Lecture Notes in Physics

Springer-Verlag Berlin Heidelberg GmbH

The Editorial Policy for Proceedings

The series Lecture Notes in Physics reports new developments in physical research and teaching – quickly, informally, and at a high level. The proceedings to be considered for publication in this series should be limited to only a few areas of research, and these should be closely related to each other. The contributions should be of a high standard and should avoid lengthy redraftings of papers already published or about to be published elsewhere. As a whole, the proceedings should aim for a balanced presentation of the theme of the conference including a description of the techniques used and enough motivation for a broad readership. It should not be assumed that the published proceedings must reflect the conference in its entirety. (A listing or abstracts of papers presented at the meeting but not included in the proceedings could be added as an appendix.)
When applying for publication in the series Lecture Notes in Physics the volume's editor(s) should submit sufficient material to enable the series editors and their referees to make a fairly accurate evaluation (e.g. a complete list of speakers and titles of papers to be presented and abstracts). If, based on this information, the proceedings are (tentatively) accepted, the volume's editor(s), whose name(s) will appear on the title pages, should select the papers suitable for publication and have them refereed (as for a journal) when appropriate. As a rule discussions will not be accepted. The series editors and Springer-Verlag will normally not interfere with the detailed editing except in fairly obvious cases or on technical matters.
Final acceptance is expressed by the series editor in charge, in consultation with Springer-Verlag only after receiving the complete manuscript. It might help to send a copy of the authors' manuscripts in advance to the editor in charge to discuss possible revisions with him. As a general rule, the series editor will confirm his tentative acceptance if the final manuscript corresponds to the original concept discussed, if the quality of the contribution meets the requirements of the series, and if the final size of the manuscript does not greatly exceed the number of pages originally agreed upon. The manuscript should be forwarded to Springer-Verlag shortly after the meeting. In cases of extreme delay (more than six months after the conference) the series editors will check once more the timeliness of the papers. Therefore, the volume's editor(s) should establish strict deadlines, or collect the articles during the conference and have them revised on the spot. If a delay is unavoidable, one should encourage the authors to update their contributions if appropriate. The editors of proceedings are strongly advised to inform contributors about these points at an early stage.
The final manuscript should contain a table of contents and an informative introduction accessible also to readers not particularly familiar with the topic of the conference. The contributions should be in English. The volume's editor(s) should check the contributions for the correct use of language. At Springer-Verlag only the prefaces will be checked by a copy-editor for language and style. Grave linguistic or technical shortcomings may lead to the rejection of contributions by the series editors. A conference report should not exceed a total of 500 pages. Keeping the size within this bound should be achieved by a stricter selection of articles and not by imposing an upper limit to the length of the individual papers. Editors receive jointly 30 complimentary copies of their book. They are entitled to purchase further copies of their book at a reduced rate. As a rule no reprints of individual contributions can be supplied. No royalty is paid on Lecture Notes in Physics volumes. Commitment to publish is made by letter of interest rather than by signing a formal contract. Springer-Verlag secures the copyright for each volume.

The Production Process

The books are hardbound, and the publisher will select quality paper appropriate to the needs of the author(s). Publication time is about ten weeks. More than twenty years of experience guarantee authors the best possible service. To reach the goal of rapid publication at a low price the technique of photographic reproduction from a camera-ready manuscript was chosen. This process shifts the main responsibility for the technical quality considerably from the publisher to the authors. We therefore urge all authors and editors of proceedings to observe very carefully the essentials for the preparation of camera-ready manuscripts, which we will supply on request. This applies especially to the quality of figures and halftones submitted for publication. In addition, it might be useful to look at some of the volumes already published. As a special service, we offer free of charge LaTeX and TeX macro packages to format the text according to Springer-Verlag's quality requirements. We strongly recommend that you make use of this offer, since the result will be a book of considerably improved technical quality. To avoid mistakes and time-consuming correspondence during the production period the conference editors should request special instructions from the publisher well before the beginning of the conference. Manuscripts not meeting the technical standard of the series will have to be returned for improvement.

For further information please contact Springer-Verlag, Physics Editorial Department II, Tiergartenstrasse 17, D-69121 Heidelberg, Germany

Guy Chavent Pierre C. Sabatier (Eds.)

Inverse Problems of Wave Propagation and Diffraction

Proceedings of the Conference
Held in Aix-les-Bains, France,
September 23–27, 1996

Springer

Editors

Guy Chavent
INRIA Rocquencourt
Domaine de Voluceau, BP 105
F-78153 Le Chesnay Cedex, France

Pierre C. Sabatier
Laboratoire de Physique Mathématique
Université Montpellier II
F-34095 Montpellier Cedex 05, France

Sponsored by D.R.E.T., U.S.A.F., and THOMSON. Organized by INRIA, and belonging to the series "SIAM-GAMM Meetings on Inverse Problems".

Cataloging-in-Publication Data applied for.

Die Deutsche Bibliothek - CIP-Einheitsaufnahme

Inverse problems of wave propagation and diffraction : proceedings of the conference, held in Aix-les-Bains, France, September 23 - 27, 1996 / Guy Chavent ; Pierre C. Sabatier (ed.). -
(Lecture notes in physics ; Vol. 486)
ISBN 978-3-662-14154-0 ISBN 978-3-540-68713-9 (eBook)
DOI 10.1007/978-3-540-68713-9

ISSN 0075-8450

Originally published by Springer-Verlag Berlin Heidelberg New York in 1997
Softcover reprint of the hardcover 1st edition 1997

Typesetting: Camera-ready by the authors/editors
Cover design: *design & production* GmbH, Heidelberg
SPIN: 10550714 55/3144-543210 - Printed on acid-free paper

Preface

Guy CHAVENT and Pierre C. SABATIER

1 Introduction

The set of lectures published here were all presented at the "Conference on inverse problems of wave propagation and diffraction", which we organized in Aix-les-Bains, France, September 23–27, 1996.

Let us first express our gratitude to the meeting scientific committee (see the list p. 373) , to our sponsors, **Délégation Générale à l'Armement**[1] **U.S. Air Force (European Office)**[2] **Thomson-Csf (Paris)**, and, last but not least, to **INRIA**, who gave us all kinds of valuable help, and, in particular, that provided by Mrs Marie-Claude Sance.

Not all the lectures which were presented at the meeting (see the list p. 374) are given here – the size of the book was limited!

Among the lectures proposed by their authors for the proceedings, unfortunately, we had to discard several excellent ones, either because they were less adequate to the study of "wave propagation and diffraction" or because they did not contain enough original developments not appearing elsewhere in published papers. We wish to thank the "invited lecturers" for their help in this selection process, but of course if there have been some mistakes, blame us; we apologize.

The aim of the meeting was to emphasize the three fundamental steps of inverse modeling: modeling the problem, analyzing it, and giving numerical solutions. It is clear that the lectures are mainly concerned with the two last steps: the modeled problem often appears as the first working assumption.

[1] "Document établi en exécution du Contrat n° 95 A 0120 passé par la Direction de la Recherche et de la Technologie - Direction Scientifique - Section Soutien à la Recherche".

[2] We wish to thank the United States Air Force European Office of Aerospace Research and Development for its contribution to the success of the Conference.

Nevertheless, the real progress in this domain over the years is shown in all the lectures by the more and more realistic character of solved inverse problems, by the caution of authors in using reasonable definitions of stability or robustness and in providing examples which rigorously prove the power of their methods in real-world problems. One may regret that a dialog between the modeling and analyzing of the inverse problem is generally missing. It is clear that a standard publication cannot cover it: such a dialog is naturally present in syntheses of attempts over several years to solve a problem, but syntheses of this kind are rare in a meeting. The cited increase of care for realism and rigor is what remains from them in standard papers.

Let us now go through the present set of lectures. Some of them (in particular invited lectures) are more general than others. All of them give an appraisal of what can be called the state of art on the two essential steps of analyzing and numerically solving the inverse problem. However, one can put most papers in one of the two following classes, which have fuzzy frontiers.

In the first class, authors *begin* with an *exact* study of inverse problems, i.e., a study starting from supposedly exact data and aiming at exact or carefully defined generalized solutions in well defined spaces. After a full understanding of the problem on these grounds, i.e., a full understanding of strong existence, uniqueness or non- uniqueness conditions for solutions and generalized solutions, they are supposed to provide stable and robust numerical algorithms for constructing them from real data. If they do it, the problem is done, and all questions of interest can be answered. Unfortunately, analyses of this class usually do not go very far into the numerical schemes and hardly match real-world needs. It is a fact that constructive methods used in fine analysis are only a starting point for constructing algorithms able to handle real data – and authors are rarely interested by going into more details in this direction.

In the "second class", authors start with a reasonable definition of generalized solutions as functions which minimize the values ϵ_i of some given cost functionals F_i and keep inside a priori bounds the values η_j of some given constraints C_j. The real-world "needs" usually require that the ϵ_i's be smaller than given values (data errors) and if this condition were enforced, analyses of this class would not be essentially different from those of the previous one. But since one usually drops any a priori bound on the ϵ_i's and only requires that these values be minima on a reasonable class of functions, quite new features usually appear in analyses of this class:

1. the mathematical tools are borrowed from optimal control theory rather than from operator theory or partial differential equations,
2. defining generalized solutions by minimizing processes makes it easier to guarantee stability and robustness, and the authors are more enthusiastic at handling real data,
3. but artificial non-uniqueness ("secondary minima") may appear,
4. and possible relations between non-uniqueness and well identified struc-

tures (coming for instance from physics, or from spectral theory) are hidden.

Thus, analyses of the first class are more appropriate in theoretical physics problems, and contain more exact functional analysis; analyses of the second class are more appropriate in applied physics and engineering, and contain more numerical analyses. To repeat, boundaries are fuzzy.

Let us now survey the papers reproduced hereafter. In Sect. 2 we go through the various topics of interest, then we focus on scattering problems. In Sect. 3, we focus on methods for reconstructing distributed parameters in media, and in particular layered media. Since many problems of applied physics and applied geophysics reduce to such reconstructions, it is not surprising that optimization techniques are more prominent there than in Sect. 2.

2 A Survey of Topics

Topics can be classed according to the nature of the waves used (acoustic, quantum, seismic, electromagnetic), to the nature of the information given (single scattering, tomography, close measurements, etc.), and to the nature of the objects to be recovered (shape, distributed parameters, location of objects, etc.). Of course, problem analyses and reconstruction methods should be adapted to the topics. Several invited lectures are particularly concerned about this general adaptation problem.

Thus, resolution and superresolution in inverse diffraction is the object of a tutorial, authoritative lecture by Bertero, Bocacci, and Piana, who also study the far-field data and the near-field data, so important now with the new microscopy techniques. Resolution is nothing but our power of disentangling information mixed by wave propagation.

In the same spirit, Mc Nally and Pike give us a very clear understanding of what amount of information is added by current a priori constraints, such as positivity or known moments, to a finite set of blurry measurements. They shatter the "great expectations" of those who believe this kind of constraints alone is sufficient to guarantee a sort of superresolution!

With Weder's lecture, we enter the important topic of quantum potential scattering. Weder deals with the new and very difficult problem of an N-body system of particles in $n \geq 2$ space dimensions with interactions given by time-dependent long-range local pair potentials – for example potentials that behave as

$$V(t, \mathbf{x}) \approx C(1+|t|)^{-\alpha}(1+|\mathbf{x}|)^{-\beta}, \ \alpha > 0, \ \frac{1}{2} < \beta \leq 1, \ \alpha + \beta > 1 \qquad (1)$$

as $|t| \to \infty$ and $|\mathbf{x}| \to \infty$.

Exact results are derived and in the inverse problem one starts from supposedly exact data. Reconstruction formulas are produced. In a well-defined class as (1), it is proved that the potential can be uniquely reconstructed from the high velocity limit of the canonical scattering operator with unperturbed evolution given by the free hamiltonian!

The same care at extending exact results to new and difficult inverse potential scattering problems appears in the paper by Boutet de Monvel and Shepelsky. One works in $\mathbb{R}$, but all difficulties due to the propagation of a transient electromagnetic field in inhomogeneous media are present. Relations between data and parameters are completely analyzed, with a resulting Gelfand– Levitan method of reconstruction which is, to our knowledge, the most general one in one-dimensional problems. Thus the two lectures above offer together a fairly good sampling of new tendencies in *exact* inverse potential scattering.

So do two other lectures for *approximate* inverse potential scattering. For Fiddy and Pommet, who manage acoustic waves, the real starting point is the Lipmann–Schwinger equation as in quantum scattering:

$$\Psi(\mathbf{r}, k\hat{\mathbf{r}}_0) = \Psi_0(\mathbf{r}, k\hat{\mathbf{r}}_0) - k^2 \int_D d\mathbf{r}' \; G_0(\mathbf{r}, \mathbf{r}') V(\mathbf{r}') \Psi(\mathbf{r}', k\hat{\mathbf{r}}_0)$$

Going to the far-field, and fixing $\hat{r}_0$, it is possible to recover approximately $W = V(\mathbf{r}) \dfrac{\Psi(\mathbf{r}, k\hat{\mathbf{r}}_0)}{\Psi_0(\mathbf{r}, k\hat{\mathbf{r}}_0)}$ from the scattering amplitude, and then to make a (nonlinear) inversion of V from W known at various $\hat{\mathbf{r}}_0$. The method is better than Born or Rytov ones but it keeps the physical features that appear in these approximate methods. It is also closer to exact methods and such a lecture give us the philosophy of modern approximate methods: to go as close as possible to exact methods without losing physical features. The lecture of Scheerschmidt obviously has the same concern. It deals with electron diffraction by crystal defects, and tries to use approximations for solving the inverse scattering problem without reconstructing the whole crystal potential. Analysis is led up to numerical calculations on real data, with rather convincing results.

While the lectures above dealt with potential scattering, the ones below now deal with the quite important topic of obstacle shape reconstruction from scattering data.

The Colton and Kress lectures offer us the great interest of complementing and updating their authoritative books on the subject. Colton's lecture deals with the "resonance region", i.e., with scattering at intermediate frequencies, where linearizations are not reasonable managements of the inverse problem. Kress' lecture reviews more widely new numerical methods and new applications of classical ones for deriving the obstacle shape from far-field data. Both produce convincing methods and examples of applications. In the same spirit, the lecture by Carfantan and Mohammed Djafari casts, for those who prefer a probabilistic presentation, many obstacle reconstruction algorithms into the unifying framework of Bayesian estimation, and analyzes the pros and cons of each approach on a theoretical level. The help provided by approximate methods in sound scattering problems is demonstrated on a simple example and in a few pages by Louis.

Two papers present complete studies of an obstacle reconstruction problem, from the theoretical analysis to the numerical results. The first one, by Rozier, Lesselier, Angell, and Kleinman, handles the case of an obstacle immersed in an acoustical wave guide and illuminated by a single harmonic source. In this approach, the obstacle boundary is parameterized in polar coordinates, and the coefficients of this parameterization are estimated by minimizing a two-term objective function that measures the data misfit and the pressure defect on the obstacle boundary. In the second lecture, by Kleinman, Van den Berg, Duchêne, and Lesselier, the "modified gradient approach" is used for solving a variety of obstacle reconstruction problems in the acoustic and electromagnetic domains. As in the previous paper, the method is based on the minimization of a two-term objective function, but now, the diffracting obstacle is represented by its characteristic function: the numerical unknown is therefore a distributed function, so that this paper could also have been presented in Sect. 3 below devoted to scattering by distributed media – but as we have said already, topics boundaries are fuzzy!

Studies of obstacle inverse scattering should also tell us when uniqueness, or non-uniqueness, is related to symmetry. Problems of this kind were adressed in the books by Colton and Kress, who draw our attention to Karp's theorem relating bijectively a sound soft spherical obstacle to the invariance of far-field amplitude under orthogonal transformations in $\mathbb{R}^3$. We are glad to be able to publish in the present book a lecture by Ha-Duong which produces the widest generalization of Karp's theorem.

The remaining lectures deal with more specialized topics on the obstacle problem. Labreuche wants to understand the workings of a popular practical method of reconstruction based on obstacle resonances (the so-called target signature). He proves that resonant frequencies *and* associated eigenfunctions uniquely determine the obstacle, whereas resonance positions only give some size estimates. To our knowledge, this is the first attempt to appraise the information of target signatures in three-dimensional cases without any symmetry (whereas there have been many in one-dimensional media or in spherical, cylindrical, etc., cases). One may object to the lack of references, in this kind of lecture, to real approaches to real data. Overwhelming information of this kind is supplied in the paper by Gerard, Guran, Maze, Ripoche, and Überall on target recognition and remote sensing. In the Haas, Rieger, Lehner lecture, the recovery of an obstacle by an adaptive iteration is described: dealing with acoustic or electromagnetic scattering, the idea is to search a closed boundary along which the tangential component of a suitable field "vanishes" in a minimum norm sense. Examples involve reasonably complicated, but smooth, scatterers. One can consider this lecture as an addendum to Kress' lecture on numerical methods.

As in the potential scattering case, we guess that this sample enables us to appraise quite well new trends in this domain.

Before going on to the problems of scattering by classical media (distributed parameters, more or less layered inhomogeneities), which will be studied in Sect. 3, we think that the best transition is offered by Natterer's lecture. It deals with acoustic scattering, distributed parameters, and tomographic soundings. It is a short lecture, showing the state of art briefly but sufficiently to get a sound idea of it, and then working out a simple method for extracting information from three-dimensional ultrasound tomography, with examples of application to real data.

3 Inverse Scattering by Distributed Media

Most of the papers in this section use, in one form or another, a least-squares approach for the definition and numerical resolution of inverse problems.

The only exceptions are:

- Two papers consider basic properties which are important for theoretical analysis and/or numerical resolution. The lecture of Bao and Symes provides conditions under which the linearized density-to-measurement map in an inverse acoustical problem is bounded. This theoretical result is important for the justification of the least-squares approach used to solve this kind of problem. Another basic result, in the lecture by Joly, concerns the development of both efficient and precise numerical tools for solving the forward problem (this is known to be one of the cornerstones in the study of the inverse problem): a new discretization approach for the obstacle is proposed, which does not generate artificial diffraction, together with new high-order finite elements, that allow mass-lumping for the Maxwell equations.
- Two papers test alternative approaches to the least-squares formulation: in their lecture, Litman, Lesselier, and Santosa adapt the level-set approach of Osher and Sethian to show convincing numerical results. A revival of the good (or bad?) old layer stripping method is proposed by Fatone, Maponi, Rignotti, and Zirilli for reconstructing the velocity in a layered half-space from surface measurements of a 2D wavefield. We have decided to include this paper for its "tour de force" in calculations, but of course limitations of the layer-stripping approach will apply in 2D as strongly as in 1D.

We come now to the six papers which are the main corpus of this section. They all correspond to the same general class of least-squares methods, where a data misfit objective function, defined through solving a state equation (forward model), is minimized by local (gradient) or global (simulated annealing) methods with respect to a distributed unknown parameter. With the exception of the last paper, they all concern the domain of reflection seismics, where one tries to image the Earth's interior from acoustic or elastic surface measurements.

The first lecture, by Cuer, recalls difficulties associated with the multimodality of the least-squares objective function in the case of a horizontally layered medium, and shows how the replacement of depth by vertical travel time partially overcomes them. An important though very technical contribution of his results is their giving an exact transparent boundary condition for a 3D source in a 1D elastic medium: it allows an efficient calculation of the forward model required for evaluating the objective function.

The second lecture, by Ernst and Herman, is devoted to eliminating from reflection data the reflection and diffraction events created by near-surface scatterers, which are considered an unwanted "noise" when one is interested in imaging the deeper structure of the Earth. These near-surface scatterers are determined by solving the corresponding inverse problem (it requires a lot of clever approximations in order to derive an efficient forward model based on modal decomposition), and the corresponding wavefield is then subtracted from the data.

The next three papers present inversion results from reflection data when the key point is an efficient forward modeler by ray tracing:

- Moser, Biryulina, and Ryzhikov use a fast 2D forward model ("recursive wavefront construction") for a distributed medium with discontinuous velocities, and apply it to the imaging inside a horst in a complex 2D medium;
- Amand and Virieux emphasize the computational efficiency by parallelizing their ray-tracing code, which allows them to invert realistic data (hundreds of shots with 96 receivers each) by simulated annealing;
- Ribodetti and Virieux define a ray tracing algorithm for the propagation of SH waves with attenuation, using a complex Lamé coefficient μ, and apply it to recovering the attenuation factor Q in a thin layer embedded inside a uniform background.

Alestra and Duceau present a nice application of the general least-squares methodology to an inverse scattering electromagnetism problem, where the unknow is the permittivity in a stratified biperiodic and 2D medium. Finally a short contribution of Scotti and Wirgin reminds an original method in obstacle reconstruction.

Once again, the papers of this section give a reasonable idea of the state of the art concerning theoretical difficulties and numerical achievements for inverse scattering by distributed media: the gap between theory and practice is still important, but steadily shrinking.

Acknowledgements

The (essential) technical work in preparing this book is due to Marie-Claude Sance (INRIA) and to Patricia Revel (Physique Mathématique, Montpellier).

It is also a pleasure to acknowledge that the meeting in Aix les Bains was part of a series of meetings on Inverse Problems originally initiated by Professors H. Engl and W. Rundell, and which was sponsored by the two scientific organizations G.A.M.M. and S.I.A.M.

We have already acknowledged on the first page the support of U.S.A.F., D.R.E.T., THOMSON, and INRIA, without which nothing would have been possible.

Contents

Resolution and Super-Resolution in Inverse Diffraction

M. Bertero, P. Boccacci and M. Piana

INFM *and Dipartimento di Fisica*
Università di Genova
Via Dodecaneso 33, 16146 Genova, Italy
e-mail: bertero@ge.infn.it

Abstract. In this tutorial paper we discuss the concept of resolution in problems of inverse diffraction. These problems have direct applications in areas such as acoustic holography and can also be considered as intermediate steps of more general problems of inverse scattering. We justify the generally accepted principle that the resolution achievable is of the order of the wavelength of the radiation used in the experiment. Moreover we indicate two cases where super-resolution, i.e. resolution beyond the limit of the wavelength, can be achieved. The first is the case of near-field data where super-resolution is possible thanks to the information conveyed by evanescent waves. The second is the case of subwavelength sources, where super-resolution is possible thanks to out-of-band extrapolation of far-field data. Simple algorithms for obtaining this result are also described.

1 Introduction

In problems of wave propagation such as those occuring in optics, acoustics, electromagnetism etc., a generally accepted principle is that the resolution achievable about the sources from observations of the scattered or emitted radiation is of the order of the wavelength λ of the radiation. In other words this means that it is only possible to recover details of the source whose linear dimensions are of the order of λ. Moreover one says that in a particular problem super-resolution is achieved if it is possible to obtain a resolution limit much smaller than λ. In this tutorial paper we discuss two problems of inverse diffraction and we use these problems for investigating two cases where super-resolution is achievable.

Inverse diffraction can be defined as the problem of determining the field distribution on a boundary surface Σ_1, from the knowledge of the field distribution on a surface Σ_2 situated within the domain where the wave propagates. Such a problem is, implicitly or explicitly, an intermediate step in a problem of inverse scattering: the recovery of the structure of the source (or obstacle) from observations of the field on a surface Σ_2 implies the recovery of the field on a surface Σ_1 surrounding the source. Then the resolution in the recovery of the source is roughly of the order of the resolution in the recovery of the field on Σ_1.

We will consider two very simple cases: in the first Σ_1 and Σ_2 are two parallel planes while in the second Σ_1 and Σ_2 are two concentric spheres. The first case is of interest both in far field acoustic holography (FAH) (Sondhi 1969) and in near field acoustic holography (NAH) (Williams and Maynard 1980) as well as in the application of holographic techniques to inverse scattering in optics (Wolf 1970), since in these applications the amplitudes are detected over planar surfaces. The second case clearly applies to experiments where the field is observed over a sphere surrounding the sources or scatterers.

In the first case it is possible to define in a precise way the so-called Rayleigh resolution limit which is proportional to λ and which corresponds to the case of far-field data. Then super-resolution is possible or by the use of *a priori* information about the source if only far-field data are available or by the use of near-field data by taking advantage of the information conveyed by evanescent waves.

In the case of spherical surfaces one can also consider the two problems, that with far-field data and that with near-field data. For the first problem the data are the values of the so-called diffraction pattern, which coincides with the scattering amplitude in the case of a scattering experiment. For the second problem the data are the values of the field amplitude over a sphere surrounding the sources. For the inverse diffraction problems corresponding to these situations it is possible to show, by investigating the behaviour of the eigenvalues of the propagation operators, that effects similar to those occuring for planar surfaces must also hold true, even if the analysis is essentially qualitative.

Finally, in the last section, we describe methods which can be used for the restoration of objects of the order of the wavelength from far-field data. We also briefly discuss the effect of different constraints on the regularized solution in these circumstances.

2 Inverse diffraction from plane to plane

Let the sources of a monochromatic field, $u(\boldsymbol{r}) = u(x_1, x_2, x_3)$, be located in the half-space $x_3 < 0$; we consider the free propagation in the half space $x_3 > 0$. Then in this region the field amplitude u is a solution of the Helmoltz equation

$$\Delta u + k^2 u = 0 \quad , \qquad x_3 > 0 \tag{2.1}$$

where k is the wavenumber, related to the wavelength λ by

$$k = \frac{2\pi}{\lambda} \quad . \tag{2.2}$$

There exists a unique solution of equation (2.1) satisfying the following conditions:

1. Sommerfeld radiation condition at infinity

$$\lim_{r\to\infty} r\left(\frac{\partial u}{\partial r} - iku\right) = 0 \quad , \quad r = \sqrt{x_1^2 + x_2^2 + x_3^2} \ ; \tag{2.3}$$

2. a boundary condition on the plane $x_3 = 0$ (the source plane)

$$u(x_1, x_2, 0) = f(x_1, x_2) \quad . \tag{2.4}$$

In general it is reasonable to assume that f is a square-integrable function.

It has been proved by Sommerfeld (Sommerfeld 1896) that there exists a unique solution of this problem, which is given by

$$u(x_1, x_2, x_3) = \int_{-\infty}^{+\infty}\int_{-\infty}^{+\infty} G^{(+)}(x_1 - x_1', x_2 - x_2', x_3) f(x_1', x_2') dx_1' dx_2' \tag{2.5}$$

where

$$G^{(+)}(r) = -\frac{1}{2\pi}\frac{\partial}{\partial x_3}\frac{e^{ikr}}{r} \tag{2.6}$$

is the (forward) Green function of the problem. However, for the discussion of the inverse diffraction problem, the so-called representation in terms of an *angular spectrum of plane waves* (Shewell and Wolf 1968) is more useful.

Let us consider the plane $x_3 = a > 0$ and let us denote by $\rho = \{x_1, x_2\}$ the position of a point in a plane orthogonal to the x_3-axis. Then the amplitude $u_a(\rho) = u(x_1, x_2, a)$ of the field on the plane $x_3 = a$ can be written as a convolution product

$$u_a(\rho) = (S_a^{(+)} * f)(\rho) \tag{2.7}$$

where $S_a^{(+)}(\rho) = G^{(+)}(x_1, x_2, a)$ acts as a *point spread function* (PSF). The Fourier transform of $S_a^{(+)}$, i.e. the *transfer function* (TF) of the system, can be computed and it is given by

$$\hat{S}_a^{(+)}(\omega) = e^{iam(\omega)} \tag{2.8}$$

where ($|\omega| = \sqrt{\omega_1^2 + \omega_2^2}$)

$$m(\omega) = \begin{cases} (k^2 - |\omega|^2)^{\frac{1}{2}} & |\omega| \le k \\ i(|\omega|^2 - k^2)^{\frac{1}{2}} & |\omega| > k \end{cases} \quad . \tag{2.9}$$

Therefore the transfer function has an oscillatory behaviour at low spatial frequencies (more precisely when $|\omega| \le k$) while it has an exponential decay, as $\exp(-a|\omega|)$, at high spatial frequencies, i.e. when $|\omega| > k$. The plane waves with spatial frequencies $|\omega| < k$ are called *homogeneous waves* while the plane waves with $|\omega| > k$ are called *evanescent waves*. In figure 1 we plot the real and imaginary part of the PSF $S_a^{(+)}$ as a function of $\rho = |\rho|$, for two values of a: $a = \lambda/5$ and $a = 5\lambda$. When $a < \lambda$ the PSF has a rather narrow central peak and small side-lobes. In fact the PSF tends to a Dirac delta function when $a \to 0$. On the other hand, when $a > \lambda$, the PSF shows oscillations with roughly equispaced zeroes, the distance between adjacent zeroes being of the order of $\lambda/2$. These different behaviours correspond to different behaviours of the transfer function $\hat{S}_a^{(+)}$. In figure 2 we plot the modulus of $\hat{S}_a^{(+)}$ as a function of $\omega = |\boldsymbol{\omega}|$, for the same values of a used in figure 1. The modulus of the transfer function is one up to $\lambda\omega = 2\pi$ and then decays exponentially for $\lambda\omega > 2\pi$. In the case $a = 5\lambda$ it is so sharp that the modulus of the transfer function is very close to a step function.

The previous analysis clearly indicates that the effect of propagation can be described in terms of a Fourier filter, more precisely a low pass Fourier filter and that two distinct spatial regions can be considered:

- *Near-field region*: corresponds to distances $a < \lambda$; in such a case the contribution of evanescent waves is important.
- *Far-field region*: corresponds to distances $a > \lambda$; in such a case the contribution of evanescent waves is negligible; one can assume that the field amplitude $u_a(\boldsymbol{\rho})$ is band-limited with a band given by

$$\mathbb{B} = \{\boldsymbol{\omega} \ , \ |\boldsymbol{\omega}| \leq k\} \ . \tag{2.10}$$

We can formulate now the problem of *inverse diffraction from plane to plane: evaluate the field amplitude $f(\boldsymbol{\rho})$ on the boundary plane $x_3 = 0$, being given the field amplitude $g(\boldsymbol{\rho}) = u_a(\boldsymbol{\rho})$ (corrupted by noise or experimental errors) on the plane $x_3 = a$.*

2.1 Inverse diffraction from far-field data

In this case evanescent waves can be completely neglected. Therefore the inverse diffraction problem is equivalent to solve a convolution equation where the PSF $S_a^{(+)}$ is a band-limited function with band $\mathbb{B}$, equation (2.10), i.e. $\hat{S}_a^{(+)}(\boldsymbol{\omega}) = \exp[iam(\boldsymbol{\omega})]$ when $|\boldsymbol{\omega}| < k$ and $\hat{S}_a^{(+)}(\boldsymbol{\omega}) = 0$ when $|\boldsymbol{\omega}| > k$. Its inverse Fourier transform $S_a^{(+)}(\boldsymbol{\rho})$ will be called the *forward propagation kernel.*

It is obvious that the solution of the problem $g = S_a^{(+)} * f$ is not unique; moreover it may not exist if g is affected by out-of-band noise. However, in such a case, the ill-posedness of the problem is not very serious and it can be cured by considering the *generalized solution*, i.e. the least-squares solution of minimal norm (Groetsch 1977).

It is very easy to prove that the generalized solution $f^\dagger(\boldsymbol{\rho})$ can be written as follows

$$f^\dagger(\boldsymbol{\rho}) = (S_a^{(-)} * g)(\boldsymbol{\rho}) \tag{2.11}$$

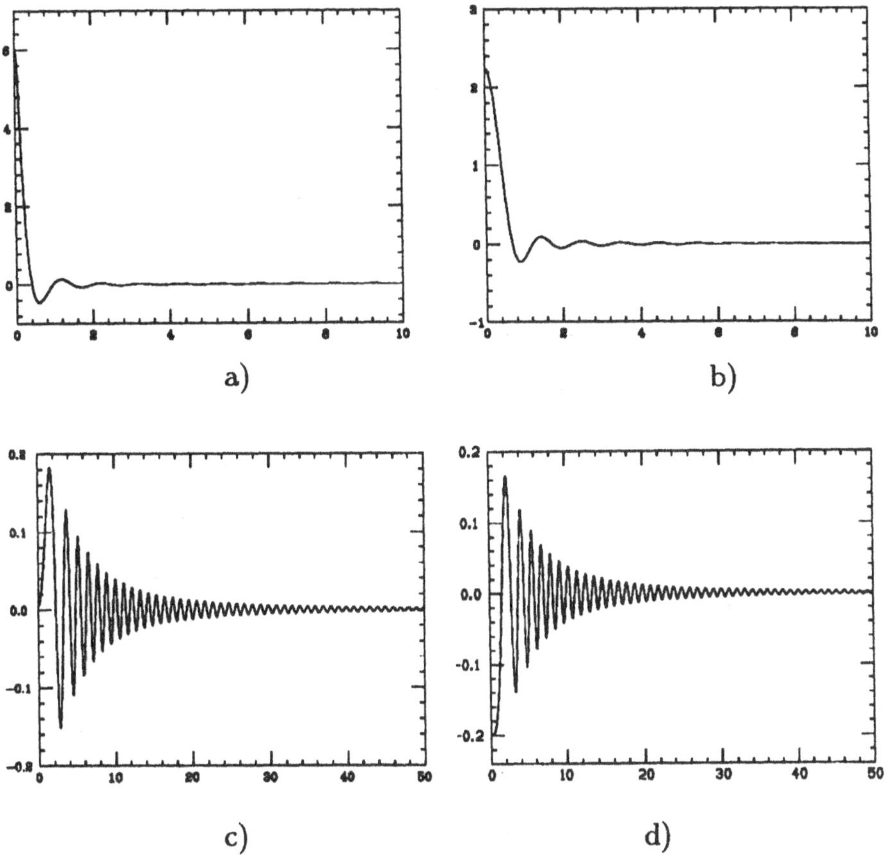

Fig. 1. Plot of the real and imaginary part of S_a^+, as a function of $x = \rho/\lambda$, for $a = \lambda/5$ (panels a) and b)), corresponding to the near-field region and for $a = 5\lambda$ (panels c) and d)), corresponding to the far-field region.

where the *backward propagation kernel* $S_a^{(-)}(\rho)$ is given by

$$S_a^{(-)}(\rho) = \frac{1}{(2\pi)^2} \int_{\mathbb{B}} e^{-iam(\omega)} e^{i\rho\cdot\omega} d\omega \quad . \tag{2.12}$$

The problem of determining the generalized solution is well-posed.

If we assume that the data are given by

$$g = S_a^{(+)} * f + w \tag{2.13}$$

where w is a term describing noise or experimental errors, then from equation (2.11) and (2.13) we obtain

$$f^\dagger = (S_a^{(-)} * S_a^{(+)}) * f + S_a^{(-)} * w \quad . \tag{2.14}$$

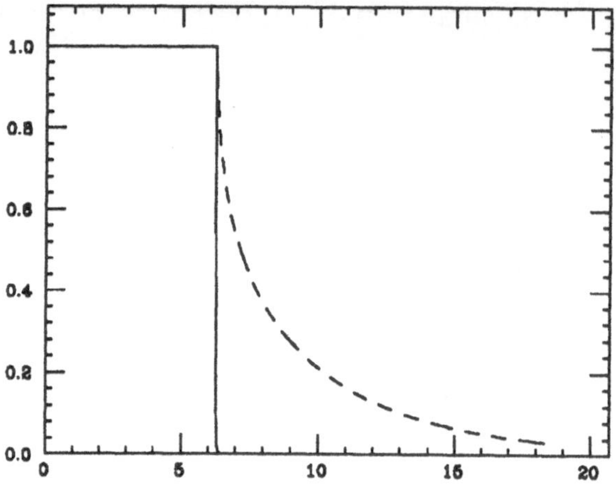

Fig. 2. Plot of the modulus of the TF $S_a^+(\omega)$, as a function of $\xi = \lambda\omega$, for the values of a of figure 1: $a = \lambda/5$ (dashed line) and $a = 5\lambda$ (solid line).

The term $S_a^{(-)} * w$ is the noise contribution to the generalized solution. Its L^2-norm is smaller than the L^2-norm of the noise: since the out-of-band noise does not contribute to $S_a^{(-)} * w$, from equation (2.12) and Parseval equality one easily derives that $\|S_a^{(-)} * w\| \le \|w\|$. As concerns the first term, the kernel $H_{\mathbb{B}} = S_a^{(-)} * S_a^{(+)}$ is the inverse Fourier transform of the characteristic function of the band $\mathbb{B}$, equation (2.10), and therefore it is given by

$$H_{\mathbb{B}}(\rho) = \frac{1}{(2\pi)^2} \int_{\mathbb{B}} e^{i\rho\cdot\omega} d\omega = \frac{k}{2\pi} \frac{J_1(k|\rho|)}{|\rho|} \quad . \tag{2.15}$$

We conclude that:

1. The generalized solution $f^\dagger$ is a noisy band-limited approximation of the boundary amplitude f.
2. The kernel $H_{\mathbb{B}}(\rho)$ has a central peak at $\rho = 0$ and is zero over circles with centre the origin and radii proportional to the zeros of the Bessel function $J_1(t)$.

The radius of the first circle is given by

$$R = 1.22\frac{\pi}{k} = 1.22\frac{\lambda}{2} \tag{2.16}$$

and this is the famous *Rayleigh resolution limit* (Born and Wolf 1980). It is the radius of the central peak of the function $H_{\mathbb{B}}(\rho)$ and it provides a measure of the smallest details of $f(\rho)$ which are recoverable. Moreover it is closely connected to the size of the band $\mathbb{B}$: the larger is the radius of the band, the smaller is the resolution distance.

In figure 3 we give two examples of restorations of binary objects obtained by means of the generalized solution $f^\dagger$. They make evident that details of the order of the wavelength are recovered, while details smaller than the wavelength are not. In fact the first object is a grid which does not contain details smaller than the wavelength, while the second one contains details of the order of $\lambda/2$. The images of the two objects are computed on the plane $a = 5\lambda$ and are contaminated by white gaussian noise (with $\sigma = 0.01$, about 1% of the maximum value of the field amplitude). The restoration of the first object, provided by equation (2.11), clearly shows the vertical and horizontal bars, which are 1λ wide. In the restoration of the second object the bars are essentially lost, in agreement with the Rayleigh criterion.

Now, the Rayleigh limit (2.16) is related to the radius k of the band of the generalized solution (2.11), more precisely it is proportional to the inverse of this radius. Therefore, in order to obtain a resolution better than the Rayleigh limit, it should be necessary to increase the band, i.e. to extrapolate the generalized solution outside $\mathbb{B}$. This is possible, in principle, if the Fourier transform of the unknown amplitude f is analytic and a sufficient condition for the analyticity of $\hat{f}$ is the boundedness of the support of f.

In order to investigate the consequences of this condition, one can proceed as follows. First recover (noisy) values of $\hat{f}(\omega)$ inside $\mathbb{B}$ by computing the generalized solution $f^\dagger(\rho)$. If we neglect the noise term, from equations (2.14) and (2.15) we obtain that

$$f^\dagger = H_{\mathbb{B}} * f \quad . \tag{2.17}$$

This convolution operator is the band-limiting operator which projects f onto the subspace of the functions whose band is interior to the band $\mathbb{B}$, equation (2.10). Next, if we have *a priori* information about the support of f and, in particular, if we know that its support is interior to some bounded domain $\mathbb{D}$ of the plane, then we can restrict the convolution operator (2.17) to the subspace of functions whose support is interior to $\mathbb{D}$. This is equivalent to introduce the following operator from $L^2(\mathbb{D})$ into $L^2(\mathbb{R}^2)$

$$(Af)(\rho) = \int_{\mathbb{D}} H_{\mathbb{B}}(\rho - \rho')f(\rho')d\rho' \quad . \tag{2.18}$$

Then extrapolation of $\hat{f}$ outside $\mathbb{B}$ is equivalent to solve the equation

$$f^\dagger = Af \quad . \tag{2.19}$$

In fact, in the absence of noise, this equation has a unique solution whose support is interior to $\mathbb{D}$ and whose Fourier transform coincides with $\hat{f}(\omega)$ over $\mathbb{B}$.

However the problem (2.19) is ill-posed. The operator $A : L^2(\mathbb{D}) \to L^2(\mathbb{R}^2)$ is a compact and injective operator. Its singular functions are related to the *generalized prolate spheroidal functions* introduced by Slepian (Slepian 1964) and its singular values are the square roots of the eigenvalues associated with these functions. By investigating the singular value spectrum of the operator (2.18) and by using the most simple regularization techniques

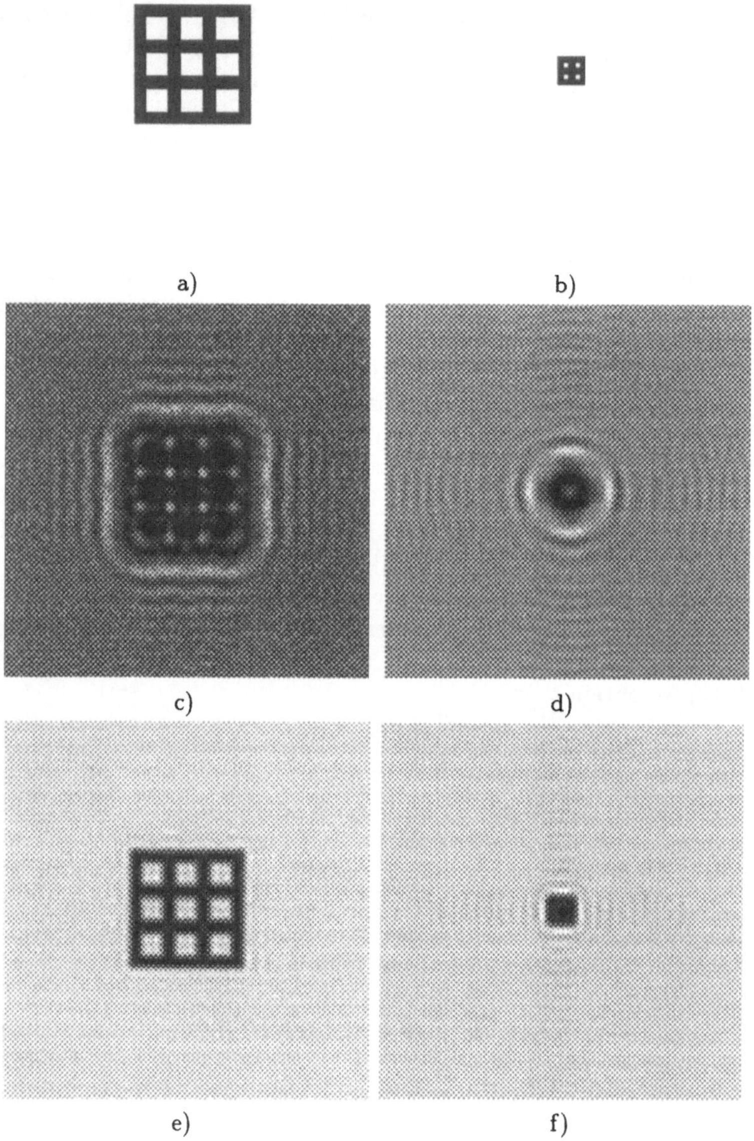

Fig. 3. Two examples of restorations obtained by means of the generalized solution (2.11). Panel a) is a grid 10λ wide with vertical and horizontal bars 1λ wide; panel b) is a grid 2.5λ wide with bars 0.5λ wide. Panels c) and d) show the modulus of the corresponding images on the plane $a = 5\lambda$. Panels e) and f) show the restorations provided by the generalized solution (2.11).

(truncated singular functions expansion) it is possible to obtain the following result (Bertero and Pike 1982): if $\mathbb{D}$ is a disc of radius d, then it is possible to obtain a significant out-of-band extrapolation (and therefore a significant improvement of resolution) for reasonable values of the signal-to-noise ratio, if the quantity $c = kd$ is not much larger than one. In other words, super-resolution is feasible when the size of the region where f is different from zero is of the order of the wavelength λ. We observe that the second object of figure 3 satisfies this condition. Such objects are sometimes referred to as *subwavelength sources.*

In general, the computation of the singular system of the operator (2.18) is difficult and therefore singular function expansions cannot be used for improving Rayleigh resolution limit in 2-D problems. However a very simple iterative method, only based on Fourier transform, was proposed by Gerchberg (Gerchberg 1974). Since it can be proved that this method is equivalent to the well-known Landweber method (De Santis and Gori 1975), it follows that it is equivalent to a filtering of the singular function expansion of the solution, which can be obtained without computing the singular system of the operator. A more general algorithm for super-resolution will be discussed in section 4.

2.2 Inverse diffraction from near-field data

As we already remarked the near-field region corresponds to distances between the two planes smaller than the wavelength λ. This condition is satisfied, for instance, in *near-field acoustic holography* (NAH) (Williams and Maynard 1980) and in *scanning near-field optical microscopy* (SNOM) (Pohl and Courjon 1993).

In such a case the information conveyed by evanescent waves allows to increase the resolution beyond the Rayleigh limit. If we consider again the integral equation $g = S_a^{(+)} * f$, where $S_a^{(+)}$ is given now by equations (2.8) - (2.9), the solution of this equation is unique but the problem is still ill-posed as a consequence of the exponential decay of $\hat{S}_a^{(+)}(\omega)$ when $|\omega| > k$. The most simple regularized solution of the problem can be obtained by a truncated Fourier transform inversion. If we have an estimate ϵ of the norm of the noise and an estimate E of the norm of the boundary amplitude f, then an estimate $\tilde{f}$ of f satisfying the bound E and reproducing the data within an error ϵ is given by

$$\hat{\tilde{f}}(\omega) = \begin{cases} e^{-iam(\omega)}\hat{g}(\omega) & |\omega| < k_{\text{eff}} \\ 0 & |\omega| > k_{\text{eff}} \end{cases} \tag{2.20}$$

where

$$k_{\text{eff}} = \max\{|\omega| \, , \, |\hat{S}_a^{(+)}(\omega)| \geq \frac{\epsilon}{E}\} \ . \tag{2.21}$$

This is a particular case of the methods, based on truncated spectral representations, investigated by Miller (Miller 1970).

Since $\epsilon/E < 1$, the condition in equation (2.21) can be replaced by the following one

$$\exp\left[a(|\omega|^2 - k^2)^{\frac{1}{2}}\right] \leq \frac{\epsilon}{E} \tag{2.22}$$

and one easily finds that

$$k_{\text{eff}} = k\left[1 + \frac{1}{(ka)^2}\log^2\left(\frac{E}{\epsilon}\right)\right] \quad . \tag{2.23}$$

The regularized solution $\tilde{f}$ is still bandlimited with a band $I\!B_{\text{eff}}$ which is a disc of radius k_{eff}:

$$\tilde{f}(\rho) = \frac{1}{(2\pi)^2}\int_{I\!B_{\text{eff}}} e^{-iam(\omega)}\hat{g}(\omega)e^{i\rho\cdot\omega}d\omega \quad . \tag{2.24}$$

By applying the Rayleigh criterion one finds that the resolution limit is now given by

$$R_{\text{eff}} = 1.22\frac{\pi}{k_{\text{eff}}} \simeq \left(\frac{ka}{\log(\frac{E}{\epsilon})}\right)^2 R \quad , \tag{2.25}$$

where the factor 1 in the r.h.s. of equation (2.23) has been neglected.

We point out that this resolution distance depends on the distance between the two planes and, in fact, it decreases quadratically when a decreases. Moreover it depends logarithmically on the signal-to-noise ratio $\frac{E}{\epsilon}$ and it decreases when $\frac{E}{\epsilon}$ increases.

In order to give an idea of the considerable improvement of resolution which can be achieved in this way, we consider the case of acoustic waves with a frequency of 3.3 kHz (we remind that the range of frequencies of acoustic waves is between 20 Hz and 20 kHz). The corresponding wavelength is about 10 cm and therefore the Rayleigh resolution distance (2.16) is about 6 cm. Now, if we assume to collect data at the distance of 1 cm from the source plane and if we also assume that the signal-to-noise ratio E/ϵ is of the order of 100, then from equations (2.23) and (2.25) we derive that $R_{\text{eff}} \simeq 0.11$ cm, with an improvement, with respect to the Rayleigh limit, by a factor 54. If we should be able to collect data at a distance of 1 mm, then we should have an improvement by a factor 5400. If $E/\epsilon = 10$ then these figures must be reduced by a factor 4 but they still imply a spectacular improvement of resolution.

3 Inverse diffraction from sphere to sphere

In the previous section we investigated the case of planar surfaces and we considered two cases of super-resolution: a) sources of the order of the wavelength in the case of far-field data; b) sources of arbitrary size in the case of near-filed data. The most significant improvement of resolution can be obtained in the second case.

It is expected that similar results apply also to other surfaces, in particular closed and bounded surfaces. It is interesting to note that, for these surfaces, we have uniqueness of the solution also in the case of far-field data. The problem, however, is still ill-posed because the solution does not exist for arbitrary data and, when it exists, does not depend continuously on the data. In order to clarify these points we investigate the case of spherical surfaces.

Assume that Σ_1 is a sphere, with centre the origin and radius a_1, containing all the sources (or scatterers) of the radiation field. Then the solution of the diffraction problem consists in determining a solution $u = u(r, \theta, \phi)$ of the Helmoltz equation (2.1) in the region $r > a_1$ satisfying Sommerfeld radiation condition (2.3) at infinity and also a boundary condition on the sphere $r = a_1$

$$u(a_1, \theta, \phi) = f(\theta, \phi) \tag{3.1}$$

where f is a given function (direct problem). The solution of this diffraction problem can be easily obtained by means of expansions in terms of the spherical harmonics $Y_{l,m}$. If we denote by $f_{l,m}$ the expansion coefficients of the boundary data f

$$f_{l,m} = \int_{\Omega} f(\theta, \phi) Y^*_{l,m}(\theta, \phi) d\Omega \quad , \tag{3.2}$$

where Ω is the unit sphere and $d\Omega = \sin\theta d\theta d\phi$, then the spherical harmonics expansion of $u(r, \theta, \phi)$ is given by

$$u(r, \theta, \phi) = \sum_{l,m} f_{l,m} \frac{h_l^{(1)}(kr)}{h_l^{(1)}(ka_1)} Y_{l,m}(\theta, \phi) \tag{3.3}$$

where the functions $h_l^{(1)}(r) = (\pi/2r)^{\frac{1}{2}} H^{(1)}_{l+\frac{1}{2}}(r)$ are the spherical Hankel functions of the first kind.

The *inverse diffraction problem* can now be formulated as follows: given the values of the field amplitude on the sphere Σ_2 with centre the origin and radius $a_2 > a_1$, determine the unknown field amplitude f on the boundary sphere a_1. This problem is, in fact, equivalent to the inversion of the following integral operator $A : L^2(\Omega) \to L^2(\Omega)$

$$(Af)(\theta, \phi) = \int_{\Omega} S^{(+)}(\theta, \phi; \theta', \phi') f(\theta', \phi') d\Omega' \tag{3.4}$$

where

$$S^{(+)}(\theta, \phi; \theta', \phi') = \sum_{l,m} \frac{h_l^{(1)}(ka_2)}{h_l^{(1)}(ka_1)} Y_{l,m}(\theta, \phi) Y_{l,m}(\theta', \phi') \quad . \tag{3.5}$$

This is a compact operator and its eigenvalues λ_l, with multiplicity $2l + 1$, are given by

$$\lambda_l = \frac{h_l^{(1)}(ka_2)}{h_l^{(1)}(ka_1)} \quad . \tag{3.6}$$

Since for large l one has

$$\lambda_l \simeq \exp\left[-(l+1)\log\left(\frac{a_2}{a_1}\right)\right] \quad , \tag{3.7}$$

the problem is severely ill-posed. However the exponential decay of the eigenvalues decreases when the ratio a_2/a_1 decreases. It follows that, if we regularize the problem by considering truncated spherical harmonics expansions, one can recover more and more terms as the sphere Σ_2 approaches the sphere Σ_1. This is an effect which is due again to evanescent waves even if a clear distinction between evanescent and homogeneous waves does not appear from the expansion (3.3). In fact, to this purpose, a much deeper analysis of the solutions of the wave equation is needed (Levi and Keller 1959).

As concerns the problem with far-field data one can now consider the asymptotic case $r \to \infty$. From the asymptotic behaviour of the spherical Hankel functions

$$h_l^{(1)}(r) \simeq (-i)^{l+1}\frac{e^{ikr}}{r} \tag{3.8}$$

one obtains the asymptotic behaviour of the field amplitude (3.3)

$$u(r,\theta,\phi) \simeq \frac{e^{ikr}}{r} g(\theta,\phi) \tag{3.9}$$

where

$$g(\theta,\phi) = \sum_{l,m}(-i)^{l+1}\frac{f_{l,m}}{h_l^{(1)}(ka_1)} Y_{l,m}(\theta,\phi) \quad . \tag{3.10}$$

The function $g(\theta,\phi)$ is usually called *diffraction pattern* and is related to the scattering amplitude in the case of scattering problems. The problem of inverse diffraction from far-field data can now be formulated as the problem of estimating the boundary function $f(\theta,\phi)$ from knowledge of the diffraction pattern $g(\theta,\phi)$. This problem, which is still ill-posed and, in fact, much more ill-posed than the problem of inverse diffraction from near-field data, can be formulated as the inversion of the integral operator

$$(Af)(\theta,\phi) = \int_\Omega S_\infty^{(+)}(\theta,\phi;\theta',\phi') f(\theta',\phi') d\Omega' \tag{3.11}$$

where

$$S_\infty^{(+)}(\theta,\phi,\theta',\phi') = \sum_{l,m}\frac{(-i)^{l+1}}{h_l^{(1)}(ka_1)} Y_{l,m}(\theta,\phi) Y_{l,m}^*(\theta',\phi') \quad . \tag{3.12}$$

This is a compact operator in $L^2(\Omega)$ and it is also injective (uniqueness of the solution of the inverse diffraction problem with far-field data). Its eigenvalues are given by

$$\lambda_l = \frac{(-i)^{l+1}}{h_l^{(1)}(ka_1)} \tag{3.13}$$

and they tend to zero much more rapidly than the eigenvalues (3.6) of the problem with near-field data. In fact their asymptotic behaviour is $|\lambda_l| \simeq \exp[-l\log(2l/eka_1)]$.

An analysis of super-resolution in the case of far-field data has not yet been performed. It is reasonable to conjecture that super-resolution can be obtained in the case where the radius a_1 of the sphere is of the order of the wavelength of the radiation. An indication in this direction has been obtained by an analysis of the inverse scattering problem in the case of Born approximation (Habashy and Wolf 1994).

4 An algorithm for super-resolution

We come back now to the problem of inverse diffraction from plane to plane, section 2, and we describe an algorithm which can be used for achieving super-resolution when *a priori* information about the support of the boundary function $f(\rho)$ is available.

At the end of section 2.1 we mentioned the Gerchberg algorithm which can be used for this purpose. However this algorithm cannot be applied directly to a convolution problem such as that described by equation (2.7). One must first estimate the Fourier transform of $f(\rho)$ over an effective band (for instance the disc of radius k, as in section 2.1, or the disc of radius k_{eff}, as in section 2.2); then one can use Gerchberg algorithm for extrapolating the Fourier transform of $f(\rho)$ outside the effective band.

We describe now an algorithm which is a generalization of the Gerchberg algorithm and does not require to solve the problem in two steps.

For generality, we consider a bounded convolution operator

$$(Af)(\rho) = (K * f)(\rho) \tag{4.1}$$

and the associated first kind equation

$$Af = g \tag{4.2}$$

where g is a given function, the data of the problem. We also assume that f belongs to the subspace of functions whose support is interior to a given and bounded domain $\mathbb{D}$. The projection operator onto this subspace is given by

$$(P_{\mathbb{D}}f)(\rho) = \chi_{\mathbb{D}}(\rho)f(\rho) \tag{4.3}$$

where $\chi_{\mathbb{D}}(\rho)$ is the characteristic function of the domain $\mathbb{D}$.

Under rather broad conditions on the PSF $K(\rho)$, the operator $AP_{\mathbb{D}}$ is compact and regularized solutions of the equation

$$AP_{\mathbb{D}}f = g \tag{4.4}$$

are provided by the *Landweber method*

$$f_{n+1} = f_n + \tau P_{\mathbb{D}}(A^*g - A^*AP_{\mathbb{D}}f_n) \tag{4.5}$$

where τ is the relaxation parameter, satisfying the usual conditions

$$0 < \tau < \frac{2}{\|A\|^2} \quad . \tag{4.6}$$

In the case $f_0 = 0$, it is easy to show by induction that all the iterates f_n satisfy the condition $P_{\mathbb{D}} f_n = f_n$. Therefore the method (4.5) is equivalent to the following *projected Landweber method*

$$f_{n+1} = P_{\mathbb{D}} f_n + \tau P_{\mathbb{D}}(A^* g - A^* A f_n) \quad . \tag{4.7}$$

From this equation one can easily derive that the algorithm can be implemented using only the Fourier transform. In fact, if $f_n(\rho)$ has been computed, then one can compute $\hat{f}_n(\omega)$ and by $\hat{f}_n(\omega)$ the function

$$\hat{h}_{n+1}(\omega) = \hat{f}_n(\omega) + \tau(\hat{K}^*(\omega)\hat{g}(\omega) - |\hat{K}(\omega)|^2 \hat{f}_n(\omega)) \quad . \tag{4.8}$$

The last step consists in computing the inverse Fourier transform of $\hat{h}_{n+1}(\omega)$, $h_{n+1}(\rho)$, and in projecting this function by means of $P_{\mathbb{D}}$ in order to obtain $f_{n+1}(\rho) = (P_{\mathbb{D}} h_{n+1})(\rho)$.

The advantage of this method is that it does not require the use of the singular functions of the operator $AP_{\mathbb{D}}$ and therefore can be easily implemented.

In figure 4 we give an example of restoration obtained by means of this method. The object in figure 4(a) is the smaller object of figure 3, i.e. a grid with size 2.5λ, therefore of the order of λ as required for achieving super-resolution from far-field data. The data are the same of figure 3(d), i.e. the noisy amplitude on the plane with $a = 5\lambda$. The restoration of this object provided by algorithm (4.7) is represented in figure 4(b) where it is evident the recovery of the four square holes, with size 0.5λ each, which are completely lost in the generalized solution of figure 3(f). The support used is just the square with size 2.5λ. If we do not have this information, it can be inferred from the data by performing inversions with different supports. We note that the super-resolution effect is just the one described at the end of section 2.1.

In (Piana and Bertero 1996) it has been pointed out that the iteration (4.7) defines a regularization algorithm. In fact, if the data g is not affected by noise and if $f_0 = 0$ the sequence $\{f_n\}_{n=1}^{\infty}$ defined by equation (4.7) converges to the unique solution of equation (4.4) in the strong topology of L^2. Moreover, in presence of noise, the algorithm is characterized by the so-called semiconvergence property, i.e. the restoration error $\|f_n - f\|$ decreases first and increases later with respect to the number of iterations. This means that in this method the number of iterations plays the role of the regularization parameter. In order to determine the optimum value of this number, several "ad hoc" criterions have been formulated in the case of real data. Things are significantly simpler when the data function is obtained synthetically, since, in this case, the theoretical model f is explicitly known and the best number of iterations can be obtained by minimizing $\|f - f_n\|$ with respect to n.

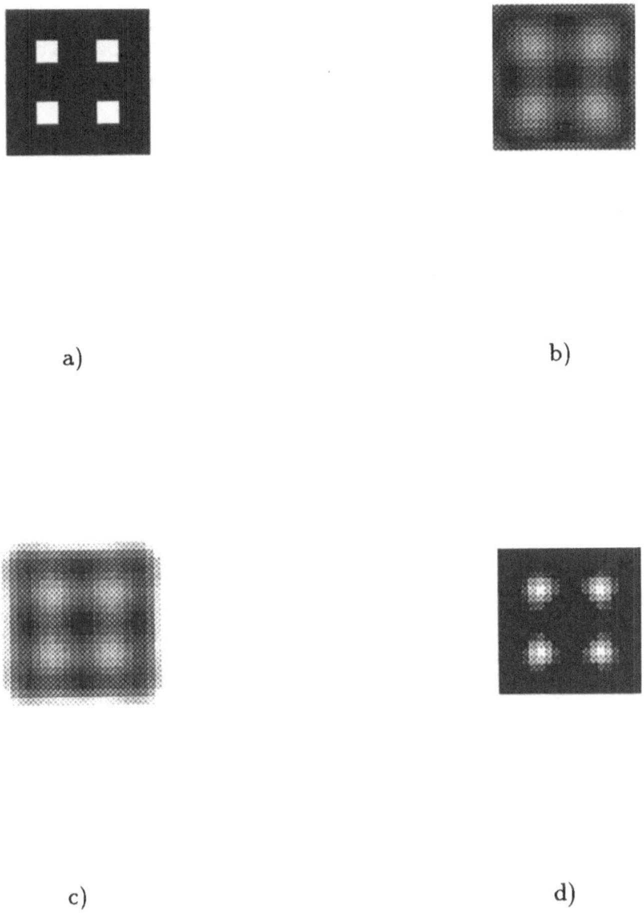

Fig. 4. Example of restoration obtained by means of the projected Landweber method. Panel a) is the smaller grid in figure 3. Panels b) and c) show the restorations provided by the method, by using respectively the constraint of compact support and the constraint of upper bound. Finally panel d) shows the effect of the combined use of the compact support and the positivity constraints.

When, as in the present example, the behaviour of this restoration error is characterized by an extremely flat minimum, it is possible to stop the iteration before the minimum is reached, without a significant loss of accuracy in the restoration. In the case of figure 4(b) it is $n = 100$.

However, in general, a notable acceleration of the projected Landweber method can be obtained by means of the so-called *preconditioning*. This procedure consists in the application of the algorithm to a modified least-squares problem. In several numerical examples regarding one dimensional models, it has been shown (Piana and Bertero 1996) that the application of preconditioning allows to obtain a gain in convergence speed up to a factor ten with no substantial modifications in the reconstructions.

The projected Landweber method can be readily generalized to the restoration of functions which belong to a closed convex subset $\mathcal{C}$ of the source space. In this case, the convex non-linear projection operator $P_{\mathcal{C}}$ can be introduced and the constrained algorithm becomes

$$f_{n+1} = P_{\mathcal{C}}(f_n + \tau A^* g - \tau A^* A f_n) \tag{4.9}$$

The method is now non-linear and the convergence of the iteration (4.9) to the generalized solution has been shown only in the weak topology. Nevertheless, numerical evidence of the strong convergence is provided by several examples.

The algorithm (4.9) can be used when it is necessary to impose upper or lower bounds on the solution. A typical lower bound is provided, for instance, by positivity. These constraints are, in general, useful in order to reduce the ringing effects which appear when linear methods are used for restoring discontinuous objects. For these constraints $P_{\mathcal{C}}$ is easily computable as well as in the case where one wishes to combine the support constraint with upper or lower bound constraints.

The object of figure 4(a) is a binary object which takes only the values 0 and 1. Therefore these values can be used as lower and upper bounds. In figure 4(c) we give the restoration obtained by means of the algorithm (4.9) after 100 iterations when only the upper bound is imposed. It is remarkable that a super-resolution effect is obtained without using the constraint on the support. This is probably due to the reduction of the ringing effects which are evident in the generalized solution (see figure 3(f)). In this example, the constraint of positivity is useless because the generalized solution of figure 3(f) does not take negative values. This is not true for the restoration of figure 4(b). Therefore in such a case positivity can be useful. In figure 4(d) we give the result obtained by combining positivity and support constraint. It is evident that the restoration is quite good.

References

Bertero M. and Pike E. R. (1982): Resolution in diffraction-limited imaging, a singular value analysis - I: The case of coherent illumination *Opt. Acta* **29** 727-746.

Born M. and Wolf E. (1980): *Principles of Optics* (Pergamon Press, Oxford).

De Santis P. and Gori F. (1975): On an iterative method for super-resolution *Opt. Acta* **22** 691-695.

Gerchberg R. W. (1974): Super-resolution through error energy reduction *Opt. Acta* **21** 709-720.

Groetsch C. W. (1977): *Generalized Inverses of Linear Operators* (Dekker, New York).

Habashy T. and Wolf E. (1994): Reconstruction of scattering potentials from incomplete data *J. Modern Opt.* **41** 1679-1685.

Levi B. R. and Keller G. B. (1959): Diffraction by a smooth object *Comm. Pure Appl. Math.* **12** 159-209.

Miller K. (1970): Least squares method for ill-posed problems with a prescribed bound SIAM *J. Math. Anal.* **1** 52-74.

Piana M. and Bertero M. (1996): Projected Landweber method and preconditioning *Inverse Problems* (in press).

Pohl D. W. and Courjon D. eds. (1993): *Near-Field Optics* (Kluwer, Dordrecht).

Shewell J. R. and Wolf E. (1968): Inverse diffraction and a new reciprocity theorem *J. Opt. Soc. Am.* **58** 1596-1603.

Slepian D. (1964): Prolate spheroidal wave functions, Fourier analysis and uncertainty - IV: Extensions to many dimensions, generalized prolate spheroidal functions *Bell. Syst. Tech. J.* **43** 3009-3057.

Sommerfeld A. (1896): Mathematische theorie der diffraction *Math. Ann.* **47** 317-336.

Sondhi M. M. (1969): Reconstruction of objects from their sound-diffraction patterns *J. Acoust. Soc. Am.* **4b** 1158-1164.

Williams E. G. and Maynard J. D. (1980): Holographic imaging without the wavelength resolution limit *Phys. Rev. Lett.* **45** 554-557.

Wolf E. (1970): Determination of the amplitude and the phase of scattered fields by holography *J. Opt. Soc. Am.* **60** 18-20.

Mathematical Programming for Positive Solutions of Ill-Conditioned Inverse Problems

B McNally[1] and E R Pike[1]

King's College London, London, WC2R 2LS

Abstract. The study of Fredholm equations of the first kind, which are ill-posed inverse problems, is an area of physics and mathematics that is currently flourishing, with applications in many areas of interest. Problems of this type are characterised by a forward relation that includes some loss of information. It is the loss of information that makes calculating the backward relation so difficult. Work is presented here which attempts to add in some of the lost information by making use of such a priori constraints as positivity and known moments. This is achieved by the method of quadratic programming, with a choice of optimisation criteria studied.

1 Introduction

Examples of Fredholm equations of the first kind range from restoration of diffraction limited optical images (E. G. Steward, 1983), to problems in high temperature superconductivity (C. E. Creffield *et al*, 1995) and experimental sizing of macromolecules (Cummins & Pike, 1974). These equations describe how the collected data (the image) is formed from the unknown solution (the object) and the blurring function (the kernel). If we first consider the one dimensional continuous case, we can denote the object by $f(x)$, the image by $g(y)$ and the blurring kernel by $K(y,x)$. Then the Fredholm equation of the first kind is:

$$g(y) = \int_{-\infty}^{\infty} K(y,x) f(x)\, dx \tag{1}$$

This equation is perfectly general. It is only out of convenience that f is described as the object, g as the image and K as the blurring kernel. For example, in the case of macromolecular sizing, f is a particle size probability distribution, and g is the first-order correlation function.

When the data is collected experimentally it can only be sampled at a finite number of points, which leads to the problem being discretised. The image is represented on N_{i} points with values $g(y_j)$ $1 \le j \le N_{\mathrm{i}}$, the object on N_{o} points with values $f(x_i)$ $1 \le i \le N_{\mathrm{o}}$, and the integral operator represented as a matrix operator $K(y_j, x_i)$ $1 \le j \le N_{\mathrm{i}}$ $1 \le i \le N_{\mathrm{o}}$. This discretised Fredholm equation of the first kind, with a suitably discretised noise vector, is represented as:

$$g(y_j) = K(y_j, x_i) f(x_i) + \eta(y_j) \tag{2}$$

Or more simply, setting the image, object and noise as vectors, and the imaging kernel as a matrix operator:

$$g = Kf + \eta \tag{3}$$

Computationally, since **f** represents a continuous function, the number of points used to represent it, N_o, has an upper limit constrained only by the available processing power. For most purposes a value of $N_o \sim 200$ is adequate to approximate the continuous case. The number of sampled data points, N_i, is dependent on the effort required to sample the data experimentally, and the amount of "blurring" present in the experiment. Typically, N_i will be much less than N_o.

1.1 The SVD System as a General Description

The most general description of (1), which also leads to a method of solving for **f**, is the singular value decomposition - widely referred to as SVD. If **K** is a compact linear operator then it maps $\mathbf{K} : X \mapsto Y$ and two sets of orthonormal basis function - $\{\mathbf{u}_k\}$ and $\{\mathbf{v}_k\}$ exist for the ranges of **K** and its adjoint $\mathbf{K}^*$ respectively. There also exist a set of singular values $\{\sigma_k\}$ that determine the mapping from X to Y such that:

$$\begin{aligned} \mathbf{K}\mathbf{u}_k &= \sigma_k \mathbf{v}_k \\ \mathbf{K}^*\mathbf{v}_k &= \sigma_k \mathbf{u}_k \end{aligned} \tag{4}$$

$$\begin{aligned} \langle \mathbf{u}_j, \mathbf{u}_k \rangle &= \delta_{j,k} \\ \langle \mathbf{v}_j, \mathbf{v}_k \rangle &= \delta_{j,k} \end{aligned} \tag{5}$$

Thus, the SVD of a system - $\{\sigma_k, \mathbf{u}_k, \mathbf{v}_k\}$ - depends on the support of the object, the support of the image and the form of the kernel. In analogy with Fourier analysis, the increasing index k corresponds to an increasing measure of spatial or temporal frequency. Because $\mathbf{u}_k$ is a basis set for X, and $\mathbf{v}_k$ is a basis set for Y, it is simple to decompose both the object and the image into a weighted sum of these basis functions,

$$\begin{aligned} \mathbf{f} &= \sum_{k=1}^{N_o} \langle \mathbf{f}, \mathbf{u}_k \rangle \mathbf{u}_k \\ \mathbf{g} &= \sum_{k=1}^{N_i} \langle \mathbf{g}, \mathbf{v}_k \rangle \mathbf{v}_k \end{aligned} \tag{6}$$

where $\langle \mathbf{a}, \mathbf{b} \rangle = a_1 b_1 + a_2 b_2 + \cdots + a_N b_N$ for **a** and **b** being N-dimensional vectors.

The standard method of recovering **f** from **g**, first performed in a simplified case by Slepian & Pollak (Slepian & Pollak, 1961) may be formulated as follows. From (6), the sampled data may also be written as:

$$\mathbf{g} = \sum_{k=1}^{N_i} \frac{\langle \mathbf{g}, \mathbf{v}_k \rangle \mathbf{v}_k \sigma_k}{\sigma_k}$$

Which, from (4), is equivalent to:

$$\begin{aligned} \mathbf{g} &= \sum_{k=1}^{N_i} \frac{\langle \mathbf{g}, \mathbf{v}_k \rangle \mathbf{K} \mathbf{u}_k}{\sigma_k} \\ &= \mathbf{K} \sum_{k=1}^{N_i} \frac{\langle \mathbf{g}, \mathbf{v}_k \rangle \mathbf{u}_k}{\sigma_k} \\ &= \mathbf{K} \mathbf{f} \end{aligned}$$

Which, when equating the last two lines, yields the inversion:

$$\mathbf{f} = \sum_{k=1}^{N_i} \frac{\langle \mathbf{g}, \mathbf{v}_k \rangle \mathbf{u}_k}{\sigma_k} \tag{7}$$

However, this solution is still very much ill-conditioned due to $\sigma_k \to 0$ as the index k increases. It is clear that (7) must have the summation stopped at some stage if there is to be any chance of a successful inversion. What happens in practice is that an index is chosen ($R < N_i$) where it is decided that experimental noise has effectively hidden the basis function weighting coefficients. This acts as a regularising parameter and gives the **T**runcated **S**ingular **V**alue **D**ecomposition solution.

$$\tilde{\mathbf{f}} = \sum_{k=1}^{R} \frac{\langle \mathbf{g}, \mathbf{v}_k \rangle \mathbf{u}_k}{\sigma_k} \tag{8}$$

In essence, SVD elucidates the underlying mathematical structure in any situation described by (1) in a way which is most effectively related to the physical problem involved. It gives a method of partitioning "noise space" from "signal space", using the singular value spectrum, which minimises the effects of physical noise, and is widely used in many problems in science and engineering.

2 Aiming for a Positive Solution

In many situations (e. g. Incoherent Imaging and Photon Correlation Spectroscopy - P. C. S.) it is known that $\mathbf{f}$ is non-negative. However, when a TSVD solution is formed from the data, it is discovered that $\tilde{\mathbf{f}}$ contains negative regions. In the noiseless case this is solely due to the summation cutoff at index R. In the presence of noise the case is worse still, due to the weights of index $k < R$ being slightly perturbed from their true values.

Another way of comparing the object, and the TSVD approximation to it, is the following:

$$\mathbf{f} = \sum_{k=1}^{R} \frac{\langle \mathbf{g}, \mathbf{v}_k \rangle}{\sigma_k} \mathbf{u}_k + \sum_{k=R+1}^{N_i} \frac{\langle \mathbf{g}, \mathbf{v}_k \rangle}{\sigma_k} \mathbf{u}_k$$

$$\tilde{\mathbf{f}} = \sum_{k=1}^{R} \frac{\langle \mathbf{g}, \mathbf{v}_k \rangle}{\sigma_k} \mathbf{u}_k + \sum_{k=R+1}^{N_i} c_k . \mathbf{u}_k \qquad (9)$$

$$c_k \equiv 0$$

It is the choice of $c_k \equiv 0$ that produces the unwanted negative regions. However, it is only because there is no reliable information about $\{c_k\}$ which leads to the arbitrary choice of setting them to zero. On the other hand, if this causes negative regions in the reconstruction, it is immediately obvious that this choice for the values of $\{c_k\}$ is incorrect. The task, then, is to try to find a choice for $\{c_k\}$ that leads to a non-negative reconstruction, with low order components equal to those of the TSVD solution.

2.1 Mathematical Programming

Mathematical programming is the general term used to describe the branch of mathematics concerned with choosing values for a set of variables, subject to various constraints placed upon them. Probably the best known subset of Mathematical Programming is called "linear programming". Problems of this type can be described by the following set of relations:

Minimise	$f = d_1 c_1 + d_2 c_2 + \cdots + d_n c_n$
	$a_{11}c_1 + a_{12}c_2 + \cdots + a_{1n}c_n = b_1$
	$a_{21}c_1 + a_{22}c_2 + \cdots + a_{2n}c_n = b_2$
Subject to the constraints	....
	$a_{m1}c_1 + a_{m2}c_2 + \cdots + a_{mn}c_n = b_m$
	$c_i \geq 0 \qquad (i = 1, \ldots, n)$

The *Simplex Method* is a procedure for solving such a set of equations. In brief, the method finds a basic feasible solution, calculates the direction that will decrease f and moves in that direction until one of the constraints is about to be violated. At that point the routine selects a new direction for decreasing f. In this manner the optimal set of $\{c_n\}$ is found that minimises f while still satisfying the linear constraints. It is a feature of linear programming, in the case of zero degeneracy, that the solution is always to be found on a vertex of the n-dimensional volume defined by the constraints.

Quadratic programming has a very similar definition to that of linear programming, but is not so straightforward to solve. Using matrix notation for the definition we have:

Minimise	$F(\mathbf{c}) \qquad \mathbf{c} \in \mathcal{R}^n$
Subject to the constraints	$\mathbf{l} \leq \left\{ \begin{array}{c} \mathbf{c} \\ \mathbf{Dc} \end{array} \right\} \leq \mathbf{u}$
Quadratic Programming	$F(\mathbf{c}) = \mathbf{e}^T\mathbf{c} + \frac{1}{2}\mathbf{c}^T\mathbf{A}\mathbf{c}$
Least Squares	$F(\mathbf{c}) = \mathbf{e}^T\mathbf{c} + \frac{1}{2}\|\|\mathbf{b} - \mathbf{A}\mathbf{c}\|\|^2$

This notation is based upon that used in (NAG, rev. 15). In all of the work so far undertaken the routine has always been used in the least squares mode, with the choice of $\{e_n\} \equiv 0$. As will be shown in the following section, the least squares mode is ideal for minimising certain aspects of the reconstruction. In the above notation, $\mathbf{c}$ represents the set of unknown basis weights, $\{c_k\}$, and $\mathbf{D}$ is an array, formed from the $\mathbf{u}$ basis functions, which along with $\mathbf{l}$ and $\mathbf{u}$ are used to ensure positivity. The matrix $\mathbf{A}$ was formed so as to implement the various optimisation choices.

2.2 What Choice for $\{c_k\}$

Borwein and Lewis (Borwein & Lewis, 1992) showed that there is a *unique* choice for $\{c_k\}$ that leads to a non-negative TSVD reconstruction with minimum L^2 norm. It may be tempting to hope that since there is a unique choice of $\{c_k\}$ that produces the minimum L^2 norm, it must be the original object. Unfortunately (fortunately for those wishing to pursue a career in inverse problems) this is not the case. The explanation is fairly simple; there is nothing to say that the original object has to be the positive, minimum L^2 norm realisation for the first R values of $\langle \mathbf{f}, \mathbf{u}_k \rangle$.

This leads to an arbitrary choice having to be made about the reconstruction. It has already been decided that $\{c_k\}$ should be chosen to ensure a non-negative TSVD solution - but that constraint is not enough to ensure a unique solution. Thus, one of many possible choices must be made for $\{c_k\}$. Let $\hat{\mathbf{f}}$ represent the TSVD solution with possibly non-zero values of $\{c_k\}$ added to ensure positivity. The various optimisation choices so far studied include:

- Minimise $\sum |\hat{\mathbf{f}}|$
- Minimise $\sum |\hat{\mathbf{f}}|^2$
- Minimise $(1-\alpha)\sum |\hat{\mathbf{f}}| + \alpha \sum |\hat{\mathbf{f}}|^2$
- Minimise $\sum |\frac{\partial \hat{\mathbf{f}}}{\partial x}|^2$
- Minimise $\sum |\frac{\partial^2 \hat{\mathbf{f}}}{\partial x^2}|^2$

It is obvious that there are an endless number of these minimisation choices; each one capable of producing a different positive $\hat{\mathbf{f}}$ that exactly fits the data to within the noise level. Indeed, for certain choices of norm - such as the minimum L^1 norm - the situation is even worse than this. While the L^2 norm is strictly convex, the L^1 is only convex. This means there is a possibility that the solution with minimum L^1 norm may be degenerate. *i. e.* there may well be an infinite number of different solutions that have exactly the same minimum L^1 norm.

2.3 Including Constraints and Optimisations

As indicated in section 2.1, the constraints for the system are specified by the system of equations:

$$\mathbf{l} \le \begin{Bmatrix} \mathbf{c} \\ \mathbf{Dc} \end{Bmatrix} \le \mathbf{u}$$

The first of these constraints, $\mathbf{l} \le \mathbf{c} \le \mathbf{u}$, imposes some constraint on the values that the missing higher order basis weights are allowed to take. In all cases so far studied there was no a priori knowledge to suggest there should be any limit on these values. Hence, $\mathbf{l}$ and $\mathbf{u}$ were set to $-\infty$ and ∞ respectively. $\mathbf{D}$ is known as the "Constraint Matrix", and is used to specify constraints on the form of the reconstruction. Examining the case of imposing positivity on point i gives, from (9):

$$\begin{aligned} \hat{f}(i) &= \sum_{k=1}^{R} \frac{\langle \mathbf{g}, \mathbf{v}_k \rangle}{\sigma_k} u_k(i) + \sum_{k=R+1}^{N_\mathrm{i}} c_k . u_k(i) \\ &= \tilde{f}(i) + \sum_{k=R+1}^{N_\mathrm{i}} c_k . u_k(i) \end{aligned}$$

Introducing the constraint of $0 \le \hat{f}(i) \le \infty$ then gives:

$$\begin{aligned} & 0 \le \hat{f}(i) \le \infty \\ \Rightarrow \quad & 0 \le \tilde{f}(i) + \sum_{k=R+1}^{N_\mathrm{i}} c_k . u_k(i) \le \infty \\ \Rightarrow & -\tilde{f}(i) \le \sum_{k=R+1}^{N_\mathrm{i}} c_k . u_k(i) \le \infty \end{aligned}$$

Recognising that every point in the reconstruction ($1 \le i \le N_\mathrm{o}$) must be positive gives $\mathbf{l} = -\tilde{\mathbf{f}}$ and $\mathbf{u} = \infty$. Thus, the constraint matrix, $\mathbf{D}$, is such that $D_{ij} = u_j(i)$. The various minimisation schemes were included in a very similar manner to this, with some aspect of the reconstruction equalling the corresponding aspect of $\tilde{\mathbf{f}}$ plus some combination of the higher order $\mathbf{u}$ basis functions.

2.4 Weakness of the Positivity Constraint

Some recent reports have claimed methods of superresolution[1] based on constraining the restoration to be non-negative. It will be shown here that the positivity constraint is *not* sufficient to accurately accurately the missing spatial frequencies.

[1] A restored image could be said to be superresolved when it contains accurate high spatial frequency components that are not detectable in the collected data.

Imagine two non-negative objects, $\hat{\mathbf{f}}_1$ and $\hat{\mathbf{f}}_2$, such that for the first R singular functions $\langle \hat{\mathbf{f}}_1, \mathbf{u}_k \rangle = \langle \hat{\mathbf{f}}_2, \mathbf{u}_k \rangle$. Both $\hat{\mathbf{f}}_1$ and $\hat{\mathbf{f}}_2$ will have identical images and TSVD solutions to within the noise level. Now suppose that $\hat{\mathbf{f}}_1$ is double peaked, and $\hat{\mathbf{f}}_2$ is single peaked. Finally, take it that $\hat{\mathbf{f}}_1$ just happens to be a L^1 minimisation, and $\hat{\mathbf{f}}_2$ the L^2 minimisation, for the first R fixed weights.

Suppose $\hat{\mathbf{f}}_2$ is imaged, the data collected, and the TSVD solution formed. This will most probably contain negative regions, so it may be desired to find a choice for $\{c_k\}$ that ensures positivity. If the L^2 minimisation criterion is chosen then the original object is recovered exactly, since $\hat{\mathbf{f}}_2$ is defined to have this property. However, if the L^1 minimisation criterion is instead chosen, the restored object is not $\hat{\mathbf{f}}_2$, but something similar to $\hat{\mathbf{f}}_1$ (identically equal to in the non-degenerate case). But it is not enough to select the L^2 choice again in the future just because it has worked so well this time. If the experiment is repeated with $\hat{\mathbf{f}}_1$ being imaged instead of $\hat{\mathbf{f}}_2$, it turns out the L^2 minimisation will return a very poor reconstruction of the original object - which would again be equal to $\hat{\mathbf{f}}_1$ for a non-degenerate problem.

3 Further Constraints in Different Situations

A proper choice of minimisation criterion is dependent on the form of the original object. However, since it is the form of the object that the process is trying to recover, the choice of minimisation criterion is not much better than a pure guess. Fortunately, some inverse problems outside the realm of imaging lend themselves to further a priori constraints beyond that of positivity.

In P. C. S. (Cummins & Pike, 1974) accurate values of the "area" of $\hat{\mathbf{f}}$, and the "centre of mass" of $\hat{\mathbf{f}}$ are known. These are the first two moments of the reconstruction. In work on high temperature superconductivity (C. E. Creffield *et al*, 1995), a reconstruction of the "Spectral Weight Function" is required from "Matsubara Green's function" - another ill-posed inverse problem involving analytic continuation. In this case it is possible to calculate exactly the first three moments of the reconstruction. Using this extra a priori information narrows down the range of possible functions that still fit the data and the a priori information to within the noise level.

3.1 An Example from High-T$_c$ Superconductivity Theory

The particular inverse problem in this situation is

$$g(y) = \int_{-\infty}^{\infty} \left[\frac{\exp(-xy)}{1 + \exp(-\beta x)} \right] f(x)\, dx \qquad y \in [0, \beta] \tag{10}$$

where $g(y)$ is the numerically calculated Matsubara Green's Function, and $f(x)$ is the Spectral Weight Function - the desired function. The "n^{th}-moment" of a function is defined as

$$\mu_n = \int_{-\infty}^{\infty} x^n f(x)\, dx \tag{11}$$

In the case studied, not only was it known that the reconstruction had to be non-negative, but also values for μ_0, μ_1 and μ_2 could be accurately precalculated. These moment constraints were built into the quadratic programming routine in much the same way as the positivity constraint. That this was able to be done relied on the linearity of the problem. Making use of (6) leads to:

$$\begin{aligned}
\int_{-\infty}^{\infty} x^n f(x)\, dx &= \int_{-\infty}^{\infty} x^n \sum_k \langle \mathbf{f}, \mathbf{u}_k \rangle \mathbf{u}_k \, dx \\
&= \int_{-\infty}^{\infty} x^n \left(\sum_{k=1}^{R} \langle \mathbf{f}, \mathbf{u}_k \rangle \mathbf{u}_k + \sum_{k=R+1}^{No} c_k \mathbf{u}_k \right) dx \\
&= \int_{-\infty}^{\infty} x^n \tilde{\mathbf{f}}\, dx + \sum_{k=R+1}^{No} c_k \int_{-\infty}^{\infty} x^n \mathbf{u}_k \, dx
\end{aligned} \tag{12}$$

This means that the nth moment of the final reconstruction is equal to the nth moment of the TSVD solution, plus a weighted sum of the nth moments of each of the higher order basis functions. In practice, if m moments are to be used as constraints, and $N_o - R$ higher order weights are to be found, the m moments of each of the $N_o - R$ functions are precalculated. These are then used as part of the general constraint matrix in the quadratic programming routine.

4 Conclusions

It has been shown that it is possible to add weighted amounts of the unmeasurable higher order $\mathbf{u}_k$ basis singular functions by using the technique of quadratic programming. This can be easily adapted to fit any linear constraints - such as known coefficients for the first N singular functions in the expansion, positivity and known moment values. Once the constraints have been met it is then necessary to specify some further optimisation condition - such as the reconstruction possessing minimum L^2 norm - and, having done this, our numerical work confirms and implements the Borwein-Lewis uniqueness theorem (Borwein & Lewis, 1992). The choice of minimisation criterion is purely arbitrary - with no one choice working well in all cases. Once one has chosen to guess unmeasurable components of the solution to make it positive, using these methods or any of the many alternative iterative nonlinear methods, there will be an infinite family of solutions which all fit the data to the same accuracy and which *may* differ widely. The utmost caution is therefore required if non-linear methods are used since any apparent increase in resolution over that of the TSVD solution could be spurious.

5 Acknowledgements

Thanks go to Geoff de Villiers of DRA Malvern for initial thoughts on the Borwein and Lewis paper, as well as constant input throughout this work.

Ben McNally also thanks the United States Department of Army for the research grant: `DAAH04-95-1-0280` under which this work was completed.

References

E. G. Steward. (1983) Fourier Optics, Ellis Horwood Limited, **1983**

C. E. Creffield, E. G. Klepfish, E. R. Pike and Sarben Sarkar. Physical Review Letters, **75,** 517-520, 1995

H. Z. Cummins and E. R. Pike, NATO Advanced Study Institute Series B: Physics, Plenum Press, New York, 1974

D. Slepian and H. O. Pollak. Bell Systems Technical Journal, **40,** 43-64 1961

J. M. Borwein and A. S. Lewis, Mathematical Programming, **57,** 15-48 1992

G. Sierksma, Linear and Integer Programming, Marcel Dekker, pp77, **1996**

Fortran NAG Library, **E04NCF,** pp15

Inverse Scattering for N-Body Systems with Time-Dependent Potentials

Ricardo Weder

Universidad Nacional Autónoma de México,Instituto de Investigaciones en Matemáticas Aplicadas y en Sistemas,Apartado Postal 20–726,DF 01000,México,
E–Mail :weder@servidor.unam.mx

Abstract. I prove that the high velocity limit of any one of the Dollard scattering operators of an N–body quantum mechanical system with long–range time–dependent pair potentials determines uniquely the potentials. I also show that in the particular case when the potentials go to zero fast enough as time goes to plus and minus infinity it is not necessary to introduce a modified Dollard time evolution and that pair potentials that decrease slowly as the interparticle distances go to infinity (for example Coulomb potentials) can be uniquely reconstructed from the high velocity limit of the canonical scattering operator with unperturbed evolution given by the free Hamiltonian.These results are obtained from reconstruction formulae with bound of the error term that I prove with a simple time–dependent method.

1 Introduction

In this paper I study the inverse scattering of an N–body quantum mechanical system of particles in $n \geq 2$ space dimensions with interactions given by time–dependent local pair potentials of long range.I prove that the high velocity limit of any one of the Dollard scattering operators determines uniquely the potentials.I also obtain a formula with bound of the error term for the constructive reconstruction of the potentials.

I prove these results in Section 2 by extending to this case the simple time–dependent method of Enss and Weder (1993),Enss and Weder (1994),Enss and Weder (1995a), Enss and Weder (1995b),and Weder (1995) where time–independent potentials are considered.In fact it is quite remarkable that due to the time–dependent nature of this method the time dependency of the potentials poses essentially no new problems.The basic physical intuition here is that during the short time interval in which a high velocity state remains in the interaction region the potential changes very little and it is approximately time independent.I can even consider pair potentials that grow to infinity in time provided that they go to zero fast enough when the corresponding interparticle distance goes to infinity.

Moreover, in the time–dependent case the border line between short– and long–range potentials is not always given by the decay of the Coulomb potential.I consider in Section 3 pair potentials that go to zero very slowly as the interparticle distances go to infinity but that go to zero fast enough as time goes to plus and minus infinity.For example potentials that behave as

$$V(t,\mathbf{x}) \approx C(1+|t|)^{-\alpha}(1+|\mathbf{x}|)^{-\beta}\ ,\alpha > 0, 1/2 < \beta \leq 1, \alpha+\beta > 1\ ,$$

as $|t| \to \infty$,and $|\mathbf{x}| \to \infty$ .An important particular case are potentials of compact support in time.This corresponds physically to (external) potentials that are turned on and then switched off after some (short)time.For this potentials it is not necessary to introduce a modified time evolution.
I also obtain in this case a formula with bound of the error term that allows me to uniquely reconstruct the potentials from the high velocity limit of the canonical scattering operator defined with the unperturbed time evolution given by the free Hamiltonian.
As is well known there are many inportant applications of scattering with time–dependent potentials,for example the charge transfer model .I discuss this literature at the end of Section 2.

There is an extensive literature on multidimensional inverse scattering for the Schrödinger equation with time–independent potentials.See for example the references mentioned in Enss and Weder (1995a) and the books by Chadan and Sabatier (1989)and Newton (1989) .However much less was known in the case of time–dependent potentials.This is perhaps so because many of the previous results for time–independent potentials where obtained with stationary methods.Starting with the work of Perla Menzala (1985) there are a number of papers that consider the inverse scattering problem for the wave equation with time–dependent potentials.See Stefanov (1989) and the references quoted there.In Ramm and Sjöstrand (1991) the uniqueness of an inverse data problem for the wave equation with time–dependent potential is proven.

2 N–Body Inverse Scattering

Let $\tilde{\mathbf{x}}_j \in \mathbf{R}^n$ and $m_j, j = 1, 2, \ldots, N$ be respectively the positions and the masses of the particles.The free Hamiltonian is given by

$$\tilde{H}_0 - \sum_{j=1}^{N}(2m_j)^{-1}\,\tilde{\mathbf{p}}_j^2\ ,\ \tilde{\mathbf{p}}_j = -i\nabla_{\tilde{\mathbf{x}}_j}\ .$$

As usual I formulate our scattering problem in the total center of mass frame and I substract the Hamiltonian of the center of mass:

$$H_{CM} = \left(2\sum_{j=1}^{N} m_j\right)^{-1}\left(\sum_{j=1}^{N}\tilde{\mathbf{p}}_j\right)^2 .$$

The free Hamiltonian is then $H_0 := \tilde{H}_0 - H_{CM}$. The space of states in the center of mass frame is a Hilbert space,$\mathcal{H}$, that in configuration space is represented by wave functions ϕ in

$$L^2(\mathbf{X})\,, \mathbf{X} = \left\{ (\tilde{\mathbf{x}}_1, \ldots, \tilde{\mathbf{x}}_N) \middle| \sum_{j=1}^{N} m_j \tilde{\mathbf{x}}_j = 0 \right\} \cong \mathbf{R}^{n(N-1)}$$

with the measure induced on $\mathbf{X}$ by the norm in $\mathbf{R}^{nN}$, $\left[\sum_{j=1}^{N} m_j \tilde{\mathbf{x}}_j^2\right]^{1/2}$. The set of momentum space wave functions,$\hat{\phi}$, is given by

$$L^2(\hat{\mathbf{X}}), \hat{\mathbf{X}} = \left\{ (\tilde{\mathbf{p}}_1, \ldots, \tilde{\mathbf{p}}_N) \middle| \sum_{j=1}^{N} \tilde{\mathbf{p}}_j = 0 \right\} \cong \mathbf{R}^{n(N-1)}$$

where I give to $\hat{\mathbf{X}}$ the dual norm induced by $\left[\sum_{j=1}^{N} (m_j)^{-1} \tilde{\mathbf{p}}_j^2\right]^{1/2}$ on $\mathbf{R}^{nN}$.The Fourier transform is a unitary operator from $L^2(\mathbf{X})$ onto $L^2(\hat{\mathbf{X}})$.H_0 is a self–adjoint operator in $L^2(\mathbf{X})$ with domain $D(H_0) = H_2(\mathbf{X})$,the second Sobolev space.For a general reference in multiparticle scattering see,e.g.,Reed and Simon (1979).

I suppose that the potential is a sum of pair potentials that are multiplication operators by real–valued functions

$$V = \sum_{j<k} V_{jk}(t, \tilde{\mathbf{x}}_k - \tilde{\mathbf{x}}_j) \ .$$

I split each pair potential into parts of short and long range depending on their decay rate at infinity and their differentiability

$$V_{jk}(t, \tilde{\mathbf{x}}_k - \tilde{\mathbf{x}}_j) = V_{jk}^s(t, \tilde{\mathbf{x}}_k - \tilde{\mathbf{x}}_j) + V_{jk}^l(t, \tilde{\mathbf{x}}_k - \tilde{\mathbf{x}}_j).$$

An operator in $L^2(\mathbf{R}^n)$ is said to be Kato–small if it is bounded with respect to the Laplacian with relative bound zero(see Kato (1976) for definitions).For any set $O \subset \mathbf{R}^n$ I denote by $F(\mathbf{x} \in O)$ the multiplication operator by the characteristic function of O.I define the following class of short–range potentials.

DEFINITION 2.1. *$\mathcal{V}_{SR}$ denotes the class of potentials*

$$V^S = \sum_{j<k} V_{jk}^s(t, \tilde{\mathbf{x}}_k - \tilde{\mathbf{x}}_j)$$

where for each fixed $t \in \mathbf{R}$, $V_{jk}^s(t, \mathbf{y})$ is Kato–small and

$$\left\| V_{jk}^s(t, \mathbf{y})(-\Delta_{\mathbf{y}} + I)^{-1} \right\| \leq C(1 + |t|)^M \qquad (1)$$

for some constants C and M.Moreover, for every $q \in \mathbf{R}$ there is a $v_0 > 0$ and a function h with $h(r) \in L^1((0,\infty))$ such that for all $r \in \mathbf{R}$

$$\left\| V^s_{jk}(r/v + q, \mathbf{y})(-\Delta_{\mathbf{y}} + I)^{-1}\, F(|\mathbf{y}| \geq |r|) \right\| \leq h(|r|) \tag{2}$$

for all $v \geq v_0$.

Condition (2) is equivalent to the existence of a $v_0 > 0$ and a function h with $h(r) \in L^1((0,\infty))$ such that for all $r \in \mathbf{R}$

$$\left\| F(|\mathbf{y}| \geq |r|) V^s_{jk}(r/v + q, \mathbf{y})(-\Delta_{\mathbf{y}} + I)^{-1} \right\| \leq h(|r|) \ .$$

This decay condition is more intuitive,but (2) is technically more convenient.

I denote by $C^u_\infty(\mathbf{R}^n)$ the space of all continuous functions that go to zero at infinity and that have continuous derivatives of all orders up to u.By $D^\alpha_{\mathbf{y}}$ I designate the derivatives with the usual multi–index notation.I define my class of long–range potentials as follows.

DEFINITION 2.2. *$\mathcal{V}_{LR}$ denotes the class of potentials*

$$V^L = \sum_{j<k} V^l_{jk}(t, \tilde{\mathbf{x}}_k - \tilde{\mathbf{x}}_j)$$

where for each fixed $t \in \mathbf{R}$, $V^l_{jk}(t, \mathbf{y}) \in C^{2u}_\infty(\mathbf{R}^n)$ and

$$\left| D^\alpha_{\mathbf{y}} V^l_{jk}(t, \mathbf{y}) \right| \leq C(1 + |t|)^\epsilon (1 + |\mathbf{y}|)^{-1-|\alpha|(\gamma-1)}, 1 \leq |\alpha| \leq 2u \ , \tag{3}$$

for some non negative constants C, ϵ, γ with $\gamma - \epsilon > 3/2$ and $u \geq 2$.

To simplify the notation later I will only use $\gamma < 2$.

The spliting of the potential into short– and long–range parts is not unique.In what follows I take one spliting and keep it fix. Note that the short–range part of the potential is allowed to grow in time as fast as any power of t in any compact region of space.On the contrary, the long–range part that contains the tail of the potential that decays slowly at infinity and that in consequence acts upon the particles during quite a long time is only allowed to grow slowly as $|t| \to \infty$ and this provided that there is enough decay as $|\mathbf{y}| \to \infty$.This is the meaning of the condition $\gamma - \epsilon > 3/2$ in (3). The interacting time–dependent Hamiltonian, defined as

$$H(t) = H_0 + V(t) \ ,$$

is a self–adjoint operator with domain $D(H(t)) = D(H_0)$. I make the assumption that there is a unitary propagator that generates the time evolution corresponding to H(t).

ASSUMPTION P. *I assume that there is a family of unitary operators in $\mathcal{H}$,$U(t,q), t, q \in \mathbf{R}$, such that*

1. *$U(t,q)$ is a strongly continuous function of $(t,q) \in \mathbf{R}^2$.*
2. *$U(t,r)U(r,q) = U(t,q)$ for all $t,q,r \in \mathbf{R}$.*
3. *$U(t,q)\,H_2(\mathbf{X}) \subset H_2(\mathbf{X})$ for all $t,q \in \mathbf{R}$ and if $\Phi \in H_2(\mathbf{R}^n), U(t,q)\Phi$ is strongly continuously differentiable in t and q and*

$$i\frac{\partial}{\partial t}U(t,q)\Phi = H(t)U(t,q)\Phi;\; i\frac{\partial}{\partial q}U(t,q)\Phi = -U(t,q)H(q)\Phi\ . \qquad (4)$$

Starting with the pioneering work of Kato (1953) there is an extensive literature on the derivation of sufficient conditions for the validity of Assumption P.See for example Kato (1970), Yajima (1987) , Yajima (1991) and the references mentioned there.The results in Yajima (1987) and Yajima (1991) contain most of the interesting applications,including moving singularities.A simple set of sufficient conditions is the following one. Suppose that each pair potential $V_{jk}(t, \tilde{\mathbf{x}}_k - \tilde{\mathbf{x}}_j)$ is Kato–small and that the operator valued functions

$$V_{jk}(t,\mathbf{y})(-\Delta_{\mathbf{y}} + I)^{-1}$$

are strongly continuously differentiable functions of $t \in \mathbf{R}$.Then it follows from TheoremX.70 and the proof of TheoremX.71 of Reed and Simon (1975) that Assumption P is satisfied.

The Dollard modified time evolution in the free channel is generated by the time–dependent Hamiltonian

$$H_D(t) = H_0 + \sum_{j<k} V^l_{jk}(t, t\,\mathbf{p}_{jk}/\mu_{jk})$$

where μ_{jk} and $\mathbf{p}_{jk}$ are, respectively, the reduced mass $\mu_{jk} = m_j m_k/(m_j + m_k)$ and the relative momentum $\mathbf{p}_{jk} = \mu_{jk}(\tilde{\mathbf{p}}_k/m_k - \tilde{\mathbf{p}}_j/m_j)$ of the particles j and k. The Dollard propagator is the following unitary multiplication operator in momentum space

$$U^D(t,q) = e^{-i(t-q)H_0} \exp\left[-i\sum_{j<k}\int_q^t dr\; V^l_{jk}(r, r\,\mathbf{p}_{jk}/\mu_{jk})\right]\ . \qquad (5)$$

Clearly, different splitings into short– and long–range parts of the pair potentials give rise to different Dollard propagators. The modified Dollard wave operators for the free channel with initial time q are defined as

$$\Omega^D_\pm(q) = s - \lim_{t\to\pm\infty} U(q,t)U^D(t,q)\ . \qquad (6)$$

The existence of the strong limits is proven below (see the argument starting with (32)).The modified Dollard scattering operator between the free channels with initial time q is given by

$$S^D(q) = \left(\Omega^D_+(q)\right)^* \Omega^D_-(q)\ . \qquad (7)$$

I reconstruct the pair potentials one by one .For any given pair of particles I introduce as in Enss and Weder (1995a) appropriate states where all particles have high relative velocity with respect to each other.I first introduce some kinematical notation.I number the particles in such a way that the given pair consists of particles one and two.I take as one n–dimensional coordinate the relative distance $\mathbf{x}$ and the corresponding relative momentum $\mathbf{p}$ of the given pair (12)

$$\mathbf{x} := \tilde{\mathbf{x}}_2 - \tilde{\mathbf{x}}_1, \ \mathbf{p} := -i\nabla_{\mathbf{x}} = \mu_{12}\left[(-i\nabla_{\mathbf{x}_2}/m_2) - (-i\nabla_{\mathbf{x}_1}/m_1)\right] \ .$$

By $\mathbf{x}_j$ and $\mathbf{p}_j$ I denote,respectively,the position and the momentum of the j–th particle,$j = 1, \ldots, N$, relative to the center of mass of the pair (12)

$$\begin{aligned} \mathbf{x}_j &:= \tilde{\mathbf{x}}_j - (m_1\tilde{\mathbf{x}}_1 + m_2\tilde{\mathbf{x}}_2)/(m_1 + m_2), \\ \tilde{\mathbf{p}}_j &:= \mu_j(\tilde{\mathbf{p}}_j/m_j - (\tilde{\mathbf{p}}_1 + \tilde{\mathbf{p}}_2)/(m_1 + m_2)) \ , \end{aligned} \tag{8}$$

where μ_j is the reduced mass of the j–th particle with respect to the center of mass of the pair (12)

$$\mu_j := m_j(m_1 + m_2)/(m_j + m_1 + m_2), \ j = 1, \ldots, N \ .$$

$\{\mathbf{x}, \mathbf{x}_3, \ldots, \mathbf{x}_N\}and\{\mathbf{p}, \mathbf{p}_3, \ldots, \mathbf{p}_N\}$are sets of N–1 independent n–dimensional configuration and momentum coordinates in the total center of mass frame. $\mathbf{p}_j/\mu_j, j = 1, \ldots, N$, is the relative velocity of particle j with respect to the center of mass of the pair (12). The relative momentum of particles j and k is

$$\mathbf{p}_{jk} = -i\nabla_{(\tilde{\mathbf{x}}_k - \tilde{\mathbf{x}}_j)}, \ j, k = 1, \ldots, N \ ,$$

and their relative velocity is

$$\frac{\mathbf{p}_{jk}}{\mu_{jk}} = \frac{\tilde{\mathbf{p}}_k}{m_k} - \frac{\tilde{\mathbf{p}}_j}{m_j} = \frac{\mathbf{p}_k}{\mu_k} - \frac{\mathbf{p}_j}{\mu_j}, \ j, k = 1, \ldots, N \ . \tag{9}$$

Let $\Phi_0 \in \mathcal{H}$ be an asymptotic configuration with product wave function of the following form in momentum space ($\hat{}$ denotes Fourier transform)

$$\Phi_0 \sim \hat{\phi}_{12}(\mathbf{p})\, \hat{\phi}_3(\mathbf{p}_3, \ldots, \mathbf{p}_N) \ , \tag{10}$$

where $\hat{\phi}_{12} \in C_0^\infty(\mathbf{R}^n)$ varies while $\hat{\phi}_3 \in C_0^\infty(\mathbf{R}^{n(N-2)})$ is a fixed function normalized to one, $||\hat{\phi}_3|| = 1$. The high–velocity state is defined as follows (see Enss and Weder (1995a))

$$\Phi_{\mathbf{v}} \sim \hat{\phi}_{12}(\mathbf{p} - \mu_{12}\mathbf{v})\, \hat{\phi}_3(\mathbf{p}_3 - \mu_3\mathbf{v}_3, \ldots, \mathbf{p}_N - \mu_N\mathbf{v}_N) \tag{11}$$

where $\mathbf{v} = v\hat{\mathbf{v}}, |\hat{\mathbf{v}}| = 1, \mathbf{v}_j = v^2\,\mathbf{e}_j$,with $\mathbf{e}_j \neq 0$,$\mathbf{e}_j \neq \mathbf{e}_k, j \neq k$, for $j, k = 3, \ldots, N$.I define also $\mathbf{v}_1 = -\mathbf{v}m/m_1$,and $\mathbf{v}_2 = \mathbf{v}m/m_2$.I denote by

$$\mathbf{v}_{jk} = \mathbf{v}_k - \mathbf{v}_j, \ v_{jk} = |\mathbf{v}_{jk}|, \ j, k = 1, \ldots, N$$

,respectively,the approximate relative velocity of the particles j and k and its absolute value.It follows from my definitions that

$$\mathbf{v}_{jk} = v^2(\mathbf{e}_k - \mathbf{e}_j) \neq 0,\ j,k = 3,\ldots,N \ ,$$

$$\mathbf{v}_{2j} = v^2(\mathbf{e}_j - \frac{m}{m_2}\frac{\hat{\mathbf{v}}}{v}) \neq 0 \,\mathrm{if}\, v > \frac{m}{m_2}(e_j)^{-1} \ ,$$

$$\mathbf{v}_{1j} = v^2\,(\mathbf{e}_j + \frac{m}{m_1}\frac{\hat{\mathbf{v}}}{v}) \neq 0, \mathrm{if}\, v > \frac{m}{m_1}(e_j)^{-1} \ ,$$

where $e_j = |\mathbf{e}_j|$.Note that

$$\Phi_{\mathbf{V}} = e^{i\mu_{12}\mathbf{V}\cdot\mathbf{X}} \prod_{j=3}^{N} e^{i\mu_j \mathbf{V}_j\cdot\mathbf{X}_j}\,\Phi_0 \ .$$

Then in the high–velocity state the approximate relative velocity of the pair (12) is v while all other particles travel with minimal velocity proporcional to v^2 relative to each other as well as relative to the particles in the distinguished pair.

THEOREM 2.3. *(Reconstruction Formula) Suppose that $V^S \in \mathcal{V}_{SR}, V^L \in \mathcal{V}_{LR}$,with $u \geq M+1$, that Assumption P is satisfied, that for every $\Phi \in H_2(\mathbf{R}^n)$ the function $t \to V_{12}^s(t,\mathbf{y})\Phi$ is strongly continuous from $\mathbf{R}$ into $L^2(\mathbf{R}^n)$ and that for some $1 \leq a \leq n$ and all $\Phi \in L^2(\mathbf{R}^n)$ the function $t \to (\frac{\partial}{\partial y_a}V_{12}^l)(t,\mathbf{y})\Phi$ is strongly continuous from $\mathbf{R}$ into $L^2(\mathbf{R}^n)$. Then for all $\Phi_{\mathbf{V}}, \Psi_{\mathbf{V}}$ as in (11)*

$$\lim_{v\to\infty} iv([S^D(q),p_a]\Phi_{\mathbf{V}},\Psi_{\mathbf{V}}) = \int_{-\infty}^{\infty} d\tau \Big[(V_{12}^s(q,\mathbf{x}+\tau\hat{\mathbf{v}})p_a\Phi_{12},\Psi_{12})$$

$$-(V_{12}^s(q,\mathbf{x}+\tau\hat{\mathbf{v}})\Phi_{12},p_a\Psi_{12})\Big] + i\int_{-\infty}^{\infty} d\tau((\frac{\partial}{\partial x_a}V_{12}^l)(q,\mathbf{x}+\tau\hat{\mathbf{v}})\Phi_{12},\Psi_{12}) \quad (12)$$

for all $q \in \mathbf{R}$.

Remark that it follows from (2) and Fatou's lemma that for all $\Phi \in L^2(\mathbf{R}^n)$

$$\int_{-\infty}^{\infty} dr \left\| V_{12}^s(q,\mathbf{y})\,(-\Delta_{\mathbf{y}}+I)^{-1}\,F(|\mathbf{y}| \geq |r|)\,\Phi \right\| \leq$$

$$\liminf_{v\to\infty} \int_{-\infty}^{\infty} dr \left\| V_{12}^s(r/v+q,\mathbf{y})\,(-\Delta_{\mathbf{y}}+I)^{-1}\,F(|\mathbf{y}| \geq |r|)\,\Phi \right\| < \infty \ . \quad (13)$$

Then for all Φ in the space of Schwarz

$$\int_{-\infty}^{\infty} d\tau \left\| V_{12}^{s}(q, \mathbf{y} + \tau\hat{\mathbf{v}})\Phi \right\| \leq \int_{-\infty}^{\infty} d\tau \left\| V_{12}^{s}(q, \mathbf{y} + \tau\hat{\mathbf{v}})(-\Delta_{\mathbf{y}} + I)^{-1} \right.$$

$$\left. F(|\mathbf{y} + \tau\hat{\mathbf{v}}| \geq |\tau/2|)\,(-\Delta_{\mathbf{y}} + I)\Phi \right\| + \int_{-\infty}^{\infty} d\tau \left\| V_{12}^{s}(q, \mathbf{y} + \tau\hat{\mathbf{v}})(-\Delta_{\mathbf{y}} + I)^{-1} \right\|$$

$$\left\| F(|\mathbf{y}| \geq |\tau/2|)\,(-\Delta_{\mathbf{y}} + I)\Phi \right\| < \infty \ , \tag{14}$$

where I used (13) and the rapid decay of Φ in configuration space.It follows that the integral in the first term in the right–hand side of (12) is well defined.Since $(\frac{\partial}{\partial y_a} V_{12}^{l})(t, \mathbf{y})$ satisfies (13) (in this case the regularization $(-\Delta_{\mathbf{y}} + I)^{-1}$ is not necessary) also the integral in the second term in the right–hand side of (12) is well defined. Formula (12) tells us that from the high–velocity limit of the commutator with p_a of the modified Dollard scattering operator with initial time q we reconstruct the scalar product in a dense set of states of the Radon (or X–ray) transform of the "derivative" with respect to x_a of the potential $V_{12}(q, \mathbf{x})$ at the same initial time q.This substanciates the remark made in the introduction that in the high–velocity limit the potentials can be considered constant during the scattering time. Moreover (12) gives us enough information to uniquely reconstruct the potential in a constructive way.

COROLLARY 2.4. *Suppose that each one of the pair potentials V_{jk} satisfies the assumptions of Theorem 2.3.Then if (the high velocity limit of) any one of the Dollard scattering operators $S^D(q)$ is known for all $q \in \mathbf{R}$ the potential is uniquely defined.*

Remark that it follows from the definitions in (6) and (7) that for all $q_1, q_2 \in \mathbf{R}$

$$S^D(q_2) = U^D(q_2, q_1)\, S^D(q_1)\, U^D(q_1, q_2) \ .$$

Then if the long–range part of the potential is a priori known and we know $S^D(q_1)$ for a fixed q_1, we know $S^D(q)$ for all $q \in \mathbf{R}$.It follows that in this case it is enough to know $S^D(q)$ for only one q to uniquely reconstruct the potential.In particular when $V^L \equiv 0$ the modified Dollard wave and scattering operators coincide with the canonical ones

$$\Omega_{\pm}(q) = s - \lim_{t \to \pm\infty} U(q, t)\, e^{-i(t-q)H_0} \ ,$$

$$S(q) = (\Omega_{+}(q))^{*}\ \Omega_{-}(q) \ ,$$

$$S(q_2) = e^{i(q_1 - q_2)H_0}\, S(q_1)\, e^{-i(q_1 - q_2)H_0}$$

and the potential V is uniquely reconstructed from the canonical scattering operator $S(q)$ at only one initial time,say $q = 0$.

To estimate the rate of convergence in (12) we have to strengthen the continuity in t of the potentials.

COROLLARY 2.5. *Suppose that the hypotheses of Theorem 2.3 are satisfied,that for each $q \in \mathbf{R}$ there are $\rho_i > 0$ and functions $Q_i(\mathbf{y}), i = 1,2$ such that for every function $g \in C_0^\infty(\mathbf{R}^n)$ the operators $Q_i(\mathbf{y})g(\mathbf{p})$are bounded and*

$$|V_{12}^s(t+q,\mathbf{y}) - V_{12}^s(q,\mathbf{y})| \leq |t|^{\rho_1} Q_1(\mathbf{y}) \ ,$$

$$\left|(\frac{\partial}{\partial y_a} V_{12}^l)(t+q,\mathbf{y}) - (\frac{\partial}{\partial y_a} V_{12}^l)(q,\mathbf{y})\right| \leq |t|^{\rho_2} Q_2(\mathbf{y}) \ ,$$

and

$$\int_0^\infty dr\,(1+r)^{\rho_i} \, \|Q_i(\mathbf{y})\, g(\mathbf{p}) F(|\mathbf{y}| \geq r)\| < \infty, \ i = 1,2$$

and that for some $0 \leq \rho_3 \leq u - M - 1$,all $g \in C_0^\infty(\mathbf{R}^n)$ and all $q \in \mathbf{R}$, there is a $v_0 > 0$ and a function h with $(1+r)^{\rho_3}\, h(r) \in L^1((0,\infty))$ such that for all $r \in \mathbf{R}$

$$\|V_{12}^s(r/v + q, \mathbf{y}) g(\mathbf{p})\, F(|\mathbf{y}| \geq |r|)\| \leq h(|r|), \ v \geq v_0 \ . \tag{15}$$

Then for all $\Phi_\mathbf{v}, \Psi_\mathbf{v}$ as in (11)

$$iv([S^D(q), p_a]\Phi_\mathbf{v}, \Psi_\mathbf{v}) = \int_{-\infty}^\infty d\tau \Big[(V_{12}^s(q, \mathbf{x} + \tau\hat{\mathbf{v}}) p_a \Phi_{12}, \Psi_{12})$$

$$-(V_{12}^s(q, \mathbf{x} + \tau\hat{\mathbf{v}})\Phi_{12}, p_a \Psi_{12}) \Big] + i \int_{-\infty}^\infty d\tau ((\frac{\partial}{\partial x_a} V_{12}^l)(q, \mathbf{x} + \tau\hat{\mathbf{v}})\Phi_{12}, \Psi_{12})$$

$$+ O(v^{-\rho_1}) + O(v^{-\rho_2}) + o(v^{-\eta}), \ \eta \leq \rho_3, \ \eta < \gamma - \epsilon - 1$$

as $v \longrightarrow \infty$.

Condition (2) implies (15) with $\rho_3 = 0$. Larger ρ means faster decay as $|\mathbf{y}| \rightarrow \infty$.

I now prepare some results that I use in the proof of Theorem 2.3 and its Corollaries. It follows from (8) that

$$\left\|(1 + |\tilde{\mathbf{x}}_k - \tilde{\mathbf{x}}_j|^2)^u \Phi_\mathbf{v}\right\| \leq C, \ j,k = 1,\ldots,N \ . \tag{16}$$

Equation (9) implies that there are functions $f_{jk} \in C_0^\infty(\mathbf{R}^n)$ such that

$$\Phi_\mathbf{v} = f_{jk}(\mathbf{p}_{jk} - \mu_{jk}\mathbf{v}_{jk})\Phi_\mathbf{v}, \ 1 \leq j < k \leq N \ . \tag{17}$$

I denote by $\tilde{U}(t,q)$ the correction term in the Dollard propagator

$$\tilde{U}(t,q) := \exp\left[-i \sum_{j<k} \int_q^t dr\, V_{jk}^l(r, r\,\mathbf{p}_{jk}/\mu_{jk}) \right] \ . \tag{18}$$

As in the proof of (4.16),(4.17) and of Proposition 3.1 of Enss and Weder (1995a) I prove that for each $q \in \mathbf{R}$ there is a constant $v_0 > 0$ such that for all $v \geq v_0$

$$\left\| (\tilde{\mathbf{x}}_k - \tilde{\mathbf{x}}_j)\, \tilde{U}(t,q) \prod_{j'<k'} f_{j'k'}(\mathbf{p}_{j'k'} - \mu_{j'k'}\mathbf{v}_{j'k'}) \left(1 + |\tilde{\mathbf{x}}_k - \tilde{\mathbf{x}}_j|^2\right)^{-1/2} \right\|$$

$$\leq C(1 + v_{jk}|t-q|)^{2+\epsilon-\gamma} \ , \tag{19}$$

$1 \leq j < k \leq N$,with ϵ and γ as in (3) and that for $\lambda > 0$ and u as in (3)

$$\left\| F\left(|\tilde{\mathbf{x}}_k - \tilde{\mathbf{x}}_j| \geq \lambda v_{jk}|t-q| \right) \tilde{U}(t,q) \prod_{j'<k'} f_{j'k'}(\mathbf{p}_{j'k'} - \mu_{j'k'}\mathbf{v}_{j'k'}) \right.$$
$$\left. \left(1 + |\tilde{\mathbf{x}}_k - \tilde{\mathbf{x}}_j|^2\right)^{-u} \right\| \leq C\left(1 + v_{jk}|t-q|\right)^{-u-\delta} , \tag{20}$$

for some $\delta > 0$ that depends on u, ϵ, and γ.

The following relations ,that I will use frequently,are obtained under translation in configuration or momentum space:

$$e^{i\mathbf{p}\cdot\mathbf{v}t} f(\mathbf{x}) e^{-i\mathbf{p}\cdot\mathbf{v}t} = f(\mathbf{x} + \mathbf{v}t) \ , \tag{21}$$

$$e^{-im\mathbf{v}\cdot\mathbf{x}} f(\mathbf{p}) e^{im\mathbf{v}\cdot\mathbf{x}} = f(\mathbf{p} + m\mathbf{v}) \tag{22}$$

for any bounded measurable function f, $m > 0$, and in particular

$$e^{-im\mathbf{v}\cdot\mathbf{x}} e^{-it\mathbf{p}^2/2m} e^{im\mathbf{v}\cdot\mathbf{x}} = e^{-i\mathbf{p}\cdot\mathbf{v}t} e^{-it\mathbf{p}^2/2m} e^{-imv^2t/2} \tag{23}$$

where $v = |\mathbf{v}|$.

LEMMA 2.6. *Suppose that $V^S \in \mathcal{V}_{SR}, V^L \in \mathcal{V}_{LR}$ with $u \geq M+1+\rho$, for some $\rho \geq 0$ and that for all $g \in C_0^\infty(\mathbf{R}^n)$ and all $q \in \mathbf{R}$ there is a $\tilde{v}_0 > 0$ and a function $\tilde{h}$ with $(1+r)^\rho \tilde{h}(r) \in L^1((0,\infty))$ such that V_{jk}^s satisfies (15) for all $v \geq \tilde{v}_0$.Then for each $q \in \mathbf{R}$ and $f_{j'k'} \in C_0^\infty(\mathbf{R}^n), 1 \leq j' < k' \leq N$, there is a function h with $(1+r)^\rho h(r) \in L^1((0,\infty))$ and a $v_0 > 0$ such that for all $v \geq v_0$*

$$\left\| V_{jk}^s(t, \tilde{\mathbf{x}}_k - \tilde{\mathbf{x}}_j) U^D(t,q) \prod_{j'<k'} f_{j'k'}(\mathbf{p}_{j'k'} - \mu_{j'k'}\mathbf{v}_{j'k'}) \left(1 + |\tilde{\mathbf{x}}_k - \tilde{\mathbf{x}}_j|^2\right)^{-u} \right\|$$

$$\leq h(v_{jk}|t-q|) \ . \tag{24}$$

Proof: Take $g \in C_0^\infty(\mathbf{R}^n)$ with $g \equiv 1$ on the support of f_{jk}.Let us denote by I the left–hand side of (24). Then $I \leq I_1 + I_2 + I_3$,where

$$I_1 = \Big\| V^s_{jk}(t, \tilde{\mathbf{x}}_k - \tilde{\mathbf{x}}_j) g(\mathbf{p}_{jk} - \mu_{jk}\mathbf{v}_{jk}) F(|\tilde{\mathbf{x}}_k - \tilde{\mathbf{x}}_j - \mathbf{v}_{jk}(t-q)|$$

$$\geq v_{jk}|t-q|\,5/8)\, e^{-i(t-q)\mathbf{p}^2_{jk}/2\mu_{jk}} g(\mathbf{p}_{jk} - \mu_{jk}\mathbf{v}_{jk}) F(|\tilde{\mathbf{x}}_k - \tilde{\mathbf{x}}_j| \leq v_{jk}$$

$$|t-q|/8)\tilde{U}(t,q) \prod_{j'<k'} f_{j'k'}(\mathbf{p}_{j'k'} - \mu_{j'k'}\mathbf{v}_{j'k'}) \left(1 + |\tilde{\mathbf{x}}_k - \tilde{\mathbf{x}}_j|^2\right)^{-u} \Big\| \, , \qquad (25)$$

$$I_2 = \Big\| V^s_{jk}(t, \tilde{\mathbf{x}}_k - \tilde{\mathbf{x}}_j) g(\mathbf{p}_{jk} - \mu_{jk}\mathbf{v}_{jk}) F\Big(|\tilde{\mathbf{x}}_k - \tilde{\mathbf{x}}_j - \mathbf{v}_{jk}(t-q)| \geq$$

$$v_{jk}|t-q|\,5/8 \Big) e^{-i(t-q)\mathbf{p}^2_{jk}/2\mu_{jk}} g(\mathbf{p}_{jk} - \mu_{jk}\mathbf{v}_{jk}) F\Big(|\tilde{\mathbf{x}}_k - \tilde{\mathbf{x}}_j| > v_{jk}|t-q|/8 \Big)$$

$$\tilde{U}(t,q) \prod_{j'<k'} f_{j'k'}(\mathbf{p}_{j'k'} - \mu_{j'k'}\mathbf{v}_{j'k'}) \left(1 + |\tilde{\mathbf{x}}_k - \tilde{\mathbf{x}}_j|^2\right)^{-u} \Big\| \, , \qquad (26)$$

$$I_3 = \Big\| V^s_{jk}(t, \tilde{\mathbf{x}}_k - \tilde{\mathbf{x}}_j) g(\mathbf{p}_{jk} - \mu_{jk}\mathbf{v}_{jk})\, F(|\tilde{\mathbf{x}}_k - \tilde{\mathbf{x}}_j - \mathbf{v}_{jk}(t-q)|$$

$$< |t-q|\,5/8) e^{-i(t-q)\mathbf{p}^2_{jk}/2\mu_{jk}}\, g(\mathbf{p}_{jk} - \mu_{jk}\mathbf{v}_{jk}) \tilde{U}(t,q)$$

$$\prod_{j'<k'} f_{j'k'}(\mathbf{p}_{j'k'} - \mu_{j'k'}\mathbf{v}_{j'k'} \left(1 + |\tilde{\mathbf{x}}_k - \tilde{\mathbf{x}}_j|^2\right)^{-u} \Big\| \, , \qquad (27)$$

where I used that $H_0 = \mathbf{p}^2_{jk}/2\mu_{jk} + \tilde{H}_0$ with a operator $\tilde{H}_0$ that commutes with $\tilde{\mathbf{x}}_k - \tilde{\mathbf{x}}_j$ (for example taking Jacobi coordinates). By (1) and (22) for $v_{jk} \geq v_0$ with v_0 large enough

$$I_1 \leq C\,(1+|t|)^M \Big\| F(|\tilde{\mathbf{x}}_k - \tilde{\mathbf{x}}_j - \mathbf{v}_{jk}(t-q)| \geq v_{jk}|t-q|\,5/8) e^{-i(t-q)\mathbf{p}^2_{jk}/2\mu_{jk}}$$

$$g(\mathbf{p}_{jk} - \mu_{jk}\mathbf{v}_{jk}) F(|\tilde{\mathbf{x}}_k - \tilde{\mathbf{x}}_j| < v_{jk}|t-q|/8) \Big\|$$

$$\leq C(1+|t|)^M \left(1 + v_{jk}|t-q|\right)^{-u-1} \, , \qquad (28)$$

where the last estimate is proven as in Lemma 2.2 in Enss and Weder (1995a) using rapid decay away from the classically allowed region (non–stationary phase). Moreover by (1),(20) and (22)

$$I_2 \leq C(1+|t|)^M \; (1 + v_{jk}|t-q|)^{-u-\delta} \, , \qquad (29)$$

and by (15) and (22)

$$I_3 \leq C \left\| V^s_{jk}(t, \tilde{\mathbf{x}}_k - \tilde{\mathbf{x}}_j) g(\mathbf{p}_{jk}) F(|\tilde{\mathbf{x}}_k - \tilde{\mathbf{x}}_j| \geq v_{jk}|t-q|\, 3/8) \right\| \leq \tilde{h}(v_{jk}|t-q|) \tag{30}$$

with $(1+r)^\rho \tilde{h}(r) \in L^1((0,\infty))$.The Lemma follows from (28)–(30) since $u \geq M+1+\rho$.

□

If $V^S \in \mathcal{V}_{SR}$ and $V^L \in \mathcal{V}_{LR}$ we have that for all $f_{j'k'} \in C_0^\infty(\mathbf{R}^n)$, $1 \leq j' < k' \leq N$

$$\left\| \left(V^l_{jk}(t, \tilde{\mathbf{x}}_k - \tilde{\mathbf{x}}_j) - V^l_{jk}(t, t\, \mathbf{p}_{jk}/\mu_{jk}) \right) U^D(t,q) \prod_{j'<k'} f_{j'k'}(\mathbf{p}_{j'k'} - \mu_{j'k'} \mathbf{v}_{j'k'}) \left(1 + |\tilde{\mathbf{x}}_k - \tilde{\mathbf{x}}_j|^2 \right)^{-u} \right\| \leq C \left(1 + v_{jk}|t| \right)^{-1-\delta} \tag{31}$$

for some $\delta > 0$.This estimate is proven using (19) and (20) as in the proof of equation (4.19) and Lemma 3.3 in Enss and Weder (1995a).

It follows from Duhamel's formula that

$$U(q,t)\, U^D(t,q) \Phi_{\mathbf{v}} = \Phi_{\mathbf{v}} + i \int_q^t dr\, U(q,r) \sum_{j<k} (V_{jk}(r, \tilde{\mathbf{x}}_k - \tilde{\mathbf{x}}_j) - V^l_{jk}(r, r\, \mathbf{p}_{jk}/\mu_{jk}))\, U^D(r,q) \Phi_{\mathbf{v}} \ . \tag{32}$$

Then by (15) with $\rho_3 = 0$,(16),(17), (24) and (31)

$$\int_{-\infty}^{\infty} dr \left\| U(q,r) \sum_{j<k} \left(V_{jk}(r, \tilde{\mathbf{x}}_k - \tilde{\mathbf{x}}_j) - V^l_{jk}(r, r\, \mathbf{p}_{jk}/\mu_{jk}) \right) U^D(r,q)\, \Phi_{\mathbf{v}} \right\| \leq \int_{-\infty}^{\infty} dr\, h(v_{jk}|r-q|) + C \int_{-\infty}^{\infty} dr\, (1 + v_{jk}|r|)^{-1-\delta} \leq \frac{C}{v_{jk}} \tag{33}$$

for some constant C. By (32) and (33) the strong limits in (6) exist if $V^S \in \mathcal{V}_{SR}, V^L \in \mathcal{V}_{LR}$ with $u \geq M+1$ and Assumption P is satisfied.Moreover, using again Duhamel's formula

$$(U(t,q)\, \Omega^D_{\pm}(q) - U^D(t,q))\, \Phi_{\mathbf{v}} = i \int_0^{\pm\infty} dr\ U(t, t+r)(H(t+r) - H^D(t+r))\, U^D(t+r,q)\, \Phi_{\mathbf{v}} \tag{34}$$

and I obtain as in (33) that

$$\left\|(U(t,q)\,\Omega_{\pm}^{D}(q) - U^{D}(t,q))\Phi_{\mathbf{v}}\right\| \le \frac{C}{v}, \quad v \ge v_0 \tag{35}$$

where C and v_0 are uniform in t.

Proof of Theorem 2.3: Since the $\Omega_{\pm}^{D}(q)$ are partially isometric $\left(\Omega_{\pm}^{D}(q)\right)^{*}\Omega_{\pm}^{D}(q) = I$ and then

$$\begin{aligned} i(S^{D}(q) - I)\Phi_{\mathbf{v}} &= i\Big((\Omega_{+}^{D}(q))^{*} - (\Omega_{-}^{D}(q))^{*}\Big)\,\Omega_{-}^{D}(q)\Phi_{\mathbf{v}} \\ &= \int_{-\infty}^{\infty} dt\, U^{D}(q,t)\,(H(t) - H^{D}(t))\,U(t,q)\,\Omega_{-}^{D}(q)\,\Phi_{\mathbf{v}} \end{aligned} \tag{36}$$

Noting that $[S^{D}(q), p_a] = [S^{D}(q), p_a - \mu_{12}v_a]$ and that $(p_a - \mu_{12}v_a)\Phi_{\mathbf{v}} = (p_a\Phi_0)_{\mathbf{v}}$ I prove:

$$iv\left([S^{D}(q), p_a]\Phi_{\mathbf{v}}, \Psi_{\mathbf{v}}\right) = \int_{-\infty}^{\infty} d\tau\, l_{\mathbf{v}}(\tau) + R(\mathbf{v}) \tag{37}$$

with $\tau = v(t-q)$ where

$$\begin{aligned} l_{\mathbf{v}}(\tau) = \Bigg[&\Big(V^{s}(\tau/v + q, \mathbf{x})U^{D}(\tau/v + q, q)(p_a\Phi_0)_{\mathbf{v}}, U^{D}(\tau/v + q, q)\Psi_{\mathbf{v}}\Big) \\ &- \Big(V^{s}(\tau/v + q, \mathbf{x})U^{D}(\tau/v + q, q)\Phi_{\mathbf{v}}, U^{D}(\tau/v + q, q)(p_a\Psi_0)_{\mathbf{v}}\Big)\Bigg] + i \\ &\Big((\frac{\partial}{\partial x_a}V_{12}^{l})(\tau/v + q, q)U^{D}(\tau/v + q, q)\Phi_{\mathbf{v}}, U^{D}(\tau/v + q, q)\Psi_{\mathbf{v}}\Big) \end{aligned} \tag{38}$$

is the leading term and the remainder is

$$\begin{aligned} R(\mathbf{v}) = v \sum_{j<k=3}^{N} \int_{-\infty}^{\infty} dt \Big\{ &\Big([V_{jk}(t, \tilde{\mathbf{x}}) - V_{jk}^{l}(t, t\,\mathbf{p}_{jk}/\mu_{jk})]U^{D}(t,q)(p_a\Phi_0)_{\mathbf{v}}\ , \\ &U^{D}(t,q)\Psi_{\mathbf{v}}\Big) - \Big([V_{jk}(t, \tilde{\mathbf{x}}_k - \tilde{\mathbf{x}}_j) - V_{jk}^{l}(t, t\mathbf{p}_{jk}/\mu_{jk})]U^{D}(t,q)\Phi_{\mathbf{v}}\ , \\ &U^{D}(t,q)(p_a\Psi_0)_{\mathbf{v}}\Big)\Big\} + v\int_{-\infty}^{\infty} dt\Big((U^{D}(t,q)\Omega_{-}^{D}(q) - U^{D}(t,q))(p_a\Phi_0)_{\mathbf{v}}\ , \\ \sum_{j<k}[V_{jk}(t, \tilde{\mathbf{x}}_k - \tilde{\mathbf{x}}_j) - &V_{jk}^{l}(t, t\,\mathbf{p}_{jk}/\mu_{jk})]U^{D}(t,q)\Psi_{\mathbf{v}}\Big) - v\int_{-\infty}^{\infty} dt\Big((U^{D}(t,q)\Omega_{-}^{D}(q) \end{aligned}$$

$$- U^D(t,q))\Phi_{\mathbf{v}}, \sum_{j<k}[V_{jk}(t,\tilde{\mathbf{x}}_k - \tilde{\mathbf{x}}_j) - V^l_{jk}(t, t\,\mathbf{p}_{jk}/\mu_{jk})]U^D(t,q)(p_a\Psi_0)_{\mathbf{v}}\Big) . \tag{39}$$

It follows from (24),(31) and (35) that

$$R(\mathbf{v}) = O(1/v) . \tag{40}$$

Moreover since $(\frac{\partial}{\partial x_a}V^l_{12})(\mathbf{x})$ satisfies (1) and (2) if follows from (24) that for $v \geq v_0$

$$|l_{\mathbf{v}}(\tau)| \leq h(\tau) \tag{41}$$

for some $h(\tau) \in L^1((0,\infty))$.Furthermore, it follows from (21)–(23) that for each fixed τ

$$\lim_{v\to\infty} l_{\mathbf{v}}(\tau) = \Big(V^s_{12}(q,\mathbf{x}+\tau\hat{\mathbf{v}})p_a\Phi_{12},\Psi_{12}\Big) - \Big(V^s_{12}(q,\mathbf{x}+\tau\hat{\mathbf{v}})\Phi_{12} , \\ p_a\Psi_{12}\Big) + i\Big((\frac{\partial}{\partial x_a}V^l_{12})(q,\mathbf{x}+\tau\hat{\mathbf{v}})\Phi_{12},\Psi_{12}\Big) \tag{42}$$

and it follows from (41) and Lebesque dominated convergence theorem that

$$\lim_{v\to\infty}\int_{-\infty}^{\infty} d\tau\, l_{\mathbf{v}}(\tau) = \int_{-\infty}^{\infty} d\tau\Big[\Big(V^s_{12}(q,\mathbf{x}+\tau\hat{\mathbf{v}})p_a\Phi_{12},\Psi_{12}\Big) \\ -\Big(V^s_{12}(q,\mathbf{x}+\tau\hat{\mathbf{v}})\Phi_{12},p_a\Psi_{12}\Big) + i\Big((\frac{\partial}{\partial x_a}V^l_{12})(q,\mathbf{x}+\tau\hat{\mathbf{v}})\Phi_{12},\Psi_{12}\Big)\Big] .$$

This and (40) yield (12).

Proof of Corollary 2.4: I prove this Corollary as in the proof of Theorem 1.2 in Enss and Weder (1995a).I give details for the reader's convenience.Let us identify any $\mathbf{z} = (z_1,z_2) \in \mathbf{R}^2$ with the vector $z_1\mathbf{c}_a + z_2\mathbf{c}_k \in \mathbf{R}^n$ with $a \neq k$ and where $\mathbf{c}_j, j = 1,\cdots,n$ denote the unit vectors along the z_j direction in $\mathbf{R}^n$. For any Φ,Ψ with $\hat{\phi},\hat{\psi} \in C_0^\infty(\mathbf{R}^n)$ I denote

$$\Phi(\mathbf{z}) := e^{-i\mathbf{p}\cdot\mathbf{z}}\Phi,\; \Psi(\mathbf{z}) := e^{-i\mathbf{p}\cdot\mathbf{z}}\Psi ,$$

and

$$f(\mathbf{z}) := \Big(V^s_{12}(q,\mathbf{x})p_a\Phi(\mathbf{z}),\Psi(\mathbf{z})\Big) - \Big(V^s_{12}(q,\mathbf{x})\Phi(\mathbf{z}),p_a\Psi(\mathbf{z})\Big) \\ +i\Big((\frac{\partial}{\partial x_a}V^l_{12})(q,\mathbf{x})\Phi(\mathbf{z}),\Psi(\mathbf{z})\Big) .$$

The function $f(\mathbf{z})$ is continuous,bounded and by the rapid decay of Φ in configuration space, (13) and a similar estimate for $(\frac{\partial}{\partial x_a} V^l_{12})(q, \mathbf{x})$, $f(\mathbf{z}) \in L^2(\mathbf{R}^2)$ .The Radon transform of $f(\mathbf{z})$ is given by

$$\tilde{f}(\hat{\mathbf{v}}, \mathbf{z}) = \int_{-\infty}^{\infty} d\tau f(\mathbf{z} + \hat{\mathbf{v}}\tau) = \lim_{v \to \infty} iv([S^D(q), p_a]\Phi_{\mathbf{v}}(\mathbf{z}), \Psi_{\mathbf{v}}(\mathbf{z})) \ .$$

Then I reconstruct the Radon transform of $f(\mathbf{z})$ from the high velocity limit of $[S^D(q), p_a]$ and inverting this Radon transform I uniquely reconstruct $f(\mathbf{z})$ (see Helgason (1984) Theorem 2.17 in Chapter I).Finally

$$(V_{12}(q, \mathbf{x})\Phi, \Psi) = i \int_0^{\infty} dz_1 \ f(z_1, 0)$$

and $V_{12}(q, \mathbf{x})$ is uniquely reconstructed as an operator and as a function for a.e. $\mathbf{x}$ in a constructive way.

Proof of Corollary 2.5 :this Corollary follows as in the proof of Theorem 2.4 in Enss and Weder (1995a) estimating the rate of convergence of $\int_{-\infty}^{\infty} d\tau \, l_{\mathbf{v}}(\tau)$ as $v \to \infty$.

If the long–range potentials are a priori known I can reconstruct the short–range potentials from the high velocity limit of the Dollard scattering operator without taking the commutator with a component of momentum.This follows as in Theorem 4.1 of Enss and Weder (1995a).I omit details.

There is an extensive literature about direct scattering with time–dependent potentials.See for example Theorem XI.28 of Reed and Simon (1979),Yafaev (1980),Yajima (1980),Kitada and Yajima (1982),Kitada and Yajima (1983), Graf (1990) , Wüller (1991),Ito (1993),Ito (1995) and the monograph Cycon et al.(1987) where further references are given.The papers Yajima (1980),Graf (1990) ,Wüller (1991),Ito (1993) and Ito (1995) study the charge transfer model for the scattering of a (light) particle under the action of moving centers of force.This type of potentials are included in my class under appropriate conditions.

3 Slowly Decreasing Potentials that Vanish Asymptotically in Time

In this Section I study the particular case of potentials that tend to zero as $t \to \pm\infty$.An important physical example is the case of potentials of compact support in time.This corresponds to a situation where a potential is turned on and then switched off after a (short) time. An interesting aspect of this case is that potentials that decay very slowly as $|\mathbf{x}| \to \infty$ are of short range in the sense that no modified Dollard free evolution is required and the canonical wave operators exist provided that the potentials go to zero fast enough as $t \to \pm\infty$.The basic physical intuition here is that since the strength of the interaction goes to zero as $t \to \pm\infty$ the potentials may go to zero as $|\mathbf{x}| \to \infty$ very slowly and still the trajectories are well approximated as $t \to \pm\infty$ by the free motion.I now define an appropiate class of potentials.

DEFINITION 3.1. *I denote by $\mathcal{V}_0$the class of potentials*

$$V^0 = \sum_{j<k} V^0_{jk}(t, \tilde{\mathbf{x}}_k - \tilde{\mathbf{x}}_j) \tag{43}$$

where $V^0_{jk}(t, \mathbf{y})$ is Kato–small,

$$\|V^0_{jk}(t,\mathbf{y})(-\Delta_{\mathbf{y}} + I)^{-1}\| \leq C$$

for some constant C.Moreover, for all $q \in \mathbf{R}$ there are positive constants C and v_0 such that

$$\left\|V^0_{jk}(t,\mathbf{y})(-\Delta_{\mathbf{y}} + I)^{-1} F\left(|\mathbf{y}| \geq v|t-q|\right)\right\| \leq C(1+|t|)^{-\alpha}(1+v|t-q|)^{-\beta} \tag{44}$$

for $v \geq v_0$, with $\alpha > 0, 1 \geq \beta > 1/2$,and $\alpha + \beta > 1$.

Equation (44) is equivalent to the more intuitive decay condition

$$\left\|F(|\mathbf{y}| \geq v|t-q|)V^0_{jk}(t,\mathbf{y})(-\Delta_{\mathbf{y}} + I)^{-1}\right\| \leq C\,(1+|t|)^{-\alpha}\,(1+v|t-q|)^{-\beta} \ .$$

LEMMA 3.2. *Suppose that $V \in \mathcal{V}_0$ with $0 < \beta \leq 1$ and that Assumption P is satisfied.Then for all $f_{j'k'} \in C_0^\infty(\mathbf{R}^n)$ and $q \in \mathbf{R}$*

$$\int_{-\infty}^{\infty} dt \left\| V^0_{jk}(t, \tilde{\mathbf{x}}_k - \tilde{\mathbf{x}}_j)\, e^{-i(t-q)H_0} \prod_{j'<k'} f_{j'k'}(\mathbf{p}_{j'k'} - \mu_{j'k'}\mathbf{v}_{j'k'}) (1 + |\tilde{\mathbf{x}}_k - \tilde{\mathbf{x}}_j|^2)^{-1} \right\| = \begin{cases} O(v_{jk}^{-\beta}), & 0<\beta<1, \\ O\left((\ln v_{jk})\,/\,v_{jk}\right), & \beta = 1\ , \end{cases} \tag{45}$$

as $v \to \infty$.

*Proof:*Let us denote the integrand in the left hand side of (45) by I.Then $I \leq I_1 + I_2 + I_3$ where $I_j, j = 1,2,3$ are defined, respectively,as in (25),(26) and (27) with V^s_{jk} replaced by V^0_{jk} and $\tilde{U}(t,q)$ replaced by 1.As in Lemma 2.6 we prove that the integrals of I_1 andI_2 give contributions to the right hand side of (45) that are of order $O(v^{-1})$.Furthermore, by (44)

$$\int_{-\infty}^{\infty} dt\, I_3 \leq C \int_{-\infty}^{\infty} (1+|t|)^{-\alpha}(1+v_{jk}|t-q|)^{-\beta} \leq C \begin{cases} v_{jk}^{-\beta}, & 0<\beta<1, \\ |\ln v_{jk}|/v_{jk}, & \beta = 1\ . \end{cases}$$

□

It turns out that only the slowly decreasing parts of the pair potentials need to vanish as time goes to plus and minus infinity.In fact I can even allow for pair potentials that grow in time, provided that they have fast enough decay as the interparticle distances go to infinity, in the sense that they belong to $\mathcal{V}_{SR}$.So I assume that

$$V = V^S + V^0$$

with $V^S \in \mathcal{V}_{SR}$ and $V^0 \in \mathcal{V}_0$.The time–dependent interacting Hamiltonian

$$H(t) = H_0 + V^S + V^0$$

is self–adjoint in $D(H(t)) = D(H_0)$.Note that Lemma 2.6 holds with $V^L \equiv 0$ and $U^D(t,q)$ replaced by $e^{-i(t-q)H_0}$.Using also (45) we prove as in Section 2 that the canonical wave operators

$$\Omega_{\pm}(q) = s - \lim_{t\to\pm\infty} U(q,t)\, e^{-i(t-q)H_0} \tag{46}$$

exist and that

$$\|(U(t,q)\,\Omega_{\pm}(q) - e^{-i(t-q)H_0})\Phi_{\mathbf{v}}\| \le C \begin{cases} v^{-\beta}, & 0<\beta<1, \\ |\ln v|/v, & \beta = 1\ , \end{cases} \tag{47}$$

where $\Phi_{\mathbf{v}}$ is defined as in (11). The canonical scattering operator is

$$S(q) = (\Omega_{+}(q))^{*}\ \Omega_{-}(q)\ . \tag{48}$$

THEOREM 3.3. *Suppose that $V^S \in \mathcal{V}_{SR}, V^0 \in \mathcal{V}_0$,that Assumption P is satisfied,that for some $1 \le a \le n$ and all $g \in C_0^\infty(\mathbf{R}^n)$*

$$\left\| \left(\frac{\partial}{\partial x_a} V_{12}^0\right)(t,\mathbf{x})\, g(\mathbf{p}) F(|\mathbf{x}| \ge r) \right\| \le C(1+r)^{-1-\delta}\ , \tag{49}$$

$0 < \delta \le 1$,and that for all $\Phi \in H_2(\mathbf{R}^n)$ the functions $t \to V_{12}^s(t,\mathbf{x})\Phi$ and $t \to (\frac{\partial}{\partial x_a} V_{12}^0)(t,\mathbf{x})\Phi$ are strongly continuous from $\mathbf{R}$ into $L^2(\mathbf{R}^n)$.Then for all $\Phi_{\mathbf{v}}, \Psi_{\mathbf{v}}$ as in (11),

$$\lim_{v\to\infty} iv\Big([S(q),p_a]\Phi_{\mathbf{v}},\Psi_{\mathbf{v}}\Big) = \int_{-\infty}^{\infty} d\tau \Big[\Big(V_{12}^s(q,\mathbf{x}+\tau\hat{\mathbf{v}})p_a\Phi_{12},\Psi_{12}\Big)$$

$$-\Big(V_{12}^s(q,\mathbf{x}+\tau\hat{\mathbf{v}})\Phi_{12},p_a\Psi_{12}\Big)\Big]+$$

$$i\int_{-\infty}^{\infty} d\tau \Big((\tfrac{\partial}{\partial x_a} V_{12}^0)(q,\mathbf{x}+\tau\hat{\mathbf{v}})\Phi_{12},\Psi_{12}\Big)\ . \tag{50}$$

Proof: The proof of Theorem 2.3 applies in this case.I replace the $V^l_{jk}(t,\mathbf{x})$ by $V^0_{jk}(t,\mathbf{x})$ and $U^D(t,q)$ by $e^{-i(t-q)H_0}$ and I eliminate the terms containing $V^l_{jk}(t, t\,\mathbf{p}_{jk}/2\mu_{jk})$.Remark that it follows from (45) and (47) that now

$$R(\mathbf{v}) = \begin{cases} O(v^{-(2\beta-1)}), & 1/2 < \beta < 1, \\ \\ O((\ln v)^2/v), & \beta = 1 \ , \end{cases} \tag{51}$$

as $v \to \infty$.Finally since $(\frac{\partial}{\partial x_a} V^0_{12}(\mathbf{x}))$ is a short–range potential the particular case of Lemma 2.6 with $\tilde{U}(t,q) \equiv I$ applies and we complete the proof as in Theorem 2.3 using Lebesque dominated convergence theorem.

COROLLARY 3.4. *Suppose that each one of the $V_{jk}(t, \tilde{\mathbf{x}}_k - \tilde{\mathbf{x}}_j), 1 \leq j < k \leq N$,satisfies the assumptions of Theorem 3.2.Then (the high velocity limit of) $S(q)$ for one $q \in \mathbf{R}$ determines uniquely the potential V.*

*Proof:*This Corollary follows as in the proof of Corollary 2.4. Recall that since

$$S(q_2) = e^{i(q_1-q_2)H_0} S(q_1) e^{-i(q_1-q_2)H_0}$$

if we know $S(q)$ for one $q \in \mathbf{R}$ we know it for all $q \in \mathbf{R}$.

COROLLARY 3.5. *Suppose that the hyphotesis of Theorem 3.3 are satisfied and that for each $q \in \mathbf{R}$ there are $\rho_i > 0$ and functions $Q_i(\mathbf{x}), i = 1,2,$ such that for every $g \in C_0^\infty(\mathbf{R}^n)$ the operators $Q_i(\mathbf{x})g(\mathbf{p})$ are bounded and*

$$|V^s_{12}(t+q,\mathbf{x}) - V^s_{12}(q,\mathbf{x})| \leq |t|^{\rho_1} Q_1(\mathbf{x}) \ ,$$

$$|(\frac{\partial}{\partial x_a} V^0_{12})(t+q,\mathbf{x}) - (\frac{\partial}{\partial x_a} V^0_{12})(q,\mathbf{x})| \leq |t|^{\rho_2} Q_2(\mathbf{x}) \ ,$$

with

$$\int_0^\infty dr\,(1+r)^{\rho_i}\, \|Q_i(\mathbf{x})g(\mathbf{p})F(|\mathbf{x}| \geq r)\| < \infty, i = 1,2$$

and that for some $0 \leq \rho_3 \leq (2\beta - 1), \rho_3 < \delta$, with β as in (44)and δ as in (49), all $g \in C_0^\infty(\mathbf{R}^n)$ and all $q \in \mathbf{R}$, there are a $v_0 > 0$ and a function h with $(1+r)^{\rho_3} h(r) \in L^1((0,\infty))$ such that for all $r \in \mathbf{R}$

$$\|V^s_{12}(r/v + q, \mathbf{x}) g(\mathbf{p}) F(|\mathbf{x}| \geq |r|)\| \leq h(|r|), \ \ v \geq v_0 \ .$$

Then for all $\Phi_\mathbf{v}, \Psi_\mathbf{v}$ as in (11)

$$iv\Big([S(q),p_a]\Phi_{\mathbf{v}},\Psi_{\mathbf{v}}\Big) = \int_{-\infty}^{\infty} d\tau\Big[\Big(V_{12}^s(q,\mathbf{x}+\tau\hat{\mathbf{v}})p_a\Phi_{12},\Psi_{12}\Big)$$

$$-\Big(V_{12}^s(q,\mathbf{x}+\tau\hat{\mathbf{v}})\Phi_{12},p_a\Psi_{12}\Big)\Big] + i\int_{-\infty}^{\infty} d\tau\Big((\frac{\partial}{\partial x_a}V_{12}^0)(q,\mathbf{x}+\tau\hat{\mathbf{v}})\Phi_{12},\Psi_{12}\Big)$$

$$+O(v^{-\rho_1}) + O(v^{-\rho_2}) + \begin{cases} o(v^{-\rho_3}), & \rho_3 < 2\beta - 1, \\ O(v^{-\rho_3}), & \rho_3 = 2\beta - 1 < 1 \ . \end{cases} \tag{52}$$

Proof: The proof follows as in Corollary 2.5 and using the bound given in (51) for the error term $R(\mathbf{v})$.Note that in (52) I do not allow for $\rho_3 = 2\beta - 1 = 1$ when $\beta = 1$ because I need that $\rho_3 < \delta \leq 1$.

Acknowledgement
I thank Prof. V. Enss and Prof. K. Yajima for information on the literature on propagators and direct scattering for Schrödinger operators with time-dependent potentials.

References

Chadan K and Sabatier P. C. (1989): *Inverse Problems in Quantum Scattering Theory* (2nd edition Springer,Berlin)

Cycon H.L.,Froese R. G.,Kirsch W. and Simon B. (1987): *Schrodinger Operators* (Springer, Berlin)

Enss V. and Weder R. (1993): Inverse potential scattering:A geometrical approach ,in: *Mathematical Quantum Theory II:Schrödinger Operators* ,Proceedings of the Summer School in Mathematical Quantum Theory,August 1993,Vancouver,B.C.,Feldman J.,Froese R. G. and Rosen L.,editors (CRM Proceedings and Lecture Notes **8** AMS,Providence (1995)),151–162

Enss V. and Weder R. (1994): Uniqueness and reconstruction formulae for inverse N–particle scattering, in:*Differential Equations and Mathematical Physics*,Proceedings of the International Conference, University of Alabama at Birmingham,March 1994,Knowles I.,editor (International Press,Boston (1995)),55–66

Enss V. and Weder R. (1995a): The geometrical approach to multidimensional inverse scattering. J. Math. Phys. **36**, 3902–3921

Enss V. and Weder R. (1995b): Inverse two–cluster scattering. Inverse Problems **12**,409–418

Graf G. M. (1990): Phase space analysis of the charge transfer model. Helv. Phys. Acta **63**,107–138

Helgason S. (1984): *Groups and Geometric Analysis* (Academic Press, Orlando)

Ito H. T. (1993): Charge transfer model and (two–cluster) → (two–cluster) three–body scattering. J. Math. Kyoto Univ. **33**,65–113

Ito H. T. (1995): Charge transfer model and three–body scattering. J. Math. Phys. **36**,115–132

Kato T. (1953): Integration of the equation of evolution in a Banach space. J. Math. Soc. Japan **5**,200–234

Kato T. (1970): Linear evolution equations of hyperbolic type I. J. Fact. Sci. Univ. of Tokyo Sect IA Math. **17**,241–258

Kato T (1976): *Perturbation Theory for linear Operators* (2nd edition,Springer,Berlin)

Kitada H. and Yajima K. (1982): A scattering theory for time–dependent long–range potentials. Duke Math. J. **49**,341–376

Kitada H. and Yajima K. (1983): Remarks on our paper A scattering theory for time–dependent long–range potentials. Duke Math. J. **50**, 1005–1016

Newton R. G. (1989): *Inverse Schrödinger Scattering in three Dimensions* (Springer,Berlin)

Perla Menzala G. (1985): Sur l'opérateur de diffusion pour l'équation des ondes avec des potentiels dépendant du temps. C.R. Acad. Sc. Paris **300**,621–624

Ramm A. G. and Sjöstrand J. (1991): An inverse problem for the wave equation. Math. Z. **206**,119–130

Reed M. and Simon B. (1975): *Methods of Modern Mathematical Physics II.Fourier Analysis,Self–Adjointness* (Academic Press,New York)

Reed M. and Simon B. (1979): *Methods of Modern Mathematical Physics III:Scattering Theory* (Academic Press,New York)

Stefanov P. D. (1989): Uniqueness for the multi–dimensional inverse scattering problem for time–dependent potentials. Math. Z. **201** ,541–559

Weder R. (1995): Multidimensional inverse scattering in an electric field. J. Func. Anal.**139**,441–465

Wüller U. (1991): Geometric methods in scattering theory of the charge transfer model. Duke Math. J. **62**,273–313

Yafaev D. R. (1980): Asymptotic completeness for the multidimensional time–dependent Schrödinger equation. Sov. Math. Dokl. **21**,545–549

Yajima K. (1980): A multi–channel scattering theory for some time–dependent Hamiltonians,charge transfer model. Comm. Math. Phys. **75**,153–178

Yajima K. (1987): Existence of solutions for Schrödinger evolution equations. Comm. Math. Phys. **110**,415–426

Yajima K. (1991): Schrödinger evolution equations with magnetic fields. J. d'Analyse Math. **56**,29–76

Inverse Scattering Approach for Stratified Chiral Media

Anne Boutet de Monvel[1,2] and Dimitri Shepelsky[2]

[1] Institut de Mathématiques de Jussieu, CNRS UMR 9994, Laboratoire de Physique mathématique et Géométrie, case 7012, Université Paris 7 Denis Diderot, 2 place Jussieu, F-75251 Paris Cedex 05

[2] B. Verkin Institute for Low temperature Physics, 47, Lenin Avenue, 310164, Kharkov, Ukraine

Abstract. Propagation of a transversely polarized time-harmonic electromagnetic plane wave normally incident on a stratified nonreciprocal chiral (bi-isotropic) slab is considered. The medium is modeled by the constitutive relations in the frequency domain. The structure of the scattering matrix is analyzed. The inverse problem of reconstruction of material characteristics of a medium is studied.

1 Introduction

In this paper we deal with the scattering problem of a transient electromagnetic field normally incident upon a stratified nonreciprocal chiral slab. Our main goal is to study the inverse scattering problem, that is given the scattering data (obtained from the measurements of the incident and the reflected fields), to determine information about material parameters of the media.

According to Maxwell's theory of electromagnetism, one can describe material media by constitutive relations that relate dielectric displacement $\overline{D}$ and magnetic induction $\overline{B}$ to electric and magnetic fields ($\overline{E}$ and $\overline{H}$, respectively).

A bi-isotropic (nonreciprocal chiral) medium is a linear medium that has the following constitutive relations (Lindell, Viitanen 1992, Kong 1986):

$$\overline{D} = \varepsilon\overline{E} + [\chi - \mathrm{i}\kappa]\overline{H} \tag{1}$$

$$\overline{B} = \mu\overline{H} + [\chi + \mathrm{i}\kappa]\overline{E} \tag{2}$$

where ε, μ, χ and κ are the permittivity, permeability, nonreciprocity parameter and chirality parameter, respectively.

Recent years have seen growing interest in wave propagation in media with complex structure (cf. Lakhtakia, V.K.Varadan, V.V.Varadan 1989, Jaggard, Engheta 1991) motivated, from one hand, by a variety of novel phenomena and characteristic features and their potential usefulness in applied electrodynamics, and, from the other hand, by impressive advances in material science and technology that make possible for such materials to be manufactured.

In the present paper, the material characteristics that enter the constitutive relations (1), (2) are assumed to vary with one direction (depth), which makes the problem one-dimensional.

We analyze the structure of the scattering matrix and transform Maxwell equations to a form suitable for the solution of the inverse problem. The uniqueness theorem on the recovering the material parameters is proven and the reconstruction formulae are given.

2 Field equations

Maxwell equations

$$\begin{aligned} \operatorname{rot} \overline{H} &= \mathrm{i}\omega \overline{D}, \\ \operatorname{rot} \overline{E} &= -\mathrm{i}\omega \overline{B} \end{aligned} \tag{3}$$

together with the constitutive relations (1), (2) describe the propagation of the harmonic (with time dependence $\exp(\mathrm{i}\omega t)$) electromagnetic waves in a bi-isotropic nonconducting medium. It is assumed that the medium parameters have the following structure:

$$\{\varepsilon, \mu, \chi, \kappa\} = \begin{cases} \{\varepsilon_1, \mu_1, \chi_1, \kappa_1\} & z < 0 \\ \{\varepsilon(z), \mu(z), \chi(z), \kappa(z)\} & 0 \le z \le l \\ \{\varepsilon_2, \mu_2, \chi_2, \kappa_2\} & z > l \end{cases} \tag{4}$$

i.e. the half-spaces $z < 0$ and $z > l$ are homogeneous (with the related constant parameters marked by the indices 1 and 2), and the slab $0 \le z \le l$ is stratified in z-direction. Assume also that a transversely polarized plane wave is incident from the half-space $z < 0$ such that $\overline{E} = (E_1, E_2, 0)$, $\overline{H} = (H_1, H_2, 0)$. Then one can rewrite Maxwell equations in the following form:

$$\frac{\partial}{\partial z}\begin{pmatrix} E \\ H \end{pmatrix} = \mathrm{i}\omega W(z)\begin{pmatrix} E \\ H \end{pmatrix} \tag{5}$$

where $E = \begin{pmatrix} E_1 \\ E_2 \end{pmatrix}$, $H = \begin{pmatrix} H_1 \\ H_2 \end{pmatrix}$,

$$W(z) = \begin{pmatrix} -(\chi + \mathrm{i}\kappa)B & -\mu B \\ \varepsilon B & (\chi - \mathrm{i}\kappa)B \end{pmatrix}(z) \tag{6}$$

with $B = \begin{pmatrix} 0 & 1 \\ -1 & 0 \end{pmatrix}$ (cf. He 1992).

To simplify the following analysis, we assume that the slab is matched to the half-spaces $z < 0$ and $z > l$, so that the real-valued (for lossless media) functions $\varepsilon(z), \mu(z), \chi(z)$, and $\kappa(z)$ are continuous (and, moreover, differentiable) on the whole axis $-\infty < z < +\infty$. For practical cases, χ and κ are comparatively small, so that $\chi^2 + \kappa^2 < \varepsilon\mu$ is also assumed.

As it was shown in (He 1992) (see also Corones, Stewart 1993), to analyze the wave propagation in complex media, and to obtain a physically clear interpretation, it is convenient to reformulate the relevant differential system in such a way that one can identify the down-going modes (propagating in the positive z-direction) and up-going modes (propagating in the negative z-direction). It can be done by diagonalizing the leading part (with respect to the spectral parameter ω) of the relevant system. The diagonalizing transformation (He 1992) is given by

$$Q = T(z) \begin{pmatrix} E \\ H \end{pmatrix} \tag{7}$$

where

$$T(z) = \frac{1}{2} \begin{pmatrix} 1 - \mathrm{i}q_1 & \mathrm{i} + q_1 & -\mathrm{i}q_2 & q_2 \\ \mathrm{i} - q_1 & 1 + \mathrm{i}q_1 & -q_2 & \mathrm{i}q_2 \\ 1 - \mathrm{i}q_1 & -\mathrm{i} - q_1 & -\mathrm{i}q_2 & -q_2 \\ -\mathrm{i} + q_1 & 1 + \mathrm{i}q_1 & q_2 & \mathrm{i}q_2 \end{pmatrix} (z) \tag{8}$$

$$q_1 = \frac{\chi}{\sqrt{\varepsilon\mu - \chi^2}}, \qquad q_2 = \frac{\mu}{\sqrt{\varepsilon\mu - \chi^2}} \tag{9}$$

and the wave equation becomes

$$\begin{aligned} \frac{\partial Q}{\partial z} &= \mathrm{i}\omega T(z) W(z) T^{-1}(z) Q + T'(z) T^{-1}(z) Q \\ &= \mathrm{i}\omega D(z) Q + V(z) Q \end{aligned} \tag{10}$$

where

$$\begin{aligned} D(z) &= \mathrm{diag}\{-\lambda_1(z), -\lambda_2(z), \lambda_1(z), \lambda_2(z)\} \\ \lambda_1 &= \sqrt{\varepsilon\mu - \chi^2} + \kappa \\ \lambda_2 &= \sqrt{\varepsilon\mu - \chi^2} - \kappa \end{aligned} \tag{11}$$

$$V(z) = T'(z) T^{-1}(z) = \begin{pmatrix} -\mathrm{i}\gamma & 0 & 0 & \overline{\gamma} \\ 0 & \mathrm{i}\overline{\gamma} & -\gamma & 0 \\ 0 & -\overline{\gamma} & -\mathrm{i}\gamma & 0 \\ \gamma & 0 & 0 & \mathrm{i}\overline{\gamma} \end{pmatrix} (z) \tag{12}$$

$$\gamma = \frac{1}{2}\{q_1' + \mathrm{i}q_2' p\}, \qquad p = \frac{\sqrt{\varepsilon\mu - \chi^2} + \mathrm{i}\chi}{\mu} . \tag{13}$$

To study (10), it is convenient to apply the transformation that make the "potential" part $V(z)$ of the equation off-diagonal. Let

$$Y = E_d(z) Q = E_d(z) T(z) \begin{pmatrix} E \\ H \end{pmatrix} \equiv T_1(z) \begin{pmatrix} E \\ H \end{pmatrix} \tag{14}$$

where

$$E_d(z) = \exp\left\{ -\int_0^z V_{\mathrm{diag}}(t)\mathrm{d}t \right\} = \mathrm{diag}\{e(z), \overline{e}(z), e(z), \overline{e}(z)\} \tag{15}$$

$$e(z) = \exp\left\{ \mathrm{i} \int_0^z \gamma(t) \mathrm{d}t \right\} . \tag{16}$$

From (10) and (14), we obtain the following equation

$$\frac{\partial Y}{\partial z} = \mathrm{i}\omega D(z) Y + V_1(z) Y \tag{17}$$

where

$$V_1(z) = \begin{pmatrix} 0 & 0 & 0 & \beta \\ 0 & 0 & -\overline{\beta} & 0 \\ 0 & -\beta & 0 & 0 \\ \overline{\beta} & 0 & 0 & 0 \end{pmatrix} (z) , \qquad \beta(z) = \overline{\gamma}(z) \exp\left\{ 2\mathrm{i} \int_0^z \mathrm{Re}\, \gamma(t) dt \right\} . \tag{18}$$

Notice that $V_1(z) = 0$ for $z < 0$ and $z > l$.

3 Scattering matrix

The Jost solutions $Y_\pm(z, \omega)$ of (17) are defined by their behaviour for large $|z|$ (that is, for our case, for z outside the interval $(0, l)$):

$$\begin{aligned} Y_-(z,\omega) &= \mathrm{diag}\left\{ \mathrm{e}^{-\mathrm{i}\omega\lambda_1(0)z}, \mathrm{e}^{-\mathrm{i}\omega\lambda_2(0)z}, \mathrm{e}^{\mathrm{i}\omega\lambda_1(0)z}, \mathrm{e}^{\mathrm{i}\omega\lambda_2(0)z} \right\} \text{ for } z < 0 \\ Y_+(z,\omega) &= \mathrm{diag}\left\{ \mathrm{e}^{-\mathrm{i}\omega\lambda_1(l)z}, \mathrm{e}^{-\mathrm{i}\omega\lambda_2(l)z}, \mathrm{e}^{\mathrm{i}\omega\lambda_1(l)z}, \mathrm{e}^{\mathrm{i}\omega\lambda_2(l)z} \right\} \text{ for } z > l . \end{aligned} \tag{19}$$

First two columns of these solutions correspond to down-going waves, whereas last two columns correspond to up-going waves. The Jost solutions are related by the scattering 4×4 matrix $S(\omega)$:

$$Y_+(z,\omega) = Y_-(z,\omega) S(\omega) . \tag{20}$$

In what follows we will denote by $A_{(i,j)}$, $i, j = 1, 2$ the corresponding 2×2 block elements of a 4×4 matrix A. From (20) we have

$$\begin{pmatrix} Y_{+(1,1)} \\ Y_{+(2,1)} \end{pmatrix} S^{-1}_{(1,1)}(\omega) = \begin{pmatrix} Y_{-(1,1)} \\ Y_{-(2,1)} \end{pmatrix} + \begin{pmatrix} Y_{-(1,2)} \\ Y_{-(2,2)} \end{pmatrix} \cdot R(\omega) \tag{21}$$

where $R(\omega) = S_{(2,1)}(\omega) \cdot S^{-1}_{(1,1)}(\omega)$. The left-hand side of (21) corresponds to the transmitted down-going wave for $z > l$, whereas the first and the second terms in the right-hand side are the incident down-going wave and reflected up-going wave, respectively. Therefore, $R(\omega)$, $\omega > 0$, is the physical reflection coefficient matrix for the slab which can be considered as input data for the inverse problem, that is the problem of reconstruction of the unknown material parameters using scattering information.

Come back for the moment to the scattering relation (20) and study the structure of the scattering matrix $S(\omega)$. Obviously, due to the specific structure of the potential matrix $V_1(z)$ in (17), the structure of $S(\omega)$ is also special. Indeed, set

$$\tilde{Y} = CYC, \tag{22}$$

where

$$C = \begin{pmatrix} 1 & 0 & 0 & 0 \\ 0 & 0 & 0 & 1 \\ 0 & 0 & 1 & 0 \\ 0 & 1 & 0 & 0 \end{pmatrix} \tag{23}$$

Then (17) and (20) yield

$$\frac{\partial \tilde{Y}}{\partial z} = \mathrm{i}\omega \tilde{D}(z)\tilde{Y} + \tilde{V}(z)\tilde{Y} \tag{24}$$

$$\tilde{Y}_{+} = \tilde{Y}_{-} \cdot \tilde{S}(\omega) \tag{25}$$

where

$$\tilde{D}(z) = \mathrm{diag}\{-\lambda_1(z), \lambda_2(z)\lambda_1(z), -\lambda_2(z)\}, \tag{26}$$

$$\tilde{V}(z) = \begin{pmatrix} 0 & \beta & 0 & 0 \\ \overline{\beta} & 0 & 0 & 0 \\ 0 & 0 & 0 & -\beta \\ 0 & 0 & -\overline{\beta} & 0 \end{pmatrix}(z) \tag{27}$$

$$\tilde{S}(\omega) = CS(\omega)C \ . \tag{28}$$

Therefore, the 4×4 scattering problem (24), (25) factors into two 2×2 scattering problems for its (1,1) and (2,2) blocks:

$$\tilde{S}(\omega) = \begin{pmatrix} \tilde{S}_{(1,1)}(\omega) & 0 \\ 0 & \tilde{S}_{(2,2)}(\omega) \end{pmatrix} \qquad \tilde{Y}_{\pm} = \begin{pmatrix} \Psi_{\pm}^{(1)} & 0 \\ 0 & \Psi_{\pm}^{(2)} \end{pmatrix} \tag{29}$$

and

$$\Psi_{+}^{(i)} = \Psi_{-}^{(i)} \cdot \tilde{S}_{(i,i)}(\omega), \quad i = 1, 2, \tag{30}$$

where 2×2 matrices $\Psi_{\pm}^{(i)}$ are the solutions of the following equations:

$$\frac{\partial \Psi_{\pm}^{(1)}}{\partial z} = \mathrm{i}\omega \begin{pmatrix} -\lambda_1(z) & 0 \\ 0 & \lambda_2(z) \end{pmatrix} \Psi_{\pm}^{(1)} + \begin{pmatrix} 0 & \beta(z) \\ \overline{\beta}(z) & 0 \end{pmatrix} \Psi_{\pm}^{(1)} \tag{31}$$

$$\frac{\partial \Psi_{\pm}^{(2)}}{\partial z} = \mathrm{i}\omega \begin{pmatrix} \lambda_1(z) & 0 \\ 0 & -\lambda_2(z) \end{pmatrix} \Psi_{\pm}^{(2)} + \begin{pmatrix} 0 & -\beta(z) \\ -\overline{\beta}(z) & 0 \end{pmatrix} \Psi_{\pm}^{(2)} \ . \tag{32}$$

The next step is to transform the problem (31), (30) to the scattering problem for Zakharov-Shabat equation (cf. Faddeev, Takhtajan 1987).

Let

$$\Phi_-^{(1)}(z,\omega) = \Psi_-^{(1)}(z,\omega)\exp\left\{\frac{1}{2}\mathrm{i}\omega\int_0^z(\lambda_1(t)-\lambda_2(t))\mathrm{d}t\right\}$$

$$\Phi_+^{(1)}(z,\omega) = \Psi_+^{(1)}(z,\omega)\exp\left\{\frac{1}{2}\mathrm{i}\omega\int_0^z(\lambda_1(t)-\lambda_2(t))\mathrm{d}t\right\}$$

$$\times\exp\left\{-\frac{1}{2}\mathrm{i}\omega\int_0^l[(\lambda_1(t)-\lambda_2(t))-(\lambda_1(l)-\lambda_2(l))]\mathrm{d}t\right\} \quad (33)$$

Then $\Phi_\pm^{(1)}$ satisfy the differential equations

$$\frac{\partial\Phi_\pm^{(1)}}{\partial z} = \mathrm{i}\omega\lambda_c(z)\begin{pmatrix}-1 & 0\\ 0 & 1\end{pmatrix}\Phi_\pm^{(1)} + \begin{pmatrix}0 & \beta(z)\\ \overline{\beta}(z) & 0\end{pmatrix}\Phi_\pm^{(1)} \quad (34)$$

where

$$\lambda_c(z) = \frac{1}{2}\{\lambda_1(z)+\lambda_2(z)\}\ , \quad (35)$$

and the boundary conditions

$$\Phi_-^{(1)}(z,\omega) = \mathrm{diag}\{\mathrm{e}^{-\mathrm{i}\omega\lambda_c(0)z}, \mathrm{e}^{\mathrm{i}\omega\lambda_c(0)z}\} \text{ for } z<0$$

$$\Phi_+^{(1)}(z,\omega) = \mathrm{diag}\{\mathrm{e}^{-\mathrm{i}\omega\lambda_c(l)z}, \mathrm{e}^{\mathrm{i}\omega\lambda_c(l)z}\} \text{ for } z>l\ .$$

The scattering relation for (34) is

$$\Phi_+^{(1)} = \Phi_-^{(1)}\cdot S_\Phi^{(1)}(\omega) \quad (36)$$

where

$$S_\Phi^{(1)}(\omega) = \tilde{S}_{(1,1)}(\omega)\cdot\exp\left\{-\frac{1}{2}\mathrm{i}\omega\int_0^l[(\lambda_1(t)-\lambda_2(t))-(\lambda_1(l)-\lambda_2(l))]\mathrm{d}t\right\}\ . \quad (37)$$

Introducing new variable

$$\xi(z) = 2\int_0^z\lambda_c(\tau)\mathrm{d}\tau \quad (38)$$

and defining

$$\hat{\Phi}_-(\xi) = \Phi_-^{(1)}(z(\xi))$$

$$\hat{\Phi}_+(\xi) = \Phi_+^{(1)}(z(\xi))\cdot\mathrm{diag}\left\{\mathrm{e}^{-\int_0^l(\lambda_c(\tau)-\lambda_c(l))\mathrm{d}\tau}, \mathrm{e}^{\int_0^l(\lambda_c(\tau)-\lambda_c(l))\mathrm{d}\tau}\right\}$$

from (34) and (36) we obtain

$$\frac{\partial\hat{\Phi}_\pm}{\partial\xi} = \frac{1}{2}\mathrm{i}\omega\begin{pmatrix}-1 & 0\\ 0 & 1\end{pmatrix}\hat{\Phi}_\pm + \begin{pmatrix}0 & u(\xi)\\ \overline{u}(\xi) & 0\end{pmatrix}\hat{\Phi}_\pm \quad (39)$$

$$\hat{\Phi}_+ = \hat{\Phi}_-\cdot\hat{S}(\omega) \quad (40)$$

where

$$u(\xi) = \frac{\beta}{2\lambda_c}(z(\xi)) \tag{41}$$

$$\hat{S}(\omega) = S_{\Phi}^{(1)}(\omega) \cdot \operatorname{diag}\left\{ e^{-i\omega \int_0^l (\lambda_c(\tau) - \lambda_c(l))d\tau}, e^{i\omega \int_0^l (\lambda_c(\tau) - \lambda_c(l))d\tau} \right\} \tag{42}$$

and

$$\begin{aligned} \hat{\Phi}_-(\xi, \omega) &= \operatorname{diag}\{e^{-i\omega\xi/2}, e^{i\omega\xi/2}\} \text{ for } \xi < 0, \\ \hat{\Phi}_+(\xi, \omega) &= \operatorname{diag}\{e^{-i\omega\xi/2}, e^{i\omega\xi/2}\} \text{ for } \xi > \xi(l). \end{aligned} \tag{43}$$

The problem (39), (40), (43) is the well-known Zakharov-Shabat scattering problem with self-adjoint potential matrix (cf. Faddeev, Takhtajan 1987). Hence, we can use known facts concerning the direct and the inverse problems for this problem:

1) the scattering matrix $\hat{S}(\omega)$ has the following structure:

$$\hat{S}(\omega) = \begin{pmatrix} a(\omega) & \overline{b}(\omega) \\ b(\omega) & \overline{a}(\omega) \end{pmatrix} \tag{44}$$

where $|a(\omega)|^2 - |b(\omega)|^2 = 1$, $a(\omega)$ is analytically extended into the half-plane $\operatorname{Im}\omega \geq 0$, and $a(\omega) = 1 + o(1)$, as $|\omega| \to \infty$, $\operatorname{Im}\omega \geq 0$;

2) the function $u(\xi)$ (and, therefore, the solution $\hat{\Phi}_\pm(\xi, \omega)$) is determined uniquely by the scattering coefficient of this system $r(\omega) = b(\omega)a^{-1}(\omega)$ given on the real line $-\infty < \omega < +\infty$.

Coming back to the problem (24), (25), from (37), (42) and (44) we have

$$\tilde{S}_{(1,1)}(\omega) = \begin{pmatrix} a(\omega) & \overline{b}(\omega) \\ b(\omega) & \overline{a}(\omega) \end{pmatrix} \cdot \operatorname{diag}\left\{ e^{i\omega \int_0^l (\lambda_1(\tau) - \lambda_1(l))d\tau}, e^{-i\omega \int_0^l (\lambda_2(\tau) - \lambda_2(l))d\tau} \right\} \tag{45}$$

Now consider $\tilde{S}_{(2,2)}(\omega)$. Writing (31) and (32) in the form

$$\frac{\partial \Psi^{(i)}}{\partial z} = A^{(i)}(z, \omega)\Psi^{(i)}, \tag{46}$$

one can see that $-B\overline{A^{(2)}(z, -\omega)}B$ coincides with $A^{(1)}(z, \omega)$ under the changing $\lambda_1(z) \leftrightarrow \lambda_2(z)$ in the diagonal part of $A^{(1)}(z, \omega)$. This fact, together with (45), implies the following relation:

$$\begin{aligned} &-B\overline{\tilde{S}_{(2,2)}(-\omega)}B = \\ &= \begin{pmatrix} a(\omega) & \overline{b}(\omega) \\ b(\omega) & \overline{a}(\omega) \end{pmatrix} \cdot \operatorname{diag}\left\{ e^{i\omega \int_0^l (\lambda_2(\tau) - \lambda_2(l))d\tau}, e^{-i\omega \int_0^l (\lambda_1(\tau) - \lambda_1(l))d\tau} \right\} \end{aligned} \tag{47}$$

Denote by $e_k(\omega)$, $k = 1, 2$ the exponent factors:

$$e_k(\omega) = \exp\left\{ i\omega \int_0^l (\lambda_k(\tau) - \lambda_k(l))d\tau \right\}. \tag{48}$$

Then (45), (47) and (28) lead to the following structure of the scattering matrix $S(\omega)$:

$$S(\omega)=\begin{pmatrix} a(\omega)e_1(\omega) & 0 & 0 & \bar{b}(\omega)e_2^{-1}(\omega) \\ 0 & \bar{a}(-\omega)e_2(\omega) & -b(-\omega)e_1^{-1}(\omega) & 0 \\ 0 & -\bar{b}(-\omega)e_2(\omega) & a(-\omega)e_1^{-1}(\omega) & 0 \\ b(\omega)e_1(\omega) & 0 & 0 & \bar{a}(\omega)e_2^{-1}(\omega) \end{pmatrix} \tag{49}$$

and, consequently, to the structure of the reflection matrix $R(\omega)$:

$$R(\omega)=S_{(2,1)}(\omega)\cdot S^{-1}_{(1,1)}(\omega)=\begin{pmatrix} 0 & -\frac{\bar{b}}{\bar{a}}(-\omega) \\ \frac{b}{a}(\omega) & 0 \end{pmatrix}\equiv\begin{pmatrix} 0 & r_2(\omega) \\ r_1(\omega) & 0 \end{pmatrix} \tag{50}$$

Here $r_1(\omega)$ coincides with the reflection coefficient $r(\omega)$ for the Zakharov-Shabat system (39)

$$r_1(\omega)=r(\omega)=b(\omega)a^{-1}(\omega), \tag{51}$$

and

$$r_2(\omega)=-\bar{r}(-\omega). \tag{52}$$

4 Inverse problem

It follows from (51) and (52) that $R(\omega)$ given for $\omega\geq 0$ determines the reflection coefficient $r(\omega)$ for the Zakharov-Shabat problem (39) on the whole line $-\infty<\omega<\infty$:

$$r(\omega)=\begin{cases} r_1(\omega), & \omega\geq 0 \\ -\overline{r_2}(-\omega), & \omega<0\ . \end{cases} \tag{53}$$

Knowing $r(\omega)$ allows a means to reconstruct the potential function $u(\xi)$ as well as the solutions $\hat{\Phi}_\pm(\xi,\omega)$ of the Zakharov-Shabat system (39) (see Appendix).

Further, if $\hat{\Phi}_\pm(\xi,\omega)$ are known, one can obtain the information about material parameters using, for example, the behaviour of $\hat{\Phi}_\pm(\xi,\omega)$ as $\omega\to 0$. Indeed, from (5) and (14) we can express $Y(z,0)$ in terms of the functions to be reconstructed:

$$\begin{aligned} Y(z,0)&=T_1(z)T_1^{-1}(0)= \\ &=\frac{1}{2}\begin{pmatrix} t(1-\mathrm{i}\delta) & 0 & 0 & t(\mathrm{i}+\bar{\delta}) \\ 0 & \bar{t}(1+\mathrm{i}\bar{\delta}) & \bar{t}(\mathrm{i}-\delta) & 0 \\ 0 & t(-\mathrm{i}-\bar{\delta}) & t(1-\mathrm{i}\delta) & 0 \\ \bar{t}(-\mathrm{i}+\delta) & 0 & 0 & \bar{t}(1+\mathrm{i}\bar{\delta}) \end{pmatrix} \end{aligned} \tag{54}$$

where

$$t(z)=\exp\left\{-\frac{\mathrm{i}}{2}\int_0^z[q_1'(\tau)-q_2'(\tau)\frac{q_1}{q_2}(\tau)+\mathrm{i}\frac{q_2'}{q_2}(\tau)]\mathrm{d}\tau\right\} \tag{55}$$

$$\delta(z)=q_1(z)+\mathrm{i}q_2(z)p(0). \tag{56}$$

From the other hand, (22), (29) and (33) imply

$$\begin{aligned} Y_{11}(z,0) &= \hat{\Phi}_{-11}(\xi,0) \\ Y_{14}(z,0) &= \hat{\Phi}_{-12}(\xi,0). \end{aligned} \tag{57}$$

Denote

$$\begin{aligned} A(\xi) &= A_R(\xi) + \mathrm{i}A_I(\xi) = \hat{\Phi}_{-11}(\xi,0), \\ B(\xi) &= B_R(\xi) + \mathrm{i}B_I(\xi) = \hat{\Phi}_{-12}(\xi,0). \end{aligned}$$

From (54) and (57), we obtain the linear system of equations for determining $\Delta_R(\xi) \equiv \operatorname{Re}\delta(z(\xi))$ and $\Delta_I(\xi) \equiv \operatorname{Im}\delta(z(\xi))$:

$$\begin{aligned} \Delta_R(B_I - A_R) + \Delta_I(B_R - A_I) &= -A_I - B_R \\ \Delta_R(B_R + A_I) + \Delta_I(-B_I - A_R) &= -A_R + B_I. \end{aligned} \tag{58}$$

The determinant Δ of this system is non-zero for any ξ because $\Delta = |A|^2 - |B|^2 = \det \hat{\Phi}(\xi,0) \neq 0$, so that we can determine Δ_R and Δ_I:

$$\begin{aligned} \Delta_R(\xi) &= -2\frac{A_I B_I + A_R B_R}{\Delta}(\xi) \\ \Delta_I(\xi) &= -\frac{(A_R + B_I)^2 + (A_I + B_R)^2}{\Delta}(\xi) \ . \end{aligned} \tag{59}$$

Theorem 1 *The reflection matrix $R(\omega)$ given for $\omega \geq 0$, added by the parameters $q_1(0)$ and $q_2(0)$ determines uniquely two functions $Q_k(\xi)$, $k = 1,2$ of the variable ξ such that*

$$\begin{aligned} Q_1(\xi)\Big|_{\xi=\xi(z)} &= q_1(z) = \frac{\chi(z)}{\sqrt{\varepsilon(z)\mu(z) - \chi^2(z)}} \\ Q_2(\xi)\Big|_{\xi=\xi(z)} &= q_2(z) = \frac{\mu(z)}{\sqrt{\varepsilon(z)\mu(z) - \chi^2(z)}} \end{aligned} \tag{60}$$

where

$$\xi(z) = \int_0^z [\lambda_1(t) + \lambda_2(t)]\mathrm{d}t = 2\int_0^z \sqrt{\varepsilon(t)\mu(t) - \chi^2(t)}\mathrm{d}t. \tag{61}$$

The statement of Theorem follows immediately from the considerations above and the equalities

$$\begin{aligned} Q_1(\xi) &= \Delta_R(\xi) + q_1(0)\Delta_I(\xi) \\ Q_2(\xi) &= q_2(0)\Delta_I(\xi) \ . \end{aligned} \tag{62}$$

Now we can make some conclusions concerning the reconstruction of the material parameters under normal incidence of exiting plane wave:

a) chirality parameter $\kappa = \kappa(z)$ does not affect the reflection matrix; moreover, it enter the scattering matrix $S(\omega)$ only in an averaged form $\int_0^l [\kappa(t) - \kappa(l)]\mathrm{d}t$.

b) the reflection matrix allows reconstruction of two (independent) functions of space variable. For example, if one of three functions $\varepsilon(z), \mu(z)$ and $\chi(z)$ is known then other two parameters are determined using the relations (60) and (61):

1) if $\mu(z)$ is known, then the dependence $\xi = \xi(z)$ is uniquely determined by the ordinary differential equation

$$2\mu(z)\mathrm{d}z = Q_2(\xi)\mathrm{d}\xi \tag{63}$$

with the initial condition $\xi(0) = 0$. The functions $\chi(z)$ and $\varepsilon(z)$ are reconstructed by the formulae

$$\chi(z) = \frac{1}{2}Q_1(\xi(z))\xi'(z) \tag{64}$$

$$\varepsilon(z) = \frac{\frac{1}{4}[\xi'(z)]^2 + \chi^2(z)}{\mu(z)} \quad ; \tag{65}$$

2) if $\chi(z)$ is known, then the equation

$$2\chi(z)\mathrm{d}z = Q_1(\xi)\mathrm{d}\xi \tag{66}$$

determines the dependence $\xi = \xi(z)$; $\mu(z)$ is given by

$$\mu(z) = \frac{1}{2}Q_2(\xi(z))\xi'(z) \ , \tag{67}$$

and $\varepsilon(z)$ is given by (65);

3) if $\varepsilon(z)$ is known then the equation

$$\frac{1 + Q_1^2(\xi)}{Q_2(\xi)}\mathrm{d}\xi = 2\varepsilon(z)\mathrm{d}z \tag{68}$$

gives $\xi = \xi(z)$; $\chi(z)$ and $\mu(z)$ are then reconstructed by (64) and (67).

Appendix

Here we briefly describe the scheme of solution of the inverse scattering problem for Zakharov-Shabat system based on the Gel'fand-Levitan-Marchenko integral equation (for details see Faddeev, Takhtajan 1987).

Given reflection coefficient $r(\omega)$, define 2×2 matrix

$$\Omega(\xi) = \begin{pmatrix} 0 & \Omega_{12}(\xi) \\ \overline{\Omega_{12}(\xi)} & 0 \end{pmatrix}$$

where

$$\Omega_{12}(\xi) = -\frac{1}{4\pi}\int_{-\infty}^{\infty} r(\omega)\mathrm{e}^{-\mathrm{i}\omega\xi/2}\mathrm{d}\omega \ .$$

Then the Gel'fand-Levitan-Marchenko integral equation (for the left end) is written in the form

$$\Gamma_-(\xi,\eta)+\Omega(\xi+\eta)+\int_0^\xi \Gamma_-(\xi,s)\Omega(s+\eta)\mathrm{d}s=0, \qquad 0\le\eta\le\xi \tag{69}$$

with respect to 2×2-matrix $\Gamma_-(\xi,\eta)$, for every fixed $\xi\in[0,l]$.

The solution $\Gamma_-(\xi,\eta)$ of (69) determines the potential function $u(\xi)$ and the solution of (39) $\hat{\Phi}_-(\xi,\omega)$ in the following way:

$$u(\xi)=2\Gamma_{-12}(\xi,\xi) \tag{70}$$

$$\hat{\Phi}_-(\xi,\omega)=E(\xi,\omega)+\int_0^\xi \Gamma_-(\xi,t)E(t,\omega)\mathrm{d}t \tag{71}$$

where $E(\xi,\omega)=\begin{pmatrix} \mathrm{e}^{-\mathrm{i}\omega\xi/2} & 0 \\ 0 & \mathrm{e}^{\mathrm{i}\omega\xi/2}\end{pmatrix}$.

Acknowledgment

During this work, D. S. benefitted from the hospitality of the Laboratory of Mathematical Physics and Geometry, University Paris-7. The partial support of C.N.R.S. is gratefully acknowledged.

References

Corones J.P., Stewart R. (1983): J. Opt. Soc. Am. A **10**, (9), 1941

Faddeev L.D., Takhtajan L.A. (1987): *Hamiltonian Methods in the Theory of Solitons* (Springer-Verlag, Berlin)

He S. (1992): J. Math. Phys. **33** (12), 4103

Jaggard D.L., Engheta N. (1991): in *Directions in Electromagnetic Wave Modelling*, Bertoni H.L., Felsen L.B., eds (Plenum, New-York).

Kong J.A. (1986): *Electromagnetic Wave Theory* (Wiley, New York)

Lakhtakia A., Varadan V.K., Varadan V.V. (1989): *Time-harmonic Electromagnetic Waves in Chiral Media*, Lecture Notes in Physics, **335** (Springer-Verlag, Berlin)

Lindell I.V., Viitanen A.J. (1992): IEEE Trans. Antennas Propag. **40**, 91

Recovery of Strongly Scattering Permittivity Distributions from Limited Backscattered Data Using a Nonlinear Filtering Technique

M. A. Fiddy and D. A. Pommet

Department of Electrical Engineering,
University of Massachusetts Lowell,
Lowell, MA 01854, USA
e-mail: fiddy@cemos.uml.edu
www.uml.edu/Dept/EE/RCs/CEMOS

1 Introduction

Imaging from scattered field data is traditionally referred to as diffraction tomography. These imaging methods, which are both numerically feasible (Fourier inversion-based) and yet mathematically rigorous, have generally required that the scattering objects be only weakly scattering, (e.g. based on the first-order Born or Rytov approximations). This severely limits their usefulness in practice. Many advances have been made based on iterative techniques, for example, to try to extend the class of objects that can be imaged. In this paper we report on a new approach that can be applied when these weakly scattering approximations are not valid. The methods remains a Fourier-based procedure, allowing well known spectral estimation and noise handling algorithms to be readily incorporated. We show how one can calculate a function using the available scattered field measurements from which an estimate of the permittivity distribution can be found by applying a nonlinear filtering technique. Most inversion techniques, including the one reported here, assume that scattered field data are available all around the object, for a set of incident field directions that circumscribe the object. Beyond this, we also consider the recovery of strongly scattering permittivity distributions from severely limited angular data, including backscattered data. This constraint occurs in many radar applications, as well as medical imaging, remote sensing and non-destructive testing. A Fourier-based method for image restoration can be directly built into the nonlinear filtering method and reconstructions using both simulated and real data have been calculated.

2 Imaging From Scattered Fields

Methods for inverse scattering, directed toward imaging a permittivity profile (or in acoustics a velocity distribution) from scattered field data, usually require that the object be weakly scattering or have a permittivity which varies spatially only slowly on the scale of the illuminating wavelength. Under these assumptions, inversion methods collectively referred to as diffraction tomography techniques, can be formulated as straightforward Fourier inversion procedures [1]. The scattered field data under these conditions, based on adopting the first-order Born and the Rytov approximations, are mapped onto the Ewald sphere in k-space and inverse transformed. These methods for diffraction tomography are both numerically feasible (Fourier based) and mathematically rigorous, but require that the scattering object only interact weakly with the incident field. These approximations are rarely applicable in practice, thus limiting their usefulness. Although more robust than the Born approximation, the Rytov approximation requires a nonlinear transformation of the data to be made prior to using an identical inversion step as that performed when the Born approximation is valid. This nonlinear step requires that the logarithm of the scattered field measurements be taken. There are numerous problems in doing this when the magnitude of the scattered field is close to zero and when the phase of the scattered field has a range that exceeds 2π. Phase unwrapping is exceedingly difficult especially in two and higher dimensional problems, in which wavefront dislocations can naturally occur in large numbers, rendering the idea of a smoothly unwrapped phase meaningless.

When the Born approximation is not valid, a Fourier based method can still be exploited and this is the key to the nonlinear filtering approach described here. The limitations of weak scattering and distorted wave inverse scattering methods are overcome by recognizing that the image recovered assuming the Born approximation to be valid, can be interpreted and filtered despite the approximation being invalid. One can invert or backpropagate the field for any given illumination direction and apply a filter to these field data in the cepstral domain; this is explained in more detail in a later section. This approach thus extracts information about the scattering function from sets of backpropagated fields, each one being interpreted as the product of the unknown permittivity distribution function with an unknown field distribution.

There have been many other developments which extend the domain of validity of the Born and Rytov approximations [1-3]. These methods, sometimes iterative in nature, either assume sufficiently weak scattering that the Born series or a modified form of it converges or they assume that some *a priori* information about the scattering object is available. With prior knowledge of a background scattering medium, one can recover small fluctuations in permittivity about this background applying distorted-wave Born and Rytov methods. The distorted-wave approach has been widely used and reported and is also of use in modeling scattering in nonlinear media [4,5], in which only small externally induced changes in the permittivity are expected. However, such methods do not provide a sufficiently general approach to solve an arbitrary inverse scattering problem, for which such prior knowledge is unknown. It is also important to point out that even when some prior knowledge about the scattering distribution is available, it can be a difficult task in its own right to compute the multiple scattering that occurs in a strongly scattering, but known, background structure. Also, in any of these imaging situations, additional difficulties arise from the limited number of field samples available and, typically, the limited number of illumination directions that can be employed. Indeed, in optical scattering, one may well need to estimate the object from scattered field intensity data by simultaneously estimating the missing phase information. These practical constraints necessitate that an image estimation or restoration technique be included in the inversion algorithm. This is readily done when the inverse scattering algorithm remains essentially Fourier-based in form.

More general methods or "exact" inversion procedures have proved extremely difficult to implement, sometimes relying on embedding the object in a medium whose permittivity is close to that of the mean of the object's permittivity, or are limited to recovering shape or surface profile information [6-8]. The method we report is potentially "exact" based on a nonlinear or homomorphic filtering. The method extends the range of validity of the existing techniques to arbitrary scatterers, i.e. without the need to specify an upper bound on the permittivity of the object. We briefly outline the principle behind the approach and will then describe some of the problems associated with its use, along with solutions to these problems.

3 Theoretical Model

Consider a scattering object having a permittivity $\varepsilon(\mathbf{r})$ which is embedded in a medium of permittivity ε_0, where $\varepsilon = \varepsilon_0[1 + V(\mathbf{r})]$. We assume the object to be bounded by a compact support D, and assume that ε_0 is the free-space permittivity. $V(\mathbf{r})$ is referred to as the scattering or object function, i.e. it represents $\varepsilon_r - 1$ and is the quantity we wish to estimate. If the scattering object possesses cylindrical symmetry and the polarization of the incident time-harmonic electromagnetic wave is along the symmetric axis of the scattering object, the depolarization term in the vector wave equation can be neglected [1].

For the case of an incident plane wave $\Psi_0(\mathbf{r}, k\hat{\mathbf{r}}_0) = e^{ik\hat{\mathbf{r}}_0\cdot\mathbf{r}}$, then from the scalar Helmholtz equation we can express the total field $\Psi(\mathbf{r}, k\hat{\mathbf{r}}_0)$ in terms of the inhomogeneous Fredholm integral equation of first kind, namely,

$$\Psi(\mathbf{r}, k\hat{\mathbf{r}}_O) = \Psi_O(\mathbf{r}, k\hat{\mathbf{r}}_O) - k^2 \int_D d\mathbf{r}' G_O(\mathbf{r}, \mathbf{r}') V(\mathbf{r}') \Psi(\mathbf{r}', k\hat{\mathbf{r}}_O)$$

$$= \Psi_o(\mathbf{r}, k\hat{\mathbf{r}}_O) + \Psi_s(\mathbf{r}, k\hat{\mathbf{r}}_O), \tag{3.1}$$

where $\Psi_o(\mathbf{r}, k\hat{\mathbf{r}}_O)$ is the scattered field resulting from the interaction of the incident wave $\Psi_s(\mathbf{r}, k\hat{\mathbf{r}}_O)$ with the scattering function $V(\mathbf{r})$, $G_o(\mathbf{r},\mathbf{r}')$ is the free-space Green's function, k is the wavenumber in free space, and $\hat{\mathbf{r}}_O$ denotes the direction of illumination. The integration in eq. (3.1) is over the support of $V(\mathbf{r})$ defined by D. Using the far-field approximation for the outgoing spherical wave $G_o(\mathbf{r},\mathbf{r}')$, we obtain

$$\Psi_S(\mathbf{r}, k\hat{\mathbf{r}}_O) = k^2 \frac{e^{ikr}}{4\pi r} \int_D d\mathbf{r}' e^{-ik\hat{\mathbf{r}}\cdot\mathbf{r}'} V(\mathbf{r}') \Psi(\mathbf{r}', k\hat{\mathbf{r}}_O) \tag{3.2}$$

When adopting the first Born approximation, the total field (or internal field) $\Psi(\mathbf{r}, k\hat{\mathbf{r}}_O)$ is replaced with the known incident field $\Psi_o(\mathbf{r}, k\hat{\mathbf{r}}_O)$ in the integral above [9]. This approximation is valid when $k|\varepsilon_r - 1|a < \pi/2$, which, as we stated earlier, is not valid for most imaging problems of interest. The parameter a is the characteristic dimension of the object, and as the extent of the object increases or the magnitude of the permittivity fluctuations increases, the first Born approximation becomes increasingly poor. The Born approximation improves however by increasing the wavelength of the illumination, but this in turn degrades the resolution of the resulting image as will be evident later.

Equation (3.2) may be written

$$\Psi_s(\mathbf{r}, k\hat{\mathbf{r}}_o) = k^2 \frac{e^{ikr}}{4\pi r} f(k\hat{\mathbf{r}}, k\hat{\mathbf{r}}_o), \tag{3.3}$$

where $f(k\hat{\mathbf{r}}, k\hat{\mathbf{r}}_O)$ is the scattering amplitude which is defined as

$$f(k\hat{\mathbf{r}}, k\hat{\mathbf{r}}_o) \equiv \int_D d\mathbf{r}' e^{-ik\hat{\mathbf{r}}\cdot\mathbf{r}'} V(\mathbf{r}') \Psi(\mathbf{r}', k\hat{\mathbf{r}}_o) \tag{3.4}$$

In the first Born approximation, the relationship between the (complex) scattering amplitude and the scattering function $V(\mathbf{r})$ becomes a Fourier transformation, namely,

$$f^{BA}(k\hat{\mathbf{r}}, k\hat{\mathbf{r}}_o) \equiv \int_D d\mathbf{r}' e^{-ik\hat{\mathbf{r}}\cdot\mathbf{r}'} V(\mathbf{r}') e^{ik\hat{\mathbf{r}}_o\cdot\mathbf{r}'} \tag{3.5}$$

One can calculate and estimate for $V(\mathbf{r})$ by performing an inverse Fourier transformation on the measured far field scattering amplitude data, $f^{BA}(k\hat{\mathbf{r}}, k\hat{\mathbf{r}}_0)$. These data are related to the object distribution by Fourier transformation when the scattering amplitude data are located on the Ewald sphere in k-space, a locus of points tangent to the k-space origin and of radius k. Inverse Fourier transforming these data is formally equivalent to backpropagating the scattered field into the object domain from the measurement space.

Since the scattered field data are mapped into k-space on circles of radius k , forward scattered data maps close to the k-space origin while backscattered data are located furthest from the origin giving high spatial frequency information about the object. As k increases, one can expect that higher spatial frequency information about the object can be recovered. That any inversion necessarily results from limited k-space coverage is to be expected and the image artifacts that might result will depend on how uniformly the data cover k-space and the signal to noise ratio.

When the Born approximation is not valid, this Fourier relationship can still be exploited and this is the key to the nonlinear approach presented in this paper. One can readily see that inverting the scattering amplitude data determines not $V(\mathbf{r})$ but rather the function $V_B(\mathbf{r}, k\hat{\mathbf{r}}_0)$ [10] which is given by

$$V_B(\mathbf{r}, k\hat{\mathbf{r}}_0) \approx V(\mathbf{r}) \frac{\Psi(\mathbf{r}, k\hat{\mathbf{r}}_0)}{\Psi_0(\mathbf{r}, k\hat{\mathbf{r}}_0)} \tag{3.6}$$

The symbol $\approx$ in the above equation recognizes the fact that the reconstruction is approximate since the Fourier transformation can only be taken for each $\hat{\mathbf{r}}_0$ = constant and limited k-space coverage will limit accuracy.

The total field can be expressed by

$$\Psi(\mathbf{r}, k\hat{\mathbf{r}}_0) = \Psi_0(\mathbf{r}, k\hat{\mathbf{r}}_0) - k^2 \int_D d\mathbf{r}' G_0(\mathbf{r}, \mathbf{r}') V(\mathbf{r}) \frac{\Psi(\mathbf{r}', k\hat{\mathbf{r}}_0)}{\Psi_0(\mathbf{r}', k\hat{\mathbf{r}}_0)} \Psi_0(\mathbf{r}', k\hat{\mathbf{r}}_0), \tag{3.7}$$

for the case of a general scattering object, and one can write the Fourier relation

$$f(k\hat{\beta}, k\hat{\alpha}) = \int_D d\mathbf{r}' e^{-ik(\hat{\beta}-\hat{\alpha})\cdot\mathbf{r}'} V(\mathbf{r}') \frac{\Psi(\mathbf{r}', k\hat{\alpha})}{\Psi_0(\mathbf{r}', k\hat{\alpha})} \tag{3.8}$$

Consequently, a first Born inversion of the scattered field data for $\hat{\mathbf{r}}_0$ = constant, generates a filtered estimate of $V(\mathbf{r})\Psi(\mathbf{r}, k\hat{\mathbf{r}}_0)/\Psi_0(\mathbf{r}, k\hat{\mathbf{r}}_0)$ from which one can attempt to estimate $V(\mathbf{r})$ directly, since it is assumed that $\Psi(\mathbf{r}, k\hat{\mathbf{r}}_0) \approx \Psi_0(\mathbf{r}, k\hat{\mathbf{r}}_0)$.

Under most normal circumstances, the field $\Psi(\mathbf{r}, k\hat{\mathbf{r}}_0)$, i.e. the field within the scattering volume D, cannot be assumed to be equal to the incident field. It is explicitly dependent on the direction of the incident plane wave which is known and so for each illumination direction used, one obtains an "image" of the function $V(\mathbf{r})\Psi(\mathbf{r}, k\hat{\mathbf{r}}_0)$. Given data from many illumination directions, a set of these "images" can be generated, one for each illumination direction, and in which V is common to each one of them but Ψ is different. The recovery of an image of V can therefore be formulated as a problem in which and ensemble of noisy images of V require processing, the "noise" being multiplicative in nature.

4 Homomorphic filtering

Since for each direction of the illuminating radiation, the $V(\mathbf{r})\Psi(\mathbf{r}, k\hat{\mathbf{r}}_0)$ product will change, a set of these single-view backpropagated reconstructions can be generated. We regard the term $\Psi(\mathbf{r}, k\hat{\mathbf{r}}_0)$ as an unwanted factor or multiplicative noise term, which contains a certain range of spatial frequencies determined by the distribution of energy of the radiation field and its effective wavelength within the object. With respect to the spatial frequency content of the scattering object, this *multiplicative* factor can be removed by homomorphic filtering techniques [11-13]. Direct Fourier (band-pass) filtering is not appropriate for multiplied signals of this kind since their spectra are convolved.

Let us consider the nature of the filtering that might be required. If one considers a weakly scattering object the internal field approximately equals the incident field and will have a characteristic spatial frequency in the direction of propagation being a plane wave. As the degree of scattering increases, the internal field will become increasingly complicated in all directions, but will retain a characteristic correlation length or minimum scale, determined by the wavelength of the radiation in the medium $V(\mathbf{r})$. As the permittivity increases, the effective wavelength of the radiation in the scatterer decreases. Thus there will be some characteristic set of spatial frequencies associated with $\Psi(\mathbf{r}, k\hat{\mathbf{r}}_0)$ inside the scatterer, information made available through backpropagation, which can be removed by filtering in the cepstral domain. Indeed, since the spatial frequency content of $\Psi(\mathbf{r}, k\hat{\mathbf{r}}_0)$ should be concentrated around some limited range of spatial frequencies, one can expect that the energy associated with these components would be located in an annular region in the spatial frequency domain, determined by the mean effective wavelength of the radiation in D, under ideal circumstances.

The cepstral filtering inversion approach is as follows. When the Born approximation is violated, one recovers the function given by equation (3.6). For each different incident direction, the product $V(\mathbf{r})\Psi(\mathbf{r}, k\hat{\mathbf{r}}_0)$ will change and the set of these single view reconstructions is generated and stored. Taking the logarithm of $V(\mathbf{r})\Psi(\mathbf{r}, k\hat{\mathbf{r}}_O)$ changes the multiplicative relationship between V and Ψ into an additive one. This then permits linear filtering techniques to be applied to the spectrum of $\log[V(\mathbf{r})\Psi(\mathbf{r}, k\hat{\mathbf{r}}_O)]$ to remove, or at least minimize, the effect of Ψ. The spectrum of $\log[V(\mathbf{r})\Psi(\mathbf{r}, k\hat{\mathbf{r}}_O)]$ is referred to as the cepstrum of $V(\mathbf{r})\Psi(\mathbf{r}, k\hat{\mathbf{r}}_O)$ [14]. This operation will modify the spatial frequency content of V over the same spectral region as that of the removed Ψ, but a second experiment at a different illumination wavelength should rectify this. In practice there are difficulties associated with taking the logarithm of the product $V(\mathbf{r})\Psi(\mathbf{r}, k\hat{\mathbf{r}}_O)$ as the phase of $\log[V(\mathbf{r})\Psi(\mathbf{r}, k\hat{\mathbf{r}}_O)]$ can be highly discontinuous if the phase delay incurred on propagation through the object exceeds 2π radians. The first Born approximation assumes that this phase delay is much less than π. The phase function will therefore be wrapped into $[-\pi,\pi]$ and abrupt discontinuities in this phase function generate unwanted harmonics in the cepstrum, making it difficult to correctly filter. A solution to this problem that avoids phase wrapping difficulties, is to make use of the differential cepstrum [15].

After summing the partial derivatives of $\log[V(\mathbf{r})\Psi(\mathbf{r}, k\hat{\mathbf{r}}_O)]$, we obtain an expression defined by the quantity *S*

$$S = \frac{1}{V\Psi}\left[\frac{\partial(V\Psi)}{\partial x} + \frac{\partial(V\Psi)}{\partial y}\right] \qquad (4.1)$$

where $\mathbf{r} = (x,y)$ in two dimensional problems. Here only derivatives of $V(\mathbf{r})\Psi(\mathbf{r}, k\hat{\mathbf{r}}_O)$ need be calculated (which can be easily obtained using a property of the Fourier transform) and the phase wrapping problem has been eliminated. A drawback in so doing is that the dc level of the function $V(\mathbf{r})\Psi(\mathbf{r}, k\hat{\mathbf{r}}_O)$ is also lost. This can be estimated from forward and backscattered data values however. One can define the differential cepstrum as the logarithmic derivative with respect to either x or y if one wishes to avoid only the phase ambiguity [11]; it is not necessary to form the derivative with respect to both spatial variables.

There is an ill-conditioning problem that arises when $V(\mathbf{r})\Psi(\mathbf{r}, k\hat{\mathbf{r}}_O)$ takes on small values. This can be remedied by multiplying both the numerator and the denominator of S by $[V(\mathbf{r})\Psi(\mathbf{r}, k\hat{\mathbf{r}}_O)]^*$ where * represents complex conjugate and then adding a small positive regularization parameter to $|V(\mathbf{r})\Psi(\mathbf{r}, k\hat{\mathbf{r}}_O)|^2$. The dominant contribution from $\Psi(\mathbf{r}, k\hat{\mathbf{r}}_O)$ is from the fundamental spatial frequency component present. For examples for which the spatial frequency content of $V(\mathbf{r})$ is lower than that of $\Psi(\mathbf{r}, k\hat{\mathbf{r}}_O)$, a simple low pass filter applied to the differential cepstrum should remove $\Psi(\mathbf{r}, k\hat{\mathbf{r}}_O)$ with only slight degradation to the recovered image of V(**r**). However, this is clearly not optimal, albeit effective in our experience. A low-pass filter also reduces spatial noise arising from the interpolation and summing of a limited number of views.

The sequence of steps required for this algorithm is shown in figure 1

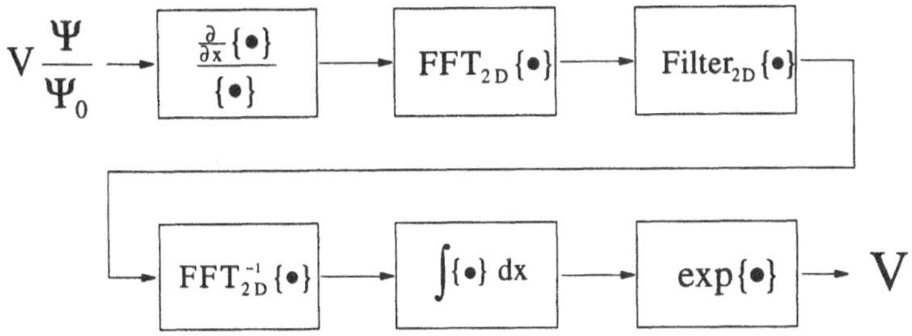

Figure 1 Differential cepstral filtering

In figure 2, we illustrate the reconstruction of a simple cylinder with the permittivities shown on the left, namely an ε of 4.0 in the central region and 2.0 in the outer annulus. The free space wavelength used was 3.0cm and the outer radius of the cylinder was 9.9cm. The exact expression for the field scattered from these concentric cylinders was calculated and used as data to check the inversion step. Following low-pass filtering of the differential cepstrum of the backpropagated field for one illumination direction. As can be seen the basic features of the original object are evident and the regions have an amplitude proportional to the original permittivity differences (not shown).

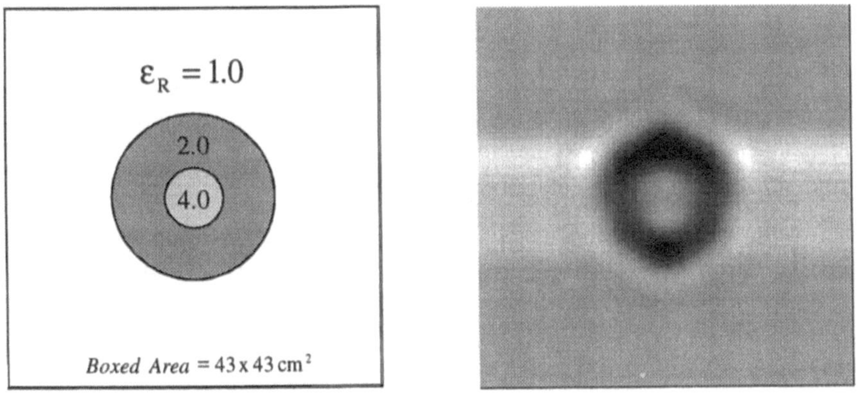

Figure 2 Differential cepstral filtering

5 Limited data points

Many important imaging techniques are based on measuring backscattered radiation. Several inverse synthetic aperture (ISAR) systems, for example, have been developed with the objective of extracting a maximum amount of information about complex targets, just from backscattered signals. Common to these methods is a stationary radar and a target whose radar aspect is varied in some specified way. In this case, an arbitrary target axis intersects the radar beam axis making an angle θ with it. As the target axis precesses around the beam axis its motion is similar to a precessing top. Varying θ between 0 and values of the order of 10 degrees yields k-space data on a spherical cap with nearly constant polarization illumination and nearly aspect independent return levels from specular scatterers. Examples of the extent of data coverage in k-space is shown in figure 3 below, for both the two and three dimensional cases.

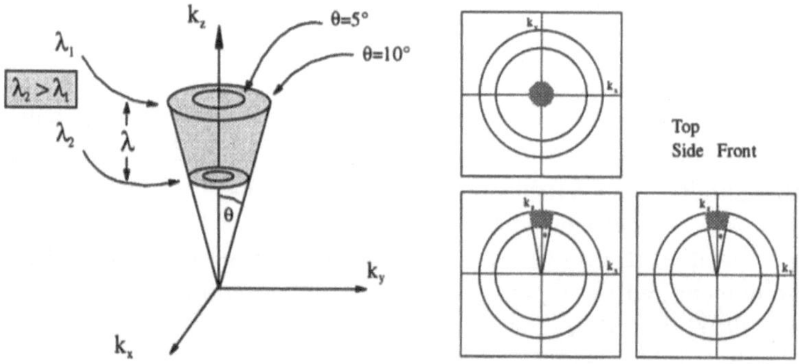

Figure 3 Limited k-space data and corresponding 2-d projection coverage.

The reconstruction of an image from such a limited k-space coverage results in considerable distortion of the image. The available data comprise a truncated conical region in k-space of angular extent 2θ with co- and cross polarization information over the full angular range of 2π.

When using these inversion methods, there is frequently a prior estimate of the scattering object. The following spectral estimation procedure exploits this prior knowledge when only limited k-space data are acquired. It takes a prior estimate, P(**r**), of the broad features of V(**r**), e.g., $V_1(\mathbf{r})$, and a set of equations of the form, expressed in 1D for convenience, is solved [19-21]:

$$f(m) = \sum_{n=-N}^{N} a_n p(m-n) \quad ; \quad m = -N, ..., N \tag{5.1}$$

The values p(m) (m = -N, ..., N) are taken from the discretized Fourier transform of P(r) and the scattering amplitude data are represented by f(m) (m = -N, ..., N). Inversion of a matrix with elements derived from p(m) allows one to solve for the coefficients a_n (n=-N,...,N). In principle, the data, f(m), need not be uniformly sampled, which means that this approach can be used to interpolate and extrapolate both nonuniformly sampled and incomplete data sets. Obtaining the coefficients, a_n allows one to define an estimator that minimizes the approximation error given by:

$$\int_{-\pi}^{\pi} dr \frac{1}{P(r)} \left| V(r) - P(r) \sum_{n=-N}^{N} a_n e^{inr} \right|^2 \tag{5.2}$$

The resulting estimator of V(**r**) is termed the PDFT estimator because of its form, namely, PDFT(r) = P(r)A(r), where A(r) is the trigonometric polynomial with coefficients, a_n, i.e. we have

$$\text{PDFT}(r) = P(r) \sum_{n=-N}^{N} a_n e^{inr} \tag{5.3}$$

If no prior knowledge is available, P(r) is a constant and the estimator reduces to the DFT of the available Fourier data. In other words, if the prior estimate, P(r), is a constant for $|r|$ ² π, then a_n= f(n) and the PDFT reduces to the discrete Fourier transform (DFT) estimator, that is usually calculated. This PDFT estimator is both continuous and data consistent, and the algorithm is easily extended to the two- or higher-dimensional case.

This estimation technique is easily regularized in the presence of noise. One can either modify the prior estimate, P(r), to take a small value outside the anticipated support of the scatterer, V(r), or one can add a small positive constant to the diagonal of the matrix, p. These can be shown to be equivalent to a Miller-Tikhonov regularization process. This is a computationally intensive algorithm because it requires the solution of a large set of linear equations. For 2M by 2M uniformly sampled scattered field data, one must invert a 2M by 2M matrix if the prior estimate, P(**r**), can be expressed as a separable function, otherwise a $(2M)^2$ by $(2M)^2$ matrix must be inverted.

6 Conclusions

We have described a method of processing backpropagated scattered field data collected from strongly scattering objects. This differential cepstral filtering approach allows direct inversion of scattered field data and incorporates a nonlinear step, required to deal directly with the nonlinear nature of the integral equation of scattering. It does not rely on linearizing methods based on the Born, Rytov, or their associated distorted-wave approximations. This filtering step has not yet been optimized but in all cases studied to date, the spatial frequency content of Ψ has been higher than that of the scattering object, allowing a simple low-pass filter to be successfully employed [22]. The filter will necessarily result in some loss of information about the target at these frequencies. On-going work involves processing more data, especially that obtained from experiments using non-cylindrically symmetric targets.

It is desirable to incorporate spectral, or in this case cepstral, extrapolation and estimation techniques into the inversion algorithm. This allows one to recover an estimate of the object even in the presence of noise and a limited data set in k-space. The estimation procedure requires some *a priori* knowledge of the object, such as its support. In the case of only having limited intensity data available to process, the PDFT can still be useful, [19,23].

An important issue that remains to be studied regarding this method, is how one should effectively use multiple view (i.e. multiple illumination direction) data over various regions in k-space and how best to filter these data in the differential cepstral domain. There also remains more research to be done in determining the optimal sampling rate necessary to adequately represent the functions involved, since taking the logarithm of a function, either implicitly or explicitly, renders it non-bandlimited.

7 Acknowledgments

MAF and DAP acknowledge the support of ONR Grant N00014-89-J-1158 and USAF contract F19628-95-C-0035. They are also grateful to their colleagues with whom they have collaborated on the work presented here, Dr. F. C. Lin, Dr. R. V. McGahan and Dr. J. B. Morris.

8 References

[1] F. C. Lin and M. A. Fiddy, "Image estimation from scattered field data," Int. J. Imaging Systems and Technology, 2, 76-95, 1990.

[2] F. C. Lin, A. Alavi, R. McGahan, and M. A. Fiddy, "Diffraction tomography in the distorted-wave Born approximation," Optical Society of America Annual Meeting, ThY29, November 4-8, 1990, Boston, MA.

[3] M. A. Fiddy, A. Alavi, F. C. Lin, and R. McGahan, "Inversion of 10GHz scattered field data using distorted-wave Born approximations," Proc. SPIE, Vol. 1351, 200-211, 1990.

[4] F. C. Lin and M.A. Fiddy, "The Born-Rytov controversy: II applications to the nonlinear and stochastic scattering problem in a one-dimensional half-space," J.O.S.A. A 10, pp1971-1983, 1993.

[5] F. C. Lin and M.A. Fiddy, "On the issue of the Born-Rytov controversy: I Comparing analytical and approximate expressions for the one-dimensional case," J.O.S.A. A9, pp1102-1110, 1992.

[6] N. Joachimowicz, Ch. Pichot and J.P. Hugonin, "Inverse scattering: an iterative numerical method for electromagnetic imaging," IEEE Trans Ant. Propag. Vol 29, Dec 1991.

[7] R.E. Kleinman and P.M. Van den Berg, "An extended range modified gradient technique for profile inversion," Proc. 1992 URSI International Symposium on Electromagnetic Theory, Sydney, Australia, August 1992.

[8] G.P.Otto, W.Chew, "Microwave scattering - local shape function imaging for improved resolution of strong scatterers," IEEE Trans. microwave Th. and Tech., Vol 42, Jan 1994.

[9] E. Wolf, "Three-dimensional structure determination of semi-transparent objects from holographic data," Optics Comm., 1, 153, 1969.

[10] M. Slaney, A.C. Kak and L.E. Larsen, "Limitations of imaging with first-order diffraction tomography," IEEE Trans Microwave Theory and Techniques MTT-32, 860, 1984.

[11] D. Raghuramireddy and R. Unbehauen, "The two-dimensional differential cepstrum," IEEE Trans. Acoustics, Speech and signal Processing, ASSP-33, 1335, 1985.

[12] W. K. Pratt, Digital Image Processing, John Wiley & Sons, New York, 1978.

[13] R. C. Gonzalez and P. Wintz, Digital Image Processing, Addison-Wesley Publishing Co., Reading, Massachusetts, 1977.

[14] A. V. Oppenheim and R. W. Schafer, Digital Signal Processing, Prentice Hall, Englewood Cliffs, New Jersey, 1975.

[15] H. Rossmanith and R. Unbehauen, "Formulas for computation of 2-D logarithmic and 2-D differential cepstrum," Signal Processing, 16, 209-217, 1989.

[16] J. B. Morris, R.V. McGahan, F.C. Lin and M. A. Fiddy, "Imaging of strongly scattering objects using cepstral filtering," Inverse Optics III, Ed. M. A. Fiddy, Proc. SPIE 2241, pp69-77, 1994.

[17] J. B. Morris, D. A. Pommet and M. A. Fiddy, "Nonlinear filtering applied to single view backpropagated images of strongly scattering objects," Proc. O.S.A. Topical Meeting on Signal Recovery and Synthesis V, March 1995.

[18] J. B. Morris, F. C. Lin, D. A. Pommet, R.V. McGahan and M. A. Fiddy, "A homomorphic filtering method for imaging strongly scattering objects," IEEE Trans. Antennas Propagation, 43, pp. 1029-1035, 1995.

[19] C.L. Byrne and M. A. Fiddy, "Estimation of Continuous Object Distribution from Limited Fourier Magnitude Data." Journal of the Optical Society of America, A4, pp. 112-117, 1987.

[20] C. L. Byrne, R. L. Fitzgerald, M. A. Fiddy, T. J. Hall, and A. M. Darling, "Image restoration and resolution enhancement," J. Opt. Soc. Am. 73, 1481-1487, 1983.

[21] C. L. Byrne and M. A. Fiddy, "Images as power spectra; reconstruction as a Wiener filter approximation," Inverse Problems, 4, 399-409, 1988.

[22] Morris, J.B., M. A. Fiddy and D. A. Pommet, "Nonlinear filtering applied to single-view backpropagated images of strong scatterers," Opt. Soc. of Amer. A, 13,July 1996.

[23] Chen, P-T, M. A. Fiddy, C-W. Liao and D. A. Pommet, "Blind deconvolution and phase retrieval using point zeros," Opt. Soc. of Amer. A, 13, July 1996.

Retrieval of Object Information from Electron Diffraction as Ill-Posed Inverse Problems

Kurt Scheerschmidt

Max Planck Institute of Microstructure Physics
Weinberg2, D-06120 Halle, Germany
Phone: +49-345-5582910, Fax: +49-345-5511223
Email: schee@mpi-msp-halle.mpg.de

Abstract. Inverse problems as direct solutions of electron scattering equations can be deduced using either an invertible linearized eigenvalue system or a discretized form of the diffraction equations. The analysis is based on the knowledge of the complex electron wave at the exit plane of an object reconstructed for single reflections by electron holography or other wave reconstruction techniques. In principle, this enables the direct retrieval of the local thickness and orientation of a sample as well as the refinement of potential coefficients or the determination of the atomic displacements, caused by a crystal lattice defect, relative to the atom positions of the perfect lattice. Considering the sample orientation as perturbation the solution is given by a generalized and regularized Moore-Penrose inverse. Extracting solely the atomic displacements the latter are given by the zeros of a function with an incompletely known Fourier spectrum. The numerical algorithms resulting from the fundamental relations imply ill-posed inverse problems.

1 Introduction

Inverse problems are difficult, always fascinating, and in most of the cases ill or improperly posed (Tichinov and Arsenin (1977), Lavrentiev (1967)). Ill or improperly posed means that one or all requirements are violated that usually characterize physics, i.e. existence, uniqueness and stability of a solution. As often occurring in many physical investigations, in the mathematical sense, the direct solution of the diffraction equations implies an inverse problem. Although the inverse problems violate especially the existence of unique and continuous solutions to arbitrary data they are of great practical importance, if the trial-and-error solution demands a large variety of possible solutions and models to be tested, mostly providing a better insight into the basic relations of the physical phenomena.

For instance, the imaging of crystal defects by high-resolution transmission electron microscopy or with the help of electron diffraction contrast technique is well known and routinely used. Though the theoretical image calculations always tend to establish standard rules of interpretation, a direct and phenomenological analysis of electron micrographs is mostly not possible, thus requiring the application of image simulation and matching techniques. Images are modelled by calculating both the interaction process of the electron beam with the almost periodic potential of the matter, and the subsequent Fourier imaging process including the microscope aberrations. The images calculated are fitted to the experiment by varying the defect model and the free parameters. This trial-and-error image matching technique is the indirect solution to the direct scattering problem applied to analyse the defect nature under investigation.

Electron holography or other reconstruction techniques (Lichte (1986), Lichte (1992), Coene et al. (1992), van Dyck et al.(1993)) permit the determination of the scattered wave function at the exit surface of the crystal directly out of the hologram or from defocus series up to the microscope information limit owing to the noise in the phase distortion. Especially the sidebands of a Fourier-transformed hologram represent the Fourier spectrum of the complete complex image wave and its conjugate, respectively, from which the object wave can be reconstructed. Thus, both the reconstructed amplitudes and phases can be compared to trial-and-error calculations (Lichte (1991), Lichte et al. (1992)).

In previous papers (Scheerschmidt and Hillebrand (1991), Scheerschmidt and Knoll (1994), Scheerschmidt and Knoll (1995a), Scheerschmidt and Knoll (1995b), Scheerschmidt (1997)) it was demonstrated that the local thickness and orientation can be calculated directly from the wave function reconstructed at the exit surface of the object instead of using trial-and-error simulation techniques. In principle, the analysis holds good also for the retrieval of the object potential, or if solely the positions of the atomic scattering centres are evaluated. The inverse problems, however, generally dealing with insufficiently measured data always require physically related information a priori. It was shown that the knowledge of both the amplitudes and phases of a sufficiently large number of plane waves scattered by the object as well as the partial knowledge of the potential of the perfect crystal structure imply the possibility of directly retrieving object information, instead of using trial-and-error simulation techniques. Two approximations are discussed to solve the resulting inverse scattering problem without reconstructing the whole crystal potential:

First, the special problem of retrieving the local sample orientation is solved on the basis of the perturbation approximation for perfect crystals, and by applying regularized and generalized matrices to invert the resulting linearized problem. The corresponding iteration procedure enables the direct analysis of the moduli and phases if a sufficient number of plane wave amplitudes can be separated yielding local thickness and bending of the object for each image pixel (Scheerschmidt and Knoll (1995b), Scheerschmidt (1997)).

Second, based on the knowledge of the reconstructed complex electron wave and using a discretized form of the diffraction equations, an alternative method is developed (Scheerschmidt and Knoll (1994), Scheerschmidt and Knoll (1995a)), yielding an algebraic equation system for the complex amplitudes and the elastic displacements. In principle, this system enables the direct retrieval of the atomic displacements, caused by a crystal lattice defect, relative to the atom positions of the perfect lattice. The equations are invertible provided the completeness of the plane waves is valid (continuity of the electron current). A special inverse problem of electron scattering is deduced considering solely those atomic displacements given by the zeros of a function with an incompletely known Fourier spectrum from the scattered electron wave of which the displacement field of a crystal lattice defect can, in principle, be retrieved.

The present paper outlines the fundamental relations for both special inverse problems describing some first numerical experiences related to the solution of the direct retrieval of local thickness and orientation. Some numerical aspects are considered as, e.g., the stability of unique inverse solutions in terms of noise, and the regularization of the problem.

2 Physical basis: Dynamical diffraction and holographic wave reconstruction

The HREM image contrast is mainly determined by two processes: First, by the electron diffraction owing to the interaction process of the electron beam with the almost periodic potential of the matter and, second, by the interference of the plane waves leaving the specimen and being transmitted by the microscope. Assuming that the object wave is reconstructed free of aberrations or under diffraction contrast conditions the influence of the microscope imaging process itself can be neglected. Thus the image contrast is solely determined by the interaction of the electrons with the object potential.

The interaction of electrons with a crystalline object is described on the basis of a periodic potential with the electron structure factors as the expansion coefficients and the Bloch-wave method for solving the high-energy transmision electron diffraction. Different formulations can be given, using Bloch wave or plane wave representations of the scattered waves, applying direct or reciprocal space expansion, and direct integration or slice techniques, which, in principle, are equivalent descriptions (van Dyck (1985), Spence and Zuo (1992), van Dyck (1989)). The object wave in terms of modified plane waves with complex amplitudes $\phi_{\mathbf{g}}$ yields

$$o(R) = \sum_{g} \phi_g e^{2\pi i((\mathbf{k}+\mathbf{g})\mathbf{R}+s_{\mathbf{g}}\mathbf{t})} \tag{1}$$

with reflections $\mathbf{g}$, excitations $s_{\mathbf{g}}$, wave vector $\mathbf{k}$, and thickness t of a parallel-sided object, $\mathbf{R} = (\mathbf{x}, \mathbf{y})$. The amplitudes $\phi_{\mathbf{g}}$ are constant with respect to z in the vacuum outside the object, which means that the plane waves are the stationary solutions of the wave equation. Within the crystal, however, the amplitudes of the modified plane waves $\phi_{\mathbf{g}}$ are z-dependent according to the Ewald pendulum solution as described by the Bloch waves, which are the stationary solution within the periodic potential.

The basic equations of the Bloch wave presentation in forward scattering approximation are given by the eigenvalue system

$$\sum_{\mathbf{h}} A_{\mathbf{gh}} C_{\mathbf{h}} - \gamma C_{\mathbf{g}} = 0, \text{ with } 2k_z A_{\mathbf{gh}} = (2\mathrm{K}.\mathbf{g} - \mathbf{g}^2)\delta_{\mathbf{gh}} - \mathbf{V}_{\mathbf{g-h}}, \tag{2}$$

yielding the amplitudes $C_{\mathbf{g}}^{(l)}$ of the l th partial wave and its "anpassung" $\gamma^{(l)}$ to the dispersion of the lattice as a function of the lattice potential (Fourier coefficient $V_{\mathbf{g}}$) as well as the relative orientation of the object with respect to the electron beam incidence **K**. With these eigenvalues and vectors, for a plane parallel perfect crystal of thickness t the complex amplitudes $\phi_{\mathbf{g}}$ of eq. (1) are directly given in matrix form by

$$\Phi = CXC^{-1}\theta \tag{3}$$

where $\Phi = [\phi_{\mathbf{g}}]$ and θ are the vectors of the amplitudes of the exit and the incident waves, respectively, and **X** represents the diagonilized scattering matrix $e^{2\pi iAt}$.

Using furthermore the deformable ion approximation a crystal lattice defect can be included by its elastic displacement field **v** as a phase shift of the Fourier spectrum of the crystal potential. The evaluation of the quantum-theoretical scattering problem using the high-energy forward scattering approximation (see, e.g., (Anstis (1989), Howie and Basinski (1968)) for the derivation and the explicit form of the equations) yields a parabolic differential equation system for vector Φ of the complex amplitudes of the elastically scattered electron waves :

$$\partial\Phi/\partial z = (\Delta + V[e^{i\mathbf{g}\mathbf{v}}])\Phi \tag{4}$$

with $\Delta = \{ik_z\nabla^2 - 2(\mathbf{k}+\mathbf{g})\nabla\}/2\mathbf{k'_z} + 2\pi(\mathbf{s_h} - \mathbf{s_g})_\mathbf{z}$, $\nabla = (\partial/\partial x, \partial/\partial y, 0)$, $k'_z = k_z + g_z + s_{\mathbf{g}}$ and the potential V=V'+iV" including the lattice potential V' and the absorption V" (one electron-optical potential approximation of inelastic scattering) as well as the diagonal matrix of the defect phase shifts.

In addition, boundary and initial conditions have to be applied: The linearized high-energy approximation directly fits $\phi_{\mathbf{g}}(\mathbf{R},\mathbf{t})$ at the crystal exit surface to $\phi_{\mathbf{g}}(\mathbf{R})$ outside, demanding $|\phi_{\mathbf{g}}(\mathbf{R},\mathbf{0})| = \delta_{\mathbf{go}}$ at the entrance surface, whereas the continuity of the derivatives has to be omitted in the linearized case. It enables one, however, to estimate the unknown displacements at the exit foil surface by using eq. (4) without potential outside and inverting eq. (4) directly at the exit surface:

$$\{V[e^{i\mathbf{g}\mathbf{v}}]\Phi = 2\Delta\Phi\}_{z=t} \tag{5}$$

Instead of boundary conditions one can assume a periodic continuation to describe large extended crystal slabs, i.e. $\phi_{\mathbf{g}}(x,y,z) = \phi_{\mathbf{g}}(x+X,y,z)$ and $\phi_{\mathbf{g}}(x,y,z) = \phi_{\mathbf{g}}(x,y+Y,z)$, with slab extensions X,Y approaching infinity.

Holography with electrons offers one of the possibilities of increasing the resolution by avoiding microscope aberrations. It also enables the complete complex object wave to be restored. Image plane off-axis holograms are recorded in a microscope which is equipped with a Möllenstedt-type electron biprism inserted between the back focal plane and the intermediate image plane of the objective lens (Lichte (1986), Lichte (1991), Lichte (1992), Lichte et al. (1992)). The object is arranged so that a reference wave outside of it is transferred through the microscope, and owing to a positive voltage of the biprism both waves mutually overlap in the image plane creating additional interference fringes. The intensity of the latter is modulated by the modulus of the object wave, whereas the fringe position is varied by the phase of the object wave. Thus the recorded interference pattern is an electron hologram from which both the modulus and the phase of the object wave can be reconstructed by optical diffraction or numerical reconstruction. The reconstruction starts with a Fourier transform of the hologram. Besides two sidebands in the central region of the Fourier spectrum the zero peak and autocorrelation occur, which is equivalent to a conventional diffractogram. The sidebands represent the Fourier spectrum of the complete complex image wave and its conjugate, respectively, from which the object wave o(x,y) can thus be reconstructed by separating, centring, and applying the inverse Fourier transform including a reciprocal Scherzer filter with damping and microscope aberrations (Lichte (1991), Orchowski et al. (1995)).

In the following it is important that, besides the whole sideband, each single reflection of sufficient intensity can be reconstructed separately (Scheerschmidt (1997)). This provides the possibility of noise reduction if suitable windows and filtering are applied and if the pixels are precisely centred to avoid additional phase shifts. The environment of the reflections included in the filtering process has to be chosen such that the information of local distortions folded with the reflections will be transferred to the reconstructed partial waves. The reconstruction of the single reflections causes modulus and phase to be distributed in the partial waves, which is the presupposition of the inverse algorithm discussed in the following.

Figure 1 demonstrates the wave and the single-reflex reconstruction using a theoretical hologram simulated for a Σ=13 (100) tilt grain boundary in gold, which is relaxed by molecular dynamics. Fig. 1(a) shows the simulated hologram, and Fig.1(c), an enlarged region with the hologram fringes extending from the central part of the boundary, with the atomic columns around the interface. The Fourier spectrum of the hologram is given in Fig.1(b), and the sideband selected for reconstruction in Fig.1(d), the pairs of the reflections are indicated with the corresponding reciprocal lattice vectors. Fig. 1(e) presents the reconstructed real space intensities of the single reflections in amplitude (AMP) and phases (PHA), separately for the two grains denoted 1 and 2, respectively: in the upper row the left grain is excited, in the lower row, the right one. The reconstructed amplitudes of the reflections can directly be interpreted as bright and dark-field images of the grain boundary.

Fig. 2 shows one sideband of the Fourier spectrum of the experimental hologram ((a), selection on the left hand side) of a Σ=13 (100) tilt grain boundary in gold ($\theta = 22.6^o$, see (Orchowski et al. (1995), Orchowski and Lichte (1996)) and preliminary common work (Orchowski et al. (1993))) and the reconstruction (b,c) of the single reflections as indicated in the spectrum of the hologram filtered through a Gaussian mask. The upper rows (b) show modulus (AMP) and phases (PHA) of the particular reflections chosen of types 000, {002}, and {220}, thus presenting the reconstruction of the corresponding amplitudes $\phi\mathbf{g}$

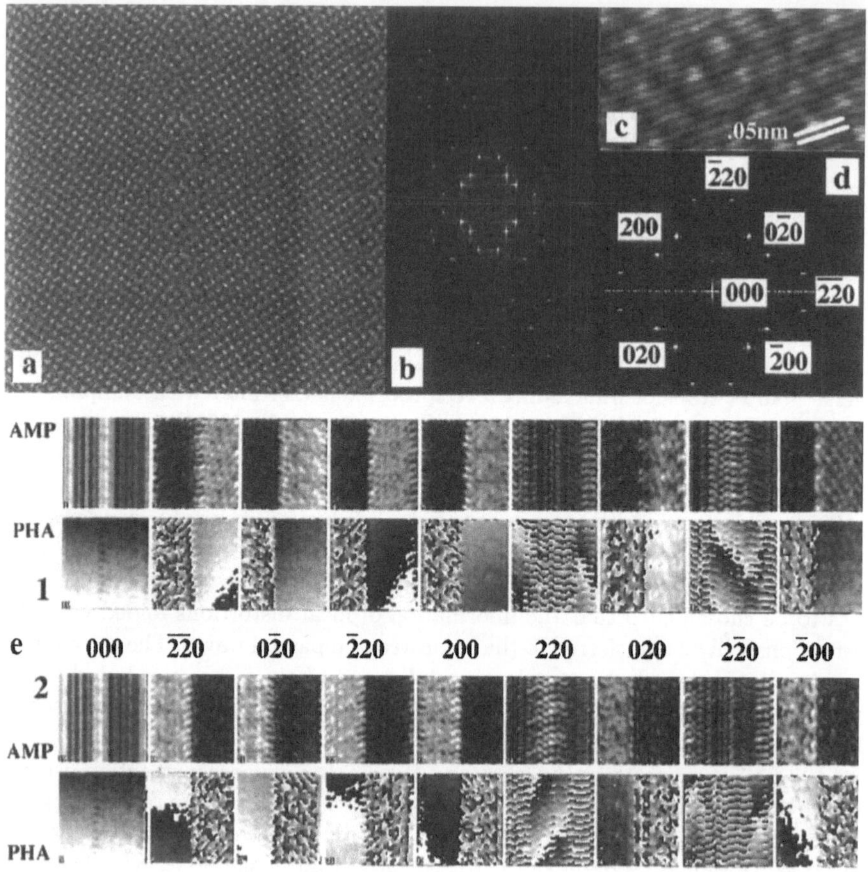

Fig. 1. Reconstruction of single reflections of a simulated MD-relaxed Σ=13 (100) Au grain boundary: (a) Theoretical hologram with the grain boundary vertically arranged in the centre, (b) Fourier spectrum of the hologram, (c) Enlarged selection of the hologram, (d) Sideband applied for reconstruction with indices of the reflections, (e) Reconstructed moduli (AMP) and phases (PHA) of the reflections of grains 1 and 2, resp.

out of the hologram. For comparison in the lower rows (c) the corresponding real (REA) and imaginary part (IMA) of the reconstructed $\phi_{\mathbf{g}}$ are presented yielding the same information, however, without the phase wrapping problem according to the multi-valued phases. The reconstruction of the higher-order reflections is impossible here because of the lower intensity of the latter and the mutual overlap of the autocorrelation and the side-band. The single reflections are denoted by 1 and 2 according to grain 1 and 2, respectively. The shift of the fringes at the grain boundary directly indicates the phase shift owing to the crystal defect. The modulation by lower frequencies is due to the local bending of the sample or to thickness oscillations.

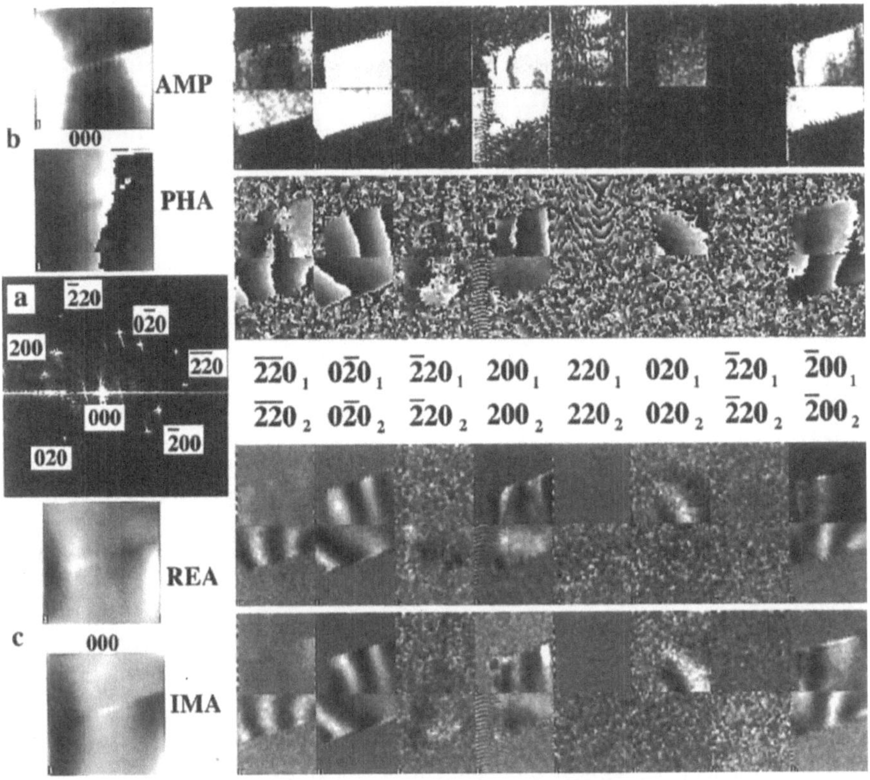

Fig. 2. Reconstruction of single reflections of a Σ=13 (100) Au grain boundary, separately for both grains denoted by 1 and 2, respectively: (a) Fourier spectrum of the hologram (0.05 nm fringes, A. Orchowski, University Tübingen (Orchowski and Lichte (1996), Orchowski et al. (1993))), with indicees of the reflections and asymmetric intensities in the sideband showing the mistilted orientation, (b) reconstructed moduli (AMP) and phases (PHA), (c) reconstructed real (REA) and imaginary (IMA) part of the partial waves.

3 Inversion by linearization and discretization

Eq. (2) can be linearized applying perturbation methods. Assuming that the eigenvalues γ are non-degenerated, and by analogy with eq. (3), the perturbation solution may read

$$\Phi = \Gamma \Xi \Gamma^{-1} \theta, \tag{6}$$

where the matrices are given by

$$\Gamma = C(1+\Delta), \Xi = \{e^{2\pi i \lambda t}\}, \text{ and } \lambda = \gamma + \Delta\{\delta_{ij}\} + \Delta^{-1}\{1/(\gamma_i - \gamma_j)\}\Delta. \tag{7}$$

As diagonal elements the perturbation matrix $\Delta_{\mathbf{gh}} = (\Delta \mathbf{K}.\mathbf{g}) + \mathbf{i}\Delta \mathbf{V}_{\mathbf{gh}}$ contains the deviation of the orientation $\Delta\mathbf{K}$ from that of the original eigenvalue

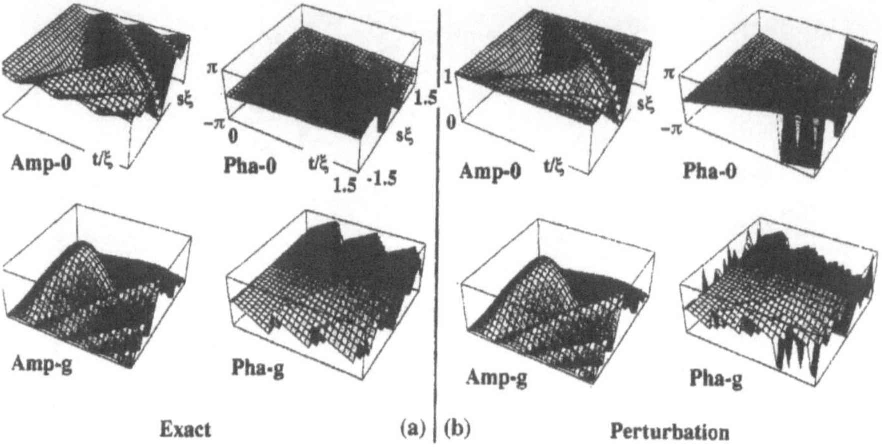

Fig. 3. Comparison of the exact two-beam solution (a) of moduli (AMP) and phases (PHA) of transmitted (0) and diffracted beam (g) with the corresponding perturbation solution (b). Differences occur for orientations with $|s\xi| > 1$, e.g., where the perturbation is no longer valid (s=Bragg deviation, t=crystal thickness, ξ=extinction distance).

system **K**. The non-diagonal elements describe a perturbation of the potential as, e.g., according to optical absorption. Fig. 3 demonstrates the validity of the perturbation solution comparing eq. (6) with the exact solution (3) of the two-beam case. As for moduli and phases, for both reflections there are remarkable deviations for $|s\xi| > 1$ almost independent of thickness t around the exact orientation of the pole $|s\xi| = 0$ of the exact two-beam excitation.

Starting from approximate values of thickness t_o and beam orientation (k_{x_o}, k_{y_o}) gained from a priori knowledge or by analysing, e.g., the asymmetry of the single reflections reconstructed from the holographically retrieved wave function, the perturbation solution is valid within certain intervals around t_o and (k_{x_o}, k_{y_o}). Eq.(6) can be expanded in a Taylor series yielding

$$\phi(t, k_x, k_y) = \phi(t_o, k_{x_o}, k_{y_o}) + (t - t_o)\partial\phi/\partial t + (k_x - k_{x_o}, k_y - k_{y_o})\nabla_k\phi. \quad (8)$$

The derivatives can directly be gained from eqs. (8) using equivalent abbreviations:

$$\partial\phi/\partial t = \Gamma\partial\Xi/\partial t\Gamma^{-1}\theta \text{ and } \nabla_k\phi = (\nabla_k\Gamma\Xi - \Gamma^{-1}\nabla_k\Gamma\Xi + \Gamma\nabla_k\Xi)\Gamma^{-1}\theta \quad (9)$$

The linearized eq.(8) together with the analytical expressions (9) enable the inverse solution:

$$(t, k_x, k_y) = \mathbf{M_{inv}}[\phi^{\mathbf{exp}} - \phi^{\mathbf{pert}}], \quad (10)$$

where the matrix is given, e.g., by the Penrose-Moore inverse $\mathbf{M_{inv}} = (\mathbf{M^TM})^{-1}\mathbf{M^T}$, which is represented analytically using the matrix of the coefficients $\mathbf{M} = (\partial\phi/\partial\mathbf{t}, \nabla_\mathbf{k}\Phi)$ of eq.(9). ϕ^{exp} are the measured data and ϕ^{pert} the solution of the perturbation equation (6) at t_o, K_o. The series expansion (8) as well as the resulting formalism (9) can be extended to include also the derivatives of deviations from potential coefficients, which are omitted here for the sake of simplicity. That means, additional unknown object parameters can be included in the retrieval procedure as far as the problem remains overdetermined with respect to the unknowns.

Algorithm (10) is the solution to the inverse problem concerning the local thickness and orientation analysis, the regularized inverse iteration can directly be applied to each pixel in the real space representation of the single reflections reconstructed from the hologram. On the suitable assumption of the basic eigenvalue system (2) and starting with suitable local thickness t_o as well as incident beam orientation (k_{xo}, k_{yo}) the values of thickness t and orientation (k_x, k_y) are probably enhanced if eq. (10) is applied to the amplitudes and phases measured of each image pixel and each reflection **g**.

Figures 4 and 5 demonstrate the applicability using the single reflection wave reconstruction of Figs. 1 and 2, respectively. In both cases, the same nine-beam eigenvalue system was used to model the diffraction behaviour. Here, no further assumption was made as to the initial thickness t_o. The best fit was revealed by searching the absolute minimum of the defect of the vector norm at an extended thickness intervall. Fig. 4 results in a flat tickness t(i,j) as assumed for the simulation of the corresponding hologram. The retrieved incident wave vector **K**(i,j) shows oscillations with the pixel numbers, caused by the bending of the lattice planes, which results from the relaxation of the grain boundary because of the additional twist component assumed. Different but small regularization parameters γ (here γ=.0001 was assumed) do not smooth the noise if solely the pixel intensity (see chapter 4) is regularized. In the case of retrieving from the experimental hologram, different initial orientations of $\mathbf{K_0}$=(.51,.71,.0) in Fig. 5a, and of $\mathbf{K_0}$=(-.28,1.21, .0) in Fig. 5b, yield very noisy results in thickness t and orientation (k_x, k_y) for the 64x64 pixels retrieved. Nevertheless, both cases show almost the same values t$\approx$ .77ξ and t=0 for the plateau of the object and the hole, respectively.

The differential equations (4) allow the diffusion-like interpretation and can be discretized using standard difference algorithms (Scheerschmidt and Knoll (1994), Scheerschmidt and Knoll (1995a)). An algebraic equation system results, which formally reads

$$\Phi(i,j,k-1) = \mathbf{F_1}\{\Phi(\mathbf{i,j,k}), \Phi(\mathbf{i}\pm\mathbf{1,j,k}), \Phi(\mathbf{i,j}\pm\mathbf{1,k}), \mathbf{v}(\mathbf{i,j,k})\} \qquad (11)$$

for the complex amplitudes Φ and the elastic displacements **v** at the (xyz)-grid points (i,j,k), (i$\pm$1,j,k), (i,j$\pm$1,k) and (i,j,k$\pm$1) representing the object. Periodic boundary conditions are assumed in x and y direction, whereas at the exit surface, a further equation is given applying the forward integration of eq. (11) outside the crystal and discretizing the symbolic equation (5).

Within the crystal the difference equations (11) are equivalent for backward (k-1) and forward (k+1) integration with respect to the beam propagation, thus being insufficient for determining both the wave amplitudes Φ(i,j,k) and the elastic displacement field **v**(i,j,k) at the grid points (i,j,k) considered. This be-

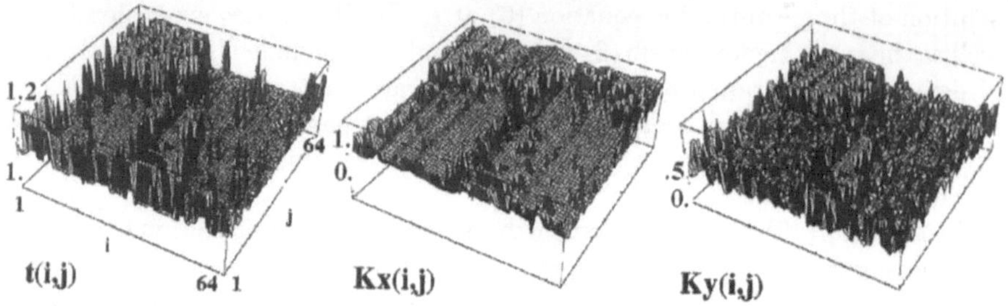

Fig. 4. Iteratively determined local sample thickness t and beam orientation (K_x, K_y) retrieved from the reconstructed reflections of the theoretical hologram in Fig. 1 for arbitrary start values of thickness and given start values of orientation without regularization.

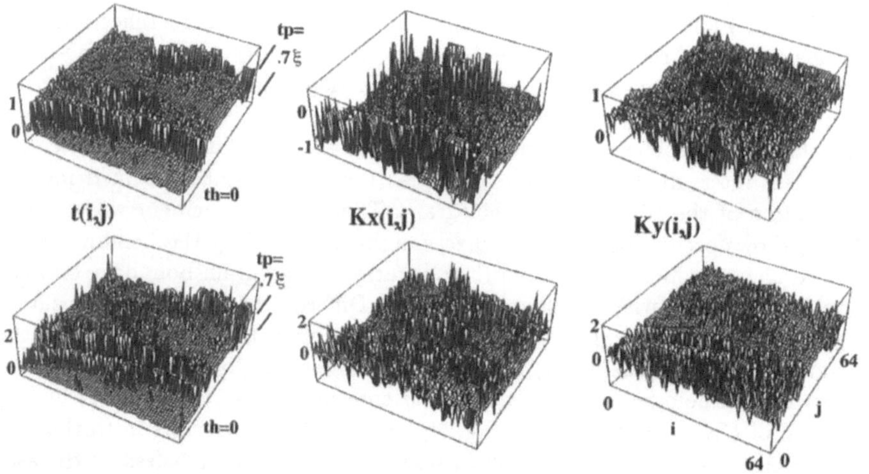

Fig. 5. Non-stabilized iteratively determined local sample thickness t and beam orientation (K_x, K_y) retrieved from the reconstructed reflections of the experimental hologram in Fig. 2 for arbitrary start values of thickness t (resulting in stable solutions th=0 in the hole, and tp $\approx .77\xi$ on the plateau) and given start values of orientation $\mathbf{K_O}$=(.51,.71,0) and $\mathbf{K_O}$=(-.28,1.21,0) for the upper and lower rows, respectively.

comes also obvious by simply numbering the unknowns and the equations at each node: for N beams, there are N unknown amplitudes and 3 unknown displacements, and N relations according to eqs. (11), using either (k-1) or (k+1). One of the difference equations, however, can be replaced as follows: While the optical potential in the reciprocal space representation is generally non-hermitian, the hermiticity of the potential V' and of the "absorption" V" yields the equation of continuity for the whole current $I = \Sigma\phi\mathbf{g}\phi\mathbf{g}^*$. The continuity equation may then read

$$\partial I/\partial z = \Phi\nabla^2\Phi^* - \Phi^*\nabla^2\Phi + 2(\mathbf{k}+\mathbf{g})\nabla\mathbf{I} - 2\Phi\mathbf{V}"[\mathbf{e}^{\mathbf{igv}}]\Phi^* \qquad (12)$$

The equation of continuity can be discretized by analogy with the discretization of the differential equations above. The differential operator, however, yields mixed terms with respect to different nodes (i,j,k) and (i±1,j±1,k):

$$\mathbf{F_2}\{\mathbf{v}(\mathbf{i},\mathbf{j}), \Phi(\mathbf{i},\mathbf{j},\mathbf{k}+1), \Phi(\mathbf{i},\mathbf{j},\mathbf{k}), \Phi(\mathbf{i}\pm 1,\mathbf{j},\mathbf{k}), \Phi(\mathbf{i},\mathbf{j}\pm 1,\mathbf{k})\} = \mathbf{0} \quad (13)$$

By analogy with the Gelfand-Levitan-algorithm (see, e.g., (Zakhariev and Suzko (1990))) an additional equation results by inverting the equation of continuity, which is a kind of completeness relation, yielding

$$\sum_{\mathbf{g}} Q_{\mathbf{g}} e^{2\pi \mathbf{g}\mathbf{v}} = 0 \quad (14)$$

for eq. (13) as well as for the additional boundary condition previously discussed. Coefficients $Q_{\mathbf{g}}$ are explicitely given in (Scheerschmidt and Knoll (1994)). Thus, in principle, the retrieval of the displacements $\mathbf{v}$ is given by the remaining inverse problem (14), implying to find the root of a function given by an incomplete Fourier transform.

The inverse problem (14) is ill-posed for two reasons: Only one equation has to be solved for the vectorial root $\mathbf{v}$(i,j,k) at node (i,j,k), thus coplanar vectors $\mathbf{g}$ leave one component unconsidered. The spectrum $Q_{\mathbf{g}}$(i,j,k) is incomplete and noisy. This results in unstable numerical solutions using standard algorithms to find the roots (viz. Newton-Raphson algorithm, genetic algorithms and neuronal networks), owing to the existence of a large number of subsidiary roots. Besides the numerical solutions, transforming eq.(14) yields iterative forms as a kind of quasi-regularization (Scheerschmidt and Knoll (1995a)), the system then refers to an overdetermined system in the same manner as discussed above.

4 Numerical aspects

The inversion proposed is based on the linearization and the fact that the problem is overdetermined with respect to the unknowns but underdetermined if the noise is included, resulting in a least square minimization of a suitable vector norm of the defect (Lois (1989), Bertero (1989)), e.g.,

$$||\Phi^{exp} - \Phi^{pert}|| = Min. \quad (15)$$

As the iteration procedure seems to be amplifying the noise, the regularizations should be further enhanced. Simple averaging of the retrieved thicknesses and orientations with values larger than a certain threshold omitted, avoids outliers and leverages, however, structural details, too, thus yielding incorrect regularization.

The stability of the procedure may be enhanced by using more general regularizations as, e.g., the Phillips regularization. The most general regularization may be of the Ivanov-Phillips-Tichonov-type (see, e.g., (Bertero (1989))),

$$||\Phi^{exp} - \Phi^{pert}||^2 + \gamma||Z||^2 = (\Phi^{exp} - \Phi^{pert})^{\dagger} C_1 (\Phi^{exp} - \Phi^{pert}) + \gamma Z^T C_2 Z = Min \quad (16)$$

While the Moore-Penrose inverse minimizes the defect, an additional constraint here allows one to weight the measured data by C_1 and to smooth the solution $Z = (t, k_x, k_y)$ by C_2. Using the Moore-Penrose or similar generalizations always allow ill-posed problems with discrete data to be transformed to well-posed, but mostly ill-conditioned, problems: The solution exists and is unique, however, mostly unstable. The generalized solution may be considered an average of the true solutions, the resulting generalized inverse including the regularization matrices may be

$$M_{inv} = (M^{\dagger}C_1M + \gamma C_2)^{-1}M^T \tag{17}$$

with the suitable regularization factor γ and matrices C_1 and C_2, respectively. The iterative solution of eq. (11) with this generalized inverse (17) yield a self-consisting approach.

The generalized approach represents the maximum-likelihood solutions if the weight matrizes C_1 are suitably chosen with respect to the reflections **g**. Gaussian distributed noise can be described by unit weights, Poisson distributed noise demands weights inversely proportional to the intensity of the reflections.

In image processing, however, the regularization is described as a procedure smoothing the pixels (i,j) (Huang (1975)) : A solution with small second derivatives with respect to neighbouring pixels tends to be more accurate. In general, any constraint C_2 which is quadratic (Huang (1975)), may be used to yield a solution resembling eqs.(10) and (17).

Assuming that the different weights can be separated without a loss of generality, the weighting C_1 is given by $W_{\mathbf{gh}}{}^{\dagger}W_{\mathbf{g'h'}}$ with W~ $|\Phi|^{\rho}$. The smoothing C_2 can be described by matrix filters with respect to the pixels (i,j). A zero-order smoothing is equivalent to outlier detection or avoiding levarages (Rousseeuw (1977)).

The regularization parameter can be bounded (Bertero (1989)), but in the physically relevant problems such bounds are too rough and should be estimated by numerical tests. To study the confidence level of the solutions the retrieved thicknesses and orientations are compared with those used in simulated holograms, which have been performed for either perfect crystals with increasing thickness and linearly varying orientation, or for the theoretical grain boundary of Figs.1 and 4 relaxed by molecular dynamics. To check the reliability and accuracy by using simulated inputs is advantageous as one can directly compare well-known numbers and thus find out the regularization parameter for the best fit. One can use different distance measures like the squared differences or the regression coefficient $r = cov(Z_{retrieve}, Z_{theory})/(\sigma_{retrieve}\sigma_{theory})$ assuming a linear hypothesis for the fit, a χ^2 test or cross correlations. Robust measures are very fast and stable: the simple sign test $s_1 = \Sigma sgn(Z_{ij} - \langle Z \rangle)/N < 1$ of all pixels or the product s_2 of neighbouring pixels, for instance allow one to detect sytematic errors, whereas weigths which are controlled by the regression coefficient between retrieved and exact data (Rousseeuw (1977)) enable the finding of outliers and leverages. No test may be considered to be perfect or superior, because always a large number of differences is reflected by only one number. Fig. 6 shows in (a) the regression coefficient r, the sign tests s_1, s_2, and different $\log(\chi^2)$ measures as function of testparameters (χ_1 from the convergence error of the retrieval procedure, χ_2 with and χ_3 without outlier detection). Figs. 6(b) and (c) represent the same $\log(\chi^2)$ measures as function of the regularization parameter γ without and including pixel smoothing, respectively, i.e. (b) with $C_1 = C_2 = I$ and (c) smoothing of the second derivative

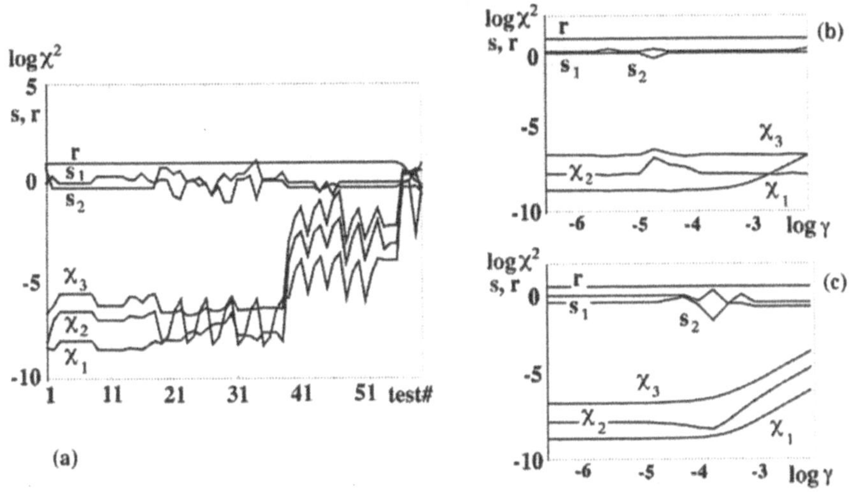

Fig. 6. Confidence tests for different measures (regression r, sign tests s_1, s_2 and $\log(\chi^2)$) of retrieved versus simulated data as function of test parameters (a, see text) or as function of the regularization parameter γ: (b) no smoothing $C_1 = C_2 = I$ and (c) smoothing of the second derivative $C_1 = I, C_2 = (\delta_{i-i_o\pm1,j-j_o\pm1} - 2\delta_{i-i_o,j-j_o})$

with $C_1 = I, C_2 = [\delta_{i-i_o\pm1,j-j_o\pm1} - 2\delta_{i-i_o,j-j_o}]$. Test 1 in (a) for comparison is calculated without any smoothing and normalization. The tests 2-18 are applied for normalizing the different reflexes and/or using different averages over the pixels and the reflexes, resp., always with $\rho = 0, \pm1, \pm2$. In the tests 19-38 additionally the weights C_1 are proportional to the amplitudes and intensities of the reflexes, resp, and $\gamma = 10^{-5}, 10^{-4}, 10^{-3}, 10^{-2}$. For the test 39-53 the regularization is related to the a priori information instead of the maximum norm itself, and $\gamma = 10^{-5}, 10^{-4}, 10^{-3}$. Clearly can be seen, that the smoothing increases the errors, whereas sytematic errors and low regression coefficients are resulting from invaluable normalization. Further systematic calculations are necessary to find out the best regularization γ, i.e., the compromise between accuracy and stability of the retrieval procedure.

5 Conclusions

Both the direct solutions (11) and (12,14), i.e. the explicit evaluation of thickness and orientation as well as the retrieval of the atomic displacements from a reconstructed electron wave function at the exit surface of an object, result in particular inverse problems of the first kind, viz. the analysis of object parameters from measured data. Thus, from the mathematical point of view the retrieval procedure is an ill-posed inverse problem requiring additional information about the periodicity of the object as the basic assumption, the thickness, the orientation and the unknown reconstructed displacements in order to make the process stable and continuous, to avoid singularities, and to restrict the manifold set of solutions possible. The procedure described has transformed these difficulties to the mathematical problem of overdetermined equation systems and of determining the roots of a function with an incomplete Fourier transform. Normalization and regularization of the solutions enable smoothing, stabilization and outlier detection.

Acknowledgements
We are grateful to the Volkswagenstiftung for financial support.

References

Anstis, G.R. (1989): Simulation techniques for reflection electron microscopy, in: Computer Simulation of Electron Microscope Diffraction and Images, Krakow, W., O'Keefe, M.A. (eds.), The Minerals, Metals and Materials Society, pp. 229–238

Bertero, M. (1989): Linear Inverse and Ill-Posed Problems, Advances in Electronics and Electron Physics **75**, 1–114

Coene, W., Janssen, G., Op de Beeck, M., van Dyck, D. (1992): Phase retrieval through focus variation for ultra-resolution in field-emission transmission electron microscopy, Phys. Rev. Letters **69**, 3743–6

van Dyck, D. (1985): Image calculation in high resolution electron microscopy: Problems, progress and prospects, in: Advances in Electronics and Electron Physics, Hawkes, P.W. (ed.), **65**, 295–355

van Dyck, D. (1989): Three-dimensional reconstruction from two-dimensional projections with unknown orientation, position and projection axis, Ultramicroscopy **30**, 435–8

van Dyck, D. , Beeck, M. Op de, Coene, W. (1993): A new approach to object wavefunction reconstruction in electron microscopy, Optik **93**, 103–7

Howie, A., Basinski, Z.S. (1968): Approximations of the dynamical theory of diffraction contrast, Philos. Mag. **17**, 1039–1063

Huang, S. (Editor) (1975), Picture Processing and Digital Filtering, Springer Vlg., N.Y.

Lavrentiev, M.M. (1967): Some Improperly Posed Problems of Mathematical Physics, Springer, Berlin, pp. 10–20

Lichte, H. (1986): Electron holography approaching atomic resolution, Ultramicroscopy **20**, 293–304

Lichte, H. (1991): Electron image plane off-axis electron holography of atomic structures, Advances in Optical and Electron Microscopy **12**, 25–91

Lichte, H. (1992): Holography - just another method of image processing?, Scanning Microscopy Suppl. **6**, 433–440

Lichte, H., Völkl, E., Scheerschmidt, K. (1991): Electron image plane off-axis electron holography of atomic structures, Advances in Optical and Electron Microscopy **12**, 25–91

Lois A.K. (1989): Inverse und schlecht gestellte Probleme, Teubner Vlg., Stuttgart

Orchowski, A., Lichte, H., Scheerschmidt, K., Scholz, R. (1993): Hochauflösende Elektronenmikroskopie zur Analyse von $\Sigma 13$ Tilt-Korngrenzen in Gold, Optik **94**, Suppl. 5, 79

Orchowski, A., Rau, W.D., Lichte, H. (1995): Electron holography surmounts resolution limit of electron microscopy, Phys. Rev. Letters **74**, 399–402

Orchowski, A., Lichte, H. (1996): High resolution electron holography of real structures at the example of a =13 grain boundary in gold, Ultramicroscopy **69**, 199–209

Rousseeuw, P.J. (1987), Robust regression and outlier detection, John Wiley & Sons, N.Y.

Scheerschmidt, K., Hillebrand, R. (1991): Image interpretation in HREM: Direct and indirect methods, Proc. 32nd Course Int. Centre of Electron Microscopy "High-Resolution Electron Microscopy - Fundamentals and applications", Heydenreich, J., Neumann, W. (eds.), Halle, p. 56–65

Scheerschmidt, K., Knoll, F. (1994): Retrieval of Object Information from Electron Diffraction, I. Theoretical preliminaries, phys. stat. sol. (a) **146**, 491–502

Scheerschmidt, K., Knoll, F. (1995): Retrieval of atomic displacements from reconstructed electron waves as an ill-posed inverse problem, Proc. Int. Workshop Electron Holography, Knoxville Tennessee USA 1994, Tonomura, A., Allard, L.F., Pozzi, G., Joy, D.C., Ono, Y.A. (eds.), Elsevier Science , p. 117–124

Scheerschmidt, K., Knoll, F. (1995): Zur Rekonstruktion von Verschiebungsfeldern aus elektronen-holographisch ermittelten Objektwellen , Optik **100**, Suppl. 6, 50.

Scheerschmidt, K. (1997): Direct retrieval of object information from diffracted electron waves, Proc. 15 Pfefferkorn Conf. on Electron Imaging and Signal Processing, May 18–22, 1996, Silver Bay, N.Y., Scanning Microscopy Suppl. **11**, submitted

Spence, J.C.H., Zuo, J.M. (1992): Electron Microdiffraction, Plenum Press, New York, pp.134–5

Tichonov, A.N., Arsenin, Y.Y. (1977): Solutions of Ill-Posed Problems, Wiley, New York, pp. 1–30

Zakhariev, B.N., Suzko, A.A. (1990): Direct and Inverse Problems, Springer Vlg., Bln.-Heidelbg.

A Linear Method for Solving Inverse Scattering Problems in the Resonance Region*

David L. Colton

Department of Mathematical Sciences, University of Delaware, Newark, DE 19716, U.S.A.

1 Introduction

Inverse scattering problems have attracted increased attention in recent years due to their appearance in a wide variety of applied areas, for example nondestructive testing, medical imaging and geophysical prospecting. Mathematically, these problems can be divided into three broad groups defined by the frequency of the probing wave: low frequency, intermediate frequency, and high frequency. In particular, the low and high frequency regimes are amenable to asymptotic methods (and hence a linearization of the inversion scheme) while at intermediate frequencies (the so-called *resonance region*) the problem is inherently nonlinear and is typically dealt with by nonlinear optimization methods. Since in many applications one is forced to work with frequencies in the resonance region (due to the conflicting needs of being able to have the probing wave penetrate deeply into the scattering object while at the same time achieving sharp resolution of the image) in recent years there has been a major effort made in solving inverse scattering problems by nonlinear optimization methods. Indeed, at the moment, *all* methods for solving inverse scattering problems in the resonance region are based on such methods. For an excellent short introduction to this approach we refer the reader to the forthcoming book by Andreas Kirsch ([6]).

In this paper we would like to briefly discuss a new approach to solving inverse scattering problems in the resonance region which totally avoids the use of nonlinear optimization methods. In fact, the scheme we are about to describe only involves solving a set of *linear* integral equations of the first kind. Since the inverse scattering problem is in fact nonlinear, something is obviously lost in such an approach. What is lost is that in our approach only the *support* of anomalies in a piecewise homogeneous background is obtained rather than the index of refraction inside the anomalies. Furthermore, the method requires being able to probe and measure the response around the

* This work was supported in part by a grant from the Air Force Office of Scientific Research.

entire scattering obstacle, thus making it impractical for some applications. However, for many applications, particularly in medical imaging and non-destructive testing, the conditions for the applicability of our method are met and the determination of the support of anomalies is all that is needed. For example, in the case of the detection and location of tumors in the body by microwaves, it is sufficient to determine if there are in fact tumors present and if so what is the support of the tumors (i.e. how big they are). The actual value of the index of refraction in the tumor is, by comparison, of little interest. Similarly, if there are flaws in a material (e.g. a crack) the fact that such flaws exist and determining an estimate of their size is the most important consideration rather than a complete reconstruction of the sound speed or refractive index inside the flaw.

Our approach for determining the support of anomalies in a material is mathematically based on the properties of solutions to interior transmission problems (c.f. [2]). Hence, in the next section we will describe the scattering problem of interest to us and how our approach to the inverse problem is naturally associated to a particular interior transmission problem. Finally, in the last section of this paper we will describe the inversion scheme itself as well as some of its peculiar characteristics.

2 Inverse Scattering and Interior Transmission Problems

Consider the two dimensional scattering problem of determining u from the equations

$$\Delta_2 u + k^2 n(x) u = 0 \quad \text{in } R^2 \tag{1}$$

$$u(x) = e^{ikx\cdot d} + u^s(x) \tag{2}$$

$$\lim_{r\to\infty} \sqrt{r}\left(\frac{\partial u^s}{\partial r} - iku^s\right) = 0 \tag{3}$$

where $x \in R^2$, $r = |x|$, $k > 0$ is the wave number and d is a vector on the unit circle Ω in R^2 giving the direction of the incident plane wave. The index of refraction n is assumed to be piecewise constant except for a compact region D (bounded by a sufficiently smooth curve ∂D) in which n is continuously differentiable and

$$m := 1 - n \tag{4}$$

has compact support. It is assumed that the *Sommerfield radiation condition* (3) holds uniformly with respect to $\hat{x} = x/|x|$. Under these conditions it is relatively easy to show that there exists a unique solution to (1)–(3) ([2]) and that the scattered field u^s has the asymptotic behavior

$$u^s(x) = \frac{e^{ikr}}{\sqrt{r}} u_\infty(\hat{x}\,;\,d) + O(r^{-3/2}) \tag{5}$$

as r tends to infinity where u_∞ is known as the *far field pattern* of the scattered field u^s.

In this paper we are concerned with the *inverse problem* of determining ∂D from a knowledge of $u_\infty(\hat{x}\,;d)$. It is also possible to use incident fields other than plane waves (e.g. point sources) and data other than far field data (e.g. near field data) but for the sake of simplicity we will not consider these cases here. The reader who is interested in such extensions should consult [3]. For the case we are interested in we have the following uniqueness theorem of Sun and Uhlmann ([9]) which shows that the *discontinuities* of n are uniquely determined by u_∞:

Theorem (Sun-Uhlmann): Let n_1, n_2 be in $L^\infty(R^2)$ and suppose $m_1 = 1 - n_1$ and $m_2 = 1 - n_2$ have compact support. If u_∞^j is the far field pattern corresponding to n_j and $u_\infty^1(\hat{x}\,;d) = u_\infty^2(\hat{x}\,;d)$ for all $\hat{x}, d \in \Omega$, then $n_1 - n_2 \in C^\alpha(R^2)$ for every $\alpha, 0 \leq \alpha < 1$.

As previously mentioned, our method for determining ∂D from u_∞ is based on transforming the problem to that of the behavior of solutions to an interior transmission problem. For the sake of simplicity, we will restrict our attention to the case when D is equal to the support of m, i.e. the background is constant ([1]). The more general case of a piecewise constant background is treated in [3]. The basic difference between the two cases is that for a constant background we must consider the *far field operator* $F : L^2(\Omega) \to L^2(\Omega)$ defined by

$$(Fg)(\hat{x}) := \int_\Omega u_\infty(\hat{x}\,;d)g(d)ds(d) \tag{6}$$

whereas for a piecewise constant background we must consider the *modified far field operator* $F_0 : L^2(\Omega) \to L^2(\Omega)$ defined by

$$(F_0 g)(\hat{x}) := \int_\Omega [u_\infty(\hat{x};d) - u_\infty^0(\hat{x}\,;d)]g(d)ds(d) \tag{7}$$

where u_∞^0 is the far field pattern corresponding to the scattering of a plane wave by the piecewise constant background (i.e. D is the empty set).

In either case, the basic idea of our method (to be described in the next section of this paper) is to try and find a superposition of plane waves such that the scattered field corresponding to this superposition is a constant multiple of a point source $\Phi(\cdot;y_0)$ located at a point $y_0 \in D\backslash\partial D$. Since the scattered field is uniquely determined by its far field pattern ([2]), it suffices to have the far field pattern corresponding to this superposition agree with a constant multiple of the far field pattern of $\Phi(\cdot;y_0)$. In particular, for the special case of a constant background, we can set

$$\Phi(x\,;y_0) = H_0^{(1)}(k|x - y_0|),\ x \neq y_0 \tag{8}$$

where $H_0^{(1)}$ is a Hankel function of the first kind of order zero. Then,, since $\Phi(\cdot; y_0)$ has a far field pattern given by $\gamma e^{-ik\hat{x}\cdot y_0}$ where

$$\gamma = \sqrt{\frac{2}{\pi k}}\ e^{-i\pi/4}, \tag{9}$$

we want to find a function $g(\cdot; y_0) \in L^2(\Omega)$ such that

$$(Fg)(\hat{x}) = e^{-ik\hat{x}\cdot y_0} \tag{10}$$

is satisfied. A short calculation using Rellich's lemma ([2]) shows that this can be done if and only if there exist functions w and v satisfying the *interior transmission problem*

$$\begin{aligned} &\Delta_2 w + k^2 n(x) w = 0 && \text{in } D \\ &\Delta_2 v + k^2 v = 0 && \end{aligned} \tag{11}$$

$$\begin{aligned} w - v &= \gamma^{-1} H_0^{(1)}(k|x - y_0|) && \text{on } \partial D \\ \frac{\partial}{\partial \nu}(w - v) &= \gamma^{-1} \frac{\partial}{\partial \nu} H_0^{(1)}(k|x - y_0|) && \end{aligned} \tag{12}$$

where ν is the unit outward normal to ∂D and v is a *Herglotz wave function* with kernel g, i.e. v is a solution of the Helmholtz equation $\Delta_2 v + k^2 v = 0$ of the form

$$v(x) = \int_\Omega e^{ikx\cdot d} g(d)ds(d)\ . \tag{13}$$

In order to proceed to the description of our inversion scheme for determining ∂D from u_∞, we will need the following two theorems concerning the interior transmission problem due to Rynne and Sleeman ([8]; see also [4]) and Colton and Potthast ([4]).

Theorem (Rynne-Sleeman): Assume that $\operatorname{Im} n(x) > 0$ for $x \in D$. Then there exists a unique solution $v, w \in H^2_{\text{loc}}(D) \bigcap L^2(D)$ to the interior transmission problem (11), (12) such that $v - w \in H^2(D)$.

Theorem (Colton-Potthast): Assume that $\operatorname{Im} n(x) > 0$ for $x \in D$. Then the solution v of the interior transmission problem can be approximated in $L^2(D)$ by a Herglotz wave function.

We will now show how these theorems can be used to derive an inversion scheme for the inverse scattering problem that is the subject of this paper.

3 The Solution of the Inverse Scattering Problem

As stated in the previous section, the basic idea of our inversion scheme is to determine g such that (10) is satisfied. If this can be done, the Herglotz wave function v with kernel g satisfies the interior transmission problem (11), (12). We would then like to conclude that, on ∂D, v becomes unbounded as y_0 tends to ∂D and hence so does $\|g\|_{L^2(\Omega)}$. If this is true then ∂D is determined by these points y_0 such that the L^2-norm of $g(\cdot\,; y_0)$ becomes unbounded. In practice, we can do this by solving (10) for y_0 on a rectangular grid known a priori to contain D and then look for those values of y_0 where $\|g\|_{L^2(\Omega)}$ is large.

Unfortunately, (10) is a (improperly posed) linear integral equation of the first kind and in general we cannot conclude that a solution $g = g(\cdot\,; y_0)$ exists. Even if a solution does exist it is not clear from (11), (12) that v (and hence $\|g\|_{L^2(\Omega)}$) becomes unbounded as y_0 tends to ∂D. To deal with these problems, we assume that $\operatorname{Im} n(x) > 0$ for $x \in D$ and proceed as follows (see [1] and [4] for details). By the Rynne-Sleeman theorem there exists a solution v, w to the interior transmission problem and by Green's formula and the trace theorem we have that for $x \in D \backslash \partial D$

$$w(x) = v(x) - \frac{ik^2}{4} \iint_D \Phi(x\,; y) m(y) w(y) dy \tag{14}$$

$$+ \frac{i}{4} \int_{\partial D} \left\{ [v(y) - w(y)] \frac{\partial}{\partial \nu(y)} \Phi(x\,; y) - \Phi(x\,; y) \frac{\partial}{\partial \nu} [v(y) - w(y)] \right\} ds(y)$$

Using

$$\int_{\partial D} \left\{ H_0^{(1)}(k|y - y_0|) \frac{\partial}{\partial \nu(y)} \Phi(x\,; y) - \frac{\partial}{\partial \nu(y)} H_0^{(1)}(k|y - y_0|) \Phi(x\,; y) \right\} ds(y) = 0 \tag{15}$$

for $x \in D \backslash \partial D$ and the boundary conditions (12), we can now conclude from (14) and (15) that

$$w(x) = v(x) - \frac{ik^2}{4} \iint_D \Phi(x\,; y) m(y) w(y) dy \tag{16}$$

for $x \in D \backslash \partial D$.

We now use the Colton-Potthast theorem to deduce that for any $\epsilon > 0$ there exists $g \in L^2(\Omega)$ such that

$$\|(Fg)(\hat{x}) - e^{-ik\hat{x}\cdot y_0}\|_{L^2(\Omega)} < \epsilon \tag{17}$$

and

$$\|v - v_g\|_{L^2(D)} < \epsilon \tag{18}$$

where v_g is the Herglotz wave function with kernel g. Assuming that v_g depends continuously on its boundary data (e.g. this is true if k^2 is not a Dirchlet eigenvalue) we can conclude that, if $\max_{x\in\partial D}|v_g(x\,;y_0)|$ remains bounded as y_0 tends to ∂D, then from (18) so does $\|v(\cdot\,;y_0)\|_{L^2(D)}$. From the integral equation (16) we now have that $w \in L^2(D)$ is bounded independently of $y_0 \in D\backslash\partial D$ and hence, using the mapping properties of volume potentials, $\|w-v\|_{H^2(D)}$ is bounded independently of $y_0 \in D\backslash\partial D$. By the trace theorem we now have that $\|w-v\|_{H^{3/2}(\partial D)}$ is bounded and this contradicts (12). Hence $v_g(x\,;y_0)$ for $x \in D$ becomes unbounded as y_0 tends to ∂D (and thus so does $\|g\|_{L^2(\Omega)}$).

From the above analysis we see that if $\operatorname{Im} n > 0$ in D then for every $\epsilon > 0$ and $y_0 \in D\backslash\partial D$ there exists a function $g(\cdot;y_0) \in L^2(\Omega)$ such that (17) is satisfied and

$$\lim_{y_0\to\partial D} \|g(\cdot\,;y_0)\|_{L^2(\Omega)} = \infty \;. \tag{19}$$

More generally, it can be shown that F is injective with dense range in $L^2(\Omega)$ ([2]). Thus ∂D can be determined by using an appropriate regularization method to solve (10) and then determining the values of y_0 for which $\|g(\cdot\,;y_0)\|_{L^2(\Omega)}$ becomes large. Note that this method is a *linear* method and makes no statement about the *values* of the index of refraction in D. The only quantity which is determined is ∂D. For numerical examples using this method we refer the reader to [1] (constant background) and [3] (piecewise constant background). Similar methods also apply to obstacle scattering ([1]) and in this case a related (but different!) method has recently been developed by Potthast ([7]). We note in passing that the inversion method described in this paper has some resemblance to the method introduced by Isakov ([5]) to prove uniqueness theorems for inverse scattering problems.

An intriguing and somewhat delicate feature of our approach for determining ∂D is that it makes a very explicit use of the improperly posed nature of the inverse scattering problem by looking for a solution of the integral equation of the first kind (10) that becomes unbounded as y_0 tends to ∂D. In particular, the regularization method used to solve (10) must allow for the fact that the solution is in fact unbounded as y_0 tends to ∂D, e.g. the penalty term in the Tikhonov regularization method should involve the derivative of g rather than g itself.

References

1. D. Colton and A. Kirsch, A simple method for solving inverse scattering problems in the resonance region, *Inverse Problems*, 12 (1996), 383–393.
2. D. Colton and R. Kress, *Inverse Acoustic and Electromagnetic Scattering Theory*, Springer-Verlag, Berlin, 1992.
3. D. Colton and P. Monk, A linear sampling method for the detection of leukemia using microwaves, submitted for publication.

4. D. Colton and R. Potthast, A simple method for inverse electromagnetic scattering from inhomogeneous media, submitted for publication.
5. V. Isakov, On uniqueness in the inverse transmission scattering problem, *Comm. Partial Diff. Equations* 15 (1990), 1565–1587.
6. A. Kirsch, *An Introduction to the Mathematical Theory of Inverse Problems*, Springer-Verlag, Berlin, 1996.
7. R. Potthast, A fast new method to solve inverse scattering problems, to appear.
8. B.P. Rynne and B.D. Sleeman, The interior transmission problem and inverse scattering from inhomogeneous media, *SIAM J. Math. Anal.* 22 (1991), 1755–1762.
9. Z. Sun and G. Uhlmann, Recovery of singularities for formally determined inverse problems, *Comm. Math. Physics* 153 (1993), 431–445.

Numerical Methods in Inverse Obstacle Scattering

Rainer Kress

Institut für Numerische und Angewandte Mathematik, Universität Göttingen, Lotzestr. 16–18, D-37083 Göttingen, Germany
e-mail : kress@math.uni-goettingen.de

1 Introduction

The inverse problem we consider is to reconstruct the shape of a scattering obstacle from a knowledge of the far field pattern for the scattering of incident time-harmonic acoustic or electromagnetic plane waves. For the sake of simplicity we confine our presentation to the inverse Dirichlet problem in two dimensions, that is, to scattering by infinitely long cylindrical sound-soft or perfectly conducting obstacles. However, much of the analysis, in principle, can be extended to other boundary conditions and also to the three-dimensional case.

Roughly speaking we can distinguish between two different approaches for the approximate solution of the inverse obstacle scattering problem. In a first group of methods the inverse obstacle problem is separated into a linear ill-posed part for the reconstruction of the scattered wave from its far field pattern and a nonlinear well-posed part for finding the location of the boundary of the scatterer from the boundary condition for the total field. In a second group of methods the inverse obstacle problem is either considered as an ill-posed nonlinear operator equation or reformulated as a nonlinear optimization problem in an output least squares sense.

Up until a few years ago research concentrated mainly on the first group of methods for the following reason: Each algorithm of the second group requires the solution of the direct scattering problem for different domains at each step of the iteration method used to arrive at an approximate solution. Hence, these methods seemed to be too costly in order to be competitive with the first group of methods and in the monograph [4] one can find an extensive treatment of two typical examples for the first group but only little material on the second approach. However, with the more recent development of computer hard and soft ware, computing time becomes a less important issue and therefore a second thought on the above argument seems to be appropriate. Consequently, in this survey we shall concentrate more on describing the basic ideas of a regularized Newton iteration and a Landweber iteration as methods of the second group of approaches. A substantial part of these methods have been developed through the interaction of the inverse scattering groups at the universities of Delaware, Erlangen and Göttingen.

2 The inverse scattering problem

We denote the cross section of the cylindrical obstacle by D and assume that $D \subset \mathbb{R}^2$ is a bounded and simply connected domain with boundary ∂D of class C^2. The simplest direct scattering problem is, given an incident field u^i, to find the total field, that is, the superposition $u = u^i + u^s$ such that u satisfies the reduced wave equation or Helmholtz equation

$$\Delta u + k^2 u = 0 \quad \text{in } \mathbb{R}^2 \setminus \bar{D} \tag{1}$$

with wave number $k > 0$, the Dirichlet boundary condition

$$u = 0 \quad \text{on } \partial D \tag{2}$$

and the Sommerfeld radiation condition

$$\lim_{r\to\infty} \sqrt{r}\left(\frac{\partial u^s}{\partial r} - iku^s\right) = 0, \quad r = |x|, \tag{3}$$

uniformly for all directions. In acoustics the Dirichlet condition (2) corresponds to scattering from a sound-soft obstacle whereas in electromagnetics it models scattering from a perfect conductor with the electromagnetic field being E–polarized. Provided the incident field u^i is an entire solution to the Helmholtz equation, then there exists a unique solution $u \in C^2(\mathbb{R}^2 \setminus \bar{D}) \cap C^{1,\alpha}(\mathbb{R}^2 \setminus D)$ to the direct scattering problem (1)–(3) for $0 < \alpha < 1$. For details we refer to [3], [4].

The Sommerfeld radiation condition (3) leads to an asymptotic behavior of the form

$$u^s(x) = \frac{e^{ik|x|}}{\sqrt{|x|}} \left\{ u_\infty \left(\frac{x}{|x|} \right) + O\left(\frac{1}{|x|} \right) \right\}, \quad |x| \to \infty. \tag{4}$$

The function u_∞, defined on the unit circle $\Omega := \{z \in \mathbb{R}^2 : |z| = 1\}$, is known as the far field pattern of the scattered wave u^s. The inverse problem we are concerned with in this survey is, given the far field pattern u_∞ of the scattered wave u^s for one incoming plane wave $u^i = e^{ik\, x \cdot d}$ with incident direction $d \in \Omega$, to determine the shape of the scatterer D. This inverse problem is nonlinear, since the solution to the direct scattering problem depends nonlinearly on the boundary ∂D and it is improperly posed, since finding the scattered wave u^s from its far field pattern u_∞ is severely improperly posed. We want to consider this inverse problem for frequencies in the resonance region, that is, for scatterers D and wave numbers k such that the wavelengths $2\pi/k$ is of a comparable size to the diameter of the scatterer. In particular, low frequency methods like impedance tomography or high frequency methods like physical or geometrical optics do not yield valid approximations in this intermediate frequency range.

For a fixed incident field u^i, the solution to the direct scattering problem defines an operator $F : \partial D \mapsto u_\infty$ which maps the boundary ∂D of the scatterer D onto the far field pattern u_∞. In terms of this operator, given a far field pattern u_∞, the inverse problem consists in solving the nonlinear and improperly posed operator equation

$$F(\partial D) = u_\infty \tag{5}$$

for the unknown boundary ∂D. Both for the theoretical investigation and the numerical solution of equation (5) a parametrization of the admissible boundary curves is required. In this survey, for the sake of simplicity, we assume the unknown scatterer to be starlike with respect to the origin. However, we wish to emphasize that the following analysis can be extended to a wider class of boundary representations. We parametrize

$$\partial D = \{r(z)\, z : z \in \Omega\}$$

with some function $r \in C^2_+(\Omega)$ where by $C^2_+(\Omega)$ we denote the set of twice continuously differentiable functions $r : \Omega \to (0, \infty)$. Then we may consider F as a mapping from $C^2_+(\Omega)$ into $L^2(\Omega)$ and will write $F(r)$ instead of $F(\partial D)$.

We proceed by summarizing some basic properties of the operator F. A first question to ask about the inverse scattering problem is uniqueness. In our formulation of the inverse problem we can state the following uniqueness result.

Theorem 1 *The far field operator $F : C^2_+(\Omega) \to L^2(\Omega)$ is injective on the ball $\{r \in C^2_+(\Omega) : k\|r\|_\infty < \zeta_0\}$ where $\zeta_0 = 2.40482\ldots$ denotes the smallest positive zero of the Bessel function J_0 of order zero.*

Proof. This extension of a classical result of Schiffer is due to Colton and Sleeman [5] (see also [4], p. 107). □

The ill-posedness of the inverse scattering problem is expressed through the following regularity property of F.

Theorem 2 *The far field operator $F : C^2_+(\Omega) \to L^2(\Omega)$ is continuous and compact.*

Proof. See Theorem 5.7 in [4]. □

The following result on the differentiability of F is of basic importance for the foundation of iterative methods for the inverse obstacle scattering problem, that is, for the iterative solution of (5).

Theorem 3 *The far field operator $F : C^2_+(\Omega) \to L^2(\Omega)$ is Fréchet differentiable. The derivative is given by*

$$F'(r)\, h = v_\infty$$

where v_∞ is the far field pattern of the solution v to the Helmholtz equation

$$\Delta v + k^2 v = 0 \quad in\ \mathbb{R}^2 \setminus \bar{D} \tag{6}$$

satisfying the Sommerfeld radiation condition and the Dirichlet boundary condition

$$v = -\nu \cdot \tilde{h}\, \frac{\partial u}{\partial \nu} \quad on\ \partial D \tag{7}$$

where $\tilde{h}(r(z)) = h(z)\,z$, $z \in \Omega$, and ν denotes the outward unit normal to ∂D.

Proof. For a proof via Hilbert space methods we refer to Theorem 5.7 in [4] and to Kirsch [10]. Proofs by boundary integral equation methods are described by Potthast [18], [19], [20] and in [13], [14]. □

Since $F'(r)\,h$ is a far field pattern, the Fréchet derivative clearly is smoothing and therefore $F'(r) : L^2(\Omega) \to L^2(\Omega)$ is a compact operator which is in agreement with Theorem 2 (see Theorem 4.19 in [4]). For the linearized operator we have the following properties which are of relevance for the application of regularization techniques to the linearized equation (5).

Theorem 4 *The Fréchet derivative $F'(r) : L^2(\Omega) \to L^2(\Omega)$ is injective and has dense range.*

Proof. We refer to [6], [9], [15] and note that the injectivity is a consequence of the boundary condition (7) and Holmgren's uniqueness theorem.

3 A regularized Newton method

We now proceed with describing the application of Newton's method to the solution of

$$F(r) = u_\infty. \tag{8}$$

In the usual fashion, the nonlinear equation (8) is replaced by the linearized equation

$$F(r) + F'(r)\,h = u_\infty \tag{9}$$

which has to be solved for h in order to improve an approximate boundary curve given by the radial function r into the new approximation given by $\tilde{r} = r + h$. Then Newton's method consists in iterating this procedure, i.e.,

$$F'(r_n)(r_{n+1} - r_n) = u_\infty - F(r_n), \quad n = 0, 1, 2, \ldots. \tag{10}$$

The question of uniqueness for the linear equation (9) is settled through Theorem 4. Since $F'(r)$ is compact, the linear equation (9) is ill-posed. Therefore regularization techniques like Tikhonov regularization or singular value cut-off have to be employed (see [12]).

For practical computations h is taken from an appropriately chosen finite dimensional subspace $W_N \subset C^2(\Omega)$ with dimension N and equation (9) is approximately solved by projecting it onto another finite dimensional subspace of $L^2(\Omega)$. The most convenient projection is given through collocation at M equidistantly spaced points $z_1, \ldots, z_M \in \Omega$. Then writing

$$h = \sum_{j=1}^{N} a_j h_j$$

where $h_1, \ldots, h_N$ denotes a basis of W_N, we have to solve the linear system

$$\sum_{j=1}^{N} a_j \,(F'(r)\, h_j)(z_i) = u_\infty(z_i) - F(r)(z_i), \quad i = 1, \ldots, M, \tag{11}$$

either by Tikhonov regularization or singular value cut-off. The Tikhonov regularization is equivalent to minimizing the penalized defect

$$\sum_{i=1}^{M} \left| \sum_{j=1}^{N} a_j (F'(r)\, h_j)(z_i) - u_\infty(z_i) + F(r)(z_i) \right|^2 + \alpha \sum_{j=1}^{N} a_j^2 \tag{12}$$

with some regularization parameter $\alpha > 0$ by solving the corresponding normal equations.

In order to set up the linear system (11), in each iteration step the direct problem for the boundary ∂D given by the radial function r has to be approximately solved for the evaluation of $F(r)(z_i)$ and in order to obtain the normal derivative $\partial u/\partial \nu$ of the total field u which enters the boundary condition (7) for the Fréchet derivative. In principle, this can be done by any numerical method for solving the exterior Dirichlet problem for the Helmholtz equation. However, we strongly recommend using a boundary integral equation approach based on a combination of a double- and a single-layer potential together with a Nyström method for the numerical solution of the integral equation as described in [4]. It is also advantageous to set the integral equation up through the use of the Green's representation formula in a manner which automatically yields $\partial u/\partial \nu$ on the boundary ∂D. In order to compute the matrix entries $(F'(r)\,h_j)(z_i)$ one has solve N additional direct problems for the same boundary ∂D and different boundary values given by (7) for the basis functions $h = h_j$, $j = 1, \dots, N$. Hence, in principle, one has to solve the same linear system as for the evaluation of $F(r)(z_i)$ for N additional different right hand sides which can be cheaply done by using an LR–decomposition. As a stopping rule for the number of iterations we suggest to use the residual $R := \|F(r) - u_\infty\|_{L^2(\Omega)}$ and terminate the iterations when the difference of the value of R for two consecutive iterations is less than a tolerance value δ or less than the noise level if working with noisy data, i.e., we suggest using the discrepancy principle.

For numerical reconstructions by regularized Newton iterations as described above using trigonometric polynomials for the approximating space we refer to Kirsch [11] and to [13], [14] for the inverse Dirichlet problem and to Mönch [16] for the inverse Neumann problem. From a numerical point of view it might be advantageous to replace the gobal trigonometric functions by functions with a more local structure. In order to illustrate this, we choose $L(t) := \exp\left(-\gamma \sin^2 \frac{t}{2}\right)$ with some $\gamma > 0$ and use approximations of the form

$$r(t) = \sum_{j=0}^{2N-1} a_j L(N(t - t_j))$$

where $t_j = j\pi/N$. For the following numerical example the boundary is given by the parametric representation

$$x(t) = (\cos t + 0.15 \sin t + 0.35 \cos 2t - 0.35, 1.2 \sin t + 0.15 \cos t)$$

with $0 \leq t \leq 2\pi$ which describes a kite-shaped starlike curve. For the solution of the boundary integral equations we used the Nyström method mentioned above with 32 grid points for the inverse algorithm and with 128 grid points and a different coupling parameter for generating the synthetic data. As initial guess for the Newton iteration we chose the unit circle. In the figures, the dashed lines give the exact boundary curve and the full lines give the reconstructions. The arrows indicate the incident directions d.

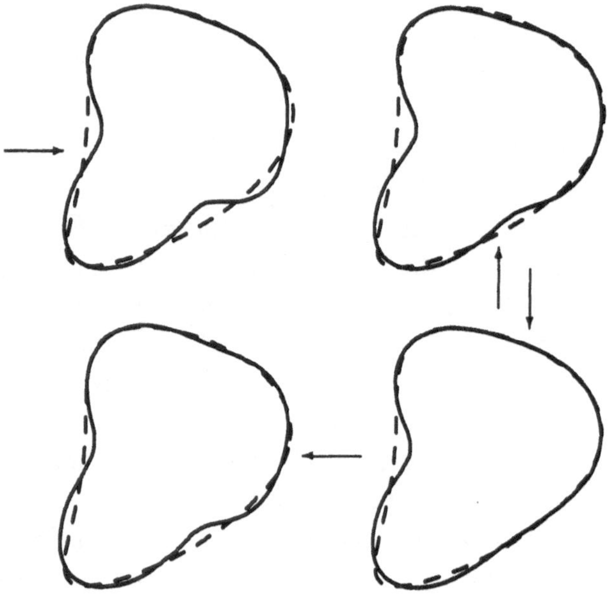

$N = 8,\ \gamma = 3,\ \alpha = 0.01$

Fig. 1. Reconstruction of kite-shaped scatterer for $k = 1$

As to be expected, using the wave number $k = 3$ yields better reconstructions in the illuminated part of the scatterer and poorer reconstructions in the shadow region.

A quasi-Newton or frozen Newton method was investigated in [15]. Related Newton schemes have been considered by Murch,

Tan and Wall [17], by Roger [22], by Tobocman [23] and by Wang and Chen [24]. The existing numerical examples provide evidence for the practicality of the regularized Newton method in inverse obstacle scattering. However further research is needed to improve on its efficiency.

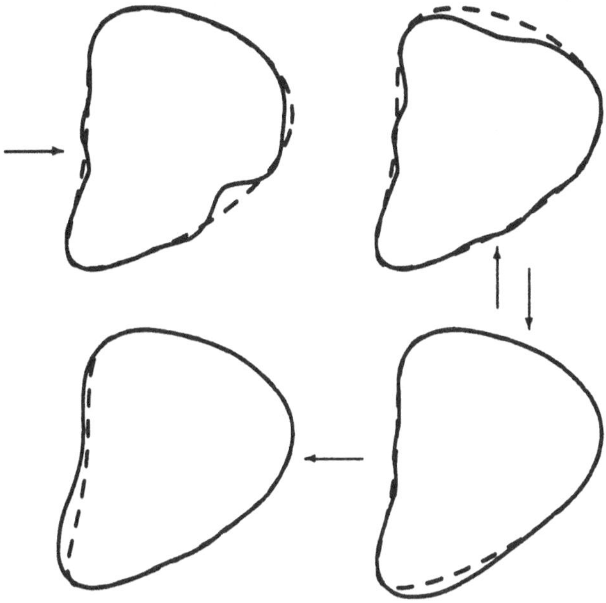

$N = 8,\ \gamma = 3,\ \alpha = 0.1$

Fig. 2. Reconstruction of kite-shaped scatterer for $k = 3$

A convergence analysis of the regularized Newton iteration for nonlinear ill-posed equations based on stopping rules via a discrepancy principle has been developed Blaschke, Neubauer and Scherzer [1]. However, these general convergence results, so far, could not be applied for obtaining convergence results for the regularized Newton method in inverse obstacle scattering.

4 Landweber iteration

The Landweber iteration has been studied and applied extensively for the solution of linear ill-posed equations (see [12]). More recently its use in the form

$$r_{n+1} = r_n - \mu[F'(r_n)]^*[F(r_n) - u_\infty], \quad n = 0, 1, 2, \ldots, \tag{13}$$

for the iterative solution of nonlinear ill-posed problems has been suggested by Hanke, Neubauer and Scherzer [7]. Here, we denote by $[F'(r)]^* : L^2(\Omega) \to L^2(\Omega)$ the adjoint operator of the Fréchet derivative $F'(r)$ and $\mu > 0$ is an appropriately chosen parameter ensuring that the iteration operator $I - \mu[F'(r)]^* F(r)$ is non expansive.

For a characterization of the explicit form of $[F'(r)]^*$ we need to make use of special solutions of the Helmholtz equation called Herglotz wave functions. These are defined by

$$w^i(x) := \int_\Omega g(d) e^{ik\, x\cdot d}\, ds(d), \quad x \in \mathbb{R}^2, \tag{14}$$

where $g \in L^2(\Omega)$ is called the kernel of w^i (c.f. [4]). Note that, by superposition, for the far field pattern w_∞ corresponding to the solution w^s of

$$\Delta w^s + k^2 w^s = 0 \quad \text{in } \mathbb{R}^2 \setminus \bar{D} \tag{15}$$

subject to the Dirichlet boundary condition

$$w^i + w^s = 0 \quad \text{on } \partial D \tag{16}$$

and the Sommerfeld radiation condition we have that

$$w_\infty(z) = \int_\Omega u_\infty(z; d) g(d)\, ds(d), \quad z \in \Omega. \tag{17}$$

Here, we indicate the dependence on the incident direction by writing $u_\infty(\cdot\,; d)$ for the far field pattern of the scattered wave $u^s(\cdot\,; d)$ for plane wave incidence $u^i(x; d) = e^{ik\, x\cdot d}$. For the proof of the following theorem we refer to [6], [9].

Theorem 5 *The adjoint operator* $[F'(r)]^* : L^2(\Omega) \to L^2(\Omega)$ *of the Fréchet derivative* $F'(r)$ *is given through*

$$[F'(r)]^*(f)(z) = -\frac{e^{-i\frac{\pi}{4}}}{\sqrt{8\pi k}}\, z \cdot a(r(z)), \quad z \in \Omega, \tag{18}$$

where

$$a := \nu \, \frac{\partial \overline{u}}{\partial \nu} \, \frac{\partial \{\overline{w^i + w^s}\}}{\partial \nu} \quad \text{on } \partial D$$

and w^i *is the Herglotz wave function with kernel* $g(d) = \overline{f(-d)}$ *and* w^s *denotes the solution of* (15)–(16).

As in the regularized Newton method, for each step of the Landweber iteration the direct scattering problem for the boundary ∂D given by the radial function r has to be solved for the evaluation of the right hand side $F(r)$. Then one has to compute the Herglotz wave function w^i with kernel g given via $f = F(r) - u_\infty$ and then solve the direct scattering problem (15)–(16) for the boundary data w^i instead of u^i. Hence, one step of the Landweber iteration is far less costly than one step of the Newton iteration. However this advantage is balanced out through a notably slower convergence of the Landweber iteration.

Numerical implementations of the Landweber iteration for inverse obstacle scattering problems have been given by Hanke, Hettlich and Scherzer [6] for sound-soft obstacles and by Hettlich [8] for sound-hard obstacles. A convergence analysis of the Landweber iteration for nonlinear ill-posed equations based on stopping rules via a discrepancy principle has been developed by Hanke, Neubauer and Scherzer [7]. However, unfortunately these general convergence results, so far, could not be employed to completely analyze the convergence behavior of the Landweber iteration in inverse obstacle scattering.

5 Starting approximations

Both the regularized Newton iteration and the Landweber iteration and most other methods for the approximate solution of inverse obstacle scattering problems rely on some a priori information for obtaining initial approximations to start the corresponding iterative procedures. In this final section we will briefly outline the principle ideas of a very simple method for finding coarse approximations for the solution to the inverse scattering problem without any use of a priori information on the obstacle which has been suggested by Colton and Kirsch [2] and which might be used to obtain initial approximations.

This method makes use of Herglotz wave functions as introduced in the previous section. Its basic idea is to try and find a Herglotz wave function w^i with kernel g, i.e., a superposition of plane waves, such that the corresponding scattered wave w^s coincides with a point source $\Phi(\cdot\,,\xi)$ located at a point ξ in the interior of the scatterer D. Here by Φ we denote the fundamental solution to the Helmholtz equation

$$\Phi(x,y) := \frac{i}{4} H_0^{(1)}(k|x-y|), \quad x \neq y,$$

in $\mathbb{R}^2$ where $H_0^{(1)}$ is the Hankel function of order zero and of the first kind. Observing that

$$\Phi(x,y) = \frac{e^{ik|x|}}{\sqrt{|x|}} \left\{ \frac{e^{i\frac{\pi}{4}}}{\sqrt{8\pi k}}\, e^{-ik\, z\cdot y} + O\left(\frac{1}{|x|}\right) \right\}, \quad |x| \to \infty,$$

where $z = x/|x|$, in view of (17) we have to find the kernel $g = g(\cdot\,;\xi)$ as a solution to the integral equation of the first kind

$$\int_\Omega u_\infty(z;d) g(d;\xi)\, ds(d) = \frac{e^{i\frac{\pi}{4}}}{\sqrt{8\pi k}}\, e^{-ik\, z\cdot\xi}, \quad z \in \Omega. \tag{19}$$

Assume that g solves equation (19). Then we have that

$$\int_\Omega u^s(x;d) g(d;\xi)\, ds(d) = \Phi(x,\xi), \quad x \in \mathbb{R}^2 \setminus \bar{D}. \tag{20}$$

Letting x tend to the boundary and using the boundary condition $u^i + u^s = 0$ on ∂D we conclude that the Herglotz wave function

$$w^i(x) = \int_\Omega g(d;\xi) e^{ik\, x\cdot d}\, ds(d), \quad x \in \mathbb{R}^2, \tag{21}$$

is a solution to the interior Dirichlet problem

$$\Delta w^i + k^2 w^i = 0 \quad \text{in } D \tag{22}$$

with boundary condition

$$w^i + \Phi(\cdot\,,\xi) = 0 \quad \text{on } \partial D. \tag{23}$$

Conversely, if the Herglotz wave function (21) solves (22)–(23) then its kernel g is a solution of (19). Hence, if a solution $g(\cdot\,;\xi)$ to the integral equation (19) of the first kind exists for all points $\xi \in D$, then from the boundary condition (23) for the Herglotz wave function we conclude that $\|g(\cdot\,;\xi)\|_{L^2(\Omega)} \to \infty$ as the source point ξ approaches the boundary ∂D.

However, in general, the solution to the interior Dirichlet problem (22)–(23) will have an extension as a Herglotz wave function across the boundary ∂D only in very special cases. Hence, the integral equation (19) will have a solution only in these special cases. Nevertheless, by making use of denseness properties of the Herglotz wave functions, it can be shown (see [2]) that approximately solving a regularized version of the integral equation (19) for ξ taken from a sufficiently fine grid in $\mathbb{R}^2$ and scanning the values for $\|g(\cdot\,;\xi)\|_{L^2(\Omega)}$ will yield an approximation for ∂D through those points where the norm of g is large. In general, this approximation will not lead to very sharply defined boundary curves. However, if necessary these approximations then could be improved by using the regularized Newton iteration or the Landweber iteration from the previous sections.

Another method which is related to the method of Colton and Kirsch and which also might be utilized for obtaining initial approximations without needing a priori information on the obstacle has been suggested by Potthast [21].

References

[1] Blaschke, B., Neubauer, A., Scherzer, O. (1996): On convergence rates for the iteratively regularized Gauss–Newton method (to appear).

[2] Colton, D., Kirsch, A. (1996): A simple method for solving the inverse scattering problems in the resonance region, *Inverse Problems* **12**, (to appear).

[3] Colton, D., Kress, R. (1983): *Integral Equation Methods in Scattering Theory*, Wiley-Interscience Publication, New York.

[4] Colton, D., Kress, R. (1992): *Inverse Acoustic and Electromagnetic Scattering Theory*, Springer-Verlag, Berlin Heidelberg New York.

[5] Colton, D., Sleeman, B.D. (1983): Uniqueness theorems for the inverse problem of acoustic scattering, *IMA J. Appl. Math.* **31**, 253–259.

[6] Hanke, M., Hettlich, F., Scherzer, O. (1995): The Landweber iteration for an inverse scattering problem, In: *Proceedings of the 1995 design engineering technical*

conferences, Vol. 3, Part C, (Wang et al, eds), 909–915, The American Society of Mechanical Engineers, New York.

[7] Hanke, M., Neubauer, A., Scherzer, O. (1995): A convergence analysis for the Landweber iteration for nonlinear ill-posed problems, *Numer. Math.* **72**, 21–37.

[8] Hettlich, F. (1996): An iterative method for the inverse scattering problem from sound-hard obstacles, In: *Proceedings of the ICIAM 95, Vol. II, Applied Analysis,* (Mahrenholz and Mennicken, eds), Akademie Verlag, Berlin.

[9] Kirsch, A. (1989): Properties of the far field operators in acoustic scattering, *Math. Meth. in the Appl. Sci.* **11**, 773–787.

[10] Kirsch, A. (1993): The domain derivative and two applications in inverse scattering theory, *Inverse Problems* **9**, 81–96.

[11] Kirsch, A. (1993): Numerical algorithms in inverse scattering theory, In: *Ordinary and Partial Differential Equations, Vol. IV,* (Jarvis and Sleeman, eds) Pitman Research Notes in Mathematics **289**, 93–111, Longman, London.

[12] Kress, R. (1989): *Linear Integral Equations,* Springer-Verlag, Berlin Heidelberg New York.

[13] Kress, R. (1994): A Newton method in inverse obstacle scattering, In: *Inverse Problems in Engineering Mechanics,* (Bui et al, eds) 425–432, Balkema, Rotterdam.

[14] Kress, R. (1995): Integral equation methods in inverse obstacle scattering, *Engineering Anal. with Boundary Elements* **15**, 171–179.

[15] Kress, R., Rundell, W. (1994): A quasi-Newton method in inverse obstacle scattering, *Inverse Problems* **10**, 1145–1157.

[16] Mönch, L. (1996): A Newton method for solving the inverse scattering problem for a sound-hard obstacle, *Inverse Problems* **12**, 309–323.

[17] Murch, R.D., Tan, D.G.H., Wall, D.J.N. (1988): Newton–Kantorovich method applied to two-dimensional inverse scattering for an exterior Helmholtz problem, *Inverse Problems* **4**, 1117–1128.

[18] Potthast, R. (1994): Fréchet differentiability of boundary integral operators in inverse acoustic scattering, *Inverse Problems* **10**, 431–447.

[19] Potthast, R. (1996): Fréchet differentiability of the solution to the acoustic Neumann scattering problem with respect to the domain, *Jour. on Inverse and Ill-posed Problems* **4**, 67–84 (1996).

[20] Potthast, R. (1996): Domain derivatives in electromagnetic scattering, *Math. Meth. in the Appl. Sci.* (to appear).

[21] Potthast, R. (1996): A fast new method to solve inverse scattering problems, *Inverse Problems* **12**, 731–742.

[22] Roger, A. (1981): Newton Kantorovich algorithm applied to an electromagnetic inverse problem, *IEEE Trans. Ant. Prop.* **AP-29**, 232–238.

[23] Tobocman, W. (1989): Inverse acoustic wave scattering in two dimensions from impenetrable targets, *Inverse Problems* **5**, 1131–1144.

[24] Wang, S.L., Chen, Y.M. (1991): An efficient numerical method for exterior and interior inverse problems of Helmholtz equation, *Wave Motion* **13**, 387–399.

An Overview of Nonlinear Diffraction Tomography Within the Bayesian Estimation Framework

Hervé Carfantan and Ali Mohammad-Djafari

Laboratoire des Signaux et Systèmes (CNRS/Supélec/UPS)
Plateau de Moulon, 91 192 Gif-sur-Yvette Cedex, France

Abstract. The Bayesian approach has been proven to give a common estimation structure to existing image reconstruction and restoration methods, in spite of their apparent diversity (Demoment 1989). The goal of this paper is to investigate diffraction tomography within the Bayesian estimation framework. A regularized solution to this ill-posed nonlinear inverse problem is defined as the maximum a posteriori estimate, introducing prior information on the object to reconstruct. Two equivalent formulations of this definition are available which lead to solution of a constrained or an unconstrained optimization problem to compute this solution. Different existing methods for solving this problem – such as *Born Iterative Method* (Wang and Chew 1989), *Newton-Kantorovitch method* (Joachimovicz et al. 1991), *Distorted Born Iterative method* (Chew and Wang 1990) and *Modified Gradient method* (Kleinman and van den Berg 1992) – are interpreted as algorithms to compute the defined solution. This common point of view allows an objective comparison between these methods, from the standpoint of their convergence properties and the solution they provide.

Introduction

Diffraction tomography consists in constructing an image representing the spatial variation of some physical properties of an inhomogeneous object (such as dielectric permittivity and conductivity for electro-magnetic waves), from a finite set of field data scattered by this object. This problem is intrinsically ill-posed and a satisfactory solution cannot be obtained from imperfect data without any introduction of a priori information on the object. The objectives of this paper are to define a regularized solution to this nonlinear inverse problem within the Bayesian estimation framework and to interpret some of the existing methods to solve this problem as algorithms to compute the defined solution.

First, we briefly present the direct model in a functional and in an algebraic framework. The algebraic framework allows a compact presentation and notably allows us to perceive strong similarities between some classical methods, which cannot be distinguished in the functional framework in which they have been proposed.

Then, we define a regularized solution within the Bayesian estimation framework. Bayes rule is a consistent way to combine information provided by the data and prior information on the solution. In this paper, we use Markov Random Fields to model this a priori information. We define the solution as the maximum a posteriori estimate; so the solution's computation requires resolution of an optimization problem.

Then some of the existing methods to solve the diffraction tomography problem are interpreted and analyzed as algorithms to compute the defined regularized solution. Among these methods are *Born Iterative Method* (Wang and Chew 1989), *Newton-Kantorovitch method* (Joachimovicz et al. 1991), *Distorted Born Iterative method* (Chew and Wang 1990) and *Modified Gradient method* (Kleinman and van den Berg 1992). Three types of methods are distinguished: the first considers successive linearizations of the forward model, the second defines the solution as the minimum of a joint criterion depending on the object and the field on the object, while methods of the third type minimize a criterion which only depends on the object.

Finally, an objective comparison between these different types of methods and the solution they provide is proposed.

1 Problem Statement

We consider an inhomogeneous 2-D object, embedded in a known homogeneous medium, illuminated with a pure harmonic Transverse Magnetic (TM) plane wave. The object is characterized by its complex contrast function $x(\boldsymbol{r}) = k^2(\boldsymbol{r}) - k_0^2$, which is related to the dielectric permittivity $\epsilon(\boldsymbol{r})$ and the conductivity $\sigma(\boldsymbol{r})$ of the object by $k^2(\boldsymbol{r}) = \omega^2\mu_0\left(\epsilon(\boldsymbol{r}) + j\sigma(\boldsymbol{r})/\omega\right)$, k_0 is the wave number of the background homogeneous medium and $\boldsymbol{r}$ denotes a position in $\mathbb{R}^2$. The direct scattering problem is modeled by the coupled integral equations:

$$y(\boldsymbol{r}_i) = \iint_{D_\mathrm{O}} \mathcal{G}(\boldsymbol{r}_i, \boldsymbol{r}')x(\boldsymbol{r}')\phi(\boldsymbol{r}')\mathrm{d}\boldsymbol{r}', \ \boldsymbol{r}_i \in D_\mathrm{M} \ , \tag{1}$$

$$\phi(\boldsymbol{r}) = \phi_0(\boldsymbol{r}) + \iint_{D_\mathrm{O}} \mathcal{G}(\boldsymbol{r}, \boldsymbol{r}')x(\boldsymbol{r}')\phi(\boldsymbol{r}')\mathrm{d}\boldsymbol{r}', \ \boldsymbol{r} \in D_\mathrm{O} \ , \tag{2}$$

where $y(\boldsymbol{r}_i), \boldsymbol{r}_i \in D_\mathrm{M}$ is the scattered field on a sensor located at $\boldsymbol{r}_i$ in the measurement area D_M, $\phi(\boldsymbol{r}), \boldsymbol{r} \in D_\mathrm{O}$ and $\phi_0, \boldsymbol{r} \in D_\mathrm{O}$ are the total and the incident field on the object area D_O, and $\mathcal{G}$ is the Green function for the homogeneous background medium.

From an algebraic viewpoint, discretization of (1–2) with a moment method (Howard and Kretzschmar 1986), leads to:

$$\boldsymbol{y} = \boldsymbol{G}_\mathrm{M}\boldsymbol{X}\boldsymbol{\phi} \ , \tag{3}$$

$$\boldsymbol{\phi} = \boldsymbol{\phi}_0 + \boldsymbol{G}_\mathrm{O}\boldsymbol{X}\boldsymbol{\phi} \ , \tag{4}$$

where $\boldsymbol{y} \in \mathbb{C}^{n_\mathrm{M}}, \boldsymbol{\phi} \in \mathbb{C}^{n_\mathrm{O}}, \boldsymbol{\phi}_0 \in \mathbb{C}^{n_\mathrm{O}}$, $\boldsymbol{X}$ is a diagonal matrix $(n_\mathrm{O} \times n_\mathrm{O})$ with the components of the vector $\boldsymbol{x} \in \mathbb{C}^{n_\mathrm{O}}$ as diagonal elements, n_O is the number of pixels of the discrete object and n_{M} is the number of measurement sensors. Note that these notations can be extended for emission from n_{S} different positions.

Formally, the total field $\boldsymbol{\phi}$ on the object can be expressed from (4) and introduced in (3). It gives an explicit relation between the contrast and the data $\boldsymbol{y} = \mathcal{A}(\boldsymbol{x})$ with:

$$\mathcal{A}(\boldsymbol{x}) = \boldsymbol{G}_{\mathrm{M}}\boldsymbol{X}\left(\boldsymbol{I} - \boldsymbol{G}_{\mathrm{O}}\boldsymbol{X}\right)^{-1}\boldsymbol{\phi}_0 \ . \tag{5}$$

The inverse problem, which we are concerned with consists in determining the contrast $\boldsymbol{x}$ from a given finite set of noisy data $\boldsymbol{y}$. Moreover, note that one can have $n_{\mathrm{O}} \gg n_{\mathrm{M}} \times n_{\mathrm{S}}$ (number of unknowns larger than number of data) so that the system of algebraic equations can be highly under-determined.

2 A Bayesian Approach for the Inverse Problem

The Bayesian inference is now a common way to handle ill-posed inverse problems in signal and image processing (Demoment 1989). We recall the main basis of the Bayesian framework before considering its application to nonlinear diffraction tomography.

2.1 General Framework

In a general Bayesian framework of parameter estimation from experimental data, the relation between the unknown parameters $\boldsymbol{x} \in \mathbb{R}^n$ or $\mathbb{C}^n$ and the data $\boldsymbol{y} \in \mathbb{R}^m$ or $\mathbb{C}^m$ can be written:

$$\boldsymbol{y} = \mathcal{A}(\boldsymbol{x}) + \boldsymbol{n} \ ,$$

where $\mathcal{A}$ models the observation mechanism (direct model) and $\boldsymbol{n}$ models errors on the measurements (measurement noise as well as modeling and discretization errors, which can often be considered additive on the data). Without particular knowledge on the errors, they are usually modeled by zero mean white Gaussian random variables, circular in case of complex quantities, with known variance σ_n^2 and independent of $\boldsymbol{x}$. These assumptions are considered hereafter.

From this modeling, the likelihood function of the unknown $\boldsymbol{x}$ for given data $\boldsymbol{y}$ can be deduced:

$$p(\boldsymbol{y} \mid \boldsymbol{x}) = \left(\frac{1}{\pi\sigma_n^2}\right)^m \exp\left(-\frac{1}{\sigma_n^2}\|\boldsymbol{y} - \mathcal{A}(\boldsymbol{x})\|^2\right) \ .$$

The a priori state of knowledge, that is before any measurement is carried out, is taken into account through a probability law:

$$p(\boldsymbol{x}) \propto \exp\left\{-\mu\mathcal{U}(\boldsymbol{x})\right\} \ ,$$

where $\mathcal{U}$ has to be chosen to enforce desired properties on the solution.

Bayes rule allows to combine information supplied by the data and prior model in the a posteriori probability law of the parameters:

$$p(\boldsymbol{x} \mid \boldsymbol{y}) = \frac{p(\boldsymbol{y} \mid \boldsymbol{x}) p(\boldsymbol{x})}{p(\boldsymbol{y})} ,$$

where, $p(\boldsymbol{y})$ is a normalizing coefficient.

From a strictly Bayesian viewpoint, the posterior law is the solution to the problem as it sums up all information available on the object. However, it is necessary to decide on a value to give to $\boldsymbol{x}$. Different estimators can be exhibited following the chosen decision rule, such as Maximum a posteriori (MAP), Maximum Marginal a posteriori (MMAP) or Posterior Mean (PM) estimators. Parameters which maximize the a posteriori law (MAP) are frequently chosen and this leads to an optimization problem. Indeed, the MAP estimate corresponds to the minimizer of the criterion:

$$\mathcal{J}(\boldsymbol{x}) = \|\boldsymbol{y} - \mathcal{A}(\boldsymbol{x})\|^2 + \lambda \mathcal{U}(\boldsymbol{x}) ,$$

where $\lambda = \sigma_n^2 \mu$ can be considered as a regularization parameter which balances between fidelity to the data and prior information.

2.2 Application to Nonlinear Diffraction Tomography

This general framework can be applied on many ways to the considered problem. We propose hereafter two distinct formulations, depending on whether the contrast $\boldsymbol{x}$ has to be estimated from the data $\boldsymbol{y}$ or both the contrast $\boldsymbol{x}$ and the field $\boldsymbol{\phi}$ on the object have to be estimated.

First Formulation : Estimation of $\boldsymbol{x}$. This formulation is straightforward. The solution is defined as the MAP estimate of $\boldsymbol{x}$:

$$\boldsymbol{x}_{\text{MAP}} = \arg\max_{\boldsymbol{x}} p(\boldsymbol{x} \mid \boldsymbol{y}) .$$

From the explicit relation (5) it corresponds to the global minimizer of the criterion

$$\mathcal{J}^{\text{MAP}}(\boldsymbol{x}) = \|\boldsymbol{y} - \mathcal{A}(\boldsymbol{x})\|^2 + \lambda \mathcal{U}(\boldsymbol{x}) , \qquad (6)$$

with

$$\mathcal{A}(\boldsymbol{x}) = \boldsymbol{G}_{\text{M}} \boldsymbol{X} (\boldsymbol{I} - \boldsymbol{G}_{\text{O}} \boldsymbol{X})^{-1} \boldsymbol{\phi}_0 .$$

Second Formulation: Joint Estimation of $\boldsymbol{x}$ and $\boldsymbol{\phi}$. The solution is defined as the joint MAP estimate of $\boldsymbol{x}$ and $\boldsymbol{\phi}$:

$$(\boldsymbol{x}, \boldsymbol{\phi})_{\text{MAP}} = \arg \max_{(\boldsymbol{x}, \boldsymbol{\phi})} p(\boldsymbol{x}, \boldsymbol{\phi} \mid \boldsymbol{y}) \ .$$

Thanks to Bayes rule, the a posteriori law can be written:

$$p(\boldsymbol{x}, \boldsymbol{\phi} \mid \boldsymbol{y}) = \frac{p(\boldsymbol{y} \mid \boldsymbol{x}, \boldsymbol{\phi}) p(\boldsymbol{\phi} \mid \boldsymbol{x}) p(\boldsymbol{x})}{p(\boldsymbol{y})} \ . \tag{7}$$

In this relation, $p(\boldsymbol{y})$ is a constant with respect to $\boldsymbol{x}$ and $\boldsymbol{\phi}$, so only the three numerator terms intervene in the MAP criterion:

- Using (3), with the considered error model, the first term can be written:

$$p(\boldsymbol{y} \mid \boldsymbol{x}, \boldsymbol{\phi}) \propto \exp\left\{ -\frac{1}{\sigma_b^2} \|\boldsymbol{y} - \boldsymbol{G}_{\mathrm{M}} \boldsymbol{X} \boldsymbol{\phi}\|^2 \right\} \ ;$$

- The second term corresponds to the probability law of $\boldsymbol{\phi}$ for a known $\boldsymbol{x}$. As $\boldsymbol{\phi}$ is the total field on the object, it is uniquely determined for a given $\boldsymbol{x}$ by (4). Thus, if δ denotes the Dirac distribution:

$$p(\boldsymbol{\phi} \mid \boldsymbol{x}) = \delta(\boldsymbol{\phi} - \boldsymbol{\phi}_0 - \boldsymbol{G}_{\mathrm{O}} \boldsymbol{X} \boldsymbol{\phi}) \ ;$$

- $p(\boldsymbol{x})$ corresponds to the prior model on the object: $p(\boldsymbol{x}) \propto \exp\{-\mu \mathcal{U}(\boldsymbol{x})\}$.

Using these expressions, the posterior probability law can be written:

$$p(\boldsymbol{x}, \boldsymbol{\phi} \mid \boldsymbol{y}) \propto \exp\left\{ -\frac{1}{\sigma_b^2} \|\boldsymbol{y} - \boldsymbol{G}_{\mathrm{M}} \boldsymbol{X} \boldsymbol{\phi}\|^2 - \mu \mathcal{U}(\boldsymbol{x}) \right\} \delta(\boldsymbol{\phi} - \boldsymbol{\phi}_0 - \boldsymbol{G}_{\mathrm{O}} \boldsymbol{X} \boldsymbol{\phi}) \ .$$

The MAP estimate of $(\boldsymbol{x}, \boldsymbol{\phi})$ corresponds to the maximum of $p(\boldsymbol{x}, \boldsymbol{\phi} | \boldsymbol{y})$, i.e. it minimizes the criterion:

$$\mathcal{J}_{\mathrm{c}}^{\text{MAP}}(\boldsymbol{x}, \boldsymbol{\phi}) = \|\boldsymbol{y} - \boldsymbol{G}_{\mathrm{M}} \boldsymbol{X} \boldsymbol{\phi}\|^2 + \lambda \mathcal{U}(\boldsymbol{x}) \ , \tag{8}$$

subject to the constraint:

$$\boldsymbol{\phi} - \boldsymbol{\phi}_0 - \boldsymbol{G}_{\mathrm{O}} \boldsymbol{X} \boldsymbol{\phi} = 0 \ . \tag{9}$$

2.3 Prior Models

Introduction of a priori information on the object is the basis of regularization. In the Bayesian framework, this information is modeled by a probability law $p(\boldsymbol{x})$ or by an energy function $\mathcal{U}(\boldsymbol{x})$.

We consider the class of Markov Random Fields (MRF) models, which is frequently used in image processing (Geman 1990) and allows the introduction of local correlations between the elements of an image. The energy function of a MRF can generally be written:

$$\mathcal{U}(\boldsymbol{x}) = \sum_{i} \sum_{i \sim j} \rho(\boldsymbol{x}_i - \boldsymbol{x}_j) \ ,$$

with $\rho(t)$ a *potential* function, and $i \sim j$ stands for neighbors pixels i and j. Note that for complex fields, ρ operate separately on real and imaginary parts of $\boldsymbol{x}$ if they are considered independent.

A large choice of such potential functions has been proposed in the literature and certain of them are summarized in Table 1 and represented Fig. 1.

Table 1. Some potential functions and their characteristics

Name	Potential function	Characteristics
L_2 norm, Gaussian	$\rho(t) = t^2$	strictly convex, scale invariant
L_1 norm, Laplacian	$\rho(t) = \|t\|$	convex, scale invariant
L_p norm	$\rho(t) = \|t\|^p, 1 < p < 2$	strictly convex, scale invariant
Hubert function	$\rho(t) = \begin{cases} \|t\|^2 \text{ if } \|t\| \leq 1 \\ 2\|t\| - 1 \text{ if } \|t\| \geq 1 \end{cases}$	convex
Truncated Quadratic	$\rho(t) = \begin{cases} \|t\|^2 \text{ if } \|t\| \leq 1 \\ 1 \text{ if } \|t\| \geq 1 \end{cases}$	non convex, implicit line process

The L_2 norm corresponds to a first order Tikhonov regularization. This kind of regularization is of significant interest when the relation between the unknown and the data is linear because a linear explicit relation between the MAP estimate and the data is then available:

$$\boldsymbol{x}_{\text{MAP}} = (\boldsymbol{A}^{\dagger}\boldsymbol{A} + \lambda \boldsymbol{W})^{-1}\boldsymbol{A}^{\dagger}\boldsymbol{y} \ ,$$

with $\boldsymbol{W}^{-1}$ the correlation matrix of the Gaussian process. However, such interest decreases for nonlinear direct models, unless linear approximations are considered.

Nonconvex potential functions, like the truncated quadratic or other models including implicit or explicit line processes, can improve considerably the reconstruction of piecewise continuous images (Künsch 1994). However, as local minima may appear in the energy function, choosing such a model generally largely increases the difficulty of computing the solution.

Convex potential functions, such as L_p norms or Hubert function, seem to be a reasonable choice for nonlinear inverse problems. They correspond to a compromise between L_2 norm and nonconvex functions, as large variations of the field are less penalized than for the L_2 norm, but more than nonconvex functions. Using such models, allows better reconstructions of piecewise continuous images than Tikhonov regularization with no difficulty increase of the solution computation.

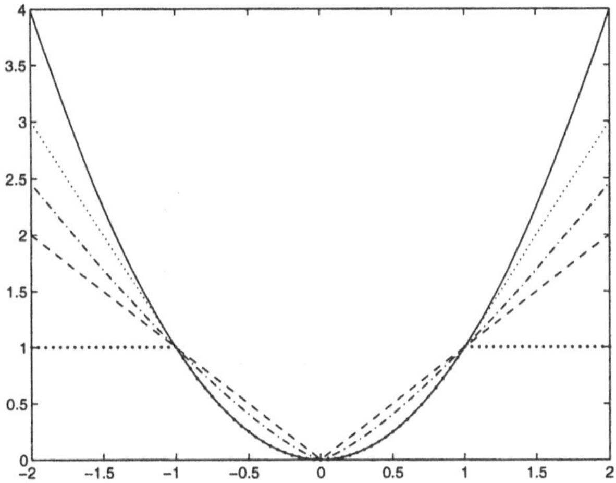

Fig. 1. 1-D representation of some potential functions: L_2 (–); L_1 (– –); L_p, $p = 1.3$ (–·); Huber function, $T = 1$ (···); Truncated Quadratic $T = 1$(· ·).

2.4 A Computational Challenge

The regularized solution has been defined as the contrast $\boldsymbol{x}$ which minimizes criterion (6) or as the contrast $\boldsymbol{x}$ and the total field $\boldsymbol{\phi}$ that jointly minimize criterion (8) under constraint (9). These two distinct formulations are equivalent in the sense that they define the same solution (for $\boldsymbol{x}$), but one may consider using different techniques to solve them.

Note that the Bayesian framework is not indispensable for defining the solution as the minimum of the criterion (6). Indeed, this criterion can be considered as a penalized least square criterion within a deterministic framework. However, the definition of the joint solution as the minimum of (8) under constraint (9) is not straightforward from deterministic arguments and other joint criteria are often proposed, which will be studied in § 4. Anyway, the Bayesian framework is not only useful to define a regularized solution to an inverse problem but also offers probabilistic tools e.g. to characterize the solution (Tarantola 1987) and to estimate some additional parameters such as the regularization parameter (Idier et al. 1996).

Due to the non-linearity of the direct problem, it is easy to show that the criteria (6) and (8) are not convex functions. Thus, even if the prior information is modeled with a convex energy function, the criteria may have local minima. From simulation experiments, appearance of local minima is closely linked to a high contrast value, a limited number of measurements and a low signal-to-noise ratio. Thus computation of the solution may be a cumbersome task, especially in these *difficult* configurations. However, the problem seems to be less difficult in more favorable configurations.

In the multiplicity of methods proposed for solving nonlinear diffraction tomography problems, we tried to establish a classification, even if not exhaustive. Three types of methods have been emphasized which can be interpreted and analyzed in terms of algorithms to compute the defined regularized solution.

3 Successive Linearizations

Methods of the *first type* consider iteratively linear approximations of the direct model, which leads to solve successively linear inverse problems. Different methods of this type have been proposed in the literature to solve the nonlinear inverse problem of diffraction tomography. As the nonlinear inverse problem is ill-posed, each linear inverse problem is ill-posed and regularization has often been introduced to stabilize the solution of each linear problem.

Before comparing these different methods, we propose a successive linearizations algorithm specifically designed to minimize the criterion (6). Finally, we study the convergence properties of such techniques.

3.1 A Successive Linearizations Algorithm to Minimize $\mathcal{J}^{\text{MAP}}$

At a given iteration n, a linear approximation of $\mathcal{A}$ has to be taken into account for $\boldsymbol{x}$ near $\boldsymbol{x}_n$. The theoretically most coherent linear approximation of $\mathcal{A}(\boldsymbol{x})$ near $\boldsymbol{x}_n$ is given by its first order Taylor series expansion:

$$\mathcal{A}(\boldsymbol{x}) = \mathcal{A}(\boldsymbol{x}_n) + \nabla_{\boldsymbol{x}}\mathcal{A}(\boldsymbol{x}_n)(\boldsymbol{x} - \boldsymbol{x}_n) + \mathcal{O}\left((\boldsymbol{x} - \boldsymbol{x}_n)^2\right)$$

(strictly speaking, one has to account for the Taylor series expansion of the real and imaginary parts of $\mathcal{A}$ to define such a relation for complex valued functions). Let $\boldsymbol{A}_n^{\text{SLMAP}} = \nabla_{\boldsymbol{x}}\mathcal{A}(\boldsymbol{x}_n)$, calculus of $\boldsymbol{A}_n^{\text{SLMAP}}$ can be done easily. If $\phi_n = (\boldsymbol{I} - G_{\text{O}}X_n)^{-1}\phi_0$ denotes the field on the object for contrast $\boldsymbol{x}_n$, and $\boldsymbol{\Phi}_n$ its corresponding diagonal matrix, $\boldsymbol{A}_n^{\text{SLMAP}}$ can be written:

$$\boldsymbol{A}_n^{\text{SLMAP}} = \boldsymbol{G}_{\text{M}}\left[\boldsymbol{I} + \boldsymbol{X}_n(\boldsymbol{I} - \boldsymbol{G}_{\text{O}}\boldsymbol{X}_n)^{-1}\boldsymbol{G}_{\text{O}}\right]\boldsymbol{\Phi}_n \ .$$

Thus minimization of $\mathcal{J}^{\text{MAP}}$ can be performed with successive linearizations of $\mathcal{A}$:

Initialize $n = 0, \boldsymbol{x}_0$.
Iterate for $n = 1, 2 \ldots$ until convergence towards a stationary point:

1. Compute the field on the object $\boldsymbol{\phi}_n$ and the matrix $\boldsymbol{A}_n^{\text{SLMAP}}$ corresponding to the linear approximation of $\mathcal{A}$ near the current solution $\boldsymbol{x}_n$.
2. Compute $\boldsymbol{x}_{n+1} = \arg\min_{\boldsymbol{x}} \mathcal{J}_n^{\text{SLMAP}}(\boldsymbol{x})$ with
$\mathcal{J}_n^{\text{SLMAP}}(\boldsymbol{x}) = ||\boldsymbol{y} - \mathcal{A}(\boldsymbol{x}_n) - \boldsymbol{A}_n^{\text{SLMAP}}(\boldsymbol{x} - \boldsymbol{x}_n)||^2 + \lambda\mathcal{U}(\boldsymbol{x})$.

Note that in such a scheme, for convex energy functions $\mathcal{U}$, all these criteria are convex functions and consequently have a unique global minimum.

3.2 The Born Iterative Method

The Born Iterative Method (BIM) has been introduced to circumvent the non-linearity, solving iteratively each of the coupled equations (1–2) (Wang and Chew 1989). Indeed, both integral equations are bilinear with respect to x and ϕ and solving each equation with respect to one of these variables leads to solution of linear equations. Using algebraic notations, the BIM scheme can be summarized:

Initialize $\boldsymbol{\phi}_n = \boldsymbol{\phi}_0$ (Born approximation).
Iterate for $n = 1, 2 \ldots$ until convergence towards a stationary point:

1. Compute $\boldsymbol{x}_{n+1}$ for field $\boldsymbol{\phi}_n$ on the object, i.e. solve the linear inverse problem: $\boldsymbol{y} = \boldsymbol{G}_{\text{M}}\boldsymbol{\Phi}_n\boldsymbol{x}$.
2. Compute the total field on the object $\boldsymbol{\phi}_{n+1}$, corresponding to contrast $\boldsymbol{x}_{n+1}$.

The linear approximation of the direct model which is accounted for in step 1. can be written:

$$\mathcal{A}(\boldsymbol{x}) \approx \mathcal{A}(\boldsymbol{x}_n) + \boldsymbol{A}_n^{\text{BIM}}(\boldsymbol{x} - \boldsymbol{x}_n), \qquad \text{with} \qquad \boldsymbol{A}_n^{\text{BIM}} = \boldsymbol{G}_{\text{M}}\boldsymbol{\Phi}_n \ .$$

It appears in calculus of $\boldsymbol{A}_n^{\text{SLMAP}}$, that $\boldsymbol{A}_n^{\text{BIM}}$ corresponds to take a zero order approximation, with respect to $\delta\boldsymbol{x}$, of $[\boldsymbol{I} - \boldsymbol{G}_0(\boldsymbol{X}_n + \delta\boldsymbol{X})]^{-1}$. This term is approximated by $[\boldsymbol{I} - \boldsymbol{G}_0\boldsymbol{X}_n]^{-1}$ so that the approximation of the BIM is coarser than the approximation of the SLMAP.

In (Wang and Chew 1989), the linear inverse problem of step 1. is solved using a zero order Tikhonov regularization on $\boldsymbol{x}$. Hence, the original BIM is equivalent to the SLMAP where $\boldsymbol{A}_n^{\text{SLMAP}}$ is replaced by $\boldsymbol{A}_n^{\text{BIM}}$, with $\mathcal{U}(\boldsymbol{x}) = ||\boldsymbol{x}||^2$ and $\boldsymbol{x} = \boldsymbol{0}$ is taken as initial solution.

3.3 The Distorted Born Iterative Method

The Distorted Born Iterative Method (DBIM) (Chew and Wang 1990) is based on a scheme similar to the BIM, using distorted wave Born approximations. At each iteration, a known inhomogeneous background medium with contrast $\boldsymbol{x}_n$ is considered, with corresponding Green function $\mathcal{G}^n$ and incident field $\boldsymbol{\phi}_n$, and an additional inhomogeneity $\delta\boldsymbol{x}$ has to be computed.

Hereafter, $\boldsymbol{\phi}_{\mathrm{O}}^n, \boldsymbol{\phi}_{\mathrm{M}}^n$ denote the field $\boldsymbol{\phi}_n$ on the object and on the measurement points respectively, $\boldsymbol{G}_{\mathrm{M}}^n$ denotes a matrix corresponding to discretization of the Green function for inhomogeneous medium $\boldsymbol{x}_n$. The DBIM scheme can then be summarized:

Initialize $\boldsymbol{x}_0 = 0, \boldsymbol{G}_{\mathrm{M}}^n = \boldsymbol{G}_{\mathrm{M}}$ and the incident fields $\boldsymbol{\phi}_{\mathrm{O}}^n = \boldsymbol{\phi}_{\mathrm{O}}^0, \boldsymbol{\phi}_{\mathrm{M}}^n = \boldsymbol{\phi}_{\mathrm{M}}^0$ (Born approximation).
Iterate for $n = 1, 2 \ldots$ until convergence towards a stationary point:

1. Compute contrast $\boldsymbol{x}_{n+1} = \boldsymbol{x}_n + \delta\boldsymbol{x}$ for the distorted wave Born approximation (field on the object $\boldsymbol{\phi}_{\mathrm{O}}^n$ and matrix $\boldsymbol{G}_{\mathrm{M}}^n$), i.e. solve the linear inverse problem: $\boldsymbol{y} + \boldsymbol{\phi}_{\mathrm{M}}^0 = \boldsymbol{\phi}_{\mathrm{O}}^n + \boldsymbol{G}_{\mathrm{M}}^n \boldsymbol{\Phi}_{\mathrm{O}}^n \delta\boldsymbol{x}$.
2. Compute incident fields $\boldsymbol{\phi}_{\mathrm{O}}^{n+1}, \boldsymbol{\phi}_{\mathrm{M}}^{n+1}$ and matrix $\boldsymbol{G}_{\mathrm{M}}^{n+1}$ corresponding to the new inhomogeneous background $\boldsymbol{x}_{n+1}$.

If discretization is performed with a moment method with pulse *basis* and *test functions*, as suggested in (Chew and Wang 1990), the update of $\boldsymbol{G}_{\mathrm{M}}^n$ can be written:

$$\boldsymbol{G}_{\mathrm{M}}^n = \boldsymbol{G}_{\mathrm{M}} + \boldsymbol{G}_{\mathrm{M}}^n \boldsymbol{X}_n \boldsymbol{G}_{\mathrm{O}} \; .$$

Note that this algebraic relation is not valid for other basis and test functions such as piecewise continuous ones, in which case the study of the DBIM in an algebraic framework is not as easy. Using algebraic notations, it can be shown that, at each iteration, the first step accounts for a linear approximation of $\mathcal{A}$ which can be written:

$$\mathcal{A}(\boldsymbol{x}) \approx \mathcal{A}(\boldsymbol{x}_n) + \boldsymbol{A}_n^{\mathrm{DBIM}}(\boldsymbol{x} - \boldsymbol{x}_n) \; , \qquad \text{with} \qquad \boldsymbol{A}_n^{\mathrm{DBIM}} = \boldsymbol{G}_{\mathrm{M}}(\boldsymbol{I} - \boldsymbol{X}_n \boldsymbol{G}_{\mathrm{O}})^{-1} \boldsymbol{\Phi}_n \; .$$

It can be shown that the approximation of the DBIM is identical to that of the SLMAP. Indeed,

$$(\boldsymbol{I} - \boldsymbol{X}_n \boldsymbol{G}_{\mathrm{O}})^{-1} = \boldsymbol{I} + \boldsymbol{X}_n (\boldsymbol{I} - \boldsymbol{G}_{\mathrm{O}} \boldsymbol{X}_n)^{-1} \boldsymbol{G}_{\mathrm{O}} \; ,$$

which can be verified by calculating the product of these matrices.

In (Chew and Wang 1990), zero order Tikhonov regularization on $\delta\boldsymbol{x}$ has been introduced to solve the linear inverse problem of step 1, i.e. it accounts for an energy function $\mathcal{U}(\boldsymbol{x} - \boldsymbol{x}_n)$ instead of $\mathcal{U}(\boldsymbol{x})$ in the SLMAP scheme. Thus the solution given by this method does not correspond to a minimum of the MAP criterion (6).

3.4 The Newton Kantorovitch Method

The BIM and the DBIM are specific to the modeling of the forward problem with coupled equations such as (2-1). The Newton-Kantorovitch method is a more general method to solve nonlinear functional equations $y = \mathcal{A}(x)$ (Roger 1981). An iterative scheme is introduced, whose iteration consists in calculating variation δx added to x_n so that $y - \mathcal{A}(x_n) = \mathcal{A}(x_n + \delta x)$. As δx is assumed to be small, $\mathcal{A}(x_n + \delta x)$ is linearized for each iteration.

A Newton-Kantorovitch Method (NKM) has been proposed to solve the problem concerned (Joachimovicz et al. 1991). The linear approximation taken into account in (Joachimovicz et al. 1991) can be written:

$$\mathcal{A}(\boldsymbol{x}) \approx \mathcal{A}(\boldsymbol{x}_n) + \boldsymbol{A}_n^{\text{NKM}}(\boldsymbol{x} - \boldsymbol{x}_n) \ , \qquad \text{with} \qquad \boldsymbol{A}_n^{\text{NKM}} = \boldsymbol{G}_{\text{M}}(\boldsymbol{I} - \boldsymbol{X}_n \boldsymbol{G}_{\text{O}})^{-1}\boldsymbol{\Phi}_n \ ,$$

which is strictly equivalent to that of the DBIM. Let us recall that this relation can be established for the DBIM when the discretization is performed with a moment method with pulse basis and test functions, while it is still valid for other functions for the NKM.

In (Joachimovicz et al. 1991), the solution of each linear inverse problem is computed using zero order Tikhonov regularization on $\delta \boldsymbol{x}$. Thus the DBIM and the NKM are strictly equivalent.

3.5 Interpretation and Analysis of the Solutions

In terms of linear approximations, the SLMAP, the DBIM and the NKM are strictly equivalent, while the BIM accounts for a coarser approximation of $\mathcal{A}$ at each iteration.

The DBIM and the NKM are identical and only differ the from the SLMAP on the way according to which the regularization is introduced. Indeed, in the DBIM and the NKM, regularization is introduced to stabilize the solution of each linear inverse problem and not to regularize the whole nonlinear inverse problem. Regularization is performed on $\delta \boldsymbol{x}$ and does not take into account any prior model on $\boldsymbol{x}$; so the provided solution does not correspond to a minimum of $\mathcal{J}^{\text{MAP}}$. Note that for such a regularization, the algorithms seem to be very sensitive to the regularization parameter. In (Joachimovicz et al. 1991) a specific adjusting method has been proposed for this parameter in the NKM. In (Chew and Wang 1990) it has been observed that the DBIM can diverge more easily than the BIM. It seems to be contradictory with the fact that the BIM accounts for a coarser approximation of $\mathcal{A}$ than the DBIM, but it can be easily understood from the fact that the DBIM does not regularize the nonlinear inverse problem satisfactorily but each linear inverse problem independently.

On the other hand, the SLMAP is a successive linearizations algorithm designed to compute a regularized solution to the nonlinear inverse problem, defined as the minimum of $\mathcal{J}^{\text{MAP}}$.

The properties of the SLMAP in terms of minimization of $\mathcal{J}^{\text{MAP}}$ can be studied. At each step, $\mathcal{J}^{\text{MAP}}$ is approximated by a convex criterion $\mathcal{J}_n^{\text{SLMAP}}$ with same value at $\boldsymbol{x}_n$ and – due to the first order Taylor series expansion – same slope at this point:

$$\mathcal{J}_n^{\text{SLMAP}}(\boldsymbol{x}_n) = \mathcal{J}^{\text{MAP}}(\boldsymbol{x}_n) \quad \text{and} \quad \nabla_{\boldsymbol{x}}\mathcal{J}_n^{\text{SLMAP}}(\boldsymbol{x}_n) = \nabla_{\boldsymbol{x}}\mathcal{J}^{\text{MAP}}(\boldsymbol{x}_n) \ .$$

The properties of such an algorithm are:

1. There exists no convergence guarantee and the algorithm could diverge.
2. If it converges, a stationary point $\boldsymbol{x}_\infty$ is reached and

$$\nabla_{\boldsymbol{x}}\mathcal{J}_n^{\text{SLMAP}}(\boldsymbol{x}_\infty) = \nabla_{\boldsymbol{x}}\mathcal{J}^{\text{MAP}}(\boldsymbol{x}_\infty) = \boldsymbol{0} \ ,$$

   so this point corresponds to a local minimum of the criterion $\mathcal{J}^{\text{MAP}}$.
3. Possible convergence and reached stationary point depend upon the initialization of the algorithm.

Note that the second property is not valid for the linear approximation taken into account in the BIM. If the BIM converges towards a stationary point, this point is not guaranteed to be a minimum of $\mathcal{J}^{\text{MAP}}$ because $\nabla_{\boldsymbol{x}}\mathcal{J}_n^{\text{BIM}}(\boldsymbol{x}_n) \neq \nabla_{\boldsymbol{x}}\mathcal{J}^{\text{MAP}}(\boldsymbol{x}_n)$. In this sense, the BIM is sub-optimal compared to the SLMAP (moreover, the SLMAP has been shown to converge more rapidly than the BIM (Carfantan and Djafari 1996)).

4 Minimization of a Joint Criterion

Some recently proposed methods – methods of the *second type* – define the solution as the minimum of criteria which account for errors on both coupled equations (3-4) with possible additional terms (Kleinman and van den Berg 1992), (Sabbagh and Lautzenheiser 1993), (Caorsi et al. 1993). In these methods, the solution is defined as the minimizer of a criterion, jointly on the contrast $\boldsymbol{x}$ and the field on the object $\boldsymbol{\phi}$, with the following generic form:

$$\boldsymbol{F}(\boldsymbol{x},\boldsymbol{\phi}) = \alpha_{\text{M}} \left\|\boldsymbol{y} - \boldsymbol{G}_{\text{M}}\boldsymbol{X}\boldsymbol{\phi}\right\|^2 + \alpha_{\text{O}} \left\|\boldsymbol{\phi} - \boldsymbol{\phi}_0 - \boldsymbol{G}_{\text{O}}\boldsymbol{X}\boldsymbol{\phi}\right\|^2 + \lambda\mathcal{U}(\boldsymbol{x},\boldsymbol{\phi}) \ . \quad (10)$$

Such a criterion is very easy to understand intuitively: it corresponds to minimizing jointly the errors on (3) and (4) and, as the problem is ill-posed, a penalization term on the unknowns is added to regularize it.

The proposed methods differ on several points:

- Criteria differ from value of parameters α_{M} and α_{O}. For example, these parameters are fixed to normalize the errors on both equations for $\boldsymbol{\phi} = \boldsymbol{0}$: $\alpha_{\text{O}} = 1/||\boldsymbol{\phi}_0||^2$ and $\alpha_{\text{M}} = 1/||\boldsymbol{y}||^2$ in (Kleinman and van den Berg 1992), while they are fixed to $1/2$ in (Sabbagh and Lautzenheiser 1993).

- There are differences on the regularization term. Originally, no regularization was introduced (Kleinman and van den Berg 1992), (Sabbagh and Lautzenheiser 1993). Then, it has been proposed to regularize both on $\boldsymbol{x}$ and $\boldsymbol{\phi}$, with $\mathcal{U}(\boldsymbol{x},\boldsymbol{\phi}) = \lambda_{\boldsymbol{x}}||\boldsymbol{x}||^2 + \lambda_{\boldsymbol{\phi}}||\boldsymbol{\phi}||^2$ in (Barkeshli and Lautzenheizer 1994) and with $\mathcal{U}(\boldsymbol{x},\boldsymbol{\phi}) = \lambda_{\boldsymbol{\phi}}||\Delta_1\boldsymbol{\phi} + (\boldsymbol{X} + k_0^2\boldsymbol{I})\boldsymbol{\phi}||^2 + \lambda_{\boldsymbol{x}}||\boldsymbol{D}_1\boldsymbol{x}||^2$ in (Caorsi et al. 1993), where Δ_1 and $\boldsymbol{D}_1$ corresponds to discretization of Laplacian and gradient operators. Finally, it has been proposed to introduce a single regularization term on $\boldsymbol{x}$, corresponding to a Markov random field with a line process in (Caorsi et al. 1995), and to a total variation penalization in (van den Berg and Kleinman 1995), which is equivalent to a L_1 regularization term.
- The methods also differ from the techniques used to compute the solution. Usual gradient type local minimization techniques has been used (Sabbagh and Lautzenheiser 1993), (Barkeshli and Lautzenheizer 1994) as well as local techniques specially designed for such a criterion (Kleinman and van den Berg 1992) and global minimization techniques such as Simulated Annealing (Caorsi et al. 1995).

4.1 Bayesian Interpretation

Recall that joint estimation of $\boldsymbol{x}$ and $\boldsymbol{\phi}$ leads to problem $\mathcal{P}_c$: minimization of criterion (8) subject to constraint (9). The constraint can be equivalently written $||\boldsymbol{\phi} - \boldsymbol{\phi}_0 - \boldsymbol{G}_{\mathrm{O}}\boldsymbol{X}\boldsymbol{\phi}||^2 = 0$, so that the Lagrangian of $\mathcal{P}_c$ can be written:

$$L(\boldsymbol{x},\boldsymbol{\phi}_0,\mu) = ||\boldsymbol{y} - \boldsymbol{G}_{\mathrm{M}}\boldsymbol{X}\boldsymbol{\phi}||^2 + \mu\,||\boldsymbol{\phi} - \boldsymbol{\phi}_0 - \boldsymbol{G}_{\mathrm{O}}\boldsymbol{X}\boldsymbol{\phi}||^2 + \lambda\mathcal{U}(\boldsymbol{x}), \tag{11}$$

with the scalar Lagrange multiplier μ. This Lagrangian looks like generic criterion (10), so that the adopted Bayesian framework gives a new way to look at it. It corresponds to the Lagrangian of the constraint optimization problems $\mathcal{P}_c$ in which the Lagrange multiplier is fixed intuitively ($\mu = ||\boldsymbol{y}||^2/||\boldsymbol{\phi}_0||^2$ (Kleinman and van den Berg 1992) or $\mu = 1$ (Sabbagh and Lautzenheiser 1993)).

Moreover, this viewpoint gives indications for regularizing such a criterion with an energy function $\mathcal{U}(\boldsymbol{x})$. Using Bayes rule for the considered model of errors on measurements, we can see on (7) that there is no need to introduce prior model on $\boldsymbol{\phi}$.

Note that in (Caorsi et al. 1995) another Bayesian interpretation has been given for this criterion. It is shown that if additive gaussian error are assumed on both coupled equations (3–4), the joint MAP estimate of $\boldsymbol{x}$ and $\boldsymbol{\phi}$ minimizes a criterion of form (10). However, it can be shown that such a criterion is obtained introducing zero mean circular Gaussian (conditionally to $\boldsymbol{x}$) noise with covariance matrix $C_{\mathrm{M}} = \sigma_2^2\boldsymbol{I} + \sigma_2^2\left[(\boldsymbol{G}_{\mathrm{M}}\boldsymbol{X})^\dagger(\boldsymbol{G}_{\mathrm{M}}\boldsymbol{X})\right]^{-1}$ on the measurements. It seems to be a very strong and unjustified hypothesis as it considers a particular correlation between the measurement errors and the unknown object.

4.2 Analysis of the Solutions

The Lagrangian theory provides a link between solutions of constrained optimization problems and saddle-point of the corresponding Lagrangian. However, in the considered case where neither the criterion nor the constraint are convex, the only available property is the following:
If $((\boldsymbol{x}, \boldsymbol{\phi}), \mu)$ is a saddle-point of Lagrangian (11), $(\boldsymbol{x}, \boldsymbol{\phi})$ is a solution to constrained optimization problem $\mathcal{P}_c$.
However, there is no guarantee that such a saddle-point exists.

Among the different methods proposed to minimize a criterion of form (10), none tries to reach a possible saddle-point of Lagrangian (11), but only a minimum of it for a fixed value of Lagrange multiplier. The given solution is then not necessarily a solution of $\mathcal{P}_c$. Moreover, note that criterion (10) is not convex, so it can have local minima. The solution computed with local minimization techniques will then possibly correspond to a local minimum of the Lagrangian, for fixed Lagrange parameters.

It is possible that the given solution corresponds to a saddle-point of the Lagrangian. If the fixed Lagrange multiplier corresponds to a maximum of the Lagrangian, the solution is solution of $\mathcal{P}_c$. It can be shown that if $(\boldsymbol{x}, \boldsymbol{\phi})$ is a local minimum of L, for μ fixed and that constraint is verified, then $\boldsymbol{x}$ corresponds to a local extremum of the unconstrained criterion $\mathcal{J}^{\text{MAP}}$. But if the constraint (9) is not verified, the solution cannot be characterized as easily.

Note that from this definition of the joint solution as the solution of $\mathcal{P}_c$, specific algorithms can be designed to compute this solution (Carfantan 1996).

5 Minimization of the MAP Criterion

From presentation of § 2, a natural idea to compute the defined solution – which corresponds to methods of the *third type* – is to minimize directly the MAP criterion (6).

Different methods have been proposed in the literature which define the solution as the minimizer of the mean square error (MSE) between experimental and simulated data, possibly taking into account a regularization penalty term (*e.g.* (Garnero et al. 1991), (Xia et al. 1994)). However, an explicit formulation of criterion (6) using the explicit algebraic relation (5) of $\mathcal{A}$ has only been proposed recently (Carfantan and Djafari 1995). Note that it is not necessary to express such a relation to try to minimize the MSE and it is sufficient to be able to simulate the forward problem. However, one can take advantage of such an expression to design specific algorithms to minimize this MSE and the criterion (6).

Different optimization techniques have been used to compute the minimum of this criterion and will not be detailed hereafter: local techniques such as conjugate gradient (Xia et al. 1994), (Lobel et al. 1996) or global ones such as Simulated Annealing (SA) (Garnero et al. 1991), Graduated non Convexity (GNC) (Carfantan and Djafari 1995), or a cheaper Iterative Conditional Mode (ICM) (Carfantan et al 1996).

Using any optimization technique to minimize $\mathcal{J}^{\text{MAP}}$ guarantees the solution to be a minimum, possibly local, of this criterion. However, in *difficult configurations*, when this criterion has local minima, a global minimization technique may be used to obtain a satisfactory solution.

6 A Comparative Study

The presented classification of existing methods allows to better compare them. One can study the number of considered unknowns, the computation cost, the convergence properties and the robustness with respect to some parameters for each type of methods. No simulation results are shown in this paper and the reader can refer to (Carfantan 1996), (Carfantan and Djafari 1996) for more details. The main conclusions of this study is presented in the following.

6.1 Number of Unknown

In methods of both first and third types, the unknown is the contrast $\boldsymbol{x} \in \mathbb{C}^{n_\text{O}}$ while in methods of second type, the contrast and the total field on the object for each incident field $\boldsymbol{\phi} \in \mathbb{C}^{n_\text{O} \times n_\text{S}}$ have to be reconstructed. So, if the number of data is increased, considering more source positions, the number of unknowns is increased as well in methods of the second type.

6.2 Computational Costs

In methods of the first type, evaluation of the criterion requires an order of $\mathcal{O}(n_\text{O} * n_\text{M} * n_\text{S})$ complex operations. However, these methods require the update of some matrices between each iteration which includes resolution of the direct problem. For the BIM, the cost of these updates is of order $\mathcal{O}({n_\text{O}}^3 + {n_\text{O}}^2 * n_\text{S} + n_\text{O} n_\text{M} n_\text{S})$ while it is of order $\mathcal{O}({n_\text{O}}^3 + {n_\text{O}}^2 * (n_\text{S} + n_\text{M}))$ for the SLMAP, the DBIM and the NKM.

The methods of the second type do not need any updates and evaluation of the criterion has a cost of order $\mathcal{O}({n_\text{O}}^2 n_\text{S} + n_\text{O} * n_\text{S} * n_\text{M})$ complex operations.

On the other hand, the third type methods require evaluations of criterion (6) whose computation order is $\mathcal{O}({n_\text{O}}^3 + {n_\text{O}}^2 n_\text{S} + n_\text{O} * n_\text{M} * n_\text{S})$. Indeed, for each evaluation of the criterion, the direct problem has to be solved. Fortunately, algorithm such as SA (Garnero et al. 1991) and ICM (Carfantan et al 1996), which update the contrast image pixel by pixel, can perform these updates without computing the whole criterion for each pixel but only once for the sweep of the whole image.

6.3 Convergence Properties

We already studied the convergence properties of each type of method which can be summarized as follows. Methods of the first type, such as the SLMAP, can diverge while the others are guaranteed to converge towards a stationary point. On the other hand when the SLMAP converges, the provided solution corresponds to a local minimum of the MAP criterion (6) (which is not true for the BIM, the DBIM and the NKM). Methods of the second type are guaranteed to converge towards a minimum, maybe local, of the criterion (10). This solution can correspond to the MAP estimate only if the constraint (9) is satisfied, which is not guaranteed by these methods. The third type methods are guaranteed to converge towards a minimum, possibly local, of $\mathcal{J}^{\text{MAP}}$.

6.4 Robustness with Respect to the Regularization Parameter

It has been experimentally established that methods of the second and third types are more robust with respect to the value of the regularization parameter λ (Carfantan 1996) than first type methods. For example, methods of the first type can give good results for a value of λ and diverge for a nearby value, while the second and third types methods are in general not very sensitive to a change of a factor ten of this parameter, on the same configuration. This is an important point to consider as no automatic adjustment of this parameter is available up to now, only the user's experience.

7 Conclusion

In this paper, we have studied diffraction tomography within the Bayesian estimation framework. It allows to consistently introduce prior information on the solution of this nonlinear ill-posed inverse problem and to define a regularized solution, the MAP estimate, with reasonable assumptions.

Different existing methods have been classified in terms of algorithms to compute the MAP estimate. Three types of methods have been distinguished. Methods of the first type correspond to successively approximating the nonlinear object/data relation with a linear one. Methods of the second type define the solution as the joint minimizer of a criterion depending on the object and on the total field on the object. Third type methods directly minimize the MAP criterion depending on the object. These methods have been compared on their convergence properties and on the solution they provide.

Three major key ideas can be emphasized:

- As regularization consists in introducing prior information on the solution, one can get benefits from introducing more advanced models than a simple L_2 (Tikhonov) one.

- A successive linearizations algorithm has been proposed to compute a regularized solution to this nonlinear inverse problem. It is both more efficient than the BIM for its linear approximation and than the NKM and the DBIM from a regularization standpoint.
- The solution given by the minimization of the (penalized) joint criterion does not correspond to the minimum of the (penalized) mean square error on the measurement.

References

Barkeshli S., Lautzenheizer R. G. (1994): An iterative method for inverse scattering problems based on an exact gradient search. *Radio Science*, vol. 29, no. 4, pp. 1119–1130

Caorsi S., Gragnani G. L., Pastorino M., Peraso A. (1993): Electromagnetic inverse scattering numerical method for non invasive diagnostic of dielectric materials. in *3rd International Conference on Electromagnetics in Aerospace Applications*, Torino, Italy

Caorsi S., Gragnani G. L., Medicina S., Pastorino M., Pinto A. (1995): A Gibbs random fields-based active electromagnetic method for noninvasive diagnostics in biomedical applications. *Radio Science*, vol. 30, no. 1, pp. 291–301

Carfantan H. (1996): Approche bayésienne pour un problème inverse non linéaire en imagerie à ondes diffractées. *PhD Thesis* Université de Paris-Sud Orsay

Carfantan H., Mohammad-Djafari A. (1995): A Bayesian approach for nonlinear inverse scattering tomographic imaging. *Proc. IEEE ICASSP*, Detroit, U.S.A., vol. IV, pp. 2311–2314

Carfantan H., Mohammad-Djafari A. (1996): Beyond the Born approximation in inverse scattering with a Bayesian approach *2nd Int. Conf. on Inverse Problems in Engineering*, Le Croisic, France

Carfantan H., Mohammad-Djafari A., Idier J. (1996): A single site update algorithm for nonlinear diffraction tomography. accepted to *IEEE ICASSP* 1997, Munich, Germany

Chew W. C., Wang Y. M. (1990): Reconstruction of two-dimensional permittivity distribution using the distorted Born iterative method. *IEEE Trans. Medical Imaging*, vol. MI-9, pp. 218–225

Demoment G. (1989): Image reconstruction and restoration : Overview of common estimation structure and problems. *IEEE Trans. Acoust. Speech, Signal Processing*, vol ASSP-37, no. 12, pp. 2024–2036

Garnero L., Franchois A., Hugonin J.-P., Pichot C., Joachimowicz N. (1991): Microwave imaging – complex permittivity reconstruction by simulated annealing. *IEEE Trans. Microwave Theory and Technology*, vol. 39, no. 11, pp. 1801–1807

Geman D. (1990): Random fields and inverse problems in imaging. *École d'Été de Probabilités de Saint-Flour XVIII - 1988*, vol. 1427, pp. 117–193, Springer-Verlag, lecture notes in mathematics

Howard A. Q. J., Kretzschmar J. L. (1986): Synthesis of EM geophysical tomographic data. *Proc. IEEE*, vol. 74, no. 2, pp. 353–360

Idier J., Mohammad-Djafari A., Demoment G. (1996): Regularization methods and inverse problems: an information theory standpoint. *2nd Int. Conf. on Inverse Problems in Engineering*, Le Croisic, France

Joachimovicz N., Pichot C., Hugonin J.-P. (1991): Inverse scattering: An iterative numerical method for electromagnetic imaging. *IEEE Trans. Ant. Propag.*, vol. AP-39, no. 12, pp. 1742–1752

Kleinman R. E., van den Berg P. M. (1992): A modified gradient method for two-dimensional problems in tomography. *J. Computational and Applied Mathematics*, vol. 42, pp. 17–35

Künsch H. R. (1994): Robust priors for smoothing and image restoration. *Annals Institute Statistical Mathematics*, vol. 46, no. 1, pp. 1–19

Lobel P., Kleinman R.E., Pichot C., Blanc-Féraud L., Barlaud M. (1996): Conjugate gradient method for solving inverse scattering with experimental data. *IEEE Trans. Ant. Propag. Magazine*, vol. 38, no. 3, pp. 48–51

Roger A. (1981): Newton-Kantorovitch Algorithm Applied to an Electromagnetic Inverse problem. *IEEE Trans. Ant. Propag.*, vol. AP-29, pp. 232–238

Sabbagh H. A., Lautzenheiser R. G. (1993): Inverse problems in electromagnetic nondestructive evaluation. *International Journal of Applied Electromagnetics in Materials*, vol. 3, pp. 235–261

Tarantola A. (1987): Inverse problem theory: Methods for data fitting and model parameter estimation. *Elsevier Science Publisher*

van den Berg P. M., Kleinman R. E. (1995): A total variation enhanced modified gradient algorithm for profile reconstruction. *Inverse Problems*, vol. 11, pp. L5–L10

Wang Y. M., Chew W. C. (1989): An iterative solution of the two-dimensional electromagnetic inverse scattering problem. *Int. J. Imaging Systems and Technology*, vol. 1, pp. 100–108

Xia J. J., Habashy M., Kong J. A. (1994): Profile inversion in a cylindrically stratified lossy medium. *Radio Science*, vol. 29, no. 4, pp. 1131–1141

Application of the Approximate Inverse to Inverse Scattering

Alfred K. Louis

Lehrstuhl für Angewandte Mathematik, Universität des Saarlandes
D-66041 Saarbrücken
e-mail: louis@num.uni-sb.de
Internet: http://www.num.uni-sb.de

1 Introduction

In this paper we propose a method for solving the inverse scattering problem for the Helmholtz equation. The nonlinear problem is separated into an ill–posed linear and a, hopefully, well–posed nonlinear problem. This technique is successfully applied by several authors, see for example Langenberg, [6], Pichot et al [18]. What is special with the proposed method here is the application of the approximate inverse introduced in [10] in order to precompute a reconstruction kernel in order to speed up the solution both of the linear and the nonlinear problem.
For applications of the elastic or electromagnetic inverse scattering problem in medical imaging or nondestructive testing see for example [3], [6], [9], [15], [17]. For the sake of simplicity we restrict our presentation here to the scalar case. In principle it is also applicable to the vector–valued case. For fixed wavelength k and incoming plane wave in direction $\Theta \in S^2$ the Lippmann–Schwinger equation serves as mathematical model. It is

$$u^s(\Theta,\eta) = \int G(|\eta - x|)u(\Theta,x)f(x)dx \qquad (1)$$

with the complex permittivity f and Green's function G. Equation (1) is valid both inside the object and outside. Hence, in a first step, we approximate the product $\Phi = uf$, which now is the solution of a linear problem. This equation is well studied, see for example [3]. Evaluating the right–hand side for η inside the object with the approximated $\Phi = uf$ results in an approximation for the scattered field u^s inside the body. Finally, dividing Φ by $u^i + u^s$ gives an approximation for the searched–for permittivity f.
The first part of this paper consists in describing an efficient method for solving the linear part by constructing a reconstruction kernel ψ_x such that the solution Φ is represented as a scalar product of the kernel ψ with the data, here u^s.

$$\Phi(x) = \langle u^s, \psi_x \rangle \ . \qquad (2)$$

We point out that no artificial discretization is needed. Hence we study operator equations $A\Phi = g$ for operators between Hilbert spaces X and Y. Approximate inverse means a solution operator which maps the data $g = u^s$ to a stable approximation of the solution of the ill - posed problem $A\Phi = g$. This inversion operator is precomputed without using the data g, see [10].
The method is based on two ideas. First, the computation of moments of the solution is stable; i.e., we compute instead of Φ the approximation $\langle \Phi, e_\gamma \rangle$ with a suitable mollifier e_γ reducing in that way the high frequency components in the solution which are mostly affected by the data noise. This can be reformulated as using a weaker topology in the space X, see [4], [8]. Examples for e_γ are given in the next section, e_γ can be a basis function for projection methods, it can be chosen such that $\langle \Phi, e_\gamma \rangle$ approximates a derivative of Φ; in wavelet language it can be a scaling function or a wavelet. Second, in the case of linear operators the computation of $\langle \Phi, e_\gamma \rangle$ is then achieved by approximating e_γ in the range of the adjoint operator A^* by the reconstruction kernel $\psi_\gamma : A^*\psi_\gamma \simeq e_\gamma$. Then

$$\langle \Phi, e_\gamma \rangle \simeq \langle \Phi, A^*\psi_\gamma \rangle = \langle A\Phi, \psi_\gamma \rangle = \langle g, \psi_\gamma \rangle := S_\gamma g \ .$$

This is the mollifier method presented in [12]. It also has been used to accelerate convergence for finite element solutions in [7]. The recently introduced method in smoothed particle hydrodynamics is based on the same ideas. Without using the possibility of precomputing a reconstruction kernel via the adjoint operator a mollification method is presented in Murio [13].
The rest of the paper is then devoted to the above mentioned application in inverse scattering.

2 Approximate Inverse For Linear Problems

In the following we assume A to be a linear, continuous operator between the Hilbert spaces X and Y. Especially we think of X as a space of functions and Y as a finite dimensional space of measurements. Hence, if necessary, we use $X = L_2(\Omega)$ for a suitable set $\Omega \subset \mathbb{R}^d$. Examples for mollifiers are

$$e_\gamma(x,y) = \frac{d}{\mathrm{vol}(S^{d-1})\gamma^d}\chi_\gamma(x-y) \tag{3}$$

where χ_γ is the characteristic function of the ball around 0 with radius γ and $\mathrm{vol}(S^{d-1})$ is the measure of the surface of the unit ball in $\mathbb{R}^d$. Here local averages of the solution are computed. With the band limiting filter

$$e_\gamma(x,y) = \left(\frac{\gamma}{\pi}\right)^d \mathrm{sinc}(\gamma(x-y)) \tag{4}$$

the high – frequency components in the solution are eliminated. Fast decaying is the kernel of the heat equation

$$e_\gamma(x,y) = (2\pi)^{-d/2}\gamma^{-d}\exp(-|x-y|^2/(2\gamma^2)) \ . \tag{5}$$

In all cases the parameter γ acts as a regularization parameter. The mollifier e_γ is not necessarily a function with mean value 1. When the essential information we need are discontinuities in Φ we can use as e_γ a function such that $\langle \Phi, e_\gamma(x,\cdot)\rangle$ approximates a derivative of $\Phi(x)$. This means that e_γ can also be a wavelet, see e.g. [13].
First we assume the equation $A^*\psi_\gamma = e_\gamma$ to be solvable. Then we put

$$\langle \Phi, e_\gamma\rangle = \langle \Phi, A^*\psi_\gamma\rangle = \langle A\Phi, \psi_\gamma\rangle = \langle g, \psi_\gamma\rangle =: S_\gamma g \ . \tag{6}$$

This is the technique to derive inversion formulas in x – ray computer tomography resulting in the so – called filtered backprojection methods, see e.g. [8], [15]. If the equation $A^*\psi_\gamma = e_\gamma$ is not solvable we approximate ψ_γ by minimizing the defect $\|A^*\psi_\gamma - e_\gamma\|$ for sufficiently smooth e_γ leading to the equation

$$AA^*\psi_\gamma = Ae_\gamma \ . \tag{7}$$

Then we get

$$\langle \Phi, e_\gamma\rangle \simeq \langle \Phi, A^*\psi_\gamma\rangle = \langle A\Phi, \psi_\gamma\rangle = \langle g, \psi_\gamma\rangle =: S_\gamma g \ .$$

It is important to mention that no artificial discretization of Φ is needed as introduced by projection methods, see e.g. [8], [15], [16]. For the numerical computation of ψ_γ the matrix AA^* needs a coarse stabilization, the fine tuning is then achieved by the choice of γ, compare [19].

3 Comparison With Other Methods

Now let A be a compact operator between the Hilbert spaces X and Y. Then it has a singular value decomposition

$$\{v_n, u_n; \sigma_n\}_n$$

where v_n, u_n are normalized and

$$Av_n = \sigma_n u_n \quad \text{and} \quad A^*u_n = \sigma_n v_n \ .$$

Regularization methods, like Tikhonov – Phillips, truncated singular value decomposition or Landweber iteration, have the form

$$T_\gamma g = \sum_n F_\gamma(\sigma_n)\sigma_n^{-1}\langle g, u_n\rangle v_n \ , \tag{8}$$

see for example [2], [5], [8]. The following result shows that these methods are special cases of the approximate inverse.

Theorem 1 *Let the regularization method T_γ in (6) be given with a filter F_γ. Then this method can be written as an approximate inverse with mollifier*

$$e_\gamma(x,y) = \sum_n F_\gamma(\sigma_n) v_n(x) v_n(y) \ . \tag{9}$$

Proof: The definition of ψ_γ as solution of $AA^*\psi_\gamma = Ae_\gamma$ in (5) leads with

$$\psi_\gamma(x) = \sum_n \sigma_n^{-1} \langle e_\gamma(x,\cdot), u_n \rangle v_n$$

to

$$\psi_\gamma(x) = \sum_n F_\gamma(\sigma_n) \sigma_n^{-1} u_n v_n(x) \ .$$

Then

$$S_\gamma g(x) = \langle g, \psi_\gamma(x) \rangle = T_\gamma g(x) \ .$$

In contrast to the Backus – Gilbert method, see [1], [19], the matrix for computing the reconstruction kernel does not depend on the reconstruction point. Also there is a the possibility to pattern the reconstruction kernel in almost any desirable way, one is not forced to approximate the delta distribution with a kernel like $|x-y|^{-2}$.

Using the smoothing property of the operator E_γ defined as $E_\gamma \Phi(x) = \langle \Phi, e_\gamma(x,\cdot)$ we can extend the concept of order optimality from the classical regularization methods, see [11].

4 Efficient Implementation

If the problem shares some invariance properties they can be used for a fast realisation of the method. Let in the following E_γ be a function of the variable y only. We can think of $E_\gamma(y) = e_\gamma(0,y)$; i.e., a mollifier concentrated around 0 which then may be shifted to arbitrary points as $e_\gamma(x,y) = E_\gamma(x-y)$.

Theorem 2 *Let $A : X \to Y$ and let T_1^x be a group representation on X and T_2^x, T_3^x be group representations on Y such that*

$$AT_1^x = T_2^x A \tag{10}$$

and

$$T_2^x AA^* = AA^* T_3^x \ . \tag{11}$$

Let w_γ be the minimum norm solution of

$$AA^* w_\gamma = AE_\gamma \ . \tag{12}$$

Then the minimum norm solution of

$$AA^*\psi_\gamma(x) = AT_1^x E_\gamma$$

is

$$\psi_\gamma(x) = T_3^x w_\gamma \ . \tag{13}$$

Proof : ¿From the invariance properties follows

$$\begin{aligned} AT_1^x E_\gamma &= T_2^x AE_\gamma \\ &= AA^* T_3^x w_\gamma \end{aligned}$$

which completes the proof.

This means that only the solution w_γ has to be computed and stored, the kernels for other reconstruction points x are found by the action of T_1^x on E_γ and by T_3^x on w_γ.

In the case of a finite number of data where $(A\Phi)_n = A\Phi(x_n)$, $n = 1,\dots,N$ for suitable points x_n the reconstruction kernel w_γ is a vector in $\mathbb{C}^N$ with $(w_\gamma)_n = w_\gamma(x_n)$. Then $w_\gamma(x_n - x)$ can be evaluated by linear interpolation between $(w_\gamma)_m$ and $(w_\gamma)_{m+1}$ with $x_m \le x_n - x < x_{m+1}$.

For $Y = \mathbb{C}^N$ and M reconstruction points the storage needs is $M \times N$ complex numbers. If the problem has invariance properties this can be dramatically reduced. If translation invariance for example holds only N complex numbers have to be stored!

5 The Inverse Scattering Problem

In the following we first consider the case of just one incident plane wave in direction Θ. We define the mapping

$$A : L_2(\Omega) \to L_2(\Gamma)$$

where we first put $\Omega = \Gamma = \mathbb{R}^3$, that means we make no use of the fact that f is compactly supported. Let G be the Green's function of the Helmholtz equation, then

$$A\Phi(\eta) = \int_\Omega G(|\eta - y|)\Phi(y)dy \tag{14}$$

which is a convolution equation. Following [3] the Operator A can be represented as

$$A\Phi(x) = \imath k \sum_{n\ge 0} h_n^{(1)}(k|x|) \sum_{m=-n}^{n} Y_m^n(x/|x|)c_{nm}$$

where $H_n^{(1)}$ are the spherical Hankel functions of the first kind and Y_m^n are the spherical harmonics. The expansion coefficients c_{nm} are given as

$$c_{nm} = \int_{S^2} Y_m^n(\vartheta) \int_0^\infty \rho^2 j_n(k\rho)\Phi(\rho\vartheta)\, d\rho\, d\vartheta$$

with the spherical Bessel functions j_n. This formula can be used to derive a singular value decomposition for the operator A.

¿From the last section we conclude that also the reconstruction is of displacement type using $T_1 = T_2 = T_3 = D^z$ where $D^z f(x) = f(x - z)$ in Theorem 2 if $e_\gamma(x, y) = D^x E_\gamma(y) = E_\gamma(y - x)$.

The mapping AA^* is generated by the kernel

$$G_2(\eta - \xi) = \int_{\mathbb{R}^3} G(|\eta - x|)G(|\xi - x|)dx\ .$$

Defining with the unitary matrix U the operator D^U as $D^U f(x) = f(Ux)$ we compute that $AD^U = D^U A$ and $AA^* D^U = D^U AA^*$, leading to another invariance and a simplification of the reconstruction kernel ψ_γ for circular symmetric $E_\gamma(|y|)$.

After solving $AA^*\psi_\gamma(x,\cdot) = Ae_\gamma(x,\cdot)$ for the reconstruction point x we put for the data given on Γ

$$\Phi_\gamma(x) = \langle u^s, \psi_\gamma(x,\cdot)\rangle_{L_2(\Gamma)} = \int_\Gamma u^s(\eta)\psi_\gamma(|x - \eta|)d\eta\ . \tag{15}$$

Evaluating the integral in (14) at the point $\eta = x \in \Omega$ we find an approximation for the scattered field inside the scatterer as

$$\begin{aligned} u_\gamma^s(x) &:= \int_{\mathbb{R}^3} G(|x - y|)\Phi_\gamma(y)dy \\ &= \int_\Gamma u^s(\eta) \int_{\mathbb{R}^3} G(|x - y|)\psi_\gamma(|y - \eta|)dyd\eta \\ &= \int_\Gamma K_\gamma(x - \eta)u^s(\eta)d\eta\ . \end{aligned}$$

We observe that also the kernel K_γ can be precomputed independent of the data. Then we put

$$f_\gamma(x) := \Phi_\gamma(x)/(u^i(x)+u^s(x)) = \frac{\int_\Gamma u^s(\eta)\psi_\gamma(|x-\eta|)d\eta}{e^{i\Theta x}+\int_\Gamma u^s(\eta)K_\gamma(x-\eta)d\eta}\ .$$

For multiple incoming plane waves we average over these values, resulting in a discretization of the integral

$$f_\gamma(x) = \frac{1}{4\pi}\int_{S^2}\frac{\int_\Gamma u^s(\Theta,\eta)\psi_\gamma(|x-\eta|)d\eta}{e^{i\Theta x}+\int_\Gamma u^s(\Theta,\eta)K_\gamma(x-\eta)d\eta}d\Theta\ . \tag{16}$$

The solution of the ill-posed linear problem is achieved by precomputing the reconstruction kernel ψ_γ and based on this also the kernel K_γ. Hence the reconstruction is realized by a fast implementation of this formula of filtered backprojection type used in x-ray tomography, compare [8], [15].

Acknowledgement

The research of the author has been supported by BMBF, grant 20M360-03LO7SAA-8, and DFG, grant Lo 310/4-1.

References

1 Backus G., Gilbert F. (1967) : Numerical application of a formalism for geophysical inverse problems. Geophys. J. R. Astron. Soc.**13** 247–76
2 Bertero M., de Mol C., Viano C. (1980) : The stability of inverse problems. In Inverse scattering in optics - ed H. P. Baltes (Berlin:Springer) pp 161–214
3 Colton D., Kress R., (1992) : Inverse Acoustic and Electromagnetic Scattering Theory, (Heidelberg:Springer)
4 Eckhardt U., (1976) Uir numerischen Behandlung inkorrekt gestellter Aufgaben. Computing**17** 193–206
5 Engl H. W., (1993) Regularization methods for the stable solution of inverse problems. Surveys on Mathematics for Industry **3** 71–143
6 Langenberg K. J., al (1996) Applied Inversion in Nondestructive Testing, to appear in : Engl, H.W., Louis, A.K., Rundell, W., (eds.): Inverse Problems in Medical Imaging and Nonsdestructive Testing, (Wien : Springer)
7 Louis A. K., (1979) Acceleration of convergence for finite element solutions of the Poisson equation. Numerische Mathematik **33** 43–53
8 Louis A. K., (1989) Inverse und schlecht gestellte Probleme. Stuttgart : Teubner)
9 Louis A. K., (199) Medical imaging : state of the art and future development. Inverse Problems **8** 709–738
10 Louis A. K., (1996) Approximate Inverse for linear and some nonlinear problems. Inverse Problems **12** 175–190
11 Louis A. K., (1997) Application of the Approximate Inverse to 3D X-Ray CT and Ultrasound Tomography, in Engl H. W., Louis, A. K., Rundell, W. (eds.) : Inverse Problems in Medical Imaging and Nondestructive Testing, (Wien:Springer)
12 Louis A. K., Maaß P (1990) A mollifier method dor linear operator equations of the first kind. Inverse Problems **6** 427–440
13 Louis A. K., Maaß P (1994) Wavelets (Stuttgart : Teubner)
14 Murio D. A., (1993) The mollification method and the numerical solution of ill-posed problems. (NewYork : Wiley)
15 Natterer F. (1986) The mathematics of computerized tomography. (Stuttgart : Teubner-Wiley)
16 Parker R. L (1984) Geophysical Inverse Theory. (Princeton : U Press)
17 Pichot Ch. al (1986) Electromagnétisme. in Série Electronique. Editions des Techniques de l'Ingénieur
18 Pichot Ch. al (1996) Gauss-Newton and gradient methods for microwave tomography, to appear in Engl, H.W., Louis A.K., Rundell W. (eds.): Inverse Problems. in Medical Imaging and Nonsdestructive Testing, (Wien : Springer)
19 Plato R. Vainikko G.(1990) On the regularization of projection methods for solving ill-posed problems. Numerische Mathematik **57** 63–79
20 Sabatier P. C (1987) Basic concepts and methods for inverse problems. in Tomography and inverse problems. ed P. C Sabatier (Bristol : Hilger) pp 471–667
21 Snieder R. (1991) An extension of Backus-Gilbert theory to nonlinear problems. Inverse Problems **7** 409–433

Reconstruction of an Impenetrable Obstacle Immersed in a Shallow Water Acoustic Waveguide

C. Rozier[1], D. Lesselier[1], T. S. Angell[2] and R. E. Kleinman[2]

[1] Laboratoire des Signaux et Systèmes, CNRS-Supélec, Plateau de Moulon, 91192 Gif-sur-Yvette, France
[2] Center for the Mathematics of Waves, University of Delaware, Newark, DE 19716, USA

Abstract. We investigate the reconstruction of the shape of a sound-soft cylindrical obstacle of smooth cross-section in a planar acoustic waveguide. This obstacle is located in the farfield of a single time-harmonic line source operating at one given frequency. Scattered pressure fields are observed on two arrays of hydrophones, one on each side of the obstacle. Using a complete family approach, the scattered field is represented as a finite sum of Green's functions whose source locations evolve with the retrieved contour. The inversion is cast as a penalized optimization problem where the unknown contour is retrieved by iterative minimization of a two-term functional. The first term measures the discrepancy between the data and the field scattered by a given obstacle, the second term measures the error in satisfying the boundary condition on its contour. After a short description of the mathematical formulation and of the needed numerical machinery, illustrative results are shown for convex and concave obstacles, low and high frequencies (few and many modes are propagated), vertical and horizontal arrays, exact and noisy data observed in the nearfield or in the farfield.

Introduction

We investigate the reconstruction of the shape of a closed, cylindrical obstacle of smooth cross-sectional contour with a known boundary condition (here, Dirichlet) which is placed in a planar acoustic waveguide. This is a model for the reconstruction of an impenetrable target in a shallow water configuration which consists of a lossless homogeneous water layer with a flat pressure-release interface with air, and a flat sound-hard sea bottom. This obstacle is located in the farfield of a single time-harmonic line source operating at one given frequency. Scattered pressure fields are observed on two receiver arrays (either vertical or horizontal), one on each side of the obstacle, either close to or far from it in terms of the wavelength in the ambient medium (water).

This inverse scattering problem is nonlinear and strongly ill-posed and, with respect to its classical free-space counterpart, it is further complicated not only by the propagation of only finitely many modes in the waveguide, the others being evanescent and not effecting the far-field pattern of the obstacle, but also by the availability of data at one single frequency which, futhermore, are aspect-limited when receivers view a limited part of the obstacle.

Generalizing the approach of [1], [2] devoted to free-space configurations, the inversion is cast as a penalized optimization problem where a contour belonging to an admissible class is constructed by iterative minimization of a two-term functional in which one term measures the discrepancy between the data and the field scattered by a given obstacle, while the other measures the error in satisfying the boundary condition on the unknown contour of the scattering obstacle. An exact contour integral formulation of the wavefield obtained by application of the Green's theorem, is employed, its discrete counterpart, derived by the Nyström method [3], providing us with synthetic data as needed. The inversion problem itself is analyzed by introducing a complete family of radiating solutions of the 2-D Helmholtz wave equation in the waveguide (they are Green's functions of the waveguide). Their line sources are initially distributed on a closed curve which is known to lie inside the obstacle. This curve is kept parallel to, and at close distance from, the contour constructed in the course of the algorithm.

In practice, a trigonometric function expansion describes the contour in polar coordinates. A finite weighted sum of Green's functions whose source locations are, as indicated before, evolving with the retrieved contour, represents the scattered field. Each Green's function involved is calculated by a summation of modes or, when the distance between source and observation point is a fraction of the wavelength, by means of a hybrid ray-mode representation [4]. Unknown coefficients of both expansions are found iteratively by a Levenberg-Marquardt solution algorithm.

The presentation is as follows. First we introduce the boundary integral formulation of the wavefield. Second, we review the complete family approach, including considerations of uniqueness of the solution of the boundary value problem, and completeness of our family of solutions according to the recent derivation of [5]. Third, the rather complex numerical machinery needed to retrieve the unknown obstacle contour from a finite number of data samples is sketched. Fourth, illustrative numerical results are shown for convex and concave obstacles, low and high frequencies (few and many modes are propagated), vertical and horizontal arrays, exact and noisy data taken in the nearfield or in the farfield. Finally, pros and cons of the method are summarized, and interesting though computationally intensive generalizations of the method are pointed out.

1 Boundary integral formulation

Let us refer to figure 1. An impenetrable cylindrical obstacle of z axis and of closed cross-section D in the $r = (x, y)$ plane is placed in a shallow water waveguide. The water layer is a homogeneous linear isotropic lossless fluid layer of thickness H, density ρ_0 and sound speed c_0 enclosed between two flat interfaces, the free air/water surface S_f at $y = 0$ and the rigid bottom S_b at $y = H$. The obstacle cross-section D has a smooth contour Γ (at least C^2) and the Dirichlet boundary condition holds (the obstacle is sound-soft). The water layer exterior to D is denoted as D_e.

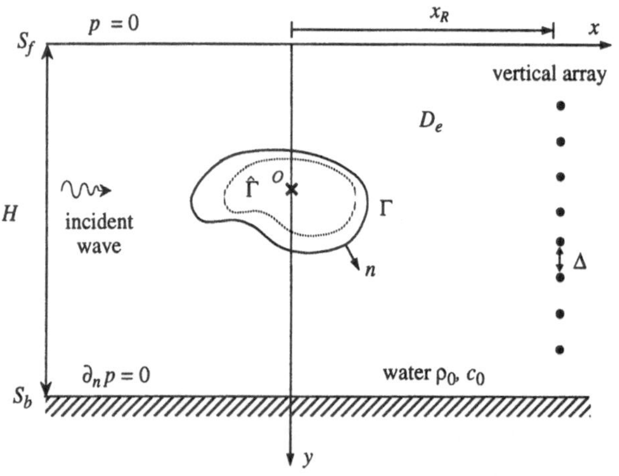

Fig. 1. The geometry of the waveguide.

A time-harmonic line source S parallel to z is placed in the farfield of D at $r_S = (x_S, y_S)$. Its operating circular frequency is ω and the $e^{-j\omega t}$ dependence of the field is dropped from now on. The resulting field is observed by means of two receiver arrays R, one on each side of the obstacle. These arrays are either vertical and then cover the whole water column, or horizontal and of finite length; in the first case both nearfield and farfield data are considered, and only farfield data in the second. One denotes by $r_R = (x_R, y_R)$ a given measurement point.

The corresponding boundary value problem reads

$$\left\{ \begin{aligned} (\Delta + k^2)\, p &= -\delta(r_s) && \text{in } D_e \\ p &= \ \ 0 && \text{on } \Gamma \\ p &= \ \ 0 && \text{on } S_f \\ \partial_n p &= \ \ 0 && \text{on } S_b \end{aligned} \right. \tag{1}$$

where p is the pressure field, where k is the (real) wavenumber in water and where the normal n is directed outside the obstacle domain, into D_e.

A radiation condition at infinity may be imposed [6], [7]. Here we simply recall that under mild geometric conditions solutions of the boundary value problem above are unique as is proved earlier by a Sturm-Liouville technique and the Green's theorem in [5].

Introducing the Green's function $G(r, r')$ of the waveguide we easily obtain a system of boundary integral equations [4]:

$$0 = \frac{1}{2}p(r) = p_0(r) - \int_\Gamma G(r, r')\,\partial_n p(r')dr', \quad r \in \Gamma \tag{2}$$

$$p(r) = p_0(r) - \int_\Gamma G(r, r')\,\partial_n p(r')dr', \quad r \in D_e \tag{3}$$

where $p_0(r)$ represents the field $G(r, r_S)$ (the *incident* field) existing in the absence of the obstacle, and where the total field $p(r)$ is equal to $p_0(r)$ + the *scattered* field $p_S(r)$.

As usual, we refer to equation (2) as the coupling equation and equation (3) as the observation equation. Solution of (2) for a given contour yields the normal derivative of the pressure field along the contour, and the pressure everywhere in the exterior D_e of the obstacle follows by straightforward integration of (3). Conversely, when the contour is unknown, both equations have to be satisfied, in some sense, from a partial knowledge of the pressure in the waveguide (the left-hand side of equation (3)).

We remark that equation (2) requires augmentation at frequencies corresponding to interior resonances. However the optimization problem which we pose below in §3 is still valid at these frequencies and moreover the numerical examples described did not suffer from such resonances.

The Green's function of the waveguide has three conventional representations [8]. It can be expanded into a ray series, or a sum of modes, or one can perform spectral integration along the k_x axis associated with the range x (since the poles lie on the real axis in the lossless case, one needs to deform the integration path into the complex k_x plane).

Closely following [4], we prefer to use a hybrid ray-mode expansion of $G(r, r')$ in order to obtain accurate numerical results at a moderate computational cost. We start from the normal modes expansion of the Green's function:

$$G(r, r') = \sum_{m=0}^{\infty} g_m(r, r') \tag{4}$$

where

$$g_m(r, r') = \frac{j}{k_m H} \sin(\beta y)\sin(\beta y')e^{jk_m|x-x'|}$$

and

$$k_m = \left\{k^2 - \left[(m + \frac{1}{2})\frac{\pi}{H}\right]^2\right\}^{\frac{1}{2}}, \quad \mathrm{Im}(k_m) \geq 0, \quad \beta = \sqrt{k^2 - k_m^2}.$$

and from the ray representation which is but the sum of the source images reflected by the impenetrable walls of the waveguide:

$$G(r, r') = \sum_{n=0}^{\infty}\left(\sum_{p=1}^{4} g_{np}(r, r')\right) \tag{5}$$

where

$$g_{np}(r, r') = -\frac{j}{4}(-1)^{n+p} \mathrm{H}_0^{(1)}(kr_{np})$$

$\mathrm{H}_0^{(1)}$ being the zero-order Hankel function of the first kind and r_{np} the distance between the observation point and the image source point resulting from the pth-type combination of n successive reflections on the walls.

In practice we use a finite number *NM* of guided modes when $| x - x' |$ is large enough (i.e., larger than a prescribed distance d). And otherwise we employ a finite series of *NR* rays, plus a remainder representation of the Green's function obtained by deformation of the integration path in the complex k_x-plane from the real axis onto the steepest-descent path [4], [9].

This calculation of the Green's function is performed both in the inversion scheme and in order to calculate synthetic data. The latter calculation is as follows. Equations (2)-(3) are discretized using the Nyström method [3], [10], which yields a complex-valued matrix system whose solution is the pressure field sought. The use of the Nyström method with trigonometric polynomials as the approximating functions has the advantage of taking into account the logarithmically singular behavior of (2) and ensuring exponential convergence with respect to the number of nodes which describe the contour Γ [3].

2 The complete family approach

The reconstruction method is based on the use of a complete family of solutions of the Helmholtz equation that satisfy the boundary conditions on the walls of the waveguide. It can be shown under mild geometric restrictions [5] that a solution p of the problem (1) together with the additional radiation conditions:

$\exists c \in \mathbb{R}^+$ such that

$$\lim_{R\to\infty} \Im\mathrm{m} \left\{ \int_{|x|=R} \bar{p} \frac{\partial p}{\partial n} \, dy \right\} = c \;\; > 0; \tag{6}$$

and

$$\lim_{R\to\infty} \int_{|x|=R} | p |^2 \, dy \;\; < \infty; \tag{7}$$

is unique and has a unique modal decomposition.

Moreover, if r_m constitutes a countably dense set of points on a curve $\hat{\Gamma}$ (cf. figure 1) completely contained in D, then the set of Green's functions with these source points is complete in $L^2(\Gamma)$. This complete family may represent the solution of (1) in the same way as was shown in [5] in the Dirichlet case.

This means, in particular, that any solution of the boundary value problem (1) (+ radiation conditions) can be approximated as closely as desired by a finite linear combination of these Green's functions.

In so doing, we get a new representation of the field scattered by an obstacle in the waveguide. And this representation is used to solve the shape inversion problem at hand (the retrieval of the contour Γ), no *inverse crime* being committed since the simulated data are obtained by solving the set of boundary integral equations (2-3).

3 The penalized inversion

As already indicated, the reconstruction method is based on the complete family approach, e.g., [11], and numerically speaking is more directly inspired from [2]. First the contour Γ is assumed to have a trigonometric expansion:

$$f(\theta) = a_0 + \sum_{n=1}^{N} a_n \cos(n\theta) + \sum_{n=1}^{N-1} a_{N+n} \sin(n\theta) \tag{8}$$

Then the corresponding scattered field, henceforth denoted as $p_S(f,r)$, is taken as a weighted sum of M Green's functions whose source points $r_m^f, m = 1, \cdots, M$ are distributed on a curve $\hat{\Gamma}$ located inside the obstacle cross-section D and set to evolve in parallel with the retrieved boundary Γ given by (8) by enforcing $\mid r_m^f(\theta) \mid = \alpha f(\theta)$. We have

$$p_S(f,r) = \sum_{m=1}^{M} c_m G(r, r_m^f) \tag{9}$$

The choice of the multiplicative factor α (which is less than 1) is somewhat arbitrary. When α is too close to 1, contours $\hat{\Gamma}$ (where the source points are lying) and Γ (where the observation points are lying) become very close to one another in terms of the wavelength in water, which may result in inaccurate or computationally costly calculations of the Green's functions. On the other hand a much smaller value of α in practice causes a premature end of the inversion procedure [12] in the sense that the cost functional (see below) to be minimized remains large.

The contour Γ should be such that the scattered field p_S fits the data on any given measurement line R (the observation equation is satisfied) and simultaneously such that the Dirichlet boundary condition $p_S + p_0 = 0$ holds on Γ (the coupling equation is satisfied).

To reach this goal we define two functionals l_1 and l_2 by:

$$l_1(f,R) = \frac{\int_R \|p_S(f,r_R) - p_S^{mes}(r_R)\|^2 dr_R}{\int_R \|p_S^{mes}(r_R)\|^2 dr_R} \tag{10}$$

which is the L_2 norm of the discrepancy between the data p_S^{mes} recorded along R and the corresponding scattered field p_S given by (9) with r taken as r_R, normalized with respect to the L_2 norm of the data; and

$$l_2(f,\Gamma) = \frac{\int_0^{2\pi} \|p_S(f,\theta) + p_0(f,\theta)\|^2 J_f(\theta) d\theta}{\int_0^{2\pi} \|p_0(f,\theta)\|^2 J_f(\theta) d\theta} \tag{11}$$

with

$$J_f(\theta) = \sqrt{f^2(\theta) + \left(\frac{df}{d\theta}\right)^2}$$

which is the L_2 norm of the discrepancy between the total field $p_S + p_0$ on the contour Γ at angle θ and its exact null value, normalized with respect to the norm of the incident field, where p_S is evaluated using (9) for f given by (8), J_f being the Jacobian of f.

Integrals in (10-11) are discretized by a trapezoidal rule, Q test points being prescribed on Γ. As for the number M of Green's functions, it is henceforth equated to Q so that source points r^f_m on $\hat{\Gamma}$ correspond point-to-point to the nodes on Γ fixed by angles θ_q. When several sets of data are recorded for the same obstacle by varying the measurement configuration, we take the cost functional to be the sum of each cost functional.

The problem is now to simultaneously determine the two sets of coefficients $\{a_n\}$ and $\{c_m\}$ which minimize the cost functional $L = l_1 + \sigma l_2$, σ being a penalty parameter favoring either the observation cost l_1 or the boundary cost l_2 in the reconstruction procedure.

This optimization problem is highly non-linear. In particular the coefficients $\{c_m\}$ depend on the $\{a_n\}$ via the points r^f_m. As for the support of the integral in (11), let us note that it should have been the obstacle contour Γ, causing further difficulty since Γ evidently depends upon the $\{a_n\}$. But, by transformation onto the unit circle, the integration contour (the unit circle) becomes independent of the unknown contour, the shape dependence being accounted for by the Jacobian of the transformation f.

Let us emphasize that the trigonometric representation (8) which is assumed for the obstacle contour constitutes strong *a priori* information about the obstacle, in addition to the already required smoothness. Indeed, the contour is constrained to be star-like with respect to the origin of the coordinate system (f,θ). This representation is convenient since it reduces the class of contours where the unknown one is sought, but it is not imposed by the underlying theory of the complete family inversion.

In practice, the coefficients $\{a_n\}$ and $\{c_m\}$ are calculated by means of a standard non-linear minimization routine which employs a Levenberg-Marquardt algorithm. The procedure is stopped when the cost functional is small enough, or when it reaches a plateau. As indicated in the above, multiple data obtained by varying source and/or receiver locations are treated by simply adding their respective cost functionals and minimizing the resulting sum.

4 Numerical results

Results are given for a hard-bottom waveguide of height $H = 100$ m and speed of the compressional waves in water $c_0 = 1500$ m/s. The operating frequency is either 30 Hz or 100 Hz, 4 or 13 modes being propagated without attenuation, respectively. The corresponding wavelength in the water is $\lambda =$ 50 m or 15 m. The boundary Γ of the obstacle (centered at 50 m depth and 0 m range) is either elliptic (horizontal semi-axis $a = 15$ m, vertical one $b = 7.5$ m) or it looks like a three-leaf clover ($f(\theta) = 10 - 3\sin(3\theta)$ m). Both perimeters are of the order of 70 m, and in the following we refer to the 10 m-radius circle as the "unit circle".

We consider a source S located on the left side of the obstacle at a large range $r_S(x_S = -10$ km, $y_S = 55$ m). Data consist in the resulting pressure field sampled by two vertical arrays R of length $L = H$ (41 ideal hydrophones equally spaced with separation $\Delta = 2.5$ m) or by two horizontal arrays of length $L = 500$ m (41 ideal hydrophones each $\Delta = 12.5$ m). The vertical arrays are located in the nearfield or the farfield of the obstacle at range $x_R = \pm 40$ m or ± 5 km. The horizontal ones are only located in the farfield, at depth $y_R = 25$ m from range -5 km to -4.5 km and 4.5 km to 5 km.

The synthetic data are calculated by directly solving the boundary integral equations (2-3) as specified in §2. Moreover in the direct solution we take at 30 Hz (resp. 100 Hz) 32 (resp. 50) equidistant nodes on Γ, which corresponds to a sampling step (arc length) less than one-tenth of the wavelength in water while in the inversion we only use 16 equiangle nodes on Γ.

The initial contour introduced in the iterative optimization is a circle of radius $a_0 = R_0$ (i.e. $\{a_n\} = 0, n \geq 1$) and we often take the "unit circle" ($R_0 = 10$ m) which is in some sense the simplest obstacle of size close to the size of the unknown obstacle. Γ is assumed to have a trigonometric expansion $f(\theta)$ given by (8) with $N = 4$ cosines and 3 sines. The number of nodes Q used to retrieve Γ is equated to the number of Green's functions $M = Q = 16$ to prevent too large a ratio between the number of unknowns and the number of data points. As for the $\{c_m\}$ they are initially set to zero (all equivalent sources are "turned off"). The parameter values are $\alpha = 0.75$ and $\sigma = 1$.

Typical data used in the inversion are shown in figure 2. Here the magnitude of the pressure field at 30 Hz recorded on the two vertical arrays and the two horizontal ones is displayed for both the elliptic and the clover contours.

Contours retrieved from data observed on the vertical arrays placed either in the nearfield or in the farfield at 30 Hz are shown in figure 3 while the influence of the type of arrays used (vertical or horizontal) is exemplified in figure 4 (in the case of farfield data only) at the same 30 Hz frequency.

Reconstructions shown in figure 3 appear to be of similar quality (this is particularly true with the clover contour) in both the nearfield and the farfield measurement configuration, even though evanescent modes are filtered out with range and only very few modes are propagated at this low frequency.

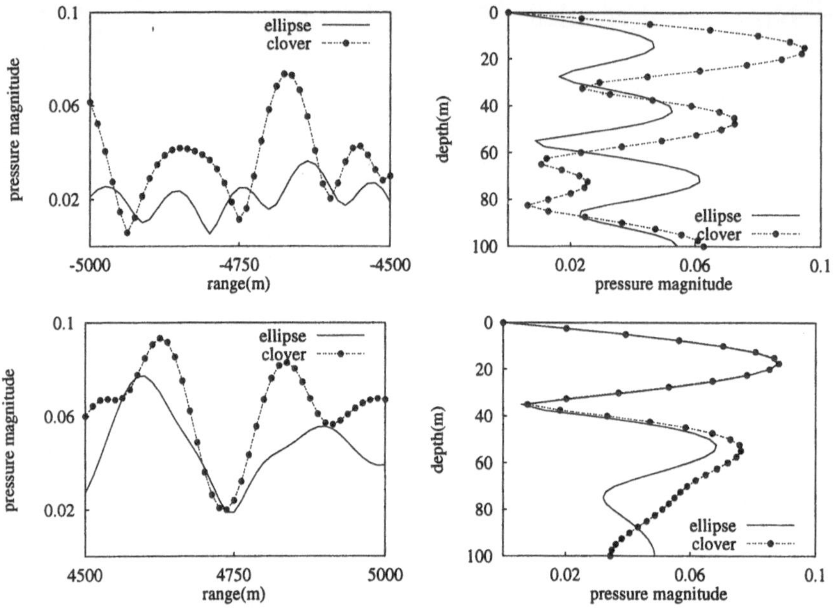

Fig. 2. Far-fields observed along the horizontal arrays (*left*) and the vertical arrays (*right*) at 30 Hz for the elliptic and clover contours. The arrays are on the left side (*top*) or on the right side (*bottom*) of the obstacle, the source is on the left side.

As for those obtained with horizontal arrays of finite length, they appear (see figure 4) only slightly less accurate than those obtained with vertical arrays even though windowing effects are expected with horizontal ones, while, in contrast, the coverage of the obstacle is complete with vertical ones (the waveguide walls are impenetrable and the arrays span the full water column).

Notice that with either horizontal arrays or vertical ones the measurement step is such that the field at the highest spatial frequency is suitably sampled as follows: (i) When measuring the field along depth at given range, this frequency corresponds to the 4-th mode ($m = 3$) and following (4) to $\beta = \frac{7\pi}{2H}$ and thus to a period equal to 9 m while the sampling step is $\Delta = 2.5$ m (about one-fourth of this period); (ii) when measuring the field along range at given depth this frequency corresponds to the 1-st mode ($m = 0$), or to $k_0 = \sqrt{k^2 - \frac{\pi^2}{4H^2}}$, thus to a period of 32 m while the sampling step is $\Delta = 12.5$ m (somewhat less than one-half of this period).

In figure 5 the results at the 100 Hz frequency are compared to those observed at 30 Hz. In both cases two vertical arrays are placed in the farfield.

At 30 Hz it is seen that the elliptic contour is not retrieved as well as the

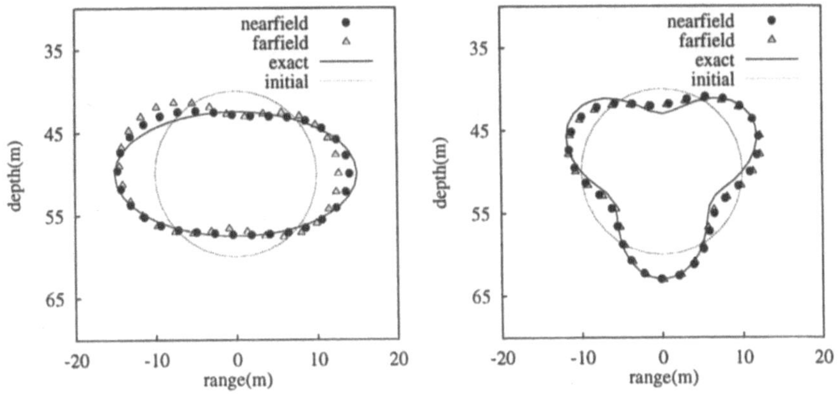

Fig. 3. Contours retrieved from exact nearfield or farfield data at 30 Hz (vertical arrays).

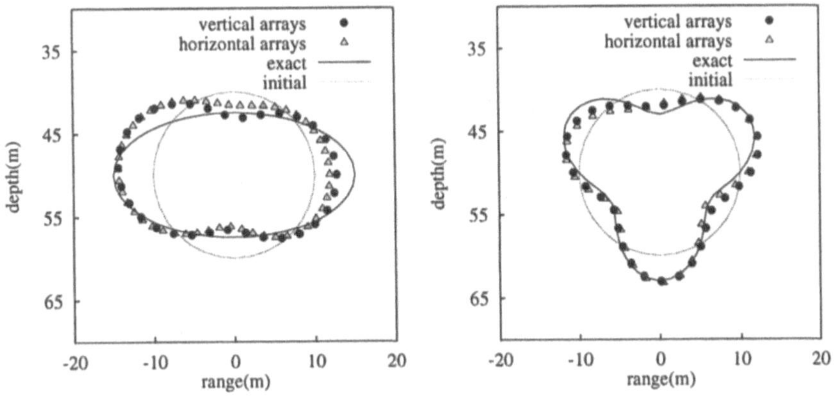

Fig. 4. Contours retrieved from exact farfield data at 30 Hz (horizontal or vertical arrays).

clover contour is, while the opposite is true at 100 Hz. Indeed, the elliptic contour cannot be represented by a finite trigonometric expansion whereas the clover contour is. So, more information appears needed if the elliptic contour is to be retrieved accurately, which here means more propagated modes (i.e., a higher frequency of operation) since we are in the farfield of the obstacle. Notice that (see figure 3) the elliptic contour is somewhat better retrieved when nearfield data are recorded since evanescent modes are now present. As for the clover contour, its concavity and its large perimeter (5 wavelengths at 100 Hz) tend to penalize its reconstruction at this high frequency.

Finally the robustness of the inversion method is considered when an uniform additive noise effecting both the real and the imaginary part of the observed pressure field is added. Typical results are shown figure 6 using two vertical arrays in the farfield and various signal-to-noise ratios *SNR*.

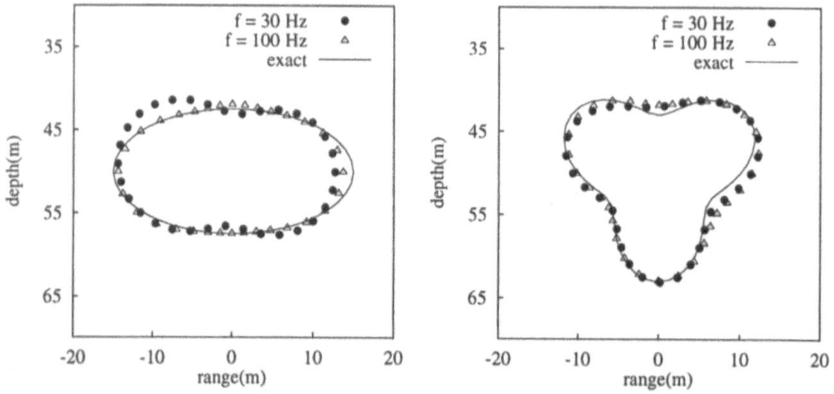

Fig. 5. Contours retrieved from exact farfield data at 30 Hz and at 100 Hz (vertical arrays).

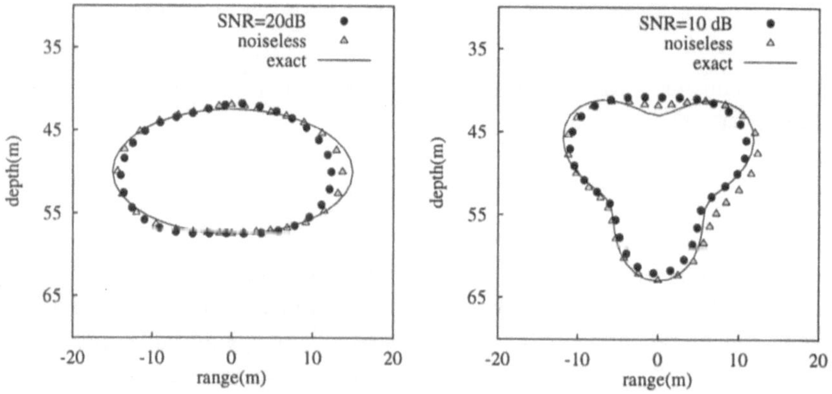

Fig. 6. Contours retrieved from noisy farfield data at 100 Hz (vertical arrays).

5 Conclusion

From the results presented here and elsewhere (abundant material is available in [13]), it appears that the complete family approach is able to retrieve with a fairly good success the contour of 2-D sound-soft convex or concave obstacles placed in a waveguide with impenetrable walls in a number of measurement configurations (near or farfield data, exact or noisy data, horizontal or vertical arrays) at a single frequency. The main drawback, in addition to the computational complexity of the method, seems to be the fact that the contour must be star-like with respect to a given point, which in particular means that one point inside the obstacle must be found by other means (for example, a backpropagation algorithm) or must be assumed beforehand.

Clearly some extensions of this method are straightforward – in addition to the introduction of multifrequency data. The introduction of a penetrable sea bottom (or a water layer whose speed of sound varies with depth) is possible, though that the Green's functions would have to be calculated in the Fourier domain, at a much higher cost, modal expansions becoming approximate. Less straightforward though still feasible in the same theoretical framework is the case of 3-D obstacles (let us refer to a somewhat similar analysis [14]). But more complicated geometries (an obstacle lying on the sea bottom, or partially buried within) require further theoretical analyses before any numerical exploitation (for example, it is required in [5] that the contour be completely confined in the waveguide).

Acknowledgments

This work was supported under a CNRS-NSF International Cooperative Grant ♯INT-9415493, a NATO Collaborative Research Grant ♯CRG-940999, and AFOSR Grant ♯F49620-96-1-0039.

References

[1] Tal J. and Leviatan Y., "Inverse scattering analysis for perfectly conducting cylinders using a multifilament current model," *Inverse Prob.*, **6**(6):1065–1074, 1990.

[2] Angell T. S., Jiang J., and Kleinman R. E., "A distributed source method for inverse acoustic scattering," to appear in *Inverse Problems*, 1996.

[3] Kress R., "Numerical solution of boundary integral equations in the time-harmonic electromagnetic scattering," *Electromagn.*, **10**(1):1–20, 1990.

[4] Lu I.-T., "Analysis of acoustic wave scattering by scatterers in layered media using the hybrid ray-mode (boundary integral equation) method," *J. Acoust. Soc. Am.*, **86**(3):1136–1142, 1989.

[5] Angell T. S., Kleinman R. E., Rozier C., and Lesselier D., *Uniqueness and Complete Families for an Acoustic Waveguide Problem.* Technical Report **96-4**, Cen-

ter for the Mathematics of Waves, University of Delaware, Newark, 1996. Submitted to *Wave Motion*.

[6] Xu Y., "The propagating solutions and far-field patterns for acoustic harmonic waves in a finite depth ocean," *Applicable Analysis*, **35**:129–151, 1990.

[7] Nosich A. I., "Radiation conditions, limiting absorption principle, and general relations in open waveguide scattering," *J. Electromagn. Waves Applic.*, **8**(3):329–353, 1994.

[8] Buckingham M. J., "Ocean-acoustics propagation models," *J. Acoustique*, **5**(3):223–287, 1992.

[9] Felsen L. B. and Kamel A. H., "Hybrid ray-mode formulation of parallel plane waveguide Green's functions," *IEEE Trans. Antennas Propagat.*, **29**(4):637–649, 1981.

[10] Nazarchuk Z. T., *Singular Integral Equations in Diffraction Theory.* Nat. Academ. Ukraine, Lviv, 1994.

[11] Angell T. S., Kleinman R. E., and Roach G. F., "An inverse transmission problem for the Helmholtz equation," *Inverse Prob.*, **3**(2):149–180, 1987.

[12] Rozier C., Lesselier D., Angell T., and Kleinman R., *Shape Retrieval of an Obstacle Immersed in Shallow Water from Single-Frequency FarFields Using a Complete Family Method.* Technical Report **96**, Center for the Mathematics of Waves, University of Delaware, Newark, 1996. Submitted to *Inverse Problems.*

[13] Rozier C., *Caractérisation d'objets cylindriques placés dans un guide d'ondes acoustique modélisant une situation de petits fonds marins.* Thèse de l'Université Paris-VII, Paris, France, 1996.

[14] Gilbert R. P. and Xu Y., *Generalized Herglotz Functions and Inverse Scattering Problem in a Finite Depth Ocean.* Technical Report **89-21**, Center for the Mathematics of Waves, University of Delaware, Newark, 1989.

Location and Reconstruction of Objects Using a Modified Gradient Approach

Ralph E. Kleinman[1], Peter M. van den Berg[2], Bernard Duchêne[3], and Dominique Lesselier[3]

[1] Center for the Mathematics of Waves, Department of Mathematical Sciences University of Delaware, Newark, Delaware 19716, USA
[2] Centre for Technical Geoscience, Laboratory of Electromagnetic Research Delft University of Technology, P.O. Box 5031, 2600 GA Delft, The Netherlands
[3] Laboratoire des Signaux et Systèmes, C.N.R.S.–SUPELEC Plateau de Moulon, 91192 Gif–sur–Yvette Cedex, France

Abstract. A large class of inverse scattering problems involves the attempt to determine the shape, location, and constitutive parameters of a bounded object or objects from a knowledge of the field scattered by the object(s) when illuminated or ensonified by a known time harmonic incident field. The fields may be electromagnetic or acoustic and while the field equations are different in each case, the inverse problem may be cast in a general framework which accommodates both phenomena and in fact may be extended to include time-harmonic inverse scattering of elastic waves. This class of problems has been attacked in a number of ways including Born-based methods [1], Newton–Kantorovich methods [2], diffraction tomography [3], and dual space methods [4]. Recently another method, a modified gradient technique has been developed [5] and used with good success in a variety of different cases. The present paper describes the essential features of the modified gradient approach and reports on recent experience in a number of specific realizations representing different physical situations and different amounts of a priori information about the scatterer.

1 Formulation

Let B denote the scattering object(s), a finite number of bounded, connected open sets in $\mathbb{R}^n$ with smooth (e.g. piecewise C^2) boundary, ∂B. Assume that B is irradiated by a number of known incident fields $u_j^{\text{inc}}(\mathbf{p})$, $j = 1, 2, \ldots, J$, where $\mathbf{p}$ is a position vector in $\mathbb{R}^n$. For each j the scattered field $u_j^{\text{S}}(\mathbf{p})$ is measured on S, a set of points (possibly an $n-1$ dimensional manifold) exterior to B ($S \cap B = \emptyset$). It should be emphasized that the field quantities may be scalar or vector quantities depending on the physical model under consideration. Denote the measured data by $f_j(\mathbf{p})$, which will be equal to $u_j^{\text{S}}(\mathbf{p})$ only in the absence of noise and measurement error. The total field induced in B by $u_j^{\text{inc}}(\mathbf{p})$ is denoted by $u_j(\mathbf{p})$, while the contrast (the difference between the constitutive parameters of B and those of the background $\mathbb{R}^n \backslash \bar{B}$) is denoted by $\chi(\mathbf{p})$ which is assumed to be a scalar valued function of position. In all the examples treated to date the contrast $\chi(\mathbf{p})$ is assumed to be independent of the incident fields $u_j^{\rm inc}(\mathbf{p})$. When $u_j^{\rm inc}$ is given for the same frequency the variation with j denotes different positions of the source or directions of the incident field, and the independence of the contrast with the respect to the incident field is evident. When the incident fields vary with frequency the contrast may vary as well. For example in Maxwell media, the contrast takes the form $(\epsilon_B + i\sigma_B/\omega)/\epsilon_0 - 1$, where ϵ_B and σ_B are the permittivity and conductivity of B, which clearly depends on the operating frequency ω. However the problems considered thus far for frequency-varying incident fields concern physical situations where the constitutive parameters are known in B, but the shape and location of B are unknown. In such cases the function $\chi(\mathbf{p})$ which is sought is the characteristic function of the domain B. For the class of problems considered here, the field $u_j(\mathbf{p})$ in B satisfies an equation (the object equation) of the form

$$u_j - G_{B_j}\chi u_j = u_j^{\rm inc} \ , \tag{1}$$

while the scattered field on S satisfies an equation (the data equation) of the form

$$u_j^{\rm S} = G_{S_j}\chi u_j \tag{2}$$

where G_{B_j} and G_{S_j} are linear operators,

$$G_{B_j} : L_2(B) \to L_2(B)\,,\ G_{S_j} : L_2(B) \to L_2(S)\ . \tag{3}$$

More generally G_{B_j} may map Sobolev spaces $H_{\rm s}(B)$ into $H_{\rm t}(B)$, and G_{S_j} may map $H_{\rm s}(B)$ into $L_2(S)$. We always use $L_2(S)$ as the range of G_{S_j}, even though the map may be much smoother, in order to be consistent with the measured data which is only assumed to be square integrable. In the realistic case that data is measured only at discrete points, we assume an L_2 interpolation. In many of the inverse problems under consideration, the position and shape of the scattering obstacle B is unknown, hence we assume that a priori information is available that restricts B to lie in a known larger bounded domain D and that $D \cap S = \emptyset$. It is this known test domain D that is then used for the object and data equations

$$\begin{gathered} u_j - G_{D_j}\chi u_j = u_j^{\rm inc}\,,\ f_j = G_{S_j}\chi u_j \\ G_{D_j} : L_2(D) \to L_2(D)\,,\ G_{S_j} : L_2(D) \to L_2(S)\ , \end{gathered} \tag{4}$$

where the contrast χ vanishes in $D\backslash\bar{B}$. This ensures that if χ is found, then not only is the contrast known in B but supp $\chi = \bar{B}$. The size of D directly depends on a priori information on B; the more that is known about B, the smaller the difference between B and D can be made. Since D will eventually be discretized, there is a clear advantage in choosing D as small as possible.

The essential features of the modified gradient method may now be given without further specification of G_{S_j} and G_{D_j}. Of course, precise definitions are needed to implement the method, and these are given in each example presented in what follows. For any $\xi \in L_\infty(D)$ and $w_j \in L_2(D)$ we may define the residual errors in the object and data equations when χ, u_j are replaced by ξ, z_j as

$$r_j = u_j^{\mathrm{inc}} - z_j + G_{D_j}\, \xi\, z_j\,,\ \rho_j = f_j - G_{S_j}\, \xi\, z_j\ . \tag{5}$$

The inverse problem is now formulated as follows:
for known incident fields $u^{\mathrm{inc}}(\mathbf{p})$, measured data f_j, test domain D and normalization constants w_{D_j} and w_{S_j}, find $\chi \in U_\infty$ and $u_j \in U_2$ to minimize the error functional

$$F(\xi, \mathbf{z}) := \sum_{j=1}^{J} \{w_{D_j} \|r_j\|^2_{L_2(D)} + w_{S_j}\ \|\rho_j\|^2_{L_2(S)}\} \tag{6}$$

where $\mathbf{z} = (z_1, \ldots, z_J)$.

The set of admissible contrasts U_∞ is a subset of $L_\infty(D)$ which incorporates any available a priori constraints. Similarly U_2 is a subset of $L_2(D)$. In the absence of any additional knowledge of χ and u_j, $U_\infty = L_\infty(D)$ and $U_2 = L_2(D)$. The minimization is carried out by constructing a sequence of approximations $\{\chi_m, u_{jm}\}$ which reduce the error functional at each step. The sequence is constructed iteratively. First a set of starting functions χ_0 and u_{j_0} are selected. This is not a trivial matter since the ultimate convergence of the sequence to the desired solution will depend on a reasonable choice of starting functions. Some of the choices that have been used are described below. Once the starting functions are defined, a sequence is generated according to the following updating scheme

$$\chi_m(\mathbf{p}) = \chi_{m-1}(\mathbf{p}) + \beta_m\, d_m(\mathbf{p})\,,\ u_{jm}(\mathbf{p}) = u_{jm-1}(\mathbf{p}) + \alpha_{jm}\, v_{jm}(\mathbf{p}) \tag{7}$$

where $d_m(\mathbf{p})$ and $v_{jm}(\mathbf{p})$ are updating directions and α_{jm}, β_m are constants. The choice of updating directions $d_m(\mathbf{p})$ and $v_{jm}(\mathbf{p})$ is given below. Once this choice is made the constants β_m and α_{jm} are found by minimizing the error functional

$$F(\chi_{m-1} + \beta_m d_m,\ u_{1m-1} + \alpha_{1m} v_{1m}, \ldots, u_{Jm-1} + \alpha_{Jm} v_{Jm})\ . \tag{8}$$

In many of the examples discussed below it is assumed that $\alpha_{jm} = \alpha_m$ for every j which reduces this algebraic optimization problem to one of finding the two, possibly complex, constants α_m, β_m.

2 Update Directions

For the update directions $d_m(\mathbf{p})$ we have used either gradient directions

$$d_m(\mathbf{p}) = g_m^{\mathrm{d}}(\mathbf{p}) := \frac{\partial}{\partial \chi} F\Big|_{\chi_{m-1},\mathbf{u}_{m-1}} \tag{9}$$

where $\frac{\partial F}{\partial \chi}\Big|_{\chi_{m-1},\mathbf{u}_{m-1}}$ denotes the gradient of F with respect to changes in the contrast, holding the field quantities constant, evaluated at χ_{m-1} and $\mathbf{u}_{m-1}$, or Polak–Ribière conjugate-gradient directions

$$d_m(\mathbf{p}) = g_m^{\mathrm{d}} + \gamma_m^{\mathrm{d}} d_{m-1} \,, \; \gamma_m^{\mathrm{d}} = \frac{< g_m^{\mathrm{d}}, g_m^{\mathrm{d}} - g_{m-1}^{\mathrm{d}} >_D}{\|g_{m-1}^{\mathrm{d}}\|_D^2} \,. \tag{10}$$

Explicitly

$$\frac{\partial F}{\partial \chi}\Big|_{\chi_{m-1},\mathbf{u}_{m-1}} = \sum_{j=1}^{J} \bar{u}_{jm-1} \left[w_{D_j} \, G_{D_j}^{\star} \, r_{jm-1} - w_{S_j} \, G_{S_j}^{\star} \, \rho_{jm-1} \right] \tag{11}$$

where $G_{D_j}^{\star}$ and $G_{S_j}^{\star}$ denote the adjoint operators mapping $L_2(D)$ into itself, and $L_2(S)$ in $L_2(D)$, respectively.

The update directions for the fields have been chosen in the simplest case as the residual field error at the previous step

$$v_{jm} = r_{jm-1} \;, \tag{12}$$

or more effectively as the gradient

$$v_{jm} = g_{jm}^{\mathrm{v}} := \frac{\partial F}{\partial u_j}\Big|_{\chi_{m-1},\mathbf{u}_{m-1}} \;, \tag{13}$$

or Polak–Ribière conjugate gradient directions

$$v_{jm} = g_{jm}^{\mathrm{v}} + \gamma_{jm}^{\mathrm{v}} \, v_{jm-1} \,, \; \gamma_{jm}^{\mathrm{v}} = \frac{< g_{jm}^{\mathrm{v}}, \, g_{jm}^{\mathrm{v}} - g_{jm-1}^{\mathrm{v}} >_D}{\|g_{jm-1}^{\mathrm{v}}\|_D^2} \;. \tag{14}$$

For each j the direction g_{jm}^{v} is the gradient of F with respect to u_j, holding the contrast and other field quantities constant, evaluated at $\chi_{m-1}, \mathbf{u}_{m-1}$. Explicitly

$$\frac{\partial F}{\partial u_j}\Big|_{\chi_{m-1},\mathbf{u}_{m-1}} = - w_{D_j} (r_{jm-1} - \bar{\chi}_{m-1} \, G_{D_j}^{\star} \, r_{jm-1}) - w_{S_j} \, \bar{\chi}_{m-1} \, G_{S_j}^{\star} \, \rho_{jm-1} \;. \tag{15}$$

3 Initialization

A number of different choices of starting functions have been employed. The simplest is the Born approximation wherein we choose

$$\chi_0 = 0\,, \; u_{j0} = u_j^{\mathrm{inc}} \; . \tag{16}$$

Better results were obtained with a slightly more sophisticated choice, a "best constant" value for χ_0 found by running the algorithm using the Born starting values but choosing the update directions in the contrast to be constant, $d_m(\mathbf{p}) = 1$, so that the updates in contrast are always constant.

An even more elaborate, but generally more effective starting guess was obtained by "back propagating" the measured data. First an initial source distribution in D is found by defining $\Phi_j = \gamma G^\star_{S_j} f_j$ where γ is chosen to minimize $\sum_{j=1}^J \|\rho_j\|$ and is found explicitly to be

$$\gamma = \frac{\sum\limits_{j=1}^{J} < f_j, G_{S_j} G^\star_{S_j} f_j >_S}{\sum\limits_{j=1}^{J} \|G_{S_j} G^\star_{S_j} f_j\|_S^2} \; . \tag{17}$$

Once Φ_j is found we define

$$u_{j0} = G_{D_j} \, \Phi_j + u_j^{\mathrm{inc}} \; , \tag{18}$$

then by equating

$$\chi_0 \, u_{j0} = \Phi_j \tag{19}$$

we may estimate χ_0 as, for example,

$$\chi_0 = \frac{\sum\limits_{j=1}^{J} \Phi_j \, \bar{u}_{j0}}{\sum\limits_{j=1}^{J} |u_{j0}|^2} \; . \tag{20}$$

4 Specific examples

A number of tests of this general method have been made with and without additional a priori information. They fall roughly into two classes, single-frequency measurements with large spatial diversity in source and receiver location and multifrequency measurements over spatially limited source and receiver locations. In addition different kinds of a priori information have been incorporated into the algorithm.

4.1 2*D* Acoustics - TM Electromagnetics

In this example the algorithm had the following explicit realization: S was a circle of radius 9λ, D was a $3\lambda \times 3\lambda$ sided square centered in S, B consisted of two distinct homogeneous square cylinders of diameter $3\lambda/4$ with $3\lambda/4$ separation. λ is the wavelength exterior to D and the contrast $\chi(\mathbf{p}) = 0.8$. Physically χ is either $[(c^2/c_D^2(\mathbf{p}))-1]$ in acoustics or $[(\tilde{\epsilon}_D(\mathbf{p})/\epsilon)-1]$ in electromagnetics, where c and c_D are the acoustic wave speeds exterior and interior to D respectively, while ϵ and $\tilde{\epsilon}_D$ are the exterior and interior complex permitivities. The operators are:

$$G_\Omega\, \chi\, u_j := \frac{ik^2}{4} \int_D H_0^{(1)}(k|\mathbf{p}-\mathbf{q}|)\, \chi(\mathbf{q})\, u_j(\mathbf{q})\, d\mathbf{q},\ \mathbf{p} \in \Omega,\ \Omega = S \text{ or } D\ , \tag{21}$$

where k is the wave number in the background, and $H_0^{(1)}$ is the zero order first kind Hankel function. The weighting constants were chosen to be the same for all j,

$$w_{D_1} = w_{D_2} = \cdots = w_{D_J} = \{\sum_{j=1}^{J} \|u_j^{inc}\|_D^2\}^{-1}$$

$$w_{S_1} = w_{S_2} = \cdots = w_{S_J} = \{\sum_{j=1}^{J} \|f_j\|_S^2\}^{-1} \tag{22}$$

and the updating constants α_{jm} were taken to be the same for every j, $\alpha_{jm} = \alpha_m$. The starting values were chosen as the "best constant" value for χ and the fields associated with this value, as described previously. Synthetic data was produced by solving the forward problem using a Galerkin method. The discretized version of the algorithm was obtained by selecting 29 equally spaced points on S, each of which served successively as a line source while all points served as receivers. The results of the iteration, both with and without noise, are shown in Fig. 1. More details may be found in [5, 6].

4.2 TE Electromagnetics

In this example we use exactly the same configuration as in **4.1**, however the different polarization implies new definitions of the operators in which the dependence on χ is no longer linear:

$$G_\Omega \chi u_j := \frac{i}{4} \int_D \frac{\chi(\mathbf{q})}{1+\chi(\mathbf{q})} \nabla_q H_0^{(1)}(k|\mathbf{p}-\mathbf{q}|) \cdot \nabla_q u_j(\mathbf{q})\, d\mathbf{q},\ \mathbf{p} \in \Omega,\ \Omega = S \text{ or } D. \tag{23}$$

To restore linearity we define $M := \chi/(1+\chi)$ and update M rather than χ. The update directions are as before, using the operator definitions given above (and their adjoints) with M replacing χ. All other definitions are as in **4.1**. Results using this algorithm are shown in Fig. 2. Additional detail may be found in [6, 7].

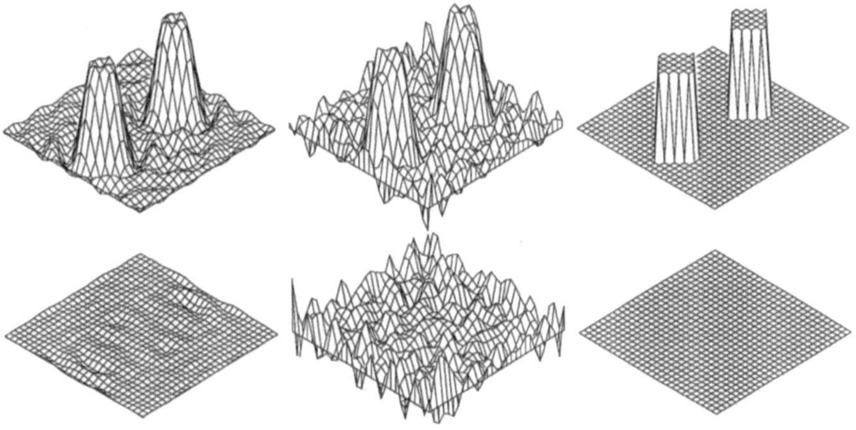

Fig. 1. Real (upper row) and imaginary (lower row) part of the contrast reconstructed after 64 iterations with the modified gradient algorithm from noiseless (left column) and noisy (middle column: 10% noise level) data compared to the exact profile (right column) - the TM case.

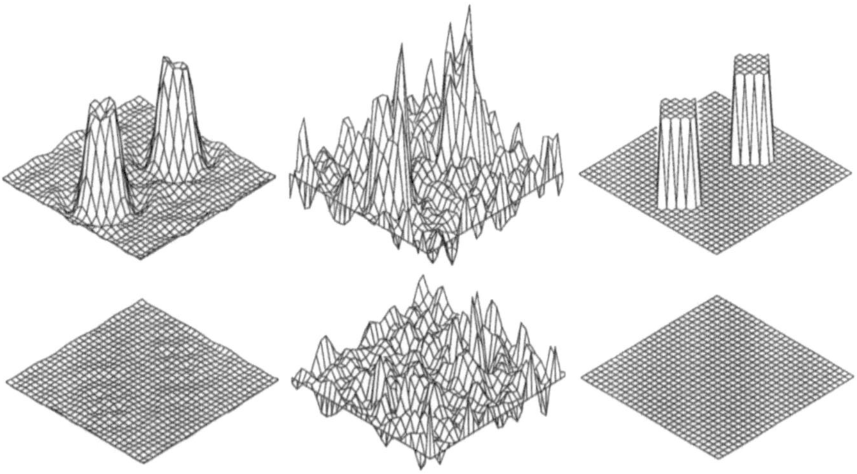

Fig. 2. Real (upper row) and imaginary (lower row) part of the contrast reconstructed after 64 iterations with the modified gradient algorithm from noiseless (left column) and noisy (middle column: 10% noise level) data compared to the exact profile (right column) - the TE case.

4.3 Total Variation

The modified gradient algorithm as given earlier has been shown to be effective and stable with respect to noise despite the absence of regularization terms usually essential in ill-posed problems. However the addition of regularizers can considerably enhance the quality of the reconstructions. There is in fact a wide choice of regularizing constraints which have been used with good effect [8, 9]. One such constraint is total variation which has been used by [10, 11, 12] and applied to the modified gradient algorithm in [13]. The essential feature is the definition of a new cost functional:

$$F_{\mathrm{TV}} = F + \omega_{\mathrm{TV}} \int_D \sqrt{|\nabla \chi|^2 + \delta^2}\, d\mathbf{q} \tag{24}$$

where F is the error functional defined previously. At the present time the penalty parameter ω_{TV} and the small parameter δ, which restores differentiability of the total variation, have been chosen only through numerical experimentation. Even with the new cost functional the update directions for the field remain unchanged whereas the gradient direction for the contrast is altered by replacing g_m^{d} defined previously by

$$g_m^{\mathrm{d}} - \omega_{\mathrm{TV}}\, \nabla \cdot \left(\frac{\nabla \chi_{m-1}}{\sqrt{|\nabla \chi_{m-1}|^2 + \delta^2}} \right) . \tag{25}$$

The positive effect of the addition of the total variation penalty term is seen in the reconstruction of the two cylinders considered in **4.1**, see Fig. 3.

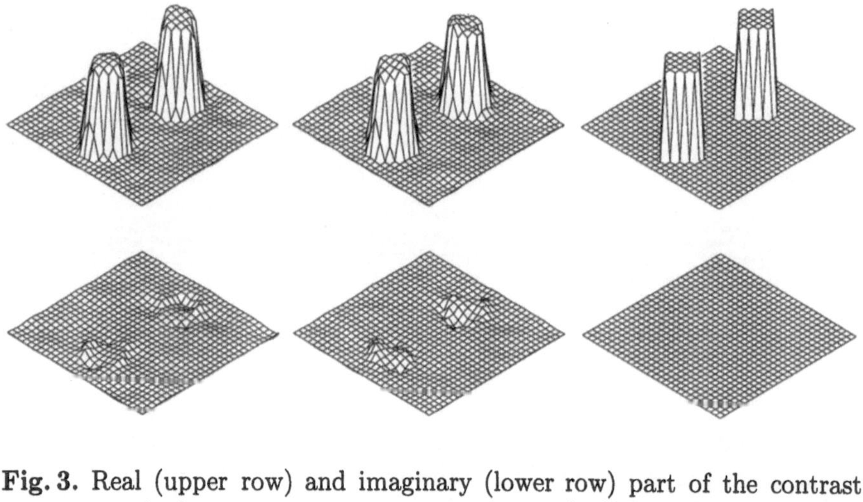

Fig. 3. Real (upper row) and imaginary (lower row) part of the contrast reconstructed after 64 iterations from noisy data (10% noise level) with the total variation constrained modified gradient algorithm in the TM (left) and TE (middle) cases compared to the exact profile (right).

4.4 Perfect Conductivity

When a priori information is available and can be incorporated directly into the algorithm, its performance can be considerably enhanced. For example in [14] it was found that the algorithm was ineffective if the contrast was too high. An upper limit of reconstructibility was found to be $kd\chi = 6\pi$, where d was the diameter of the test domain. However if it is known a priori that the contrast is large positive imaginary (e.g. a metallic conductor), then this information may be incorporated by replacing χ by $\mathrm{i}\zeta^2$, ζ real. The algorithm is slightly altered since ζ rather than χ is updated. The most significant effect of this change is that the gradient of F with respect to ζ, evaluated at the $(m-1)^{\mathrm{st}}$ step is now

$$g_m = -2\,\zeta_{m-1}\,\Im[g_m^{\mathrm{d}}] \tag{26}$$

where g_m^{d} was defined previously. Observe that the gradient vanishes if $\zeta_{m-1} = 0$, in which case the contrast remains zero. Hence, zero may not be used as a starting value. An illustration is provided in the case of TM electromagnetics as in **4.1** with the following changes: S is now a circle of sufficiently large radius so that the far field approximation may be used; B is taken to be a circle with radius a such that $ka = \pi$ centered asymmetrically with respect to D. The operators are the same for every j. They read:

$$G_D\,\chi\,u_j = -\frac{k^2}{4}\int_D \zeta^2(\mathbf{q})\,u_j(\mathbf{q})\,H_0^{(1)}(k|\mathbf{p}-\mathbf{q}|)\,d\mathbf{q},\ \mathbf{p}\in D \tag{27}$$

$$G_S\,\chi\,u_j = i\int_D e^{-ik\hat{\mathbf{p}}\cdot\mathbf{q}}\,\zeta^2(\mathbf{q})\,u_j(\mathbf{q})\,d\mathbf{q},\ \hat{\mathbf{p}}\in S \tag{28}$$

where $\hat{\mathbf{p}}$ is the unit vector in the direction of observation. The initial values in this case were obtained by back propagation as described earlier. The data were obtained from the series solution of the perfectly conducting cylinder problem and the discretized version of the algorithm had D subdivided into 31×31 subsquares and 30 incident plane waves equally spaced on the unit circle with these 30 directions also serving as receiver directions for each incident wave. The results of the algorithm are shown in Fig. 4 where it is seen that the shape and location are well reproduced. More detail is found in [15].

4.5 Spatially Limited Data

In this and the following examples it is assumed that the contrast in the object is known but the location and shape is unknown so that χ represents the characteristic function of the scatterer which, because it is non-negative, is replaced by ζ^2, ζ real. In this first example of this class of spatially limited data it is assumed that the scatterer is a void in a homogeneous lossy medium (concrete) which is illuminated by a plane wave using three different frequencies (7, 10, and 13 GHz) and that the scattered field is measured on a line

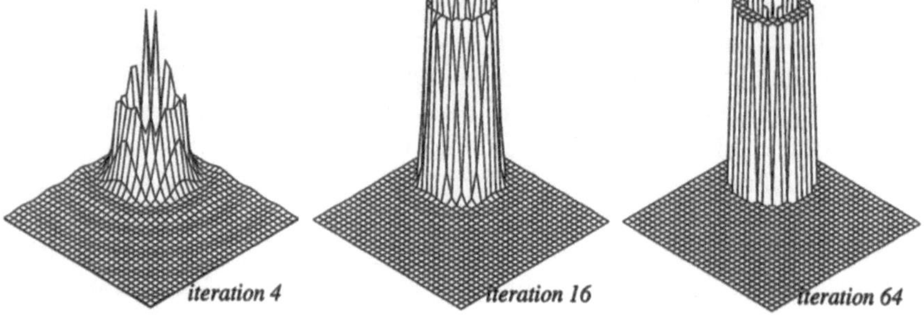

Fig. 4. Reconstruction of an impenetrable circular cylinder with perfect conductivity a priori information in the modified gradient algorithm.

perpendicular to the direction of incidence. Thus S is a line segment 32 cm long, 4.5 cm distant from the center of B, and 64 receivers are located on it; D is 1.13 cm sided square, B is a circle of radius 0.43 cm and the operators are

$$G_{\Omega_j}\chi u_j = \frac{i}{4}\int_D \zeta^2\,(k_{D_j}^2 - k_j^2)u_j(\mathbf{q})\;H_0^{(1)}(k_j|\mathbf{p}-\mathbf{q}|)\;d\mathbf{q},\;\mathbf{p}\in\Omega,\;\Omega = S\;or\;D, \tag{29}$$

where k_{D_j} and k_j are the wave numbers in D and in the background, respectively, for different frequencies. The weights ω_{D_j} and ω_{S_j} are chosen as in **4.1** to be independent of j but α_{jm}, the constants in the field updates, are now allowed to differ for each j. This increases the dimension of the algebraic optimization problem at each step. Results obtained using this algorithm are shown in Fig. 5. The starting values were those obtained by backpropagation. The stability with respect to noise is evident. More detail is found in [16].

4.6 Binary Contrast

In this set of examples we again consider spatially limited frequency diverse data for reconstructing the characteristic function of the scatterer with two major differences. First the background medium is no longer homogeneous but instead consists of two dissimilar homogeneous half spaces with the scatterer embedded in one of them. An extensive bibliography on this problem may be found in [17]. Secondly the fact that the contrast is not smooth ($\chi = 0$ or 1) is taken into account. If the contrast is updated in the gradient direction, as in the previous example, the updated contrast will no longer be a characteristic function. In fact the gradient does not exist in the space of characteristic functions. To account for this, we approximate the characteristic function by a smooth function of the form

$$\chi(\mathbf{p}) = [1 + \exp\left(-\frac{\tau(\mathbf{p})}{\theta}\right)]^{-1}\;, \tag{30}$$

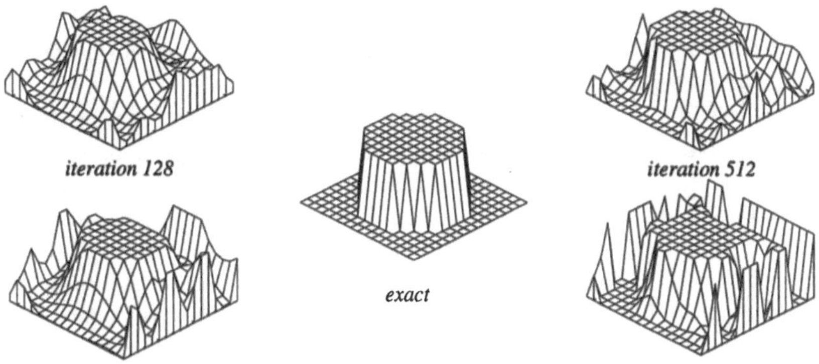

Fig. 5. The results obtained with the modified gradient algorithm in the case of microwave imaging of a void in a homogeneous concrete medium with noiseless (upper row) and noisy (lower row: 30% level random noise) data.

where τ is a real valued function of position and θ is a real positive parameter which controls the rate at which χ changes from 0 to 1. $\tau(\mathbf{p})$ is the function which is updated in the algorithm,

$$\tau_m(\mathbf{p}) = \tau_{m-1}(\mathbf{p}) + \beta_m\, d_m(\mathbf{p}) \ . \tag{31}$$

The operators are now

$$G_{\Omega_j}\chi u_j := \int_D G(\mathbf{p},\mathbf{q})\, \chi(\mathbf{q})\, (k_{D_j}^2 - k_j^2)\, u_j(\mathbf{q})\, d\mathbf{q},\ \mathbf{p}\in\Omega,\ \Omega = S \textit{ or } D \ , \tag{32}$$

where $G(\mathbf{p},\mathbf{q})$ is the Green's function for the unperturbed problem (see e.g. [17] for an explicit definition), k_{D_j} is the wave number in D at the jth frequency whereas k_j is the wave number in the half space containing D, and $\chi(\mathbf{q})$ is the approximation to the characteristic function defined above. With these definitions of the operators, the algorithm is as given before with the gradient of the error functional with respect to changes in τ given by

$$g_m^{\mathrm{d}} := 2\,\frac{d\chi}{d\tau}\bigg|_{\tau=\tau_{m-1}} \cdot R_e \sum_{j=1}^{J} \bar{u}_{jm-1} \left\{ w_{D_j}\, G_{D_j}^{*}\, r_{jm-1} - w_{S_j}\, G_{S_j}^{*}\, \rho_{jm-1} \right\} \ . \tag{33}$$

Here the weights are given by

$$\omega_{D_j} = \{\|u_j^{\mathrm{inc}}\|_D^2\}^{-1} \text{ and } \omega_{S_j} = \{\|f_j\|_S^2\}^{-1} \tag{34}$$

and the constants α_{jm} in the field updates are allowed to vary with j. This binary contrast formulation has been employed in a number of different situations and we present three examples here.

The first involves a local perturbation of a two fluid medium with incident acoustic waves normal to the interface at frequencies ω_j = 500, 700, 1100, 1400, 1700, and 2000 kHz. The perturbation B is an 0.8 mm sided square centered at 1 mm depth within a test domain D which is a 2 mm sided square. S is a line segment 1.5 mm above the interface. The acoustic velocities were $c_D = c_1 = 1470\,\mathrm{m/s}$ and $c_2 = 1800\,\mathrm{m/s}$ where c_1 denotes the velocity in the upper fluid layer and c_2 the velocity in the lower half space which contains the perturbation. In the discretized version, D is decomposed into 20×20 subsquares and the field was measured at 64 points 0.4 mm apart on S. Results of the inversion algorithm both with and without noise in the data are shown in Fig. 6.

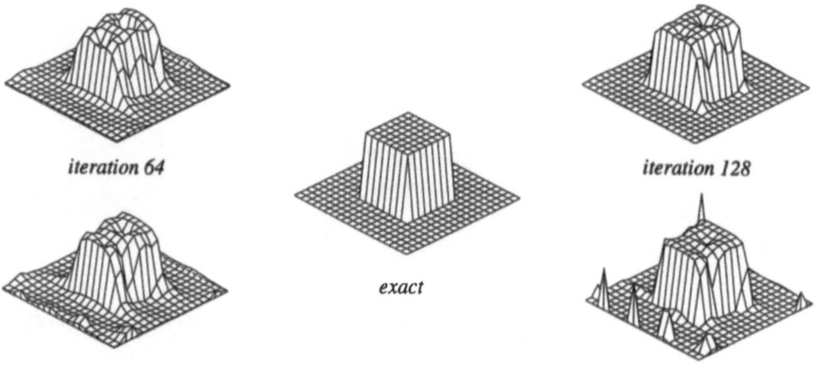

Fig. 6. The results obtained with the binary modified gradient algorithm in the case of the ultrasonic imaging of a local perturbation in a two-fluid medium with noiseless (upper row) and noisy (lower row: 20% level random noise) data. The interface is located on the left hand side of the pictures.

The second example concerns eddy current non-destructive testing of a defect in the surface of a metal-air interface. The incident fields are electromagnetic line sources (TM polarization) placed in air near the metal interface. Six frequencies between 10 and 349 kHz were employed. The defect is a 0.3 mm sided square void with one side on the air metal interface. The test domain D is a 0.5 mm sided square. The measurement domain S is a line segment 1.5 mm above the interface. In the discretized version D was again divided into 20×20 square pixels and the field was measured at 64 points, 0.3 mm apart, on S. Figure 7 shows the results of the binary version of the modified gradient algorithm and compares the performance with that of the original algorithm without introducing the particular form of the characteristic function employed here.

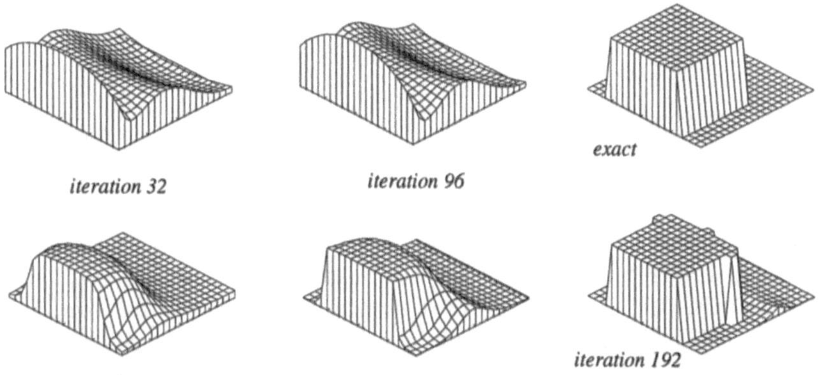

Fig. 7. Comparison of the modified gradient algorithm (upper row) and its binary version (lower row) in the case of eddy current imaging of a void in a metallic block. The air-metal interface is located on the left hand side of the pictures.

Another example of this binary constraint in a two layer medium involves a metallic structure immersed in sea water illuminated by low frequency electromagnetic waves (TM polarization). In this case B is a 9 m sided square, D is a 30 m sided square. D is again divided into 20×20 square pixels and S is a line segment 1.5 m above the interface which has 64 equally spaced receivers, 3 m apart. Six operating frequencies between 10 and 207 Hz were employed. Two different choices of conductivity within B were used, $\sigma_B = 10^7\,\mathrm{S/m}$ and $\sigma_B = 80\,\mathrm{S/m}$. The approximation $k_{D_j}^2 - k_j^2 \approx \mathrm{i}\omega_j\mu\sigma_B$ was employed in the operators. Results of this binary version of the algorithm are shown in Fig. 8. These examples demonstrate the effectiveness of this version of modified gradient algorithm in a number of relatively complex physical situations. More detail may be found in [18].

4.7 Blind Reconstruction

A final example of the modified gradient approach is provided from experimental data obtained at the Ipswich test site of Rome Laboratory, Hanscom Air Force Base. The physical scattering experiment involves TM electromagnetic scattering of an object in free space at a single frequency (10 GHz). A priori information was supplied that the target was perfectly conducting,

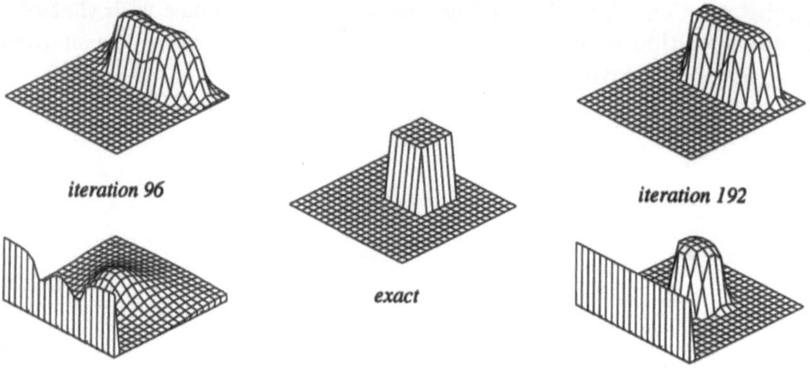

Fig. 8. The results obtained with the binary modified gradient algorithm in the case of low frequency electromagnetic imaging of an immersed metallic structure with two different choices of σ_B (upper row: $\sigma_B = 10^7$ S/m, lower row: $\sigma_B = 80$ S/m). The air-water interface is located on the left hand side of the pictures.

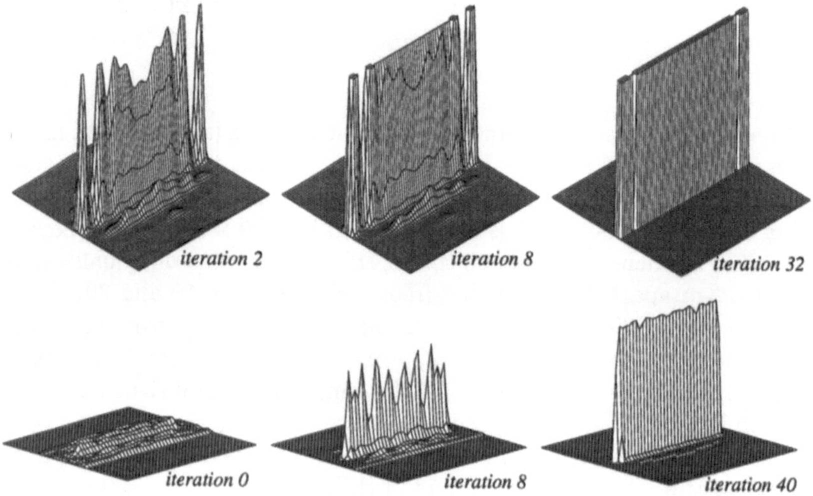

Fig. 9. Blind reconstruction of a mystery object from experimental data with two different versions of the modified gradient method: the perfect conductivity version **4.4** (upper row: D = $12.6 \times 12.6\ cm^2 = 63 \times 63$ pixels) and the binary version **4.6** (lower row: D = $16 \times 16\ cm^2 = 40 \times 40$ pixels).

symmetric about the x and y axes, but the actual shape of the target was not revealed until after the reconstructions were completed. Two different forms of the modified gradient algorithm were employed, the version described in **4.4** and the binary version described in **4.6**. Figure 9 show the results of these two approaches, both of which successfully reconstructed the object, later revealed to be a strip 4λ (12 cm) wide and 0.106λ (0.32 cm) thick. More detail on this example may be found in [19, 20].

5 Acknowledgments

This work was supported under a CNRS–NSF International Cooperative Grant #INT9415493, a NATO Collaborative Research Grant #CRG940999, and AFOSR Grant #F49620-96-1-0039.

References

[1] Tijhuis A. G.: Born–type reconstruction of material parameters of an inhomogeneous lossy dielectric slab from reflected–field data. Wave Motion **11**, 151–173 (1989)

[2] Roger A.: A Newton–Kantorovich algorithm applied to an electromagnetic inverse problem. IEEE Trans. Antennas Propagat. **AP–29**, 232–238 (1981)

[3] Tabbara W., Duchêne B., Pichot Ch., Lesselier D., Chommeloux L., Joachimowicz N.: Diffraction tomography: Contribution to the analysis of applications in microwaves and ultrasonics. Inverse Problems **4**, 305–331 (1988)

[4] Colton D., Kress R.: *Inverse Acoustic and Electromagnetic Scattering Theory* (Springer, Berlin Heidelberg, 1992)

[5] Kleinman R. E., van den Berg P. M.: A modified gradient method for two–dimensional problems in tomography. J. Comp. Appl. Math. **42**, 17–35 (1992)

[6] van den Berg P. M., Kleinman R. E.: Gradient Methods in Inverse Acoustic and Electromagnetic Scattering. *IMA Proc. Large Scale Global Optimization* (Springer, Berlin Heidelberg, to appear)

[7] Wen Lixin, Kleinman R. E., van den Berg P. M.: Modified gradient profile inversion using H-polarized waves. *Digest of IEEE-AP Symp.*, Newport Beach, CA, 1598–1601 (1995)

[8] Charbonnier P., Blanc-Féraud L., Aubert G., Barlaud M.: Deterministic edge-preserving regularization in computed imagery. IEEE Trans. Image Processing (to appear)

[9] Aubert G., Lazavoaia L.: A variational method in image recovery. SIAM J. Num. Anal. (to appear)

[10] Ruden B., Osher S., Fatemi C.: Nonlinear total variation based noise removal algorithm. Physica **60D**, 259–268 (1992)

[11] Dobson D. C., Santosa F.: An image-enhancement technique for electrical impedance tomography. Inverse Problems **10**, 317–334 (1994)

[12] Acar R., Vogel C. R.: Analysis of bounded variation penalty methods for ill-posed problems. Inverse Problems **10**, 1217–1229 (1994)

[13] van den Berg P. M., Kleinman R. E.: A total variation enhanced modified gradient algorithm for profile reconstruction. Inverse Problems **11**, L5–L10 (1995)

[14] Kleinman R. E., van den Berg P. M.: An extended range modified gradient technique for profile inversion. Radio Sci. **28**, 877–884 (1993)

[15] Kleinman R. E., van den Berg P. M.: Two-dimensional location and shape reconstruction.' Radio Sci. **29**, 1157–1169 (1994)

[16] Belkebir K., Kleinman R. E., Pichot Ch.: Microwave imaging- location and shape reconstruction from multifrequency scattering data. IEEE Trans. Microwave Theory (to appear) Tech.

[17] Lesselier D., Duchêne B.: Wavefield inversion of objects in stratified environments; from back propagation schemes to full solutions. *Review of Radio Science 1993–1996*, W. R. Stone Ed. (Oxford University Press, New York, 1996), 235–268

[18] Souriau L., Duchêne B., Lesselier D., Kleinman R. E.: Modified gradient approach to inverse scattering for binary objects in stratified media. Inverse Problems **12**, 463–481 (1996)

[19] van den Berg P. M., Coté M. G., Kleinman R. E.: "Blind" shape reconstruction from experimental data. IEEE Trans. Antennas Propagat. **AP–43**, 1389–1396 (1995)

[20] Duchêne B., Lesselier D.: On modified gradient solution methods using the binary aspect of the unknown electromagnetic parameters and their application to the Ipswich data. IEEE Antennas Propagat. Mag. **38**, 45–47 (1996)

Generalizations of Karp's Theorem to Elastic Scattering Theory

Ha-Duong Tuong

Université de Technologie de Compiègne, B.P. 20.529, 60205 Compiègne Cedex (France)
e-mail : tuong.ha-duong@utc.fr

Abstract. Karp's theorem states that if the far field pattern corresponding to the scattering of a time-harmonic acoustic plane wave by a sound-soft obstacle in $\mathbf{R}^2$ is invariant under the group of rotations, then the scatterer is a circle. The theorem is generalized to the elastic scattering problems and the axisymmetric scatterers in $\mathbf{R}^3$.

1 Introduction

In Martin and Dassios (1993), the authors proved some generalizations of Karp's theorem to elastic scattering theory, using uniqueness theorems for the inverse problem. They raised the question of a direct proof of this result (without uniqueness theorem), and discussed some difficulties in this proof. This paper is an answer to this question. In acoustic scattering theory, the author had already generalized the Karp's theorem to scatterers with any invariant group (Ha Duong (1996)), but the proof was quite different. The point here is an association of an idea of Kirsch and Kress in their proof of Karp's theorem for acoustic scatterers, using a superposition of plane incident waves, and an appropriate exploitation of the so called Atkinson-Wilcox expansion for elastic waves. In doing this, we also prove a more general version of Karp'p theorem, valid for axisymmetric scatterers. On the other hand, contraryly to Martin and Dassios (1993), we have only to suppose an invariant hypothesis for one type of incident waves (either pressure or shear waves) to obtain the invariance of the scatterer.

The main idea is first presented for acoustic problems in section 2. Applying this idea to elastic problems, besides a first complication due to the vector nature of the two far field patterns, a technical difference appears between the cases of P-incident and S-incident waves, compelling us to a more involved proof in the last case. These cases will be presented respectively in section 4 and 5. Section 3 is devoted to some recalls and notations concerning the Atkinson-Wilcox theorem for elastic scattering theory.

2 The Acoustic Problem

We consider the acoustic scattering problem of a time harmonic plane wave by an obstacle D in $\mathbf{R}^3$. To avoid technical difficulties, we suppose that D be an open bounded set with a regular border $\Gamma = \partial D$ and $\Omega = \mathbb{R}^3 \backslash \bar{D}$ is a connected exterior domain. The unitary normal vector $\boldsymbol{n}$ on Γ points to the exterior of D. The scatterer can be either sound-soft or sound-hard. Thus, the scattered wave is solution of the following Helmholtz equation problem :

$$\begin{cases} \Delta u + k^2 u = 0 & \text{in } \Omega \\ \mathcal{B}u = -\mathcal{B}u^{in} = g & \text{in } \Gamma \\ \frac{\partial u}{\partial n} - iku = 0(\frac{1}{r^2}) & \text{when } r = |x| \to \infty \end{cases}$$

where $\mathcal{B}$ is either the Dirichlet or the Neumann boundary condition. It is well-known that u has the following asymptotic behaviour

$$u(x) = \frac{e^{ik|x|}}{|x|}(F(\hat{x}) + O(\frac{1}{|x|})), \tag{1}$$

where $\hat{x} = \frac{x}{|x|}$ is the observation direction. The function F is called the far-field pattern of u

We shall prove the following generalization of Karp's theorem :

Theorem 1 *:*
Let $\boldsymbol{e}$ *be any unit vector in* $\mathbf{R}^3$ *and* U *the subgroup of orthogonal transformations leaving invariant* $\boldsymbol{e}$*. The scatterer* $\boldsymbol{D}$ *is U-invariant if and only if the far-field pattern* $\boldsymbol{F}$ *is.*

That means

$$D^Q = \{Qx; x \in D\} := D, \ \forall Q \in U$$

where D^Q is the image of D under Q, if and only if

$$F(Q\hat{x}; Q\alpha) = F(\hat{x}; \alpha), \ \forall Q \in U, \ \forall x, \alpha \in S^2 \tag{2}$$

where $F(\hat{x}; \alpha)$ is the far-field pattern under the incident

$$u^I(x; \alpha) = e^{ik\alpha . x}$$

Proof:

It is clear that only the proof of the *if* part is needed, the *only if* part resulting from the property of the laplacian !

Let us consider the following superposition of plane incident waves :

$$u^I(x) = \int_{S^2} u^I(x; \alpha)\, ds(\alpha) = \frac{4\pi sin(k|x|)}{k|x|}$$

(Funk-Hecke's formula, cf. Colton and Kress (1992)). Then, the resulting far field is the integral of $F(\hat{x};\alpha)$, and the invariance property (2) yields

$$F(Q\hat{x}) = F(\hat{x}),\ \forall \hat{x} \in S^2,\ \forall Q \in U \tag{3}$$

Now, one can recover the scattered wave u from F by the Atkinson-Wilcox's expansion (Atkinson (1949), Wilcox (1956)) :

$$u(x) = \frac{e^{i\kappa r}}{r} \sum_{n \geq 0} \frac{F_n(\kappa, \hat{x})}{r^n} \tag{4}$$

where the series is uniformly convergent for $(\hat{x}, r) \in S^2 \times \{r \geq a'\}$, for all $a' > a$ where $D \subset \{x \leq a\}$, and the functions F_n verify the following identities

$$\text{i) } F_0 = F \tag{5}$$

$$\text{ii) } 2i\kappa F_{n+1} = nF_n + \frac{1}{n+1}\Delta^* F_n \tag{6}$$

where $n \geq 0$ and Δ^* is the Laplace-Beltrami operator on S^2.

Taking a system of axis with e as the 3rd unit vector, and using the associated spherical coordinates :

$$\left.\begin{array}{l} \hat{x}_1 = sin\theta cos\varphi \\ \hat{x}_2 = sin\theta sin\varphi \\ \hat{x}_3 = cos\theta \end{array}\right\} \quad (0 \leq \theta \leq \pi, 0 \leq \varphi < 2\pi)$$

the invariance property (3) is simply expressed as

$$F(\hat{x}) = \tilde{F}(\theta)$$

for some function $\tilde{F}$. Now, since

$$\Delta^* = \frac{1}{sin^2\theta}\frac{\partial^2}{\partial\varphi^2} + \frac{1}{sin\theta}\frac{\partial}{\partial\theta}\left(sin\theta\frac{\partial}{\partial\theta}\right),$$

it follows that all the functions F_n are solely functions of θ, then independent of $\hat{x}_1$ and $\hat{x}_2$. From this and the expansion (4), one sees that for $\{|x| \geq a'\}$, u is a function of $(|x'|, x_3)$, where $x' = (x_1, x_2)$. The same is then true for its analytical continuation into $\Omega \cap \{|x| \leq a'\}$. Applying the boundary condition on $\Gamma = \partial D$, one gets the relation $(|x'| = constant)$ on the intersection of Γ with any horizontal plan $\{x_3 = constant\}$. QED

3 The Elastic Problem

We consider an homogeneous isotropic elastic medium, with density $\rho > 0$ and Lamé coefficients $\lambda, \mu > 0$. Consider the following scattering problem:

$$\begin{cases} \mathcal{A}_\omega \boldsymbol{u} = \mu \Delta \boldsymbol{u} + (\lambda+\mu)\nabla \text{ div } \boldsymbol{u} + \rho\omega^2 \boldsymbol{u} = 0 & \text{in } \Omega \quad (7) \\ \mathcal{B}\boldsymbol{u} = -\mathcal{B}\boldsymbol{u}^{inc} = \boldsymbol{g} & \text{in } \Gamma \quad (8) \\ |\boldsymbol{u}| = O(\frac{1}{r}); & \\ |\sigma(\boldsymbol{u}).\hat{x} - i\boldsymbol{W}\boldsymbol{u}| = o(\frac{1}{r}) \text{ when } r = |x| \to \infty & (9) \end{cases}$$

where $\boldsymbol{u}^{inc}$ is the incident wave, $\boldsymbol{W}$ is defined by (cf.Kupradze et al. (1979))

$$\boldsymbol{W}\boldsymbol{u} = (\lambda + 2\mu)\kappa_p(\boldsymbol{u}.\hat{x})\hat{x} + \mu\kappa_s(\boldsymbol{u} - (\boldsymbol{u}.\hat{x})\hat{x}),$$

and the boundary conditions (8) is either the Dirichlet ($\mathcal{B}\boldsymbol{u} = \boldsymbol{u}$, scattering by inclusions) or the Neumann one ($\mathcal{B}\boldsymbol{u} = \boldsymbol{\sigma}(\boldsymbol{u}).\boldsymbol{n}$, cavities problems).

The problem is well-posed and a simple application of Green's formula allows us to show that $\boldsymbol{u}$ can be extended to a distribution $\boldsymbol{u}_0$ of $\mathbf{R}^3$, null in D, verifying

$$\mathcal{A}_\omega \boldsymbol{u}_0 = T_\Gamma(\boldsymbol{u}) \tag{10}$$

where $T_\Gamma(\boldsymbol{u})$ is a distribution with support on Γ defined by Maxwell-Betti's formula

$$< T_\Gamma(\boldsymbol{u}), \boldsymbol{\phi} >= \int_\Gamma (\sigma(\boldsymbol{\phi})\boldsymbol{n}.\boldsymbol{u} - \sigma(\boldsymbol{u})\boldsymbol{n}.\boldsymbol{\phi})\, ds, \forall \boldsymbol{\phi} \in \mathcal{D}(\mathbb{R}^3)^3. \tag{11}$$

It is well-known that the elastic scattered wave has the following asymptotic behaviour at infinity:

$$\boldsymbol{u}(x) = \frac{e^{i\kappa_p|x|}}{|x|}\boldsymbol{F}^P(\hat{x}) + \frac{e^{i\kappa_s|x|}}{|x|}\boldsymbol{F}^S(\hat{x}) + O(\frac{1}{|x|^2}) \tag{12}$$

where the *elastic far field patterns* $\boldsymbol{F}^P(\theta)$ and $\boldsymbol{F}^S(\theta)$ are respectively a vector parallel and a vector orthogonal to θ. More precisely, using (10) and applying the classical decomposition of elastic waves to the fundamental tensor of Navier equation (7), it was proved in Alves and Ha Duong (1996) that the elastic scattered wave can be recovered from the far-field patterns by the following Atkinson-Wilcox expansion formula

$$\boldsymbol{u}(x) = \frac{e^{i\kappa_p r}}{r}\sum_{n\geq 0}\frac{\mathbb{F}_n^p(\hat{x})}{r^n} + \frac{e^{i\kappa_s r}}{r}\sum_{n\geq 0}\frac{\mathbb{F}_n^s(\hat{x})}{r^n} \tag{13}$$

where the series converge absolutely and uniformly as for the expansion (4), and where the $\mathbb{F}_n^W$ $(n \geq 1)$ can be obtained by recurrence from $\mathbb{F}_0^W = \boldsymbol{F}^W$, $(W = P, S)$ by the formula:

$$2i\kappa_W \mathbb{F}_{n+1}^W = n\mathbb{F}_n^W + \frac{1}{n+1}\Delta^* \mathbb{F}_n^W \quad \forall n \geq 0, \;\; W = P, S \tag{14}$$

This first order recurrence is the same as in the acoustic Atkinson-Wilcox expansion, while in Dassios (1988), only second order (and much more complicated) recurrence formulas were obtained.

In the following, we will note by $F^W(.;\beta,\alpha)$ the far-field patterns of the scattered wave under an incident plane wave $\boldsymbol{u}^I(x;\beta,\alpha) = \beta e^{i\kappa\alpha.x}$ with α as the incidence direction and β as the polarization direction, which satisfy the condition

$$\beta = \alpha \text{ if } \kappa = \kappa_P \text{ and } \beta \perp \alpha \text{ if } \kappa = \kappa_S \tag{15}$$

Also, the subscript Γ will be added to F^W when needed. Now, from the isotropy of the medium, one has

$$\boldsymbol{F}^W_{Q\Gamma}(Q\theta; Q\beta, Q\alpha) = Q\boldsymbol{F}^W_\Gamma(\theta;\beta,\alpha),$$
$$\forall Q \in U_0 \;\; \forall\theta \in S^2, \text{ and } \forall\; \beta,\alpha \text{ satisfying (15)}$$

where U_0 is the orthogonal group. Then, if the scatterer is invariant with respect to a subgroup U of U_0, one gets

$$\boldsymbol{F}^W_\Gamma(Q\theta; Q\beta, Q\alpha) = Q\boldsymbol{F}^W_\Gamma(\theta;\beta,\alpha) \tag{16}$$
$$\forall Q \in U \;\; \forall\theta \in S^2, \text{ and } \forall\; \beta,\alpha \text{ satisfying (15)}$$

Naturally, we want to prove the converse of that. Let us begin with the simpler case of a P-incident wave.

4 The P-invariance theorem

Theorem 2 *If for all P-wave incidence, the far-field patterns* $\boldsymbol{F}^P$ *and* $\boldsymbol{F}^S$ *are invariant with respect to the orthogonal group, then* Γ *is a sphere.*

Proof

Let $\boldsymbol{u}^{I,P}$ be a superposition of the plane incident P-waves, defined by

$$\boldsymbol{u}^{I,P}(x) = \int_{S^2} \boldsymbol{u}^I(x;\alpha,\alpha)\,ds(\alpha) \;=\; \int_{S^2} \alpha e^{i\kappa_P\alpha.x}\,ds(\alpha) \tag{17}$$

Using spherical coordinates to calculate the last integral, one gets the following formula, just like the Funk-Hecke formula

$$\boldsymbol{u}^{I,P}(x) = \frac{4i\pi\hat{x}}{\kappa_P^2|x|^2}(sin\kappa_P|x| - \kappa_P|x|cos\kappa_P|x|) \tag{18}$$

On the other hand, by the invariance hypothesis, one gets with $W = P, S$

$$QF^{W,P}(\hat{x}) = F^{W,P}(Q\hat{x}) \;\; \forall Q \in U_0 \;\; \forall\hat{x} \in S^2 \tag{19}$$

Now, since $F^{P,P}(\hat{x}) = C(\hat{x})\hat{x}$, where C is a scalar function, (19) implies that

$$C(\hat{x}) = C(Q\hat{x}) \ \forall Q \in U_0 \ \forall \hat{x} \in S^2,$$

which means that C must be a constant. While since $F^{S,P}(\hat{x})$ is orthogonal to $\hat{x}$, it must be null. Indeed, if we take two different Q transforming one fixed x_0 to a same x, a rotation around a line passing by O and a symmetry with respect to a plane, we obtain from $F^{S,P}(\hat{x}_0)$ two opposite images for $F^{S,P}(\hat{x})$. So that, using the Atkinson-Wilcox expansion, one sees that the scattered wave is of the form

$$\boldsymbol{u}(x) = C\frac{e^{i\kappa_P|x|}}{|x|}\hat{x} \tag{20}$$

Writing the homogeneous boundary condition for $\boldsymbol{u}^{I,P} + \boldsymbol{u}$, one finds an equation for $|x|$ to satisfy on Γ. This is obvious for the Dirichlet condition. For the Neumann one, we can write

$$\boldsymbol{v}(x) = \boldsymbol{u}^{I,P}(x) + \boldsymbol{u}(x) = \varphi(|x|)x$$

and after some calculations,

$$\boldsymbol{\sigma}(\boldsymbol{v}).n(x) = 2\mu\hat{x}\varphi'(|x|) + n(x)\{(3\lambda + 2\mu)\varphi(|x|) + \lambda|x|\varphi'(|x|)$$

And the same conclusion follows.

Remarks

1/ Incidentally, one sees how to create an incident wave such that the scattered wave (by a sphere) is a pure pressure wave. The author don't know if this result was known, neither if it is possible to do that for other geometry of the scatterer. The question of generating a pure pressure or shear outgoing wave is much easier if we were working with elastic waves created by sources rather than with scattering by obstacles, that is, if we consider the solutions of Navier equation with a right hand side and no boundary conditions in all space. See (Alves and Ha Duong (1996)).

2/ We could use the same method in section 2 to prove an invariance theorem for axisymmetric scatterer. However, one can see later that such an invariance theorem will be introduced naturally in the S-incidence case, so that, to avoid repetition, we reserve the treatment of axisymmetric scatterer for this case.

3/ It is clear from the proof that our theorem concerns the invariant property for incident waves with a fixed frequency.

5 The S-invariance theorem

Theorem 3 *If for all S-wave incidence, the far-field patterns* $\boldsymbol{F}^P$ *and* $\boldsymbol{F}^S$ *are invariant with respect to the orthogonal group, then* Γ *is a sphere.*

Proof

1/ We first note that it is *NOT* possible to use the superposition of all S-incident plane waves

$$\boldsymbol{u}^{\boldsymbol{I}}(x;\beta,\alpha) = \beta e^{i\kappa_S\alpha.x} \text{ with } \beta \perp \alpha$$

because when one integrates this $\boldsymbol{u}^{\boldsymbol{I}}(x;\beta,\alpha)$ on $\{(\alpha,\beta) \in S^2 \times S^2; \beta \perp \alpha\}$, the result is null ! We chose indeed the following S-incident plane waves

$$\boldsymbol{u}^{\boldsymbol{I}}(x;\alpha) = (\beta_0 \wedge \alpha)e^{i\kappa_S\alpha.x}$$

where β_0 is a fixed unit vector. The vector product $\beta_0 \wedge \alpha$ is no more an unit vector, but clearly $\boldsymbol{u}^{\boldsymbol{I}}(.;\alpha)$ remains a shear plane wave. Now, the superposition of $\boldsymbol{u}^{\boldsymbol{I}}(.;\alpha)$ gives the following incident wave

$$\boldsymbol{u}^{\boldsymbol{I},\boldsymbol{S}}(x) = \int_{S^2} \beta_0 \wedge \alpha e^{i\kappa_S\alpha.x}\, ds(\alpha)$$

and as in (17 and 18), one gets

$$\boldsymbol{u}^{\boldsymbol{I},\boldsymbol{S}}(x) = \varphi(|x|)\ \beta_0 \wedge \hat{x}$$

where

$$\varphi(|x|) = \frac{4i\pi}{\kappa_S^2|x|^2}(sin\kappa_S|x| - \kappa_S|x|cos\kappa_S|x|)$$

2/ Now, what about the invariance of the far field patterns with these incident waves ? We note that since

$$F^W(\hat{x}) = \int_{S^2} F^W(\hat{x};\alpha,\beta_0 \wedge \alpha)\, ds(\alpha)$$

then

$$QF^W(\hat{x}) = \int_{S^2} F^W(Q\hat{x};Q\alpha,Q(\beta_0 \wedge \alpha))\, ds(\alpha)$$

From an elementary algebra result :

$$Q(\beta_0 \wedge \alpha) = (detQ)(Q\beta_0 \wedge Q\alpha)$$

Thus, we have only the following invariance property

$$QF^W(\hat{x}) = F^W(Q\hat{x})\ \forall\hat{x} \in S^2,\ \forall Q \in U\ (W = P \text{or} S)$$

where U designates the set of orthogonal transformations Q which leave invariant the vector β_0, and with $detQ = 1$, i.e. for the subgroup of rotations around β_0. This is why we have to deal with the axisymmetric case here !
3/ Let's consider a system of axis with β_0 as the 3rd vector of the basis. Then, from the invariance property (16), the P-far field pattern is of the form

$F^P(x) = C(x_3)\hat{x}$ where C is a scalar function. Using spherical coordinates, one can write

$$F^P(\hat{x}) = \lambda(\theta)\begin{pmatrix} sin\theta cos\varphi \\ sin\theta sin\varphi \\ cos\theta \end{pmatrix},$$

and a simple calculation yields

$$\Delta^*(F^P)(\hat{x}) = \tilde{\lambda}_1(\theta)\hat{x} + \tilde{\mu}_1(\theta)e_3$$

From this and the recurrence formula (14), it follows easily that the P-coefficients of the Atkinson-Wilcox expansion can be written as

$$F_n^P(\hat{x}) = \lambda_n(x_3)\hat{x} + \mu_n(x_3)e_3.$$

So that the P-part of the scattered wave is of the form

$$u^P(x) = \varphi_1(|x'|, x_3)\hat{x} + \varphi_2(|x'|, x_3)e_3$$

(first, out of a sphere containing the scatterer by the Atkinson-Wilcox expansion, then out of the scatterer itself, by an analytical continuation argument as in theorem 1).

Now, if one writes the S-far field pattern as

$$F^S(\hat{x}) = \begin{pmatrix} F_{01}(\theta,\varphi) \\ F_{02}(\theta,\varphi) \\ F_{03}(\theta,\varphi) \end{pmatrix},$$

then, from the invariance property

$$\begin{cases} F_{01}(\theta,\varphi) = cos\varphi F_{01}(\theta,0) - sin\varphi F_{02}(\theta,0) = cos\varphi\lambda(\theta) - sin\varphi\mu(\theta) \\ F_{02}(\theta,\varphi) = sin\varphi F_{01}(\theta,0) + cos\varphi F_{02}(\theta,0) = sin\varphi\lambda(\theta) + cos\varphi\mu(\theta) \\ F_{03}(\theta,\varphi) = F_{03}(\theta,0) = \gamma(\theta), \end{cases}$$

(where the functions λ and μ satisfy

$$sin\theta\lambda(\theta) + cos\theta\gamma(\theta) = 0$$

because of the orthogonality of $F^S(\hat{x})$ with $\hat{x}$, but this relation shall not be used in the following). Consequently, one gets

$$\Delta^*(F^S)(\hat{x}) = cos\varphi\tilde{X}_1(\theta) + sin\varphi\tilde{Y}_1(\theta) + \tilde{Z}_1(\theta)$$

where $\tilde{X}_1, \tilde{Y}_1$ are orthogonal vectors in the $\{e_1, e_2\}$ plan and $\tilde{Z}_1$ is proportionnal to the vector e_3. Using again the recurrence formula (14), one sees that the S-coefficients of the Atkinson-Wilcox expansion can be written as

$$F_n^S(\hat{x}) = cos\varphi X_n(\theta) + sin\varphi Y_n(\theta) + Z_n(\theta)$$

where X_n, Y_n are now two vectors no more necessaryly orthogonal in the $\{e_1, e_2\}$ plan and Z_n is proportionnal to the vector e_3. We obtain finally the S-part of the scattered wave as

$$u^S(x) = x_1\chi_1(|x'|, x_3) + x_2\chi_2(|x'|, x_3) + \chi_3(|x'|, x_3)e_3$$

where χ_1 and χ_2 are vectors in the $\{e_1, e_2\}$ plan, and χ_3 a scalar function.

We conclude as in theorem 1 that the scatterer is invariant with respect to the subgroup U of rotations around β_0. (As for the P-incidence case, it is obvious for the Dirichlet boundary condition, while the calculations are more involved in the Neumann case). Since the vector β_0 is given arbitrarily, the theorem is proved.

Remarks

1/ As it was already noted, the above proof yields the following theorem for axisymmetric scatterer :

Theorem 4 *:*
Let e *be any unit vector in* $\mathbf{R}^3$ *and* U *the subgroup of orthogonal transformations leaving invariant* e. *The scatterer* D *is* U*-invariant if and only if the far-field patterns* F^P *and* F^S *are, for S-incident plane waves* $\beta e^{i\kappa_S\alpha.x}$ *with one frequency* κ *and all* (α, β) *such that* $\beta \perp \alpha$.

A similar theorem can be formulated for P-incident waves.

2/ It is also worth noting that we have only to suppose the invariance hypothesis for one type of plane incident waves (either P or S waves), while in Martin and Dassios (1993), the authors obtained their conclusion with the invariance hypothesis for both types of plane incident waves.

3/ Finally, it appears clearly in the proof that it cannot be generalized to lesser symmetric scatterers. In this case, a different approach, like the one in Ha Duong (1996) for acoustic scattering problems, should be adopted.

References

Alves C., Ha Duong T.(1996): On the far field amplitude for elastic waves. *(To appear)* Proceedings of the Sommerfeld'96 Workshop in Modern Mathematical Methods in Diffraction Theory and Its Applications in Engineering, Freudenstadt, Germany

Atkinson (1949): On Sommerfeld's radiation condition. Philos. Mag. **40**, 645-651

Colton D., Kress A. (1992): *Inverse Acoustic and Electromagnetic Scattering Theory.* (Springer-Verlag, Berlin Heidelberg)

Dassios G.(1988): The Atkinson-Wilcox Expansion Theorem for Elastic Waves. Quartly Appl. Math. **46**, 285-299

Ha Duong T.(1996) : An Invariance Theorem in Acoustic Scattering Theory. Inverse Problems **12**, 627-632

Isakov V.(1990): On uniqueness in the inverse transmission scattering problem. Comm. Part. Diff. Eq. **15**, 1565–1587

Karp S.N. (1962) : Amplitudes and Inverse Scattering Theory. (Electromagnetic Waves, R.E. Langer ed., University of Wiscosin Press, Madison), 291-300

Kirsch A., Kress R. (1993): Uniqueness in inverse obstacle scattering. Inverse Problems **9**, 285–299

Kupradze S., Gegelia T.G., Basheleishvili M.O., Burchuladze (1979) : *Three Dimensional Problems of the Mathematical Theory of Elasticity and Thermoelasticity.* (North Holland, Amsterdam)

Martin P., Dassios G. (1993) : Karps Theorem in Elastodynamic Inverse Scattering. Inverse Problems **9**, 97-111

Wilcox C.H. (1956) : A generalization of theorems of Rellich and Atkinson. Proc. Amer. Math. Soc. **7**, 271-276

Inverse Obstacle Scattering Problem Based on Resonant Frequencies *

Christophe Labreuche

Thomson CSF, Laboratoire Central de Recherche
Domaine de Corbeville, 91404 Orsay cedex, FRANCE.
e-mail: labreuch@thomson-lcr.fr

Abstract. Radar identification is often based on the study of the so-called resonant frequencies. But no inverse problem can emerge from this since the resonant frequencies do not uniquely determine the obstacle. We propose to consider also the eigenfunctions associated with the resonant frequencies. We first show the uniqueness of our inverse problem : the resonant frequencies *and* the associated eigenfunctions uniquely determine the obstacle. Then the stability of this problem is shown. Finally some numerical examples of the inversion are given.

1 Introduction

The problem we shall consider here is the identification of remote targets via radars. Classically, target detection is based on the Radar Cross Section (RCS) which indicates how much a target radiates in a given solid angle. The RCS of an object depends on both the altitude of the object and the orientation of the object. Accordingly, it is almost impossible to interpret the RCS for the identification of targets.

An alternative approach has been developed in the early 1970s, namely the Singularity Expansion Method (SEM) [Baum 1976]. It consists in computing the resonant frequencies of a target from the transient response. The scattered field measured on the antennas is composed of two parts: first we have the direct reflection of the incident field on the obstacle. After this direct reflection, some waves are sticking around the obstacle. These are the so-called *"creeping waves"* and correspond to the second part of the signal: the scattered field has then an exponential decay with respect to the time. The rate of the decay in time can be interpreted by some complex numbers called *"resonant frequencies"*. It turns out that the resonant frequencies (or poles) do not depend neither on the altitude of the target nor on the orientation of the target. Henceforth, poles appear as a powerful tool of characterization of a target. Unfortunately, from a theoretical point of view, there

* This work has been carried out while the author was visiting the University of Delaware, USA.

is no uniqueness result on the reconstruction of an obstacle from its poles. In fact, it is quite clear that the poles of an obstacle do not uniquely determine the shape of this obstacle. Thus the use of resonant frequencies for the inverse obstacle problem does not seem to be attractive any more. However, engineers are using poles for identification in the following way: if the target is assumed to belong to a restricted class (or catalog) of scatterers (for instance the class of all airplanes), then one can identify the target just by comparing the poles of the target with that of all the objects of the class. The main limitation of this method lies in the fact that it cannot be applied to squadrons of airplanes. The resonant frequencies of a cluster of different obstacles depend on the shape of each obstacle as well as on the relative locations of the obstacles. Since these relative locations are arbitrary, the case of multiple-targets cannot be taken into account in a catalog.

The problem of the identification of a target can be viewed as an inverse problem using the transient response. Basically, there is only one shape reconstruction technique in the time domain: the tomography. The frequencies used in remote radar detection typically belong to the slab [100 MHz, 400 MHz]. Hence the resolution of the signal in term of tomography is bracketed between 0.75 m and 3 m. Clearly, this is not enough for the reconstruction of missiles or airplanes of medium size.

Instead of studying the whole transient response, we propose here to focus directly on some particular creeping waves. More precisely, the idea of this work is to consider one resonant frequency k and one associated eigenfunction (e.g. the field scattered by a special creeping wave). This eigenfunction is solution to the Helmholtz equation at the frequency k. The use of resonant frequencies enables us to transform the inverse problem in the time domain into an inverse problem in the frequency domain. In the frequency domain, there exist some numerical methods that reconstruct the obstacle when the frequency belongs to the resonant region of the obstacle. By adapting one of these methods to our problem, we will be able to reconstruct (and thus identify) the target.

In section 2, we define the notion of resonant frequency and eigenfunction. We justify how these two quantities can be deduced from the transient response in Section 3. In addition, we review the theoretical results on the repartition of resonant frequencies that can be useful for our inverse problem. Then the inverse problem is stated in section 4. To justify the introduction of this inverse problem, we first show its uniqueness: one resonant frequency plus one associated eigenfunction uniquely determine the obstacle. The following section is devoted to showing the stability of this inverse problem. The method we use can be carried over almost unchanged to the classical inverse obstacle problem referred in [Colton et al. 1992]. In particular, we improve the stability estimate obtained in [Isakov 1994]. Finally, we show in section 6 how to use the method described in [Angell et al. 1995] to our case. We give some numerical examples that indicates the potentiality of this method for radar identifications.

2 Definition of the Resonant Frequencies

Let Ω_i be an open bounded domain and Ω its complement in $\mathbb{R}^N$. The outward normal to $\Gamma = \partial\Omega_i$ is denoted by $\mathbf{n}$. We assume that Γ is twice differentiable.

Let $w(t, \mathbf{x})$ be the acoustic field at the time t and the location $\mathbf{x} \in \mathbb{R}^N$. Then w satisfies the wave equation in the open space Ω :

$$\begin{cases} \frac{\partial^2 w(t,\mathbf{x})}{\partial t^2} - \Delta w(t, \mathbf{x}) = 0 \ , \ \text{in } \mathbb{R}^+ \times \Omega \\ w(t, \mathbf{x}) = 0 \ , \ \text{on } \mathbb{R}^+ \times \Gamma \\ w(0, \mathbf{x}) = f_0(\mathbf{x}) \text{ in } \Omega \\ \frac{\partial w(0,\mathbf{x})}{\partial t} = f_1(\mathbf{x}) \text{ in } \Omega \ . \end{cases} \tag{1}$$

$f = \begin{pmatrix} f_0 \\ f_1 \end{pmatrix}$ is the initial value for this Cauchy problem. Here we have considered the Dirichlet boundary condition on Γ.

After the incident wave has been reflected on a non-trapping obstacle, the energy around the obstacle as a function of the time decays exponentially when the dimension N is odd. More precisely, in each compact set $B \subset \mathbb{R}^N$ (N odd), we have [Lax et al. 1967]

$$\int_B \left| w(t, \mathbf{x}) - \sum_{j=1}^{J} e^{-\imath k_j t} \ u_j(\mathbf{x}) \right|^2 d\mathbf{x} \leq C(J) \ e^{\Im(k_{J+1})t} \tag{2}$$

where k_j are the resonant frequencies and u_j are non-trivial solutions to

$$\begin{cases} \Delta u_j + {k_j}^2 u_j = 0 \ , \ \text{in } \Omega \\ u_j = 0 \ , \ \text{on } \Gamma \\ u_j \text{ outgoing} \end{cases} \tag{3}$$

There are a countable number of poles. Moreover, their imaginary part is negative. In (2), we have assumed that the poles are numbered in such a way that

$$0 < \Im(k_1) \leq \Im(k_2) \leq \ \cdots \ \rightarrow -\infty \ .$$

Note that this is possible at least when the obstacle is non-trapping [Lax et al. 1967]. From the exponential decay of w and the Huygens' principle, u_j is shown to have an exponential growth

$$u_j(\mathbf{x}) \overset{|\mathbf{x}| \to \infty}{\sim} \frac{e^{\imath k_j |\mathbf{x}|}}{|\mathbf{x}|^{(N-1)/2}} \ u_j^\infty(\hat{\mathbf{x}}) \ \ , \ \hat{\mathbf{x}} = \frac{\mathbf{x}}{|\mathbf{x}|} \ , \tag{4}$$

where u_j^∞ is called the far field of u_j. In (3), the requirement that u_j is *"outgoing"* corresponds to the fact that the creeping waves are leaving the obstacle. We now make this more explicit. To this end, let us consider for $\Im(k) \geq 0$ the following problem

$$\begin{cases} \Delta u + k^2 u = 0 \ , \ \text{in } H^1_{loc}(\Omega) \\ u^s = g \ , \ \text{in } H^{1/2}(\Gamma) \\ \lim_{r\to\infty} r^{\frac{N-1}{2}} \left(\frac{\partial u}{\partial r} - \imath k u\right) = 0 \end{cases} \tag{5}$$

It is readily seen that this problem has a unique solution whenever $\Im(k) \geq 0$ [Colton et al. 1992]. Hence the mapping $g \mapsto u$ defines an operator $R(k)$. This operator is holomorphic in $\Im(k) \geq 0$ and has a meromorphic extension to $\Im(k) < 0$ [Poisson 1992]. Using integral equations [Colton et al. 1983], one can explicit the extension of $R(k)$. The operator $R(k)$ has at most a countable number of poles (resonant frequencies) in $\Im(k) < 0$. Consequently, the resonant frequencies are the complex numbers k_j for which the problem (3) has non-trivial solutions. In problem (3), by *"outgoing"* is meant an extended solution defined by $R(k)$.

The relation (2) which states that one can expand the solution for large time in term of the singularities of the problem is quite classical in physics.

3 Repartition of the Resonant Frequencies

Assume that the field $w(t, \mathbf{y})$ is measured by only one radar (located at the point $\mathbf{y}$) in a fixed range of time $t \in [T_0, T_1]$. By virtue of (2), only the resonant frequencies can be computed. The factors $u_j(\mathbf{y})$ alone do not give further information. The number J of poles one can estimate depends on the quality of the measurements (level of noise, $\cdots$). $k_1, \cdots, k_J$ are computed by minimizing the defect

$$\int_{T_0}^{T_1} \left| w(t, \mathbf{y}) - \sum_{j=1}^{J} c_j e^{-\imath k_j t} \right|^2 dt \ .$$

If the field $w(t, \mathbf{y})$ is known without error for all $t \geq T_0$ then one can clearly compute all the resonant frequencies.

As already seen in the introduction, we cannot reconstruct an obstacle from its resonant frequencies (see [Zworsky 1994] for some hints on this point). Nevertheless, since the creeping waves are travelling around the obstacle, we expect that the resonant frequencies characterize some geometrical quantities of the obstacle. Many researches have been carried out on the repartition of the poles (see [Melrose 1995] for an overview). We give two lemmas that state a relation between the repartition of the poles and some geometrical quantities of the obstacle.

Lemma 1 (Theorem 5.5 in [Lax et al. 1969]).
The counting function on the purely imaginary axis is defined by

$$N_{\mathrm{I}}(\sigma) = \#\{k \text{ pole },\ k \in \imath\mathbb{R} \text{ and } |k| < \sigma\}\ .$$

In odd space dimensions, there are infinitely many poles on the purely imaginary axis, and more precisely

$$\liminf_{\sigma\to\infty} \frac{N_{\mathrm{I}}(\sigma)}{\sigma^{N-1}} \geq \frac{1}{(N-1)!}\left(\frac{R_1}{\gamma_0}\right)^{N-1},$$

where R_1 is the radius of the largest sphere contained in Ω_i, R_2 is the radius of the smallest sphere containing Ω_i, and γ_0 is a known constant. Moreover, if Ω_i is star-shaped, then

$$\limsup_{\sigma\to\infty} \frac{N_{\mathrm{I}}(\sigma)}{\sigma^{N-1}} \leq \frac{1}{(N-1)!}\left(\frac{R_2}{\gamma_0}\right)^{N-1}.$$

Lemma 2 (Theorem 2.4 in [Lax et al. 1971]).
Assume that R_2 is the radius of the smallest sphere containing Ω_i. Then in three dimensions, there is no pole in the disk of center $-\frac{\imath}{2R_2}$ and radius $\frac{1}{2R_2}$.

Let us assume that the following limit exists

$$R_\Gamma = \gamma_0\left[(N-1)!\lim_{\sigma\to\infty}\frac{N_{\mathrm{I}}(\sigma)}{\sigma^{N-1}}\right]^{\frac{1}{N-1}}.$$

Then Lemma 1 shows that $R_1 \leq R_\Gamma \leq R_2$. For a given repartition of the poles, let ρ be the smallest number such that there is no pole in the disk of center $-\frac{\imath}{2\rho}$ and radius $\frac{1}{2\rho}$. Then Lemma 2 yields $R_2 \geq \rho$. Hence from R_Γ and ρ, one cannot directly give an upper bound of R_2. However, in practice, if we are interested only in a restricted class of scatterers, R_Γ and ρ are expected to give a good idea of the size of the obstacle. Furthermore, since the radar provides the location of the object, we assume we can a priori bound the object in a fixed sphere.

This will be very useful for the study of the inverse problem we consider now: the reconstruction of the shape of an obstacle using several resonant frequencies and associated eigenfunctions. In practice, if the scattered field is measured on a sphere S_{meas} for $t \in [T_0, T_1]$ then the poles $k_1,\cdots,k_J$ and the value of the associated eigenfunctions $u_1,\cdots,u_J$ on S_{meas} are determined by minimizing the functional

$$\int_{T_0}^{T_1}\int_{S_{\mathrm{meas}}}\left|w(t,\mathbf{x}) - \sum_{j=1}^{J} e^{-\imath k_j t}\ u_j(\mathbf{x})\right|^2 dt\, d\mathbf{x}\ ,$$

where J is fixed in advance. In next section, we will show that only one resonant frequency and one associated eigenfunction are enough to reconstruct the obstacle. Since the sphere $S_{\mathbf{meas}}$ is supposed to be far away from the target (for remote radar applications), u_1 on $S_{\mathbf{meas}}$ is almost equal to the far field up to a multiplicative factor. Hence from now on, we assume that we have measured one resonant frequency and the far field of one associated eigenfunction.

4 Uniqueness of the Inverse Problem

There is a classical result about uniqueness, which states that for the Dirichlet inverse obstacle scattering problem using plane waves, a finite number of far fields for different frequency (where this finite number depends only on the size of the obstacle) uniquely determine the scatterer. The first proof of uniqueness in inverse obstacle scattering was due to Schiffer and had been improved in [Colton et al. 1992]. This is the starting point for the proof of the uniqueness in our case.

Theorem 3. *Let $k \in \mathbb{C}$ (with $\Im(k) < 0$) and $u^\infty \in L^2(S)$ with $u^\infty \not\equiv 0$. Assume that $\Omega_i^{(1)}$ and $\Omega_i^{(2)}$ are two sound-soft obstacles having the same resonant frequency k and such that u^∞ is the far field of two eigenfunctions associated to k for the two obstacles $\Omega_i^{(1)}$ and $\Omega_i^{(2)}$. This means that for $j = 1, 2$ there exists a function $u_j \in H^1_{\mathrm{loc}}\left(\mathbb{R}^N \backslash \overline{\Omega_i^{(j)}}\right)$ such that*

$$\begin{cases} \Delta u_j + k^2 u_j = 0 \ , \ \text{in } \mathbb{R}^N \backslash \overline{\Omega_i^{(j)}} \\ u_j = 0 \ , \ \text{on } \partial \Omega_i^{(j)} \\ u_j \text{ outgoing with } u_j(\mathbf{x}) \overset{|\mathbf{x}| \to \infty}{\sim} \frac{e^{\imath k |\mathbf{x}|}}{|\mathbf{x}|^{(N-1)/2}} u^\infty(\hat{\mathbf{x}}) \ , \ \hat{\mathbf{x}} = \frac{\mathbf{x}}{|\mathbf{x}|} \end{cases} \tag{6}$$

Then $\Omega_i^{(1)} = \Omega_i^{(2)}$.

The space of all the eigenfunctions associated to a pole is a linear vectorial space of finite dimension. We would like to make it clear that in last theorem any non-zero eigen-far field u^∞ provides the stated uniqueness.

Proof of Theorem 3: The proof is done by contradiction: we assume that $\Omega_i^{(1)} \neq \Omega_i^{(2)}$. Let S be the unit sphere. Moreover, we set $\Gamma_j = \partial \Omega_i^{(j)}$. The unbounded component of $\mathbb{R}^N \backslash \left(\overline{\Omega_i^{(1)}} \cup \overline{\Omega_i^{(2)}}\right)$ is denoted by Ω^*. We also set $\Omega_* = \mathbb{R}^N \backslash \overline{\Omega^*}$. Let Ω_0 be a connected component of $\Omega_* \backslash \overline{\Omega_i^{(1)}}$. The boundary of Ω_0 is $\partial \Omega_0 = \Gamma^- \cup \Gamma^+$ where $\Gamma^- = \partial \Omega_0 \cap \Gamma_1$, and $\Gamma^+ = \partial \Omega_0 \cap \Gamma_2$.

Let u_j be defined by (6). From Theorem 2.13 in [Colton et al. 1992], the far field uniquely determines the field outside a big ball containing the obstacles, and thus by analytic continuation the field in the exterior domain ([Colton et al. 1983]). Hence $u_1 \equiv u_2$ in Ω^*. The rest of the proof is organized in three steps.

- First suppose that $\overline{\Omega_i^{(1)}}$ and $\overline{\Omega_i^{(2)}}$ are disjoint. Let u be the restriction of u_2 to $\Omega_i^{(1)}$. u is clearly a solution to the Helmholtz equation in $\Omega_i^{(1)}$. Moreover, since $u_1 = u_2$ in Ω^*, we conclude that $u = u_1 = 0$ on Γ_1. Since k is a resonant frequency, the imaginary part of k is strictly negative. Hence, by Lax-Milgram's lemma, the interior problem

$$\begin{cases} \Delta u + k^2 u = 0 \ , \ \text{in } \Omega_i^{(1)} \\ u = 0 \ , \ \text{on } \Gamma_1 \end{cases}$$

  is well-posed, so that $u \equiv 0$ in $\Omega_i^{(1)}$. By analytic continuation, u_2 vanishes identically in $\mathbb{R}^N \backslash \overline{\Omega_i^{(2)}}$, which contradicts the fact that $u^\infty \not\equiv 0$.
- Thus $\overline{\Omega_i^{(1)}} \cap \overline{\Omega_i^{(2)}} \neq \emptyset$. Here we assume that $\Omega_i^{(1)} \cap \Omega_i^{(2)} \neq \emptyset$. Thus $\Omega_0 \neq \emptyset$. Let u be now the restriction of u_1 to Ω_0. The function u satisfies the Helmholtz equation in Ω_0 and we have $u = 0$ on Γ^-. In addition, u_1 is equal to u_2 in Ω^*. Consequently, since $\partial\Omega_0 \cap \partial\Omega^* = \Gamma^+$, $u = u_2 = 0$ on Γ^+. Hence u is solution to the homogene Dirichlet problem in Ω_0. With the same argument as before, we attain a contradiction.
- It remains to consider the last case when $\Omega_i^{(1)}$ and $\Omega_i^{(2)}$ are just tangent. We again have the same contradiction by using the arguments of last step to the domain $\Omega_i^{(1)}$.

This concludes the proof. ■

5 Stability of the Inverse Problem

Inverse obstacle scattering problems are ill-posed in the sense that a small error in the measurement may imply a large error in the reconstruction. This is contrary to the idea of continuity (i.e. stability). The numerical resolution of a problem can be reasonably performed only if the problem is stable. Otherwise, the combination of the initial error on the data and the roundoff errors may overwhelm the final result, leading to something which has nothing to do with the real solution.

In fact, by adding some a-priori information, the reconstruction becomes stable. The first result on stability for inverse obstacle problems has been shown by V. Isakov [Isakov 1992, Isakov 1994] who proved that stability holds if we assume that the obstacle lies inside a fixed compact set.

In this section, we restrict ourself to the two dimensional case. We denote by S the unit circle. We will write a point $\mathbf{x} \in \mathbb{R}^2$ using either its Cartesian coordonates $(\mathbf{x}_1, \mathbf{x}_2)$ or its polar coordonates (θ, r). In the following, if $\theta \in [0, 2\pi]$, we will denote by $\hat{\theta}$ the unit vector $(\cos\theta, \sin\theta)$. We will show the stability of our inverse problem among all the obstacles whose boundary belongs to a reasonable class Λ of closed surfaces. Let us denote by Λ the class of all analytic boundaries Γ such that

(i) Γ is star-shaped with respect to the origin and is given by

$$\Gamma = \left\{ r(\theta)\hat{\theta} \ , \ \theta \in [0, 2\pi] \right\}$$

(ii) the radius r satisfies $R_1 \leq r(\theta) \leq R_2$, $\forall \theta \in [0, 2\pi]$, and

$$\|r\|_{C^2([0,2\pi])} := \max_{\theta \in [0,2\pi]} |r(\theta)| + \max_{\theta \in [0,2\pi]} |r'(\theta)| + \max_{\theta \in [0,2\pi]} |r''(\theta)| \leq C_r$$

(iii) the curvature $\mathcal{C}(\theta)$ at the point $r(\theta)\hat{\theta}$ satisfies

$$0 \leq \mathcal{C}(\theta) \leq \mathcal{C}_{max} \quad , \quad \forall \theta \in [0, 2\pi] \ .$$

In the following, C_Λ will stand for any constant which depends only on the class Λ, i.e. on the constants R_1, R_2, C_r and $\mathcal{C}_{max}$. More generally, we denote by C a generic constant, and by η a generic number satisfying $0 < \eta < 1$. We also introduce a compact fixed set Ξ of $\mathbb{C}^- = \{z \in \mathbb{C}, \Im(z) < 0\}$.

Remark that, as far as uniqueness is concerned, the restriction to the class Λ implies that only the second case in the proof of Theorem 3 has to be considered. We now give the stability theorem.

Theorem 4. *Let $\Omega_i^{(1)}$ and $\Omega_i^{(2)}$ be two obstacles whose boundaries (denoted by Γ_1 and Γ_2 respectively) belong to the class Λ. Let $k_j \in \Xi$ be a resonant frequency associated with $\Omega_i^{(j)}$. Let $u_j^\infty \in L^2(S)$ be such that there exists a function u_j solution to*

$$\begin{cases} \Delta u_j + {k_j}^2 u_j = 0 \ \text{ in } \mathbb{R}^2 \setminus \overline{\Omega_i^{(j)}} \\ u_j = 0 \ \text{ on } \Gamma_j \\ u_j \text{ is outgoing , with } u_j(\mathbf{x}) \overset{|\mathbf{x}| \to \infty}{\sim} \dfrac{e^{ik_j|\mathbf{x}|}}{\sqrt{|\mathbf{x}|}} u_j^\infty(\hat{\mathbf{x}}) \\ \|u_j^\infty\|_{L^2(S)} = 1 \end{cases} \tag{7}$$

We denote by $d(\Gamma_1, \Gamma_2)$ the Hausdorff distance $\sup_{\mathbf{x}_1 \in \Gamma_1} \inf_{\mathbf{x}_2 \in \Gamma_2} |\mathbf{x}_1 - \mathbf{x}_2|$. Then there exists a constant $C(k_1, u_1^\infty, \Lambda, \Xi)$ which depends only on k_1, u_1^∞, Λ and Ξ, and a real number $0 < \eta(k_1, u_1^\infty, \Lambda, \Xi) < 1$ depending only on k_1, u_1^∞, Λ and Ξ, such that for any bounded domain $\Omega_i^{(2)}$ we have

$$d(\Gamma_1, \Gamma_2) \leq C(k_1, u_1^\infty, \Lambda, \Xi) \quad \epsilon^{\eta(k_1, u_1^\infty, \Lambda, \Xi)\kappa(\epsilon)} \ , \tag{8}$$

$\epsilon := |k_1 - k_2| + \|u_1^\infty - u_2^\infty\|_{L^2(S)}$, and $\kappa(\epsilon)$ is given by

$$\kappa(\epsilon) = \frac{\mathrm{lx}_{\frac{e|k_1|R_2}{2}}\left(\frac{1}{\epsilon}\right)}{\log \frac{1}{\epsilon}} \ ,$$

where for y and x real, the real number $\mathrm{lx}_y(x)$ is defined as the solution z of

$$\left(\frac{z}{y}\right)^z = x \ .$$

One can easily show that the right hand side of (8) tends to zero when ϵ tends to zero. In [Isakov 1994], the following stability estimate for the inverse obstacle problem described in [Colton et al. 1992] is shown

$$d(\Gamma_1, \Gamma_2) \leq \frac{C(\Lambda)}{\kappa(\epsilon)|\log \epsilon|} .$$

Our estimate is better since we get rid of the logarithm function. But on the other hand, the constants we have in Theorem 4 depend on k_1 and u_1^∞. This is not a flaw and is enough in practice.

First we have to prove that (7) has a sense. The problem (7) corresponds to (3) plus a normalization condition on u_j^∞. The space of all the eigenfunctions associated to a pole forms a vectorial space of finite dimension. The norm of the eigenfunctions can be arbitrarily large or small. Hence, if k_j is a pole of Γ_j, the set of all solutions to (7) is not empty. To have some stability, we need to have a substantial grip on the norm of the eigenfunction u_j. This is why we have introduced the normalization condition on the far field u_j^∞.

Lemma 5. *For any domain $\Omega_i^{(j)}$ (with $\Gamma_j = \partial\Omega_i^{(j)} \in \Lambda$) and any solution u_j to (7) (for some $k_j \in \Xi$), we have the uniform bound*

$$\|u_j\|_{C^{2+\eta}(B_R\backslash\Omega_i^{(j)})} < C_{R,\Lambda,\Xi} ,$$

for some $0 < \eta < 1$ and $R > R_2$. B_R is the ball of center 0 and radius R.

The proof of this lemma is almost similar to that of Lemma 2 in [Isakov 1992] and thus is not reproduced here.

We use the same notation as in Section 4. Let Ω_0 be the connected component of $\overline{\Omega_*\backslash\Omega_i^{(1)}}$. We assume that the Hausdorff distance $d(\Gamma^-, \Gamma^+)$ is attained in Ω_0. $\partial\Omega_0$ is composed of two parts: $\Gamma^- := (\partial\Omega_0 \cap \Gamma_1)$ and $\Gamma^+ := (\partial\Omega_0 \cap \Gamma_2)$. Since Γ_1 and Γ_2 belong to Λ, $\Gamma^\pm$ are described by $\Gamma^\pm = \left\{r_\pm(\theta)\hat{\theta} \, , \, \theta \in [\theta_a, \theta_b]\right\}$ for some θ_a and θ_b.

Lemma 6. *We have*

$$d(\Gamma^-, \Gamma^+) \leq \max_{\theta \in [\theta_a, \theta_b]} (r_+(\theta) - r_-(\theta)) \leq \left(1 + \sqrt{\frac{\pi}{2}} \frac{\|r_-'\|_{C^0([\theta_a,\theta_b])}}{R_1}\right) d(\Gamma^-, \Gamma^+).$$

This lemma is a simple consequence of the definition of the Hausdorff distance. Its proof is omitted.

Let us fix $P \geq 2$. We set $\rho = d(\Gamma^-, \Gamma^+)$. By the above lemma, it is clear that the set $\Omega_P := \left\{(\theta, r) \in \Omega_0 \, , \, r_+(\theta) - r_-(\theta) \geq \frac{\rho}{P}\right\}$ is not empty. Ω_P is probably not connected: $\Omega_P = \cup_l \Omega_{P,l}$, where $\Omega_{P,l}$ is a closed and connected. Among the sets $\Omega_{P,l}$, there is one (labeled Ω_m) in which $\max_{\theta \in [\theta_a, \theta_b]}(r_+(\theta) - r_-(\theta))$ is attained. Hence $\Omega_m = \left\{(\theta, r) \in \Omega_0 \, , \, \theta \in \left[\tilde{\theta}_a, \tilde{\theta}_b\right]\right\}$, for some $\tilde{\theta}_a > \theta_a$ and $\tilde{\theta}_b < \theta_b$. For the proof of Theorem 4, we need the following lemma.

Lemma 7. *Assume that $w \in H_0^1(\Omega_0) \cap C^1(\Omega_0)$ satisfies $\Delta w + {k_1}^2 w = f$ in Ω_0, with $f \in L^2(\Omega_0)$. Then $w \in H^2(\Omega_m)$ and*

$$\|w\|_{H^2(\Omega_m)} \leq \frac{C_{\Xi,\Lambda}}{\rho}\left(\|f\|_{L^2(\Omega_0)} + \|w\|_{H^1(\Omega_0)}\right) .$$

Proof : This result is essentially a classical regularity result on weak solutions of elliptic equations. Let $P' > 2P$. We write $\Omega_m = \Omega_m^{P'} \cup \Omega_m^c$, where we have set $\Omega_m^{P'} = \{\mathbf{x} \in \Omega_m\,,\ \mathrm{dist}\,(\mathbf{x}, \partial\Omega_0) \geq \frac{\rho}{P'}\}$ and $\Omega_m^c = \Omega_m \backslash \Omega_m^{P'}$.

- A bound of the H^2 norm of w in $\Omega_m^{P'}$ is provided by Theorem 8.8 in [Gilbarg et al. 1977]:

$$\|w\|_{H^2(\Omega_m^{P'})} \leq \frac{C_\Lambda}{\rho}\left(\|F\|_{L^2(\Omega_0)} + \|w\|_{H^1(\Omega_0)}\right) .$$

  The factor $\frac{C_\Lambda}{\rho}$ correspond the $C^1(\Omega_0)$ norm of a cut-off function $\xi \in C_c^1(\Omega_0)$ that is equal to 1 in $\Omega_m^{P'}$. Since $\mathrm{dist}\,(\Omega_m^{P'}, \partial\Omega_m) = \frac{\rho}{P'}$, such a choice of ξ is possible.
- We define a cover of Ω_m^c : $\Omega_m^c = \bigcup_l \left(B(\mathbf{z}_l, \frac{3\rho}{2P'}) \cap \Omega_m^c\right)$, where the union is finite and $\mathbf{z}_l \in \partial\Omega_m \cap \partial\Omega_0$. Each set $B(\mathbf{z}_l, \frac{3\rho}{2P'}) \cap \Omega_m^c$ is transformed into a subset of $\mathbb{R}_+^2 = \{(x_1, x_2) \in \mathbb{R}^2\,,\ x_2 > 0\}$. From Theorem 8.12 in [Gilbarg et al. 1977], we have

$$\|w\|_{H^2(\Omega_m^c)} \leq \frac{C_\Lambda}{\rho}\left(\|F\|_{L^2(\Omega_0)} + \|w\|_{H^1(\Omega_0)}\right) .$$

  The factor $\frac{C_\Lambda}{\rho}$ now comes from the fact that the radius of the ball is $\frac{3\rho}{2P'}$.

■

Lemma 8. *Let $\Gamma_m^\pm := \Gamma^\pm \cap \partial\Omega_m$. For all $u \in H^1(\Omega_0)$, we have*

$$\|u\|_{L^2(\Gamma_m^-)} \leq \frac{C_\Lambda}{\sqrt{\rho}}\|u\|_{H^1(\Omega_m)} .$$

Proof : First consider $u \in C^1(\overline{\Omega_0})$. For any positive function $w \in C^1(\overline{\Omega_m})$, we have for $r_-(\theta) \leq r \leq r_+(\theta)$

$$w(\theta, r_-(\theta)) \leq w(\theta, r) + \left|\int_{r_-(\theta)}^{r} \frac{\partial w(\theta,t)}{\partial r} dt\right| \leq w(\theta, r) + \int_{r_-(\theta)}^{r_+(\theta)} \left|\frac{\partial w(\theta,t)}{\partial r}\right| dt .$$

Multiplying this by $\sqrt{r_-(\theta)^2 + r'_-(\theta)^2}$ and integrating over Ω_m, we get

$$\int_{\tilde\theta_a}^{\tilde\theta_b} (r_-(\theta) - r_+(\theta))\, w(\theta, r_-(\theta)) \sqrt{r_-(\theta)^2 + r'_-(\theta)^2}\, d\theta$$

$$\leq \int_{\tilde\theta_a}^{\tilde\theta_b}\int_{r_-(\theta)}^{r_+(\theta)} w(\theta, r) \frac{\sqrt{r_-(\theta)^2 + r'_-(\theta)^2}}{r} r\, d\theta dr$$

$$+ \int_{\tilde\theta_a}^{\tilde\theta_b}\int_{r_-(\theta)}^{r_+(\theta)} (r_-(\theta) - r_+(\theta)) \left|\frac{\partial w(\theta,t)}{\partial r}\right| \frac{\sqrt{r_-(\theta)^2 + r'_-(\theta)^2}}{t} t\, d\theta dt$$

Thus thanks to the definition of Ω_m and to Lemma 6

$$\frac{\rho}{P}\left\|\sqrt{w}\right\|^2_{L^2(\Gamma_m^-)} \leq C_\Lambda \left\|\sqrt{w}\right\|^2_{L^2(\Omega_m)} + C_\Lambda\, \rho \left\|\sqrt{\frac{\partial w}{\partial r}}\right\|^2_{L^2(\Omega_m)} .$$

Let us apply this to $w = |u|^2$. Since $\left|\frac{\partial}{\partial r}|u|^2\right| \leq \left|\nabla |u|^2\right| \leq |\nabla u|^2 + |u|^2$, the stated inequality holds for $u \in C^1(\overline{\Omega_0})$. By Theorem 1.4.2.1 in [Grisvard 1985], $C^1(\overline{\Omega_0})$ is dense in $H^1(\Omega_0)$ without any assumption on Ω_0. Thus this relation is also true for $u \in H^1(\Omega_0)$. ■

Lemma 9. *We have*

$$\text{meas}\,(\Gamma_m^{\pm}) \geq C_\Lambda \sqrt{d(\Gamma^-, \Gamma^+)}\ .$$

This lemma is a consequence of a classical result giving the link between the distance between two successive zeros of a function and the extremum of the function in this slab.

Lemma 10. *Let u_1 be a solution to (7) for $j = 1$. Then there exists a constant $C_{u_1} > 0$ and an integer M_{u_1} such that for any interval $\gamma \subset \Gamma_1$*

$$\left\|\frac{\partial u_1}{\partial \mathbf{n}}\right\|_{L^2(\gamma)} \geq C_{u_1}\ \text{meas}\,(\gamma)^{M_{u_1}+\frac{1}{2}}\ .$$

Proof : Assume that $\Gamma_1 = \left\{ r_1(\theta)\hat{\theta}\ ,\ \theta \in [0, 2\pi] \right\}$.

- The solution u_1 satisfies $u_1 = 0$ on Γ_1. Therefore by the Cauchy-Kowalewska theorem, $\frac{\partial u_1}{\partial \mathbf{n}}$ cannot vanish on $\gamma \subset \Gamma_1$, with meas $(\gamma) > 0$. Moreover since $\frac{\partial u_1}{\partial \mathbf{n}}$ is analytic on Γ_1, $\frac{\partial u_1}{\partial \mathbf{n}}$ vanishes (at most) at a finite number of points. Set $v(\theta) := \frac{\partial u_1}{\partial \mathbf{n}}(\theta, r_1(\theta))$ the value of $\frac{\partial u_1}{\partial \mathbf{n}}$ on Γ_1. The function v is periodic and analytic on $[0, 2\pi]$. By periodicity, v is extended to $[-2\pi, 0]$. $|v|$ has only a finite number p of local maximums (attained for θ equal to some values $e_1, \cdots, e_p$) on $[0, 2\pi]$. We set $e_0 = e_p - 2\pi$ and $I_i := [e_{i-1}, e_i]$. Since $v(e_{i-1})$ and $v(e_i)$ are two successive local maximums, the minimum of $|v|$ over I_i is attained at a point $\theta_i \in]e_{i-1}, e_i[$. Since v is analytic, the Taylor expansion of v at the point θ_i converges for all θ.

$$v(\theta) = \sum_{n=M_i}^{\infty} \frac{v^{(n)}(\theta_i)}{n!}(\theta - \theta_i)^n\ ,$$

  where M_i is the smallest integer $n < \infty$ for which $v^{(n)}(\theta_i) \neq 0$. Such an integer exists since $\frac{\partial u_1}{\partial \mathbf{n}}$ cannot vanish identically. Hence we conclude that there exists $C_i > 0$ such that for all $\theta \in I_i$

$$|v(\theta)| \geq C_i |\theta - \theta_i|^{M_i}\ .$$

This comes from the fact that θ_i is the minimum of $|v|$ over I_i. Let $M := \max_{i\in\{1,\cdots,p\}} M_i$. Then there exists $C > 0$ such that

$$\forall i \in \{1, \cdots, p\} \quad \forall \theta \in I_i \quad |v(\theta)| \geq C|\theta - \theta_i|^M \ . \tag{9}$$

- We give without proof two intermediate results. For a fixed integer L, and for some positive numbers $x_1, \cdots, x_L$, we have

$$\sum_{l=1}^{L} {x_l}^J \geq \frac{1}{L^{J-1}} \left(\sum_{l=1}^{L} x_l \right)^J \ . \tag{10}$$

For $J \in \mathbb{N}^*$ and $x > y \geq 0$, we have

$$x^J - y^J \geq \frac{1}{2^{J-1}} (x-y)^J \ . \tag{11}$$

- Let $\gamma = \left\{ r_1(\theta)\hat{\theta} \ , \ \theta \in G \right\} \subset \Gamma_1$, where G is an interval of $[e_0, e_p]$. By (9),

$$\int_G |v|^2 d\theta = \sum_{i=1}^{p} \int_{G\cap I_i} |v|^2 d\theta \geq \sum_{i=1}^{p} \int_{G\cap I_i} C^2(\theta - \theta_i)^{2M} d\theta \ .$$

Let $G_i := G \cap I_i$. As the intersection of two intervals, G_i is an interval, say $[a, b]$.

$$\int_{G_i} |v|^2 d\theta \geq \int_a^b C^2(\theta - \theta_i)^{2M} d\theta = \frac{C^2}{2M+1} \left[(b-\theta_i)^{2M+1} - (a-\theta_i)^{2M+1} \right] .$$

By (10) with $L = 2$ and $J = 2M + 1$, and (11) with $J = 2M + 1$

$$\int_{G_i} |v|^2 d\theta \geq \frac{C^2}{2M+1} \frac{\text{meas}\,(G_i)^{2M+1}}{2^{2M}} \ .$$

Summing over i and using (10) with $L = p$ and $J = 2M + 1$, we get

$$\int_G |v|^2 d\theta \geq \frac{C^2}{2M+1} \frac{\text{meas}\,(G)^{2M+1}}{(2p)^{2M}} \ .$$

Now we write

$$\int_\gamma \left| \frac{\partial u_1}{\partial \mathbf{n}} \right|^2 d\gamma \geq R_1 \int_G |v|^2 d\theta \geq \frac{R_1 C^2}{2M+1} \frac{\text{meas}\,(G)^{2M+1}}{(2p)^{2M}} \ .$$

Since $\text{meas}\,(\gamma) \leq C_\Lambda \text{meas}\,(G)$, we have the stated inequality. ∎

Proof of Theorem 4: The proof is split into four parts. The first two steps are similar to what is done in [Isakov 1994]. The rest of the proof is quite different to [Isakov 1994].

– Step 1: From the far field pattern to a sphere S_R.
From Lemma 5, u_1 and u_2 can be uniformly bounded. In [Bushuyev 1996], the far field $\mapsto$ middle field mapping is proved to be stable in three dimensions. This work can be carried over to the two dimensional case with complex frequencies. Thus from [Bushuyev 1996] we have for $R > 2R_2$

$$\|u_1 - u_2\|_{L^2(S_R)} \le C_{R,\Lambda,\Xi}\ \epsilon^{\kappa(\epsilon)} \tag{12}$$

– Step 2: From S_R onto $\partial\Omega^*$.
We denote by ϵ_1 the right hand side of (12). One can estimate the error of $u_1 - u_2$ on $\partial\Omega^*$ by using the same arguments as in Lemma 5 in [Isakov 1994]. In particular, there exists $0 < \eta < 1$ such that

$$\|u_1 - u_2\|_{C^0(\partial\Omega^*)} \le C_{R,\Lambda,\Xi}\ {\epsilon_1}^{\eta}\ . \tag{13}$$

– Step 3: Upper bound of $\|\frac{\partial u_1}{\partial \mathbf{n}}\|_{L^2(\Gamma_m^-)}$.
We denote by ϵ_2 the right hand side of (13). The Hausdorff distance $\rho := d(\Gamma_1, \Gamma_2)$ between Γ_1 and Γ_2 is attained in either one connected component of $\Omega_* \backslash \overline{\Omega_i^{(1)}}$ or in one of $\Omega_* \backslash \overline{\Omega_i^{(2)}}$.

(i) Assume that it is attained on a connected component of $\Omega_* \backslash \overline{\Omega_i^{(1)}}$, say Ω_0. u_1 is solution to the Helmholtz equation inside Ω_0. On $\partial\Omega_0$ $g = u_{1|\partial\Omega_0}$ satisfies $g = 0$ on Γ^- and $g = u_1 - u_2$ on Γ^+. Since $\Gamma^+ \subset \partial\Omega^*$, we have by (13) that $\|g\|_{C^0(\Gamma^+)} \le \epsilon_2$. The C^s norm of a function g over a non-smooth boundary $\partial\Omega_0$ is defined by

$$\|g\|^*_{C^s(\partial\Omega_0)} := \min_{g^* \in C^s(\Omega_0),\, g^*_{|\partial\Omega_0} = g} \|g^*\|_{C^s(\overline{\Omega_0})}\ .$$

In our case, we clearly have $\|g\|^*_{C^0(\partial\Omega_0)} \le \|g\|_{C^0(\Gamma^+)} \le \epsilon_2$. This combined with Lemma 5 and some interpolation inequalities (see Theorem 1.1.1 in [Isakov 1991]) shows that there exists $0 < \eta < 1$ such that $\|g\|^*_{C^2(\partial\Omega_0)} \le C_{R,\Lambda,\Xi}{\epsilon_2}^{\eta}$. From Theorem 1.1.1 in [Isakov 1991], one can extend g defined on $\partial\Omega_0$ to u_1^* defined in Ω_0 such that $\|u_1^*\|_{C^2(\Omega_0)} \le C\|g\|^*_{C^2(\partial\Omega_0)}$, where C is independent of the domain Ω_0. Hence

$$\|u_1^*\|_{H^2(\Omega_0)} \le C_\Lambda \|g\|^*_{C^2(\partial\Omega_0)} \le C_{R,\Lambda,\Xi}\ {\epsilon_2}^{\eta}\ . \tag{14}$$

The function $w := u_1 - u_1^*$ belongs to $H_0^1(\Omega_0)$ and satisfies $\Delta w + {k_1}^2 w = f$ in Ω_0, where $f = -(\Delta + {k_1}^2)u_1^*$. Hence for all $v \in H_0^1(\Omega_0)$, we have $a(w,v) = \int_{\Omega_0} f\overline{v}$, where $a(w,v) := \int_{\Omega_0}(\nabla w \cdot \nabla\overline{v} - {k_1}^2 w\overline{v})$. Hence by Cauchy-Schwarz inequality

$$C_{k_1}\|w\|^2_{H^1(\Omega_0)} \le \Re(-i\overline{k_1}a(w,w)) \le |k_1|\|f\|_{L^2(\Omega_0)}\|w\|_{L^2(\Omega_0)}\ ,$$

where $C_{k_1} = |\Im(k_1)|\min(1, |k_1|^2)$. Using (14), we find

$$\|w\|_{H^1(\Omega_0)} \le C_\Xi \|f\|_{L^2(\Omega_0)} \le C_{R,\Lambda,\Xi}\ {\epsilon_2}^{\eta}\ . \tag{15}$$

The function u_1 belongs to $C^2(\Omega_0)$. So does $w = u_1 - u_1^*$. Combining Lemma 7 with (14) and (15), we have

$$\|u_1\|_{H^2(\Omega_m)} \le C_{R,\Lambda,\Xi} \frac{\epsilon_2{}^\eta}{\rho} . \tag{16}$$

We denote by $\Gamma_m^\pm$ the surface $\Gamma^\pm \cap \partial\Omega_m$. Furthermore, we set $\mathbf{U} := \nabla u_1$. Lemma 8 is now applied to each component of $\mathbf{U}$. Using (16) and the relation $|\frac{\partial u_1}{\partial \mathbf{n}}| \le |\mathbf{U}|$, we finally get

$$\left\|\frac{\partial u_1}{\partial \mathbf{n}}\right\|_{L^2(\Gamma_m^-)} \le \|\mathbf{U}\|_{L^2(\Gamma_m^-)} \le \frac{C_\Lambda}{\sqrt{\rho}} \|u_1\|_{H^2(\Omega_m)} \le C_{R,\Lambda,\Xi} \frac{\epsilon_2{}^\eta}{\rho^{\frac{3}{2}}} . \tag{17}$$

(ii) Assume now that the Hausdorff distance is attained on a connected component of $\Omega_* \backslash \overline{\Omega_i^{(2)}}$, again denoted by Ω_0. Then, switching the indexes 1 and 2 (also + and −) in last point, one can show as for (17) that

$$\left\|\frac{\partial u_1}{\partial \mathbf{n}}\right\|_{L^2(\Gamma_m^-)} \le C_{R,\Lambda,\Xi} \frac{\epsilon_2{}^\eta}{\rho^{\frac{3}{2}}} . \tag{18}$$

– Step 4: Lower bound of $\|\frac{\partial u_1}{\partial \mathbf{n}}\|_{L^2(\Gamma_m^-)}$ and conclusion of the proof. By Lemma 10, that there exists a constant $C_{u_1} > 0$ and an integer M_{u_1} such that

$$\left\|\frac{\partial u_1}{\partial \mathbf{n}}\right\|_{L^2(\Gamma_m^-)} \ge C_{u_1} \operatorname{meas}(\Gamma_m^-)^{M_{u_1}+\frac{1}{2}} .$$

By Lemma 9 and (17), (18)

$$C_{u_1} C_\Lambda d(\Gamma_1, \Gamma_2)^{\frac{2M_{u_1}+1}{4}} \le \left\|\frac{\partial u_1}{\partial \mathbf{n}}\right\|_{L^2(\Gamma_m^-)} \le C_{R,\Lambda,\Xi} \frac{\epsilon^{\eta\kappa(\epsilon)}}{d(\Gamma_1,\Gamma_2)^{\frac{3}{2}}} .$$

We take $R = 3R_2$, so that $C_{R,\Lambda,\Xi}$ only depends on Λ and Ξ. Thus it is labeled $C_{\Lambda,\Xi}$. Therefore we have the require estimate with $\eta(k_1, u_1^\infty, \Lambda, \Xi) = \frac{4\eta}{2M_{u_1}+7}$. ■

6 Numerical Reconstruction of the Obstacle

Let u be a non-zero solution to (6) for the resonant frequency k and the far field u^∞. We wish to reconstruct Γ from k and u^∞. The proof of uniqueness by Schiffer is constructive and leads to the Kirsch-Kress method [Colton et al. 1992]. The method is basically designed for the inverse obstacle scattering problem using plane waves, and can be extended to our problem. This method consists in two steps :

– Step 1: Computation u from u^∞.
The field u is sought in the form of a single layer potential on a fixed artificial boundary γ:

$$u(\mathbf{x}) = \left(\tilde{V}_\gamma \phi\right)(\mathbf{x}) := \int_\gamma G(\mathbf{x},\mathbf{y})\phi(\mathbf{y})\, d\gamma(\mathbf{y}) \ , \tag{19}$$

where the potential ϕ has to be determined and G is the outgoing fundamental solution of the Helmholtz equation. The closed surface γ must lie inside the obstacle Ω_i [Colton et al. 1992]. In two dimensions, we have $\left(\tilde{V}_\gamma \phi\right)(\mathbf{x}) \overset{|\mathbf{x}|\to\infty}{\sim} \frac{e^{ik|\mathbf{x}|}}{\sqrt{|\mathbf{x}|}} \left(F_\gamma \phi\right)(\hat{\mathbf{x}})$, where

$$\left(F_\gamma \phi\right)(\hat{\mathbf{x}}) = \frac{1}{2ik}\sqrt{\frac{-ik}{2\pi}} \int_\gamma e^{-ik\hat{\mathbf{x}}\cdot\mathbf{y}}\, \phi(\mathbf{y})\, d\gamma(\mathbf{y}) \ . \tag{20}$$

Therefore ϕ is determined by solving the equation

$$F_\gamma \phi = u^\infty \quad \text{on } S \ . \tag{21}$$

As pointed out in [Colton et al. 1992], since the kernel of F_γ is analytic, this equation is severely ill-posed. This equation is solved by minimizing the functional

$$\mu_1(\phi; \Upsilon, \alpha) := \|F_\gamma \phi - u^\infty\|^2_{L^2(S)} + \alpha \|\varphi\|^2_{L^2(\gamma)} \tag{22}$$

over all $\phi \in L^2(\gamma)$. We notice the Tikhonov regularization term $\|\varphi\|^2_{L^2(\gamma)}$.
– Step 2: Determination of Γ from u.
The relation (19) now gives u in the unbounded component of $\mathbb{R}^2\backslash\gamma$. The Dirichlet boundary condition (i.e. $u = 0$ on Γ) enables us to characterize Γ :

$$\Gamma = \left\{\mathbf{x} \in \mathbb{R}^N \ / \ u(\mathbf{x}) = 0\right\} \ . \tag{23}$$

From the uniqueness Theorem 3, there is only one closed curve on which u vanishes. This step is well-posed but non-linear, and is solved by minimizing the defect

$$\mu_2(\Upsilon; \varphi) := \left\|\tilde{V}_\gamma \varphi\right\|^2_{L^2(\Upsilon)} \tag{24}$$

over all the closed curves Υ. There is no regularization term in μ_2 since this step is well-posed.

The Angell-Jiang-Kleinman method [Angell et al. 1995] corresponds to several iterations of the Kirsch-Kress method. In fact, the Angell-Jiang-Kleinman method features an update of the reference curve γ from the reconstruction Υ given by step 2. The idea is that the computation of the field is more accurate if the reference curve γ is parallel to the exact boundary Γ. Henceforth, after the computation of ϕ and $\Upsilon = \left\{r(\theta)\hat{\theta}, \theta \in [0, 2\pi]\right\}$, the surface γ is set to a curve parallel to Υ, say $\gamma = \left\{\frac{3}{4}r(\theta)\hat{\theta}, \theta \in [0, 2\pi]\right\}$. Then the two steps are done again with the new reference curve γ, leading to new values of ϕ and Υ. The reference curve γ is then updated. And so on.

In Figures 1 and 2, we show two examples of reconstructions. The use of an adaptative interior curve γ improves a lot the numerical results (compared to the Kirsch-Kress method in which the interior curve is fixed). We also notice the stability of the algorithm to some noise added on the data.

In many experimental devices, the measurement is available only in a limited aperture. This is in particular the case of radar applications since the antennas can be put only on the ground. We are here interested in half aperture. We obtained good results in this case. For real frequencies and incident plane waves, limited-aperture problems are very tough to solve. The eigen-solutions associated with resonant frequencies correspond to creeping waves which go around the obstacle. Consequently, even if the information is obtained only in a limited-aperture, these waves contain the information about the whole obstacle.

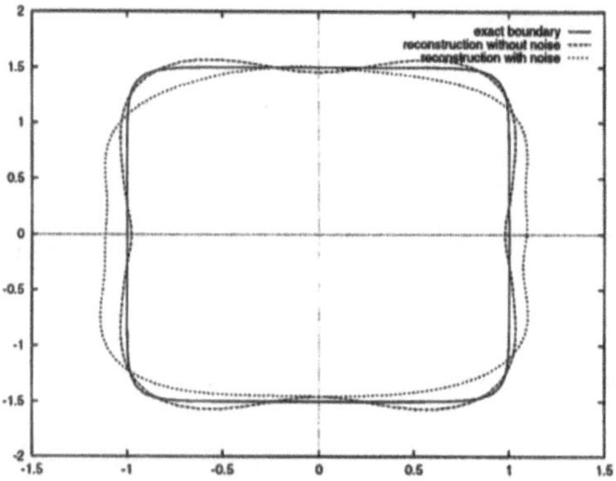

Fig. 1. The Rounded rectangle obstacle. We present the results obtained with the resonant frequency $k = 1.547 - 1.347i$ and the parameter $\alpha = 10^{-4}$ in μ_1. 5% of noise in both the resonant frequency and the far field is added for the second reconstruction. The error in the first reconstruction is $d(\Gamma, \Upsilon) = 7.3\%$, and the error in the second reconstruction is $d(\Gamma, \Upsilon) = 18.7\%$.

References

Angell T., Jiang X., Kleinman R. (1995): On a Numerical Method for Inverse Acoustic Scattering. Technical Report No. 95-9 of Department of Mathematical Sciences, University of Delaware, USA.

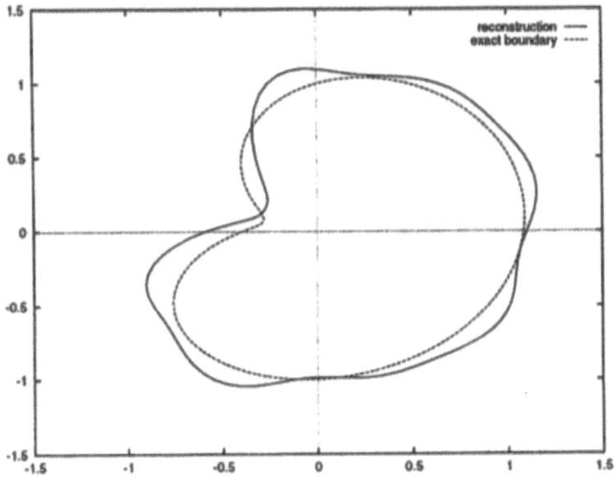

Fig. 2. The *"bean-shaped"* obstacle. We used the pole $k = 0.464 - 1.371i$ and the parameter $\alpha = 10^{-4}$. 5% of random noise on both the resonant frequency and the eigen-far field are added. The error is $d(\Gamma, \Upsilon) = 16.5\%$.

Baum C. (1976): The Singularity Expansion Method. In *Transient Electromagnetic Fields*. Ed. Felsen L. Springer-Verlag, 129-179.

Bushuyev I. (1996): Stability of the Recovering the Near Field Wave from the Scattering Amplitude. preprint of the University of Wichita, USA.

Colton D., Kress R. (1983): *Integral Equation Methods in Scattering Theory* (Wiley-Interscience Publication).

Colton D., Kress R. (1992): *Inverse Acoustic and Electromagnetic Scattering Theory* (Springer-Verlag, Berlin Heidelberg New York).

Gilbarg D., Trudinger N. (1977): *Elliptic Partial Differential Equations of Second Order* (Springer-Verlag, Berlin).

Grisvard P. (1985): *Elliptic Problems in Nonsmooth Domains* (Monographs and Studies in Mathematics 24, Pitman, Boston).

Isakov V. (1991): *Inverse Source Problems* (AMS Mathematical Monographs Series, Number 34, Providence RI).

Isakov V. (1992): Stability Estimates for Obstacles in Inverse Scattering. J. Comp. Appl. Math. **42**, 79-88.

Isakov V. (1993): New Stability Results for Soft Obstacles in Inverse Scattering. Inverse Problems **9**, 535-543.

Lax P., Philipps R. (1967): *Scattering Theory* (Academic press).

Lax P., Philipps R. (1969): Decaying Modes for the Wave Equation in the Exterior of an Obstacle. Comm. Pure Appl. Math. **22**, 737-787.

Lax P., Philipps R. (1971): On the Scattering Frequencies of the Laplace Operator for the Exterior Domain. Comm. Pure Appl. Math. **25**, 85-101.

Melrose R. (1995): *Geometric Scattering Theory* (Stanford lectures, Cambridge University press).

Poisson O. (1992): *Calculs des Pôles Associés à la Diffraction d'Ondes Acoustiques et Élastiques en Dimension 2* (Thesis of the University of Paris IX Dauphine, France).

Zworski M. (1994): Counting Scattering Poles. In *Spectral and Scattering Theory*. Ed. Ikawa, Marcel Dekker, New York, 301-331.

New Developments in the Application of Inverse Scattering to Target Recognition and Remote Sensing

A. Gérard[1], A. Guran[2], G. Maze[3], J. Ripoche[3], H. Überall[4]*

[1] LMP, CNRS 867, Université Bordeaux I, F-33405 Talence, France (U.E.)
[2] HEDCO 216, University of Southern California, Los Angeles, CA 90089-1211, USA
[3] LAUE, CNRS 1373, Université du Havre, F-76610 Le Havre, France (U.E.)
[4] Department of Physics, Catholic University of America, Washington, DC 20064, USA

Abstract. Various approaches to the solution of the inverse scattering problem are discussed here, and illustrated by selected examples. Inverse scattering, having originated with quantum mechanical scattering problems, has more recently become of interest in acoustic and electromagnetic areas, in geophysics as well as in oceanography. These topics will be described here both based on general approaches, or more specifically as based on the use of target resonances, or of surface waves on the target.

1 Introduction

The inverse scattering problem, i.e. the determination of the character of a scattering object or medium when the incident signal is known and the scattered signal has been measured, had its start in quantum mechanics[1-3] and later on became of interest in radar scattering[4-6], and in geophysical prospecting[7, 8]. In acoustic[9-12] and elastic-wave scattering[13-14] the interest in this problem is or more recent date. While the most general solution of the inverse scattering problem has spawned a veritable subfield of applied mathematics[15-18] (this also extends to nondestructive testing[19], tomography[20, 21] and imaging[16, 22, 23]) often concerned with the solution of integral equations[24-25], it has been found of advantage by us[9-11, 13, 26-29] and by others[5, 12, 14, 30-36], to combine purely mathematical approach with an application of well-identified physical phenomena (the target resonances, in this case), and the use there of in order to make additional headway towards the solution of the inverse problem, or even to provide shortcuts for the latter[29]. The present discussion selects a number of representative examples from both the general mathematical solutions of the problem (determination of electromagnetic medium parameters, or acoustic identification of layers), and the physically-oriented approach based on the Resonance Scattering Theory or RST[37-40] (Derem's "acoustic resonance spectroscopy"[41], the Le Havre experiments[12] on multilayer characterization by a combined observation of resonance spectra and pulse returns; and Gérard's[42] Generalized Debye Series approach allowing access to the resonances of individual layers[43], of obvious importance for the geophysical

* Visiting Professor into 1 and 3

problem). Finally, the application of acoustic surface waves on the target, being intimately related to the resonances, is mentioned for a determination of material parameters via a measurement of their dispersion curves[44, 45], or via time-frequency analysis[46].

2 Non-resonant inversion techniques

2.1 Acoustic Scattering

The aim of the inverse scattering formalism is to reconstruct, from a measured set of scattering data, as much information on the scattering object as possible, given a limited amount of data (which may cause problems). In addition, the ill-posed nature of the inverse problem can lead to great sensitivity of the solution to small changes of the parameters, or to fluctuations in the measured data. Fourier expansions employed will add to this sensitivity. We mention here two examples from acoustic scattering illustrating, or trying to remedy, these problems.

Bonnet[47] has considered the problem of reconstructing the given surface velocity of a vibrating cylinder from theoretical scattering data obtained from (Green's) integral equation using the boundary element method. Methods to stabilize the solution of an ill-posed problem are, among others, the Tikhonov regularization[48], or stochastic inversion[49]. Bonnet employs the latter, in particular Gaussian inversion, treating the data as random variables. Figure 1 shows the results of his inversion for the normal surface velocity U vs. distance z in the axial direction: the inverted U, bracketed by standard deviations, as compared to the exact value (heavy solid curve).

Tobocman[50] has considered the effect of Fourier vs. wavelet expansion on the reconstruction of an experimental ultrasonic pulse reflected from human tissue, from its filtered Fourier transform or from a stripped wavelet analysis; the former reconstructed pulse (dotted) is compared with the actual pulse (solid line) in Fig. 2. For the case of wavelet analysis, however, the reconstruction is found to agree much more closely with the actual pulse than the Fourier one. In addition, the wavelet approach is shown to lead to great computational simplifications.

2.2 Tomography

Examples are presented here from the area of electromagnetics[21, 51-53], although similar techniques can also be employed in ultrasonics[21]. The problem here is the reconstruction of e.g. the permittivity and conductivity profiles of a dielectric slab from measured transmission and reflection data. He and Ström[51] consider a point source above an inhomogeneous dissipative slab, and reconstruct its one-dimensional permittivity (ε) and conductivity (σ) profiles mainly from two-sided reflection data. Fields are split into downgoing and upgoing waves represented by time integrals, and for each the incident are reflected fields are expressed in terms of reflection and transmission kernels, either for the entire slab or for subregions of the same ("invariant imbedding"). The imbedding equations can be solved for ε and σ in terms of the scattering data by iteration. Figure 3 shows the assumed ε-profile of the slab (solid), reconstructed from clean data (dotted), after smoothing noisy data (dashed), and after exempting peaks in the kernels from smoothing (circles). It is seen that tomography of dielectric objects is well under control, at least in one dimension.

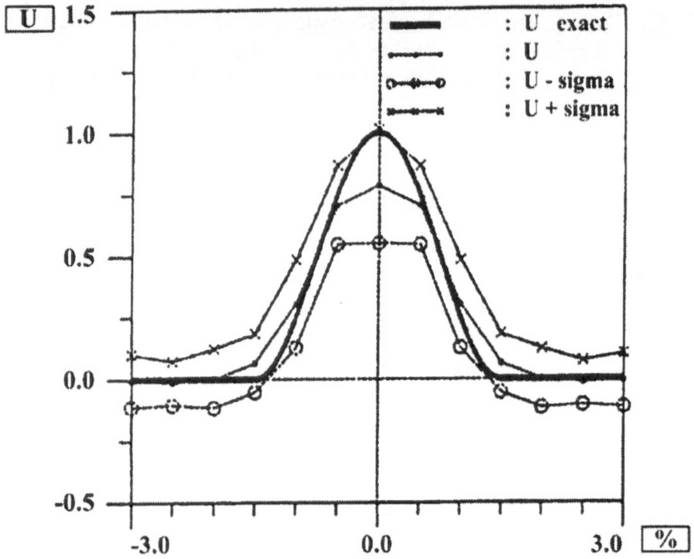

Fig. 1. Surface velocity U of vibrating cylinder: exact and reconstructed values (with bounds). From Ref. 47.

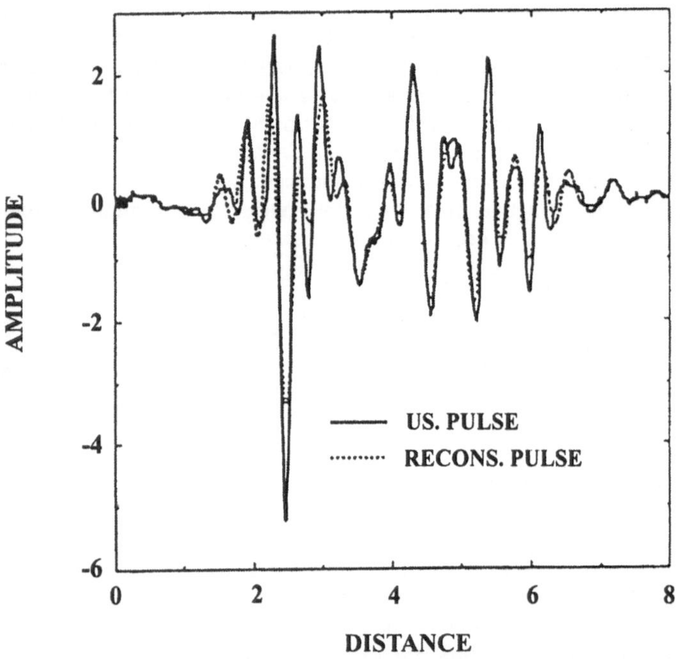

Fig. 2. Reflected ultrasound pulse (solid) and filtered Fourier transform reconstruction (dotted). From Ref. 50.

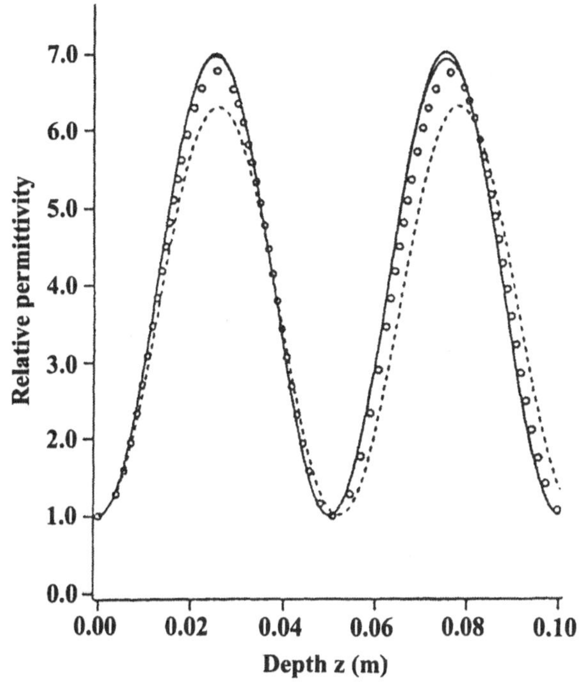

Fig. 3. Permittivity profile of dielectric slab: exact (solid), and reconstructed. From Ref. 51.

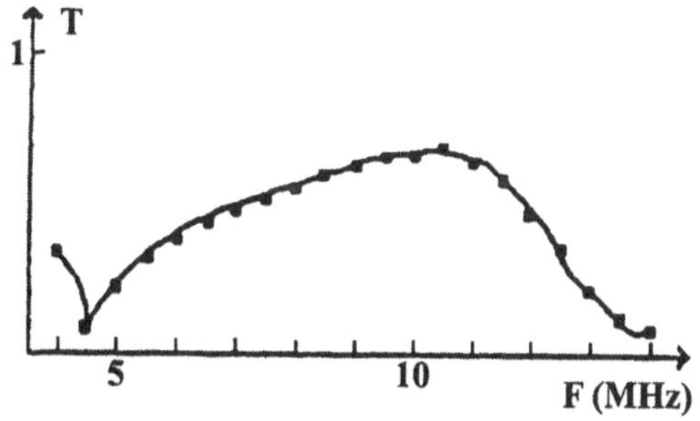

Fig. 4. Acoustic transmission coefficient for PVF2 film on glass, measured (dots) and recovered (solid). From Ref. 54.

2.3 The Inverse Problem of Layered Structures

Layers have been analyzed acoustically both in the laboratory[54-56], and by using field data concerning layered ocean floor sediments[57]. In the mentioned laboratory experiments, the acoustic parameters of a polyvinylidene fluoride (PVF2) film, which is used in the assembly of surface acoustic wave transducers, are determined from transmission coefficient (T) measurements. The PVF2 film is pasted on a glass substrate, with all this being immersed in water. Seven film parameters must be determined: its thickness, density, compressional and shear speed, two components of the stiffness tensor, and the power of the latters' frequency dependence. The calculated and measured values of T can be brought into agreement by least-square minimization, which can be done in several steps: at normal incidence, the reflection coefficient R is independent of absorption, which permits a termination of three parameters (thickness, density, and compressional speed) only from measurements of T. Subsequently, a measurement at non-normal incidence is performed obtaining the remaining parameters, with an adjustment of the three preceding ones. Figure 4 shows T as a function of frequency as calculated using the final values of the parameters (solid curve), and as measured (points), with a precision of a few percent of agreement.

Layers on the seafloor can be identified by acoustic means[57, 58]. Reference 57 has developed two approaches measuring wide-band reflected signals at normal incidence on a multilayered seabottom at various frequencies. Normal incidence causes the dependence on shear waves to cease, so that only (complex) compressional speeds in each layer, and the layer thicknesses can be determined. Multiply reflected pulses occur in addition to those corresponding to direct time-of-arrival. Their first approach, based on the Simplex scheme[59], compares in a least-square sense experimental data to a set of numerical ones and modifies the parameters of the theoretical model until a good fit between experiment and theory is obtained. No parameters need to be known a priori, but the time of flight obtained for an echo pulse is used as an input parameter for the iteration. Their second algorithm, termed "Bottom-3", requires eight measured (complex) reflection coefficients at eight different frequencies, with which the analytic expressions for the (normal incidence) reflection coefficients can be reduced to one nonlinear equation for one layer thickness divided by its complex sound speed. The many possible solutions (due to the ill-posedness of the problem) must be reduced here by imposing bounds on the bottom parameter values dictated by physical reasonableness. Laboratory data were obtained for a bottom model with one sediment and one subbottom halfspace. Figure 5a shows the measured echo signal sequence indicating reflections from the water-sediment and sediment-subbottom interfaces, plus two multiple reflection echoes. The inverse algorithms led to the reflection coefficients vs. frequency in Fig. 5b: dotted: Bottom-3; dashed: Simplex results, as compared to the measured reflection coefficient (solid curve).

3 Resonant Inversion Techniques

3.1 Acoustic Reflection from Layers

While the preceding chapter has presented results of inverse methods not specifically based on any physical features expected in the results, such features - the resonances - will now be discussed as a means for facilitating solutions of the inverse scattering

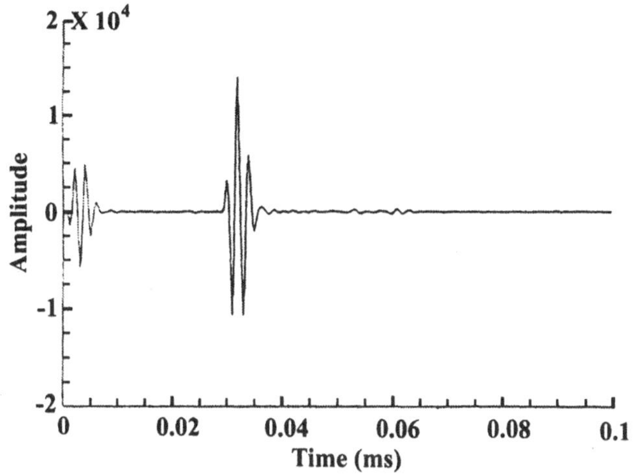

Fig. 5a. Measured signal, reflected from a seafloor model in a tank. From Ref. 57.

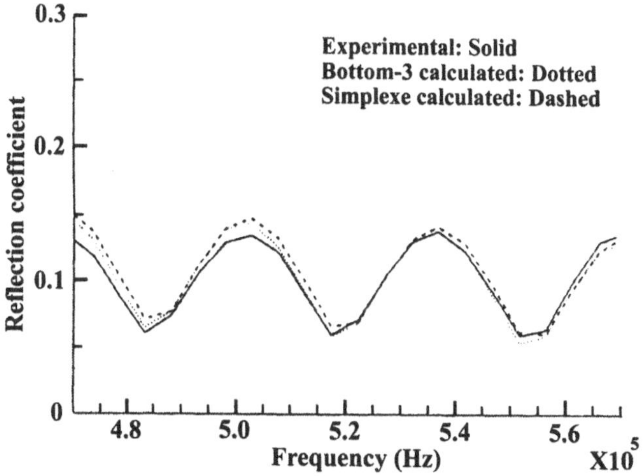

Fig. 5b. Reflection coefficient of seafloor model: measured (solid) and recovered. From Ref. 57.

problem. In radar scattering, the "Singularity Expansion Method" (SEM)[30] is based on the observation that the excitation of the normal modes of a scatterer generates sinusoidally decaying transient signals (i.e. complex exponentials); the general signal thus consists of a series of complex exponentials known as a Prony series[60]. The same is the case for acoustic scattering. Use of the eigenmode resonances for a solution of the inverse problem was first proposed in radar scattering by Moffat and Mains[5]. In acoustics, we have advocated such a use for various purposes[27], with results detailed in Reference 28. One relevant case is, as in the preceding example, the inverse solution for a (fluid) sediment layer on a subbottom which was solved analytically both for the steady-state[10] and the transient case[61], using both normal and oblique incidence.

Short-pulse laboratory experiments solving the inverse scattering problem of a structure with several layers have been carried out at the University of Le Havre[12, 62], based in part on the layer resonances. The structure, all immersed in water, consisted of a polystyrene plastic layer and an aluminum layer separated from each other by a thin water layer. Due to its strong impedance contrast, the latter only played the role of a connector, while the resonances in the two other layers' reflection coefficients (each apparently quite unaffected by the other layer, as were the resonances) were very prominent in the echoes. They were caused by the excitation of guided Lamb waves in the two layers. The inverse procedure consisted in several steps: First, the longitudinal phase velocities in the two solids were determined by a bistatic measurement of the individual plate reflection coefficients out to the critical angles. The thicknesses of the plates were determined from the spectra of the pulse echoes reflected by the individual solid layers, which showed the resonances of various Lamb modes. Their frequencies led to the layer thicknesses via a standing wave argument. The shear wave phase speeds for the plates resulted from a similar standing-wave consideration after separating the two plate wave propagations by their different decay constants. Finally, the plate densities were obtained from the absolute values of the two plate reflection coefficients at normal incidence. Figure 6 shows the guided-wave resonance spectrum at oblique incidence (5°) in which Lamb wave resonances in plastic (A_{ip}, S_{ip}, $i = 1,2,..$) and in aluminum (A_{iA},S_{iA}) are identified.

3.2 Geophysical Layer Resonances

Just as in the case of acoustics of the ocean floor where resonances are prominent in bottom reflection[10, 61] and even in the propagation loss (including shear wave resonances[63]), the same is the case in geophysical subsurface-structure propagation[14] although the corresponding resonance phenomena there apparently have not yet received all the attention they deserve. Remote sensing of underground layers is of importance for prospecting purposes. The usual Thomson-Haskell approach[64] would in general not be very useful for solving the inverse problem (except in the special case of normal incidence[57]), since a multiplicity of parameters would have to be adjusted to get a fit to measured reflection data. Gérard has developed a multilayer reflection/transmission approach[42, 65], valid for any separable geometry, which via a study of layer resonances, furnishes access to the remote sensing problem[43], while simultaneously providing physical insight via the individual reflection/transmission coefficients at layer interfaces. Details of this approach are described in Ref. 43; suffice it to say here that the individual resonances of the mth layer are obtained as the solutions of the equation

$$\det(1 - R_{m,m+1} S_m) = 0$$

where S_m is the (diagonal) matrix of outgoing P and SV wave amplitudes, and R the reflection matrix between the mth and the (m + l) st layer. It is seen that this equation furnishes the resonances of each individual layer (of course modified from those of a free layer by layer coupling) separately, and these can be separately identified in the overall backscattered signal amplitude (such as shown in the preceding discussion of the experiments of Refs. 12 and 62) for purposes of layer identification. If the multiply-reflected amplitudes are desired, one may expand using the Cayley-Hamilton theorem,

$$(1 - R_{m,\,m+1}S_m)^{-1} = \sum_{p=1}^{+\infty} (R_{m,\,m+1}S_m)^{p-1} \, .$$

This is the generalized Debye series[65], which leads to a representation of the solution in the half-space overlying the layered structure, in terms of the various reflection and transmission coefficients at the interfaces of the multilayer system. While the preceding discussion concerns resonances in layers only, the resonances in the scattering from other geophysical inhomogeneities have been discussed by Dubrobskii and Morochnik[14].

3.3 Surface-Wave and Time-Frequency Analysis

It is well known that families of resonances (such as shown e.g. in Fig. 6) are caused by the resonant reinforcement of multiply self-overlapping, phase-matching internal waves. In the case of plates (Fig. 6), these are waves that repeatedly bounce between the plate faces, thereby generating the Lamb waves and (at the resonant frequencies), their resonances[61]. In the case of finite-size scattering objects, they are (dispersive) surface waves that peripherally encircle the object over its surface, mainly in its interior. Phase matching of surface waves was first shown in Ref. 66 to generate the resonances, at resonance frequencies identical with the "natural frequencies" of its normal vibrations.

Ultrasonic waves encircling cylinders ("surface acoustic", or SAW waves) have been employed for the characterization and nondestructive testing of cylindrical objects[67, 68]. In a recent experimental study[45], the wear on the cladding of a nuclear fuel rod has been tested using circumferentially propagating ultrasonic pulses and their echo returns, as well as their resonance spectra. The latter are shown in Fig. 7, (a) near the end of the rod, and (b) towards the middle, indicating larger wear there. The larger attenuation of higher-order return pulses also indicated the amount of wear.

A series of studies were carried out measuring the dispersion curves of Rayleigh wave phase speeds in a chromium-plated steel layer[69], and of Scholte-Stoneley wave speeds in a polyvinyl chloride (PVC) layer glued to an aluminum plate[70, 71], all immersed in water. The experimental model, in the latter case, consisted of a 10mm thick PVC layer which had been prepared in such a way that the velocity profiles of its compressional and its shear bulk waves varied linearly from one face to the other, thereby simulating a consolidated ocean-floor sediment layer which had these properties. A numerical inverse-scattering model was developed, based on the conjugate-gradient formalism, in order to recover these velocity profiles (the other properties of PVC being assumed known) from measurements of the Scholte-Stoneley wave dispersion curve on the structure. Scholte-Stoneley waves on plates are

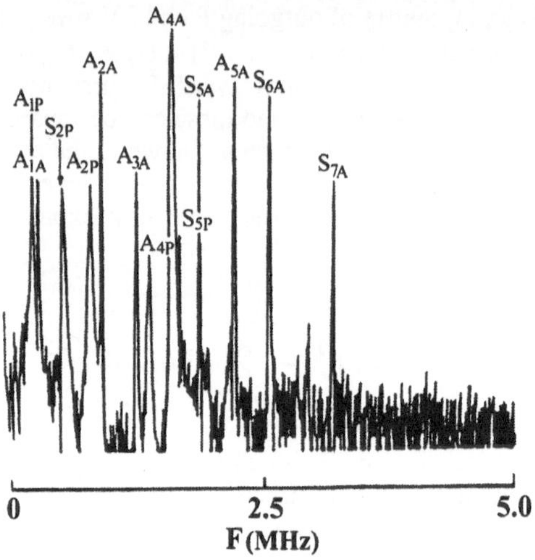

Fig. 6. Observed resonance spectrum of Lamb waves from a three-layer model in a tank. From Ref. 12.

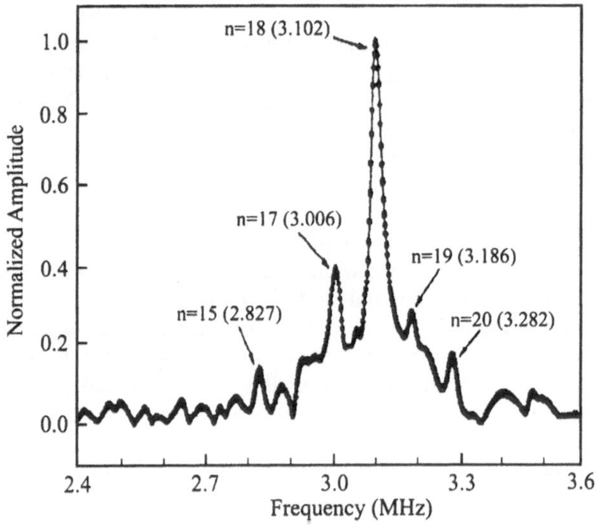

Fig. 7a. Resonance spectrum from peripheral waves on a nuclear fuel rod. From Ref. 45.

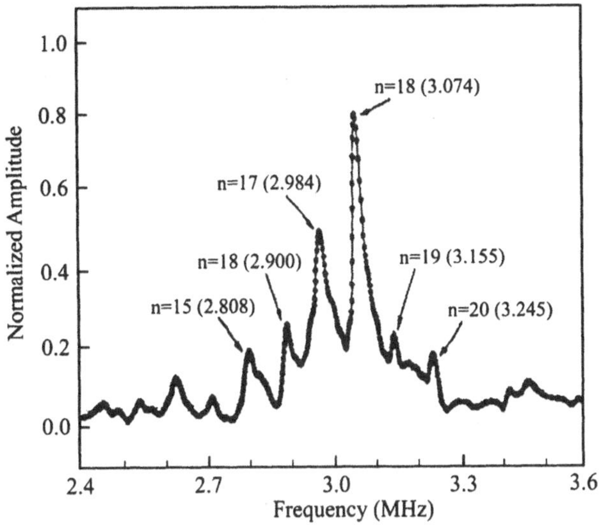

Fig. 7b. Resonance spectrum from a worn portion of a nuclear fuel rod. From Ref. 45.

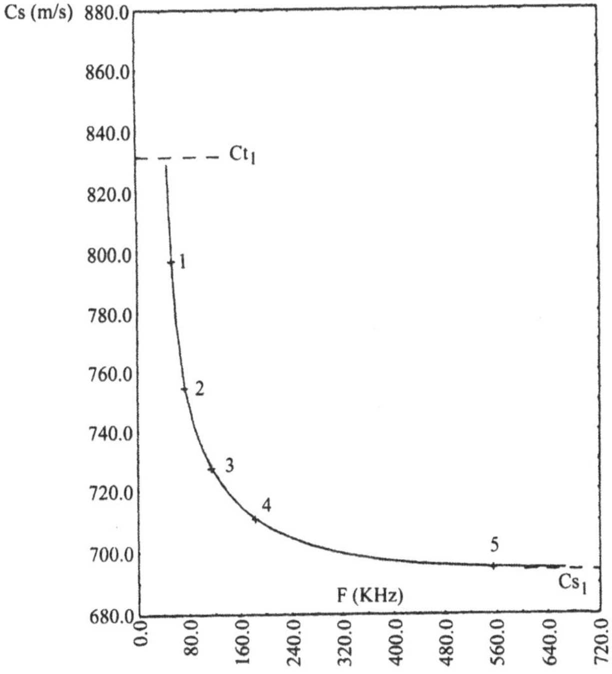

Fig. 8. Dispersion curve of Scholte-Stoneley wave on a layered seafloor model. From Ref. 71.

described in detail in Ref. 43; a calculated dispersion curve of their phase speed C_S for the present experimental model is shown in Fig. 8. A least-square adjustment criterion between experimental values of C_S and those calculated assuming linear bulk speeds in PVC served to determine the bulk speeds to 4% or better.

It is interesting to note that a model similar to the one just described has also been studied experimentally by Ref. 72 using short pulses, and employing for its analysis and inversion the Wigner-Ville distribution. For a time signal x(t), its Wigner-Ville distribution is given by

$$W_x(t,\vartheta) = \int_{-\infty}^{+\infty} x(t + \tau/2)\, x^*(t - \tau/2) \exp(-2i\pi\vartheta t)\, dt ,$$

having the property that the frequency content of an echo pulse can be obtained, and hence the dispersion relations of its different components can be found, from a single measurement using one echo pulse. This leads, e.g., to the dispersive time-frequency curve for W_X, for the sediment model mentioned above, shown in Fig. 9. With the group delay time

$$\tau_g(\vartheta) = \int_{-\infty}^{+\infty} t\, W_x(t,\vartheta)\, dt \Big/ \int_{-\infty}^{+\infty} W_x(t,\vartheta)\, dt ,$$

a (group) dispersion curve for the Scholte-Stoneley wave may then be obtained from Fig. 9, and used for the inversion (determination of the linearly increasing velocity profiles in PVC) of the model.

The Wigner-Ville distribution was earlier used for analyzing acoustic scattering, e.g. from a thin spherical elastic (10% thick brass) shell[46,73]. Figure 10 shows a corresponding time-frequency plot for the Scholte-Stoneley wave echo; the ridges in W_X are clearly frequency dependent and their spacing, at a given frequency, immediate leads via τ_g to the peripheral-wave dispersion curves. In the study of Ref.73, the Wigner-Ville plots for scattering by the swim bladder of isolated fish have also been obtained, and may be used for species classification.

3.4 Uncertainties in Inversion Procedures

The above-mentioned radar scattering response, consisting of a series of decaying sinusoids, naturally represents a Prony-series expansion. The identification of a radar target by the location pattern of its complex-frequency SEM poles is then attempted by expanding the measured time response of a target into a Prony series whose complex exponents represent this pole pattern. This approach is described by Dudley[35], who cautions, however, against the numerical uncertainties involved in the Prony inversion, especially when the data are noisy. Figure 11a shows, here for the example of acoustic scattering from a rigid sphere[74], the pole pattern in the Laplace-s plane ($s = i\omega$), calculated exactly (Δ), and recovered from Prony analysis of synthetic data of calculated echo pulses (other symbols, keeping different number of terms N in the series). The mere fact of time limitations imposed on the synthetic time series (since all real data are time limited) leads to a deteriorated pole identification (Fig.

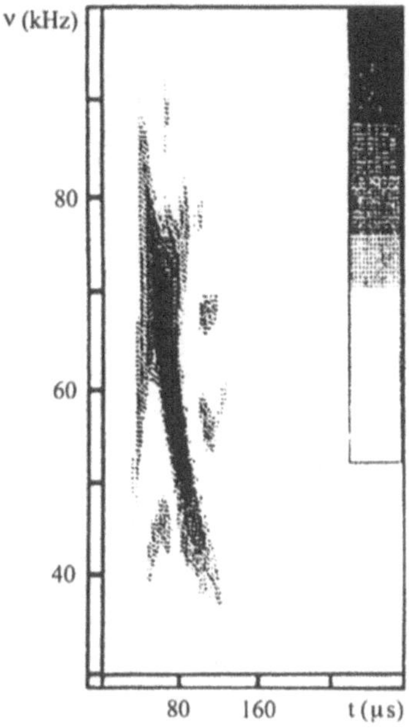

Fig. 9. Time-frequency distribution of return signal from a layered seafloor model. From Ref. 72.

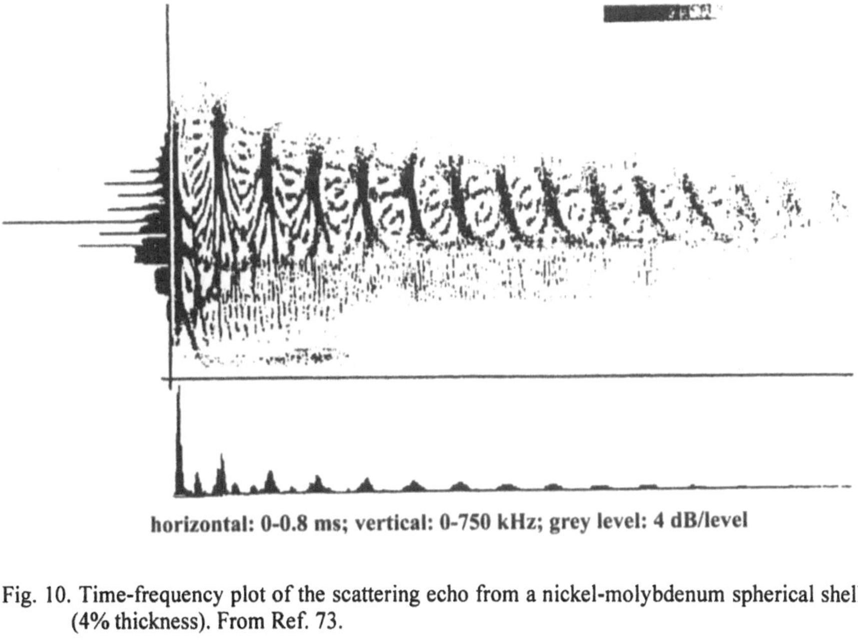

horizontal: 0-0.8 ms; vertical: 0-750 kHz; grey level: 4 dB/level

Fig. 10. Time-frequency plot of the scattering echo from a nickel-molybdenum spherical shell (4% thickness). From Ref. 73.

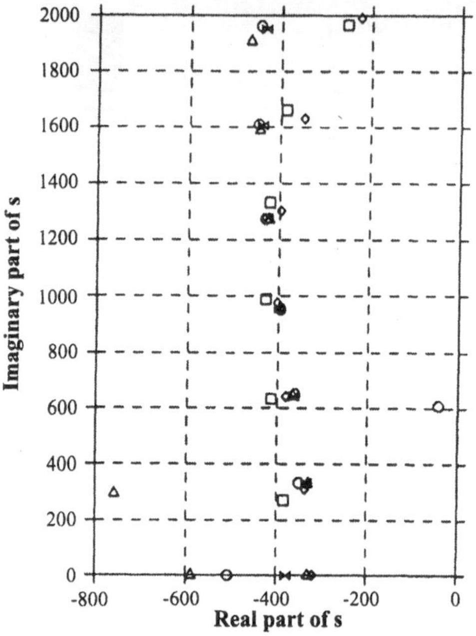

Fig. 11a. Pole pattern in Laplace plane for acoustically rigid sphere (Δ), and recovered by 20 Prony terms (❑) or more. From Ref. 74.

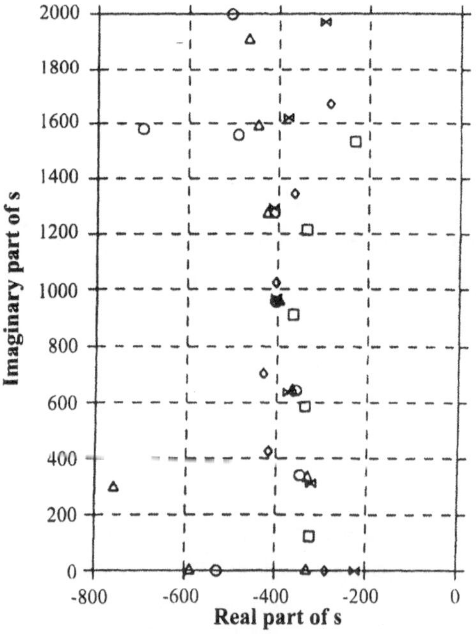

Fig. 11b. Pole pattern in Laplace plane for acoustically rigid sphere (Δ), and recovered by 20 Prony terms (❑) or more, signal terminated early. From Ref. 74.

11b). If even a small amount of Gaussian noise is added to the data, the deterioration becomes much worse[35]. In view of this, Goodman and Dudley have embarked on constructing better inversion algorithms[75]: the Prony series, although mathematically correct, is inefficient and requires an inordinate number of terms for convergence. Another alternative for improving identification beyond the Prony series is due to Sabio[76], in which a "synthetic echo response" is compiled from the data, and Fourier (or wavelet) analyzed to create a set of spectral coefficients ("spectral template"). The spectral coefficients from other data are correlated with the template, and a "target-declaration threshold" is chosen for the correlation coefficient. This method was successfully applied to the target recognition of a set of dipoles, and of a utility truck.

4 Conclusions

Inverse scattering methods, and inverse methods in general, have for their purpose the recognition of targets, and more so, their identification and characterization by remote means. We have reviewed here the applications of this approach in a variety of fields: acoustics and elastodynamics, electromagnetic scattering and tomography, geophysics, and oceanography, in a series of examples chosen from recent research which show that the field of inverse scattering has progressed well into applications of more and more practical nature. Inverse scattering approaches that make use of the target resonances have been specially stressed here, since they were introduced into the field of acoustics by some of the present authors; but more of the related work of ours, not specially discussed here, can be found described in Ref. 28.

References

1. P. C. Sabatier, "Asymptotic properties of the potentials in the inverse scattering problem at fixed energy", J. Math. Phys. 7, 1515-1531, (1966).
2. K. Chadan and P. C. Sabatier, Inverse Problems of Quantum Scattering Theory, 2nd ed., Springer Verlag, Berlin, Heidelberg and New York (1989).
3. P. C. Sabatier, "Exotics of the Schrödinger Problem on the line", in Quantum Inversion Theory and Applications (H. V. Geramb, ed.), Springer Verlag, Berlin, Heidelberg and New York (1994).
4. A. G. Rejpar, A. A. Ksinski, and L. J. White, "Object identification from multi-frequency radar returns", Radio Electron. Eng. 45, 161-167, (1975).
5. D. L. Moffat and R. K. Mains, "Detection and discrimination of radar targets", IEEE Trans. Antennas Propagat., AP-23, 358-367, (1975).
6. J. D. Young, "Radar imaging from ramp response signatures", IEEE Trans. Antennas Propagat., AP-24, 276-282, (1976).
7. See, e.g., J. C. d'Arnaud Gerkens, Foundation of Exploration Geophysics, Elsevier, Amsterdam (1989).
8. M. T. Taner, Long Period Multiples and their Suppression, Seiscom Data, Inc., Technical Report (1975).
9. R. Fiorito, W. Madigosky, and H. Überall, "Acoustic resonances and the determination of the material parameters of a viscous fluid layer", J. Acoust. Soc. Am. 69, 897-903, (1981).

10. A. Nagl, H. Überall, and W. R. Hoover, "Resonances in acoustic bottom reflection and their relation to the ocean bottom properties", IEEE Trans. Geosc. Rem. Sensing GE-20, 332-337, (1982).
11. J. D. Alemar, P. P. Delsanto, E. Rosario, A. Nagl, and H. Überall, "Spectral analysis of the scattering of acoustic waves from a fluid cylinder III: Solution of the inverse scattering problem", Acustica 61, 14-20, (1986).
12. O. Lenoir, J. L. Izbicki, P. Rembert, G. Maze, and J. Ripoche, "Acoustic scattering and inverse problem: Determination of physical parameters of a multilayered structure", Inverse Prob. 7, 369-378, (1991).
13. P. P. Delsanto, J. D. Alemar, E. Rosario, A. Nagl, and H. Überall, "Spectral analysis of the scattering of elastic waves from a fluid-filled cylinder", Materials Eval. 46, 1000-1005, (1988).
14. V. A. Dubrovskii and V. S. Morochnik, "The nonstationary scattering of elastic waves on a spherical inclusion", Izvestiya, Earth Physics 25, 87- 93, (1989), in Russian.
15. D. Colton and R. Kress, Inverse acoustic and electromagnetic theory, Springer Verlag, Berlin, Heidelberg and New York, (1992).
16. M. Bertero and E. R. Pike (eds.), Inverse Problems in Scattering and Imaging, Adam Hilger, Bristol (1992).
17. K. J. Langenberg, "Introduction to the Special Issue on Inverse Problems", Wave Motion 11, 99-112, (1989).
18. A. Kirsch (ed.), Special Issue on Inverse Problems in Scattering Theory, J. Comp. Appl. Math. 42, 1-155, (1992).
19. S. K. Datta (ed.), Elastic Wave Propagation and Ultrasonic Nondestructive Evaluation, Elsevier, Amsterdam (1990).
20. P. C. Sabatier (ed.), Basic Methods of Tomography and Inverse Problems, Adam Hilger, Bristol (1986).
21. W. Tabbara, B. Duchêne, Ch. Pichot, D. Lesselier, L. Chammeloux, and N. Joachimowicz, "Diffraction Tomography: Contribution to the analysis of some applications in microwaves and ultrasonics", Inverse Prob., 4, 305-331, (1988)
22. W. M. Boerner et al., (eds.), Inverse Methods in Electromagnetic Imaging, D. Reidel, Dordrecht (1985).
23. W. M. Boerner and H. Überall (eds.), Radar Target Imaging, Springer Verlag, Berlin, Heidelberg and New York (1994).
24. R. Burridge, "The Gelfand-Levitan, the Marchenko, and the Gopinath-Sondhi integral equations of inverse scattering theory, regarded in the context of inverse impulse response problems", Wave Motion 2, 305-323, (1980).
25. D. Colton and R. Kress, Integral Equation Methods in Scattering Theory, John Wiley, New York, (1983).
26. P. J. Moser and H. Überall, "Complex eigenfrequencies of axisymmetric perfectly conducting bodies: Radar spectroscopy", Proc. IEEE, 71, 171-172, (1983).
27. H. Überall, P .J. Moser, J. D. Murphy, A. Nagl, G. Igiri, J. V. Subrahmanyam, G. C. Gaunaurd, D. Brill, P. P. Delsanto, J. D. Alemar and E. Rosario, "Electromagnetic and acoustic resonance scattering theory", Wave Motion, 5, 307-329, (1983).
28. H. Überall and A. Guran, "Inverse scattering based on the resonances of the target", in Acoustic Interactions with Submerged Elastic Structures (Series on Stability, Vibration, and Control of Structures, 5) A. Guran et al. (eds.), World Scientific, Singapore and New Jersey, Vol. 2 (1996).

29. P. C. Sabatier, "From exact to approximate inverse scattering theory", in Electromagnetic Interactions (Series on Stability, Vibration, and Control of Structures, 5), A. Guran et al. (eds.), World Scientific, Singapore and N. J. (1996).
30. C. E. Baum, "On the singularity expansion method for the solution of electromagnetic interaction problems", Interaction Note 88, Air Force Weapons Lab., Kirtland Air Force Base, Albuquerque, NM 1971.
31. A. J. Berni, "Target identification by natural resonance estimation", IEEE Trans. Aerosp. Elec. Syst. AES-ll, 147-154, (1975).
32. M. L. Van Blaricum and R. Mittra, "A technique for extracting the poles and residues of a system directly from its transient response", IEEE Trans. Antennas Propagat. AP-23, 777-781, (1975).
33. D. G. Ramm, "Extraction of resonances from transient fields", IEEE Trans. Antennas Propagat. AP-33, 223-226, (1985).
34. M. A. Morgan, "Scatterer discrimination based upon natural resonance annihilation", J. Electrom. Waves Applic. 2, 481-502, (1988).
35. D. G. Dudley, "Progress in identification of electromagnetic systems", IEEE Antennas Propagat. Soc. News lett., 30, 5-11, (1988).
36. C. E. Baum, E. J. Rothwell, K. M. Chen, and D. P. Nyquist, "The Singularity Expansion Method and its application to target identification", Proc. IEEE, 79, 1481-1492, (1991).
37. L. Flax, L. R. Dragonette, and H. Überall, "Theory of elastic resonance excitation by sound scattering", J. Acoust. Soc. Am., 63, 723-731, (1978).
38. H. Überall (ed.), Acoustic Resonance Scattering, Gordon and Breach, New York, (1992).
39. H. Überall, G. Maze, and J. Ripoche, "Experiments and analysis of sound-induced structural vibrations", in Wave Motion, Intelliqent Structures, and Nonlinear Mechanics (Series on Stability, Vibration, and Control of Structures, 1), A. Guran and D. J. Inman (eds.), World Scientific, Singapore and New Jersey (1995).
40. N. GESPA, La Diffusion Acoustique (B. Poirée, ed.), Editions CEDOCAR, Paris, (1987).
41. A. Derem, "Relations entre les formations des ondes de surface et l'apparition de résonances dans la diffusion acoustique", Revue du CETHEDEC 58, 43-79, (1979).
42. See, e.g., A. Gérard and H. Überall, "Generalized Debye series for acoustic scattering from objects of separable geometric shape", preprint.
43. H. Überall, A. Gérard, A. Guran, J. Duclos, M. El Hocine Khelil, X. L. Bao, and P. K. Raju, "Acoustic scattering resonances: relation to external and internal surface waves", Appl. Mech. Rev., 49, S63-S71, (1996).
44. J. D. Kaplunov, E. V. Nol'de, and N. D. Veksler, "Determination of parameters of elastic layer by measured dispersion curves of zero-order Lamb-type waves", Proc. Estonian Acad. Sci. Phys. Math., 41, 39-48, (1992).
45. M. S. Choi, J. S. Joo, and J. P. Lee,"Evaluation of a nuclear fuel rod by ultrasonic resonance scattering", in New Perspectives on Problems in Classical and Quantum Mechanics, A. W. Saenz and P. P. Delsanto (eds.), Gordon and Breach, New York (1996).
46. P. Flandrin, J. Sageloli, J. P. Sessarego, and M. Zakharia, "Application of time-frequency analysis to the characterization of surface waves on elastic targets," Acoust. Lett., 10, 23-28, (1986).

47. M. Bonnet, "A numerical investigation for a source inverse problem in linear acoustics", J. Acoustique, 4, 307-334, (1991).
48. A. Tikhonov and V., Arsenine, Méthodes de Résolution de Problèmes Mal Posés, Editions Mir (1976).
49. A. Tarantola, Inverse Problem Theory, Elsevier, Amsterdam (1987).
50. W. Tobocman, "Application of wavelet analysis to inverse scattering", in Acoustic Interactions with Submerged Elastic Structures (Series on Stability, Vibration, and Control of Structures, 5), A. Guran et al. (eds.), World Scientific, Singapore and New Jersey, Vol. 1, (1996).
51. S. He and S. Ström, "The electromagnetic inverse problem in the time domain for a dissipative slab and a point source using invariant imbedding: Reconstruction of the permittivity and conductivity", J. Comp. Appl. Math., 42, 137 (1992).
52. G. Kristensson and R. J. Krueger, "Direct and inverse scattering in the time domain for a dissipative wave equation I, II", J. Math. Phys. 27, 1667-1683, (1986).
53. A. G. Tijhuis, "Iterative determination of permittivity and conductivity profiles of a dielectric slab in the time domain", IEEE Trans. Antennas Propagat., AP-29, 239 (1981).
54. M. Mortaki, B. Chenni, J. Duclos, and M. Leduc, "Transmission coefficient and acoustical parameters of polyvinylidene flouride film", Proceed. European Conf. on Underwater Acoustics (M. Weydert, ed.), Luxemburg, Sept. 1992, Elsevier, London and New York, 189-192, (1992).
55. M. Mortaki, B. Chenni, J. Duclos, and M. Leduc, "The Newton-Raphson and Conjugate Gradient algorithms for the acoustic characterization of thin film", Proceed. Ultrasonics International 93, Vienna, Austria, June 1993, Butterworth, London, 827-830, (1993).
56. M. Mortaki, B. Chenni, J. Duclos, and M. Leduc, "Characterization of layers or films by an immersion method and estimation of the imprecision of the evaluated parameters", Proceed. Second European Conference on Underwater Acoustics (L. BjØrnØ, ed.), European Commission, 443-448, (1994).
57. L. BjØrnØ, J. S. Papadakis, P. J. Papadakis, J. Sageloli, J. P. Sessarego, S. Sun, and M. I. Taraoudakis, "Identification of seabed data from acoustic reflections: Theory and experiment", Acta Acustica 2, 359-374, (1994).
58. S. D. Rajan, J. F. Lynch, and G. V. Frisk, "Perturbative inversion methods for obtaining bottom geoacoustic parameters in shallow water", J. Acoust. Soc. Am., 82, 998-1017, (1987).
59. P. B. Ryan, R. L. Barr, and H. D. Todd, "Simplex techniques for non-linear optimization", Analyt. Chemistry, 52, 1460-1467, (1980).
60. R. Prony, "Essai expérimental et analytique, etc.", Paris, J. de l'Ecole Polytechnique 1, cahier 2, 24-76, (1795).
61. A. Nagl, H. Überall, and K. B. Yoo, "Acoustic Exploration of ocean floor properties based on the ringing of sediment layer resonances", Inverse Prob., 1, 99-110, (1985).
62. O. Lenoir, J. L. Izbicki, P. Rembert, G. Maze, and J. Ripoche, "Acoustic scattering from an immersed plane multilayer: Application to the inverse problem", J. Acoust. Soc. Am. 91, 601-612, (1992).
63. S. J. Hughes, D. D. Ellis, D. M. F. Chapman, and P. R. Staal, "Low- frequency acoustic propagation loss in shallow water over hard-rock seabeds covered by a thin layer of elastic-solid sediment", J. Acoust. Soc. Am. 88, 283-297, (1990).

64. See, e.g.' D. L. Folds and C. D. Loggins, "Transmission and reflection of ultrasonic waves in layered media", J. Acoust. Soc. Am., 62, 1102-1109, (1977).

65. A. Gérard, "Diffraction d'ondes de cisaillement par une sphère élastique", C. R. Acad. Sci. Paris, 278, 1055-1057, (1974).

66. H. Überall, L. R. Dragonette, and L. Flax, "Relation between creeping waves and normal modes of vibration of a curved body", J. Acoust. Soc. Am., 61, 711-715, (1977).

67. P. B. Nagy M. Blodgett, and M. Golis, "Weephole inspection by circumferential creeping waves", NDT&E International, 27, 131-142, (1994).

68. U. Kawald, C. Desmet, W. Lauriks, C. Glorieux, and J. Thoen, "Investigation of the dispersion relations of surface acoustic waves propagating on a layered cylinder", J. Acoust. Soc. Am., 99, 926-930, (1996).

69. J. Pouliquen, A. Defebvre, and B. Chenni, "Velocity dispersion of Rayleigh waves on chromium-plated steel", Proceed. 12th International Congress on Acoustics, Toronto, Canada, July 1986, 2, G5-2., (1986)

70. A. Defebvre, J. Pouliquen, and L. M. Moukala, "Caracterisation de milieux plans stratifiés par ondes de Stoneley-Scholte", Rapport de synthèse, Faculté Libre des Sciences, Laboratoire d'Acoustique-Ultrasons, Lille, Oct. 1989.

71. A. Defebvre, J. Pouliquen, and L. M. Mukala, "Contribution to Sea-Botto characterization from Stoneley-Scholte waves celerity dispersion in special cases of media having linear sound celerity profiles", Proceed. Ultrasonics International 93, Vienna, Austria, June 1993, Butterworth, London, 591-594, (1993).

72. J. Guilbot and M. Zakharia, "Caractérisation d'un modèle de milieu sédimentaire marin par étude de la dispersion d'une onde d'interface", J. de Physique IV, Colloque C1, suppl. J. de Physique III, vol. 2, C1-845-848, (1992); J. Guilbot and M. E. Zakharia "Tank experiments on a sediment small-scale model (continuously layeréd medium): Profile inversion via surface acoustic waves", Proceed. European Conf. on Underwater Acoustics (M. Weydert, ed.), Luxembourg, Sept. 1992, Elsevier, London and New York, 543-546, (1992).

73. M. E. Zakharia, F. Magand, J. Sageloli and J. P. Sessarego, "Time- frequency approaches for sonar target descriptions: Application to fisheries", NATO ASI on Acoustic Signal Processing for Ocean Exploration, Madeira, Aug. 1992. NATO, 541-546, (1993).

74. R. R. Weyker and D. G. Dudley, "Identification of resonances on an acoustically rigid sphere", IEEE J. of Ocean Engin. OE-12, 317-326, (1987).

75. D. M. Goodman and D. G. Dudley, "An output error model and algorithm for electromagnetic system identification", Circuits Sys. Sig. Process, 6, 471-505, (1987).

76. V. Sabio, "Spectral correlation of wideband target resonances", SPIE '95 Conference, Orlando, FL, in Signal Processing, Sensor Fusion, and Target Recognition IV (I. Kadar and V. Libby, eds.), SPIE Vol., 2484, 567-573, (1995); V. Sabio", Spectral correlation of wideband target resonances", SPIE '96 Conference, Orlando, FL, in Algorithms for synthetic Aperture Radar Imagery III (E. G. Zelnio and D. J. Douglass, eds.), SPIE Vol., 2757, 145-151, (1996).

Inverse 3D Acoustic and Electromagnetic Obstacle Scattering by Iterative Adaptation

Martin Haas, Wolfgang Rieger, Wolfgang Rucker, and Günther Lehner

Institut für Theorie der Elektrotechnik, Universität Stuttgart, Pfaffenwaldring 47, D-70550 Stuttgart, Germany

Abstract. The inverse three-dimensional time-harmonic scattering problem of reconstructing the starlike and smooth boundary Λ of an impenetrable obstacle from its far field scattering data, for both, the acoustic and electromagnetic case, is considered. An approach, based on a method proposed by Kirsch and Kress [2], that employs weak a priori knowledge by choosing an auxiliary curve Γ inside the searched boundary Λ is used. Initial reconstructions are improved using an iteration scheme to adapt the internal surface Γ by exploiting information on the reconstruction Λ of the previous step. The adaptation algorithm yields significant improvements on Λ, provided a reasonable first reconstruction may be obtained.

1 Introduction

In mathematical physics, scattering of a time-harmonic incident wave from an impenetrable obstacle constitutes an exterior boundary value problem for the Helmholtz equation. Let D_- denote the interior of a simply connected obstacle, $\Lambda = \partial D_-$ its smooth boundary, and $D_+ = \mathbb{R}^3 / \bar{D}_-$ the region outside the obstacle. All time dependencies are given by the factor $\exp(-i\omega t)$ with angular frequency ω, which, for simplicity, is omitted in the sequel. We assume that the origin of the coordinate system is contained in D_- and that there is a homogeneous, isotropic and non-absorbing medium in D_+. Throughout the paper, the unit normal vector $\mathbf{n}$ to any closed surface is always directed outward into the exterior domain of the surface.

In the acoustic case, we may choose a scalar velocity potential u to fully describe all scattering quantities. Let k denote the wave number and Δ the scalar Laplace operator. Then, under suitable assumptions [2], the (linearized) acoustic scattering problem is defined by

$$(\Delta + k^2)\, u(\mathbf{r}) = 0\,, \quad \mathbf{r} \in D_+ \,, \tag{1}$$

where the total field u is given by superposition of the known incident field and the scattered field,

$$u(\mathbf{r}) = u^{\mathrm{inc}}(\mathbf{r}) + u^{\mathrm{s}}(\mathbf{r})\,, \quad \mathbf{r} \in D_+ \,. \tag{2}$$

The scattered field u^{s} has to comply with the *Sommerfeld radiation condition* for all directions $\hat{\mathbf{r}} = \mathbf{r}/|\mathbf{r}| \in \Omega$, Ω denoting the unit sphere,

$$\lim_{r\to\infty} r \left\{ \frac{\partial u^{\mathrm{s}}(\mathbf{r})}{\partial r} - iku^{\mathrm{s}}(\mathbf{r}) \right\} = 0\,, \quad r = |\mathbf{r}| \,, \tag{3}$$

in order to represent a *radiating* solution. The interaction between wave and surface of the obstacle is modeled by a boundary condition. A perfectly *soft* surface leads to a Dirichlet problem with homogeneous boundary condition

$$u(\mathbf{r}) = 0\,, \quad \mathbf{r} \in \Lambda\ , \tag{4}$$

whereas a perfectly *hard* surface leads to a Neumann problem with homogeneous boundary condition since the obstacle's rigidity implies a vanishing normal derivative of the total field,

$$\frac{\partial u(\mathbf{r})}{\partial n} = 0\,, \quad \mathbf{r} \in \Lambda\ . \tag{5}$$

In the electromagnetic case, we choose the electric field $\mathbf{E}$ as the quantity of special interest. Since in D_+ we have neither currents nor charges, from the time-harmonic Maxwell equations

$$\left.\begin{aligned} \operatorname{curl}\mathbf{H}(\mathbf{r}) &= -\mathrm{i}\omega\varepsilon\mathbf{E}(\mathbf{r}) \\ \operatorname{curl}\mathbf{E}(\mathbf{r}) &= +\mathrm{i}\omega\mu\mathbf{H}(\mathbf{r}) \end{aligned}\right\}\,, \quad \mathbf{r} \in D_+ \tag{6}$$

with electric permittivity ε and magnetic permeability μ [6], we arrive at the homogeneous vector Helmholtz equation for divergence free solutions $\mathbf{E}$,

$$(\boldsymbol{\Delta} + k^2)\,\mathbf{E}(\mathbf{r}) = \mathbf{0}\,, \quad \operatorname{div}\mathbf{E}(\mathbf{r}) = 0\,, \quad \mathbf{r} \in D_+\ . \tag{7}$$

Again, the total field may be seen to be a superposition of the incident and the scattered field,

$$\mathbf{E}(\mathbf{r}) = \mathbf{E}^{\mathrm{inc}}(\mathbf{r}) + \mathbf{E}^{\mathrm{s}}(\mathbf{r})\,, \quad \mathbf{r} \in D_+\ . \tag{8}$$

Let $\eta = \sqrt{\mu/\varepsilon}$ denote the wave impedance, and $\hat{\mathbf{r}} = \mathbf{r}/|\mathbf{r}|$. Then, similar as in the acoustic case, the *Silver-Müller radiation condition*,

$$\lim_{r\to\infty} r\,(\mathbf{H}^{\mathrm{s}}(\mathbf{r}) \times \hat{\mathbf{r}} - \mathbf{E}^{\mathrm{s}}(\mathbf{r})/\eta) = \mathbf{0} \quad \forall\, \hat{\mathbf{r}} \in \Omega\,, \quad r = |\mathbf{r}|\ , \tag{9}$$

ensures the solution to be a radiating one. Since the obstacle is assumed to be infinitely conductive the tangential component $\mathbf{n} \times \mathbf{E}$ of the total electric field needs to vanish on the surface Λ,

$$\mathbf{n} \times \mathbf{E} = \mathbf{0} \quad \text{on } \Lambda\ . \tag{10}$$

Our scattering data for inversion were simulated by numerically solving the respective forward problems. All boundary integral equations were evaluated by using the boundary element method (BEM). We did not take any special measure to handle the case of internal resonances, but in practice we never encountered any problems referring thereto.

In order to rule out so called "inverse crimes" [2], we used different basic equations and/or numerical procedures for direct and inverse problems.

The direct electromagnetic problems were solved using so called vectorial *current basis functions* (CBFs) [8] for linear triangular elements to model the surface current density and by applying the Galerkin method to equation (18), s. section 3, whereas for the inverse electromagnetic problem we used nodal quadratic elements and the collocation method.

In the acoustic case we used the double-layer equation (12) for the direct sound-soft problem and the single-layer equation (15) for the direct sound-hard problem. Both types of acoustic inverse problems were solved using the single-layer approach, i.e. only for the sound-hard problems we used the same basic equation and numerical procedure for the forward and inverse problem, however, and this is true for all of our examples, we used much finer meshes for the direct calculations than for the inverse ones.

The relative error in percent, used to specify the deviation of a disturbed function $\tilde{f}$ from its original f, is defined by $e := \|\tilde{f} - f\|_{L^2} \,/\, \|f\|_{L^2} * 100$.

To solve the inverse problem of finding the obstacle's boundary from its known boundary condition type and its far field scattering data, we used an iterative reconstruction technique that is based on a method proposed by Kirsch and Kress [2]. In an earlier paper [3], we worked out the details and demonstrated the applicability of the method for two-dimensional problems, where no distinction between acoustic and electromagnetic case is necessary. The present paper deals with the extension to three dimensions.

In the following two sections we briefly describe the approaches to solve the direct obstacle scattering problems for both the acoustic and electromagnetic case. Since the problems under consideration are defined in the outer region D_+, which is infinitely extended, integral equation formulations are most suitable since they allow not only theoretical considerations but also serve as a base for numerical solution schemes. Section 4 gives some aspects of the reconstruction method and in section 5 we comment on the regularization technique used and the problem of finding reasonable regularization parameters. Section 6 follows with details to the adaptation technique. Finally, section 7 closes with numerical examples.

2 Acoustic Scattering from Impenetrable Obstacles

Mathematically, the solution to both, the Dirichlet- and the Neumann problem could be given either as the field of a single-layer on Λ, or, alternatively, as the field of a double-layer on Λ [1]. Since, for numerical reasons, we preferred to have integral equations of the second kind rather than equations of the first kind, we chose to use a double-layer representation for the Dirichlet problem and a single-layer representation for the Neumann problem.

The *far field pattern* u^s_∞ we used for inversion is defined on the unit sphere Ω as in [2] by the asymptotic behavior of the scattering solution,

$$u^s(\mathbf{r}) = \frac{\exp(ikr)}{r} \left\{ u^s_\infty(\hat{\mathbf{r}}) + O\left(\frac{1}{r}\right) \right\} , \quad r \to \infty \,. \tag{11}$$

This means that far from the scatterer, locally, the scattered wave behaves like a plane wave with decay $1/r$, propagating in direction $\hat{\mathbf{r}}$.

Due to the jump relation of the double-layer potential the second kind integral equation

$$(\mathcal{I}+\mathcal{K})\tau := \tau(\mathbf{r}) + 2\int_{\Lambda} \tau(\mathbf{r}')\frac{\partial\Phi(\mathbf{r},\mathbf{r}')}{\partial n'}\,\mathrm{d}s' = 2\,u^{\mathrm{s}}_{+}(\mathbf{r})\,, \quad \mathbf{r}\in\Lambda \tag{12}$$

holds for the double-layer density τ, where the fundamental solution Φ of the scalar Helmholtz equation is given by $\Phi = 1/(4\pi)\exp(\mathrm{i}k|\mathbf{r}-\mathbf{r}'|)/|\mathbf{r}-\mathbf{r}'|$ and where u^{s}_{+} denotes the outer limit of the scattered field u^{s} towards the surface Λ. According to (2) and (4), $u^{\mathrm{s}}_{+} = -u^{\mathrm{inc}}(\mathbf{r})$ for $\mathbf{r}\in\Lambda$ has to be chosen. After having solved (12) the scattered field at any point in D_{+} for a sound-soft obstacle is given by

$$u^{\mathrm{s}}(\mathbf{r}) = \mathcal{W}\tau := \int_{\Lambda} \tau(\mathbf{r}')\frac{\partial\Phi(\mathbf{r},\mathbf{r}')}{\partial n'}\,\mathrm{d}s'\,, \quad \mathbf{r}\in D_{+}\ , \tag{13}$$

and the corresponding far field pattern with $\Phi_{\infty} = 1/(4\pi)\exp(-\mathrm{i}k\hat{\mathbf{r}}\mathbf{r}')$ by

$$u^{\mathrm{s}}_{\infty}(\hat{\mathbf{r}}) = \mathcal{F}_{\tau}\tau := \int_{\Lambda} \tau(\mathbf{r}')\frac{\partial\Phi_{\infty}(\hat{\mathbf{r}},\mathbf{r}')}{\partial n'}\,\mathrm{d}s'\,, \quad \hat{\mathbf{r}}\in\Omega\ . \tag{14}$$

Due to the jump relation for the normal derivative of the single-layer potential, the Neumann problem leads to the second kind integral equation

$$(\mathcal{I}-\mathcal{K}')\sigma := \sigma(\mathbf{r}) - 2\int_{\Lambda} \sigma(\mathbf{r}')\frac{\partial\Phi(\mathbf{r},\mathbf{r}')}{\partial n}\,\mathrm{d}s' = -2\frac{\partial}{\partial n}u^{\mathrm{s}}_{+}(\mathbf{r})\,, \quad \mathbf{r}\in\Lambda \tag{15}$$

for the single-layer density σ where according to (2) and (5) again $u^{\mathrm{s}}_{+} = -u^{\mathrm{inc}}(\mathbf{r})$ for $\mathbf{r}\in\Lambda$ has to be chosen. After having solved (15) the scattered field at any point in D_{+} for a sound-hard obstacle is given by

$$u^{\mathrm{s}}(\mathbf{r}) = \mathcal{V}\sigma := \int_{\Lambda} \sigma(\mathbf{r}')\Phi(\mathbf{r},\mathbf{r}')\,\mathrm{d}s'\,, \quad \mathbf{r}\in D_{+}\ , \tag{16}$$

and the corresponding far field pattern by

$$u^{\mathrm{s}}_{\infty}(\hat{\mathbf{r}}) = \mathcal{F}_{\sigma}\sigma := \int_{\Lambda} \sigma(\mathbf{r}')\Phi_{\infty}(\hat{\mathbf{r}},\mathbf{r}')\,\mathrm{d}s'\,, \quad \hat{\mathbf{r}}\in\Omega\ . \tag{17}$$

3 Electromagnetic Scattering from Perfectly Conductive Obstacles

The electromagnetic scattered field can be given either in terms of electric surface currents, or in terms of magnetic surface currents, which are equivalent to a double-layer of electric currents, on Λ. Since the obstacle is thought to be electrically infinitely conductive, electric surface currents represent the physical sources of the field, while the alternative of assuming magnetic surface currents is of a purely formal nature. Determining the electric surface currents from the tangential component of the electric field $\mathbf{E}^{\mathrm{s}}(\mathbf{r})$ on Λ leads to a boundary integral equation of the first kind [1], whereas the latter variant leads to an integral equation of the second kind due to the jump relation for the electric field $\mathbf{E}^s(\mathbf{r})$ at a magnetic surface current. Again, for numerical reasons, we preferred the second kind integral equation. Thus, for the magnetic surface current density $\mathbf{m}$, there holds

$$\mathbf{m}(\mathbf{r}) + 2\int_\Lambda \mathbf{n_r} \times \operatorname{curl}_\mathbf{r}\{\Phi(\mathbf{r},\mathbf{r}')\,\mathbf{m}(\mathbf{r}')\}\,\mathrm{d}s' = -2\,\mathbf{n_r} \times \mathbf{E}^s_+(\mathbf{r})\,, \quad \mathbf{r} \in \Lambda\,, \tag{18}$$

where $\mathbf{E}^s_+$ denotes the outer limit of the scattered electric field $\mathbf{E}^s$ towards the surface Λ. Knowing the layer $\mathbf{m}$, the scattered field is given by

$$\mathbf{E}^s(\mathbf{r}) = \mathcal{E}\mathbf{m} := -\operatorname{curl}\int_\Lambda \Phi(\mathbf{r},\mathbf{r}')\,\mathbf{m}(\mathbf{r}')\,\mathrm{d}s'\,, \quad \mathbf{r} \in D_+\,, \tag{19}$$

and the corresponding far field $\mathbf{E}^s_\infty$, defined in complete analogy to (11), reads

$$\mathbf{E}^s_\infty(\hat{\mathbf{r}}) = \mathcal{F}_\mathbf{m}\mathbf{m} := -\hat{\mathbf{r}} \times ik\int_\Lambda \Phi_\infty(\hat{\mathbf{r}},\mathbf{r}')\mathbf{m}(\mathbf{r}')\,\mathrm{d}s'\,, \quad \hat{\mathbf{r}} \in \Omega\,. \tag{20}$$

4 The Reconstruction Method

According to a method described in [2], an auxiliary closed surface Γ inside Λ is chosen, s. Fig. 1. This initial choice of Γ requires some a priori information about the size and the location of the obstacle.

In the acoustic case, given the far field $u^s_\infty(\hat{\mathbf{r}})$, we seek to represent the scattered field by a single-layer on the interior curve Γ by taking (17) as integral equation for the single-layer density σ on Γ,

$$\int_\Gamma \Phi_\infty(\hat{\mathbf{r}},\mathbf{r}')\sigma(\mathbf{r}')\,\mathrm{d}s' = u^s_\infty(\hat{\mathbf{r}})\,, \quad \hat{\mathbf{r}} \in \Omega\,. \tag{21}$$

Having computed σ, the near field $u^s(\mathbf{r})$ is given by (16), Λ being replaced by Γ, and the total field by superposing incident and scattered field. The boundary Λ is now found by searching a closed curve Λ along which the boundary condition is matched in a minimal residual norm sense, i.e. for the sound-soft obstacle Λ outside Γ is searched such that

$$F_1 = \|B_1\|^2_{L^2(\Lambda)} \stackrel{!}{=} \min\,, \quad B_1(\mathbf{r}) = \left(\mathcal{V}\sigma + u^{\mathrm{inc}}\right)(\mathbf{r})\,, \quad \mathbf{r} \in \Lambda\,, \tag{22}$$

whereas the sound-hard obstacle requires the normal derivative of the total field to vanish, i.e. for Λ outside Γ

$$F_2 = \|B_2\|^2_{L^2(\Lambda)} \stackrel{!}{=} \min\,, \quad B_2(\mathbf{r}) = \left(\frac{\partial}{\partial n}\left(\mathcal{V}\sigma + u^{\mathrm{inc}}\right)\right)(\mathbf{r})\,, \quad \mathbf{r} \in \Lambda\,. \tag{23}$$

Given the electromagnetic far field $\mathbf{E}^s_\infty(\hat{\mathbf{r}})$, in analogy to the acoustic case, a far field to near field transformation is performed by taking (20) as integral equation for $\mathbf{m}$ on Γ,

$$-\hat{\mathbf{r}} \times ik\int_\Gamma \Phi_\infty(\hat{\mathbf{r}},\mathbf{r}')\mathbf{m}(\mathbf{r}')\,\mathrm{d}s' = \mathbf{E}^s_\infty(\hat{\mathbf{r}})\,, \quad \hat{\mathbf{r}} \in \Omega\,, \tag{24}$$

to represent the reconstructed field by magnetic surface currents on Γ. The obstacle's boundary is now found by requiring that, for Λ outside Γ,

$$F_3 = \|B_3\|^2_{L^2(\Lambda)} \stackrel{!}{=} \min , \quad B_3(\mathbf{r}) = \left|\mathbf{n} \times \left(\mathcal{E}\mathbf{m} + \mathbf{E}^{\mathrm{inc}}\right)\right|(\mathbf{r}) , \quad \mathbf{r} \in \Lambda . \tag{25}$$

Since only starlike boundaries are considered, Λ may be represented by a radial function of the usual polar coordinates θ and φ in the form

$$\Lambda \; : \; \mathbf{r}(\theta,\varphi) = \hat{\mathbf{r}}(\theta,\varphi) r_\Lambda(\theta,\varphi) = \hat{\mathbf{r}} r_\Lambda(\hat{\mathbf{r}}) \; . \tag{26}$$

As the authors in [2], we used a truncated Fourier series of spherical harmonics for $0 \leq \theta \leq \pi$ and $0 \leq \varphi < 2\pi$,

$$r_\Lambda(\theta,\varphi) = \sum_{n=0}^{N} \sum_{m=0}^{n} P_n^m(\cos\theta) \left\{a_{nm}\cos(m\varphi) + b_{nm}\sin(m\varphi)\right\} \; , \tag{27}$$

to represent a finite dimensional set of starlike surfaces. To numerically minimize the functionals in (22), (23) and (25), a discrete set of collocation points on Λ is necessary. We used points on Λ in predefined directions $\hat{\mathbf{r}}_l, l = 1, \ldots, M$ such that they were about equally distributed on the unit sphere. After discretization, for $j = 1, 2, 3$, the functionals (22), (23) and (25) read

$$F_j(a_{00}, \ldots, a_{NN}, b_{00}, \ldots, b_{NN}) = \sum_{l=1}^{M} \left|B_j\left(\hat{\mathbf{r}}_l r_\Lambda(\theta_l,\varphi_l)\right)\right|^2 \stackrel{!}{=} \min \; . \tag{28}$$

The F_j are nonlinear functionals of $(N+1)^2$ unknown coefficients ($b_{n0} \equiv 0$). We used a modified Levenberg-Marquardt algorithm (routine LMDIF1 from MINPACK, jacobian approximated by finite differences) for the numerical minimization of the F_j.

5 Regularization and Parameter Choice

The discretization of the ill-posed integral equations (21) and (24) leads to ill-conditioned linear systems. We compared conjugate gradients stopped after a suitable number of iterations [4], truncated SVD (singular value decomposition), and Tikhonov regularization to obtain regularized solutions. The results did not depend very much on the method used.

In contrast, it is crucial to find reasonable regularization parameters. For all of our examples presented in section 7, we used the CGLS-algorithm (conjugate gradients least squares) from [9] to solve the overdetermined linear systems. Our parameter choice strategy was twofold, depending on whether we had "rather exact" far field data (as good as our forward solvers could simulate them) or whether they were contaminated by some (artificial) noise. We were led to do so by L-curve plots [5] for various examples and adaptation steps, showing logarithmically the relation between the norms of the layer densities $\|\sigma\|$ and $\|\mathbf{m}\|$ versus the norms of the corresponding discrepancies $\|\mathcal{F}_\sigma \sigma - u^{\mathrm{s}}_\infty\|$ and $\|\mathcal{F}_\mathbf{m}\mathbf{m} - \mathbf{E}^{\mathrm{s}}_\infty\|$,

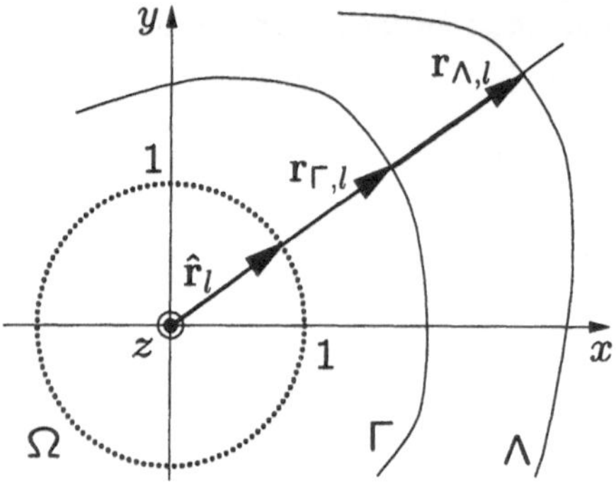

Fig. 1. unit circle Ω, inner auxiliary surface Γ, and reconstruction Λ

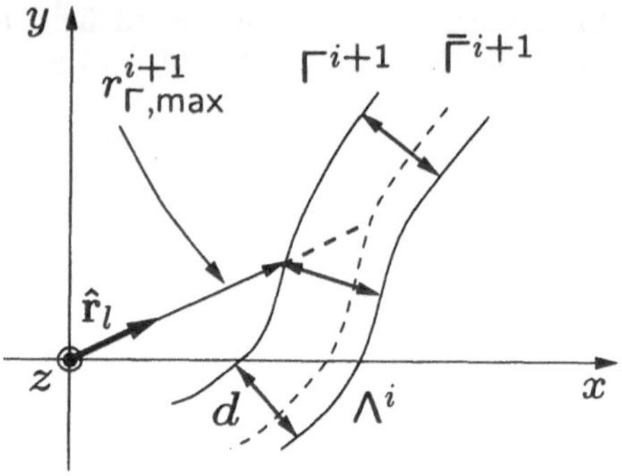

Fig. 2. Modification process to get a new auxiliary curve Γ^{i+1} from $\bar{\Gamma}^{i+1}$

respectively, as a function of the regularization parameter, i.e. in our case for different CG-iteration steps.

With "exact" far field data, we never reached the "corner" of the L-curve with a reasonable number (< 8000) of CG-iteration steps. Since we got stable solutions for any feasible number of iterations, we used a simple predefined sequence $J^{(i)}$ of CG-iteration steps ($^{(i)}$ denoting the adaptation index), depending on the adaptation step i of the auxiliary curve Γ^i by

$$J^{(i)} = J^{(0)} \cdot 2^i , \quad i = 0, 1, 2, \ldots, I-1 \text{ and } J^{(I)} = J^{(I-1)} , \tag{29}$$

that means, we doubled the number of CG iterations with every adaptation step, except for the final one. It seems reasonable to allow larger residual norms $\|\mathcal{F}_\sigma \sigma - u^s_\infty\|$ and $\|\mathcal{F}_\mathbf{m} \mathbf{m} - \mathbf{E}^s_\infty\|$, i.e. to heavier regularize the solutions, during initial adaptation steps and to decrease it with increasing adaptation steps, since for the initial internal curves a good reconstruction of Λ cannot be obtained anyhow. Of course, the arbitrarily chosen exponential increase of iteration steps according to (29) is just one of many possibilities and by no means crucial for the final result. First of all, it saves computation time, compared with the case of a constant number of CG-iterations, since our CGLS-procedure accounts for most of the CPU-time of the total reconstruction.

In the case of erroneous far field data, we used the L-curve criterion [5]. For every adaptation step, we performed 800 CG-iterations to have enough points representing the L-curve. Then we chose the CG-iteration number belonging to the "corner" of the L-curve.

6 The Adaptation Algorithm

After having computed a reconstruction Λ^i with interior curve Γ^i, an improved reconstruction Λ^{i+1} may be obtained with a more suitable interior curve Γ^{i+1}.

To obtain Γ^{i+1}, we first define

$$\bar{r}_\Gamma^{i+1} = r_\Gamma^i + \beta\left(r_\Lambda^i - r_\Gamma^i\right), \quad 0 < \beta < 1 \tag{30}$$

with adaptation factor β. Of course, to leave open the possibility for $r_\Lambda^{i+1}(\hat{\mathbf{r}}_l)$ to decrease in the following step $i+1$, a distance d between Γ^{i+1} and Λ^i must be kept. Therefore, after having evaluated (30), we take

$$\tilde{r}_\Gamma^{i+1}(\hat{\mathbf{r}}_l) = \min\left(\bar{r}_\Gamma^{i+1}(\hat{\mathbf{r}}_l), r_{\Gamma,\max}^{i+1}(\hat{\mathbf{r}}_l)\right) \quad \text{for all } \hat{\mathbf{r}}_l\,, \quad l = 1,\ldots,M \tag{31}$$

as new auxiliary curve. $r_{\Gamma,\max}^{i+1}$ is determined in direction $\hat{\mathbf{r}}_l$ such that according to Fig. 2 a distance d normal to Λ^i remains between Γ^{i+1} and Λ^i. Since the set $\{\tilde{r}_\Gamma^{i+1}(\hat{\mathbf{r}}_l)\}$ obtained from (31), in general, does not belong to the surface space defined in (27), it is subsequently smoothed out by interpolation using representation (27) for Γ. To this end, coefficients a_{nm}, b_{nm} are determined by overdetermined collocation such that with $r_\Gamma(\hat{\mathbf{r}}_l)$ according to (27) and $\tilde{r}_\Gamma(\hat{\mathbf{r}}_l)$ from (31)

$$\sum_{l=1}^{M}(r_\Gamma(\hat{\mathbf{r}}_l) - \tilde{r}_\Gamma(\hat{\mathbf{r}}_l))^2 \stackrel{!}{=} \min\ . \tag{32}$$

The final set $\{r_\Gamma^{i+1}(\hat{\mathbf{r}}_l)\}$ is computed with coefficients from (32). Of course, this interpolation procedure may slightly vary the adjusted minimal distance d between Λ^i and Γ^{i+1} again.

From a theoretical point of view, there might be an objection against this adaptation technique. Indeed, it cannot be guaranteed that our basic assumption, namely Γ^i being inside the original obstacle, is valid throughout all adaptation steps. But we found in practice that this lack does not affect reconstructions negatively.

7 Examples

In order to simplify comparisons between different methods from different authors, we used examples from [2], that were used by other authors as well, e.g. [7]. Our goal was first to compare electromagnetic and acoustic reconstructions, and second to achieve best possible results from "exact" far field data with a minimum of probing cost, i.e. with one incident wave only. Of course it's unrealistic to use exact data in view of practical applications, but nevertheless one has to know what can be achieved with highly accurate data.

All of our reconstructions were performed using one incident plane wave of the form

$$u^{\text{inc}}(\mathbf{r}) = u_0 \exp(i\mathbf{k}\mathbf{r}) \quad \text{or} \quad \mathbf{E}^{\text{inc}}(\mathbf{r}) = \mathbf{E}_0 \exp(i\mathbf{k}\mathbf{r})\,, \quad \mathbf{r} \in \mathbb{R}^3 \tag{33}$$

for the acoustic and electromagnetic case, respectively, with $k = |\mathbf{k}| = 2$. All far field data were given at 1148 different directions, about equally spaced on Ω, and

$M = 128$ collocation points on Λ were used to minimize the functionals (22), (23) and (25). The degree of the series (27) was always $N = 6$, the minimal distance d was $d = 0.25$, the initial guess for Γ^0 was a sphere with radius $R = 0.4$ and the adaptation factor β was chosen to be $\beta = 0.6$. The number of adaptation steps was 6 in the case of "exact" data and 10 for contaminated data to see whether a stable reconstruction was obtained. The reconstruction error is denoted with e_{rec} in the figures.

Although all of our examples were rotationally symmetric, we did not exploit this property to simplify and improve reconstructions.

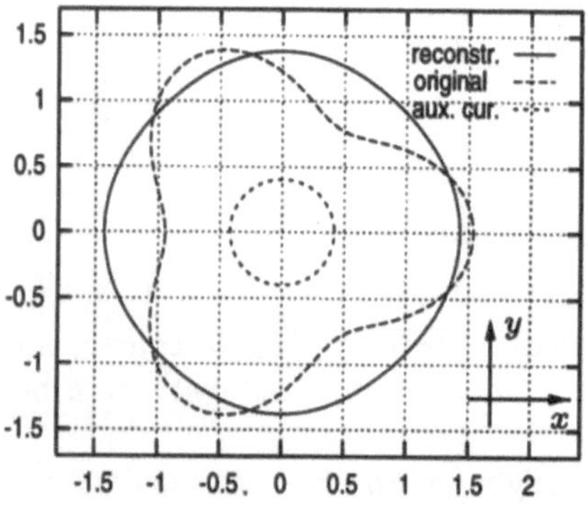

Fig. 3. Initial reconstruction compared to the original and auxiliary curve, $e_{\mathrm{rec}} = 20.2\%$.

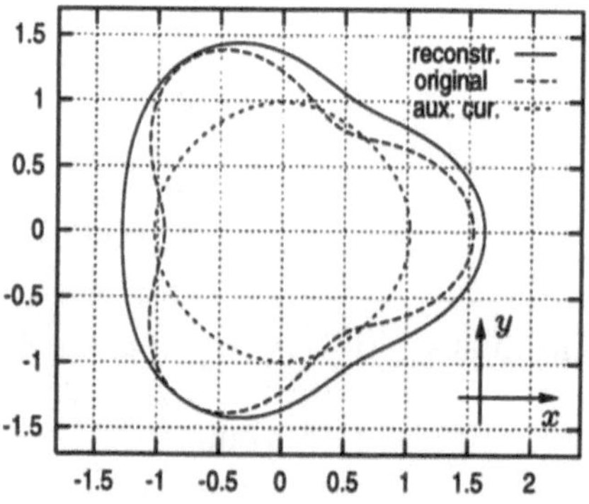

Fig. 4. Reconstruction compared to the original after 1 adaptation step, $e_{\mathrm{rec}} = 16.0\%$.

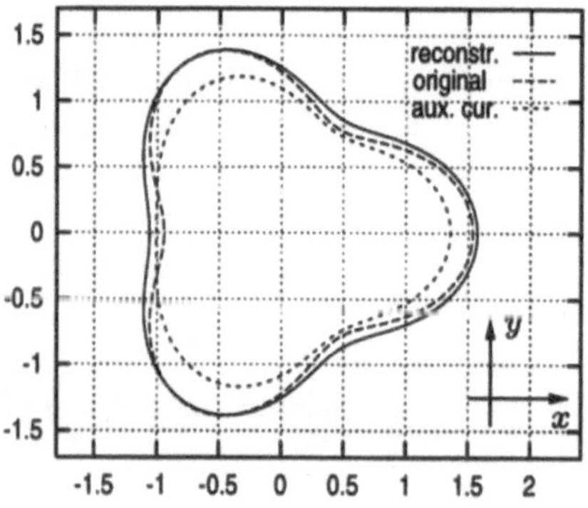

Fig. 5. Reconstruction compared to the original after 2 adaptation steps, $e_{\mathrm{rec}} = 7.6\%$.

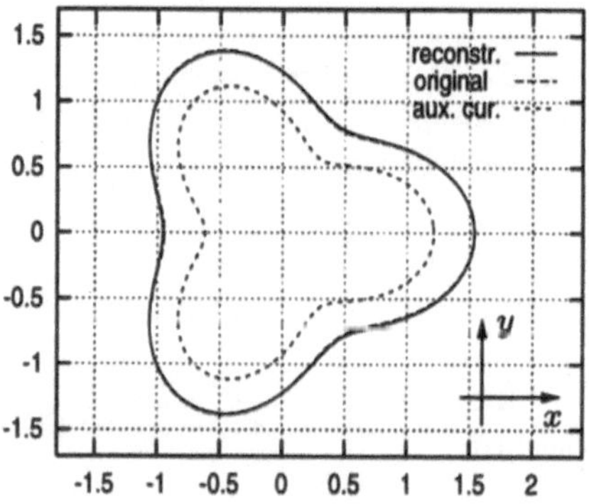

Fig. 6. Final Reconstruction, original and auxiliary curve after 6 adaptation steps, $e_{\mathrm{rec}} = 1.4\%$.

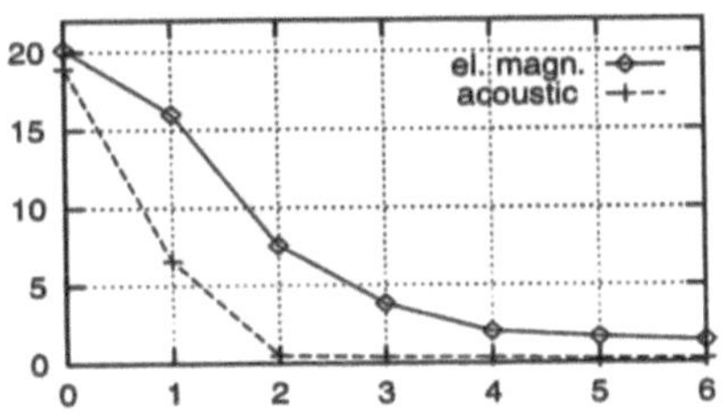

Fig. 7. Acoustic sound-soft and electromagnetic reconstruction error in % versus the adaptation step.

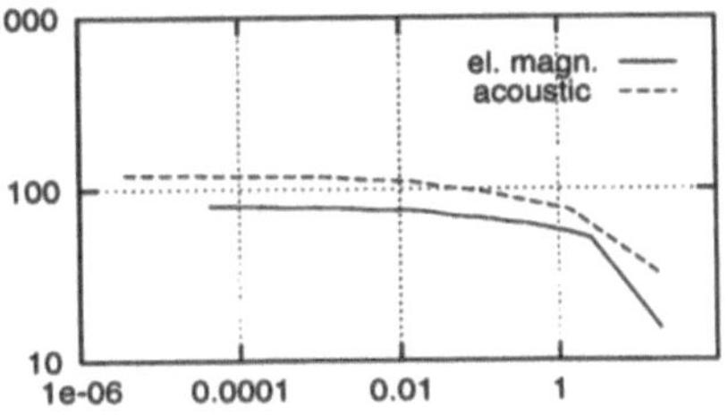

Fig. 8. $\|\mathbf{m}\|$ and $\|\sigma\|$ (ordinate) versus the residual norms $\|\mathcal{F}_{\mathbf{m}}\mathbf{m} - \mathbf{E}^{s}_{\infty}\|$ and $\|\mathcal{F}_{\sigma}\sigma - u^{s}_{\infty}\|$ (abscissa) for all CG-iteration steps.

7.1 Reconstruction of an "Acorn" from Exact Far Field Data

In our first example we reconstruct an acorn as it is mentioned in [2], given by

$$r = \frac{3}{5}\left(\frac{17}{4} + 2\cos(3\alpha)\right)^{\frac{1}{2}}, \quad 0 \le \alpha \le \pi , \tag{34}$$

(α denoting the angle between $\mathbf{r}$ and the x-axis) from both acoustic sound-soft, acoustic sound-hard and electromagnetic scattering data with $\mathbf{k} = \{0,0,2\}^T$ and $\mathbf{E}_0 = \{1,0,0\}^T$.

Fig. 3 to 6 show the same sectional view (xy-plane) of the reconstruction from electromagnetic data for different adaptation steps. The results from acoustic data, which are not shown, were better throughout all adaptation steps, s. Fig. 7. Fig. 8 shows parts of the L-curves, which did not yet reach their "corners" for $J_a^{(6)} = 3200$ and $J_e^{(6)} = 6400$ CG-steps (J_a, J_e: number of CG-iteration steps for acoustic and electromagnetic reconstructions, respectively).

For acoustic sound-soft data we got a final error of $e_{\text{rec}} = 0.24\%$, for sound-hard data of $e_{\text{rec}} = 1.1\%$, and for the electromagnetic reconstruction we had $e_{\text{rec}} = 1.4\%$.

7.2 Reconstruction of the "Acorn" from Erroneous Far Field Data

Fig. 9 and 10 show the initial and final reconstruction from electromagnetic data with 1% noise added. Especially concave parts of the obstacle suffer from less accurate data, as can be seen by a comparison of Fig. 6 with Fig. 10. For acoustic sound-soft data we got a final error of $e_{\text{rec}} = 6.0\%$, for sound-hard data of $e_{\text{rec}} = 3.2\%$, and for the electromagnetic reconstruction we had $e_{\text{rec}} = 3.9\%$.

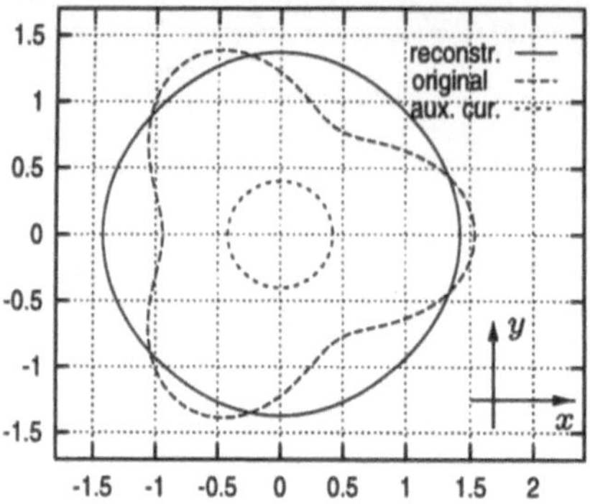

Fig. 9. Initial reconstruction, error of reconstruction is $e_{\mathrm{rec}} = 20.4\%$.

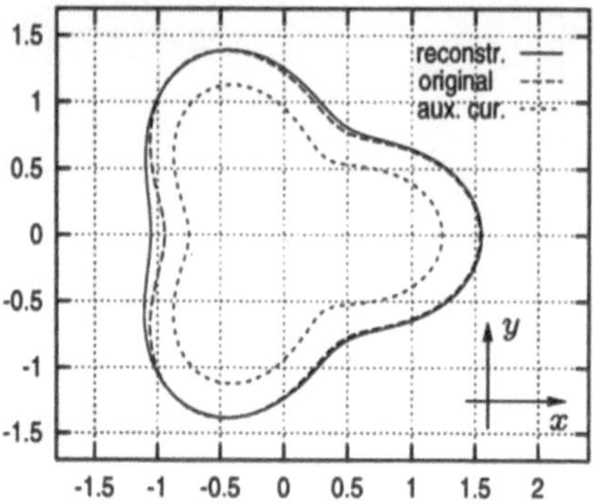

Fig. 10. Final reconstruction after 10 adaptation steps, error diminished to $e_{\mathrm{rec}} = 3.9\%$.

7.3 Reconstruction of a "Peanut" from Exact Far Field Data

Fig. 11 and 12 show the initial and final reconstruction of a peanut from electromagnetic data, as it is used in [2] and [7], given by

$$r = \frac{3}{2}\left(1 - \frac{3}{4}\sin^2\alpha\right)^{\frac{1}{2}} , \quad 0 \le \alpha \le \pi \ . \tag{35}$$

Again, we used $\mathbf{k} = \{0, 0, 2\}^T$ and $\mathbf{E}_0 = \{1, 0, 0\}^T$. The result is even better as for the acorn in section 7.1 what is probably due to a higher symmetry of the peanut and less concave parts. For acoustic sound-soft data we got a final error of $e_{\mathrm{rec}} = 0.37\%$, for sound-hard data of $e_{\mathrm{rec}} = 0.27\%$, and for the electromagnetic reconstruction we had $e_{\mathrm{rec}} = 0.62\%$.

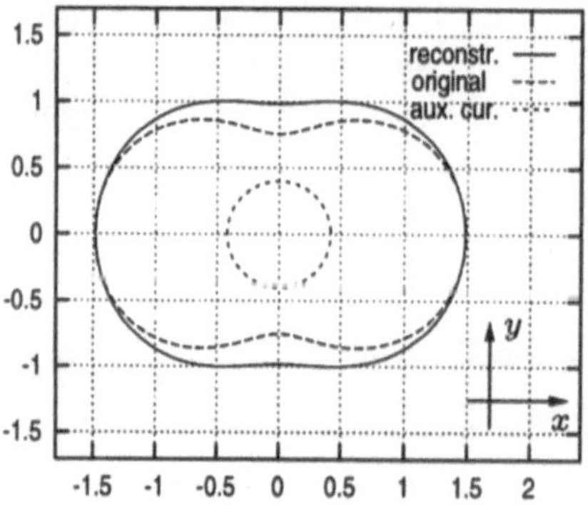

Fig. 11. Initial reconstruction, error of reconstruction is $e_{\mathrm{rec}} = 15.7\%$.

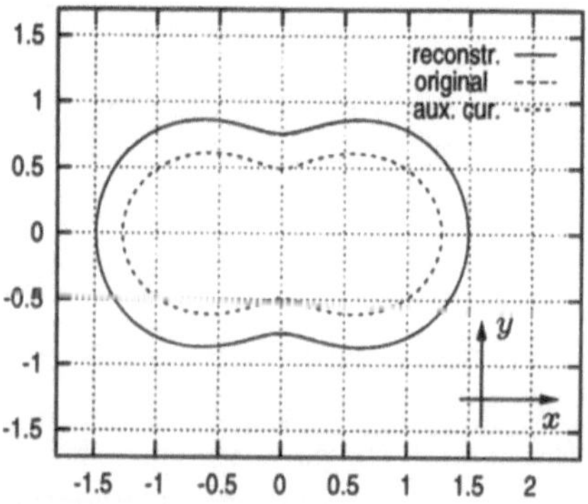

Fig. 12. Final reconstruction after 6 adaptation steps, error diminished to $e_{\mathrm{rec}} = 0.62\%$.

8 Conclusions

An iteration scheme has been presented to reconstruct impenetrable 3D scatterers with smooth and starlike boundaries from their far field scattering data.

The examples show that in both the acoustic and electromagnetic case the adaptation technique does much improve reconstructions, compared to the case of not adapting the internal surface, even when reconstructions from erroneous data are performed.

The fact that, for data "free" of errors, the acoustic reconstructions consistently turn out somewhat better than the electromagnetic ones is probably, at least partly, due to the fact that the acoustic far field data are more precise than the electromagnetic data.

The adaptation algorithm allows to start the reconstruction process with a rather poor initial guess of the obstacle's shape, i.e. a small sphere, in general, will be sufficient.

References

1 Colton, D. and Kress, R. *Integral Equation Methods in Scattering Theory.* Pure and Applied Mathematics. John Wiley & Sons, 1983.

2 Colton, D. and Kress, R. *Inverse Acoustic and Electromagnetic Scattering Theory.* Applied Mathematical Sciences 93. Springer, Berlin Heidelberg New York, 1992.

3 Haas, M. and Lehner, G. Inverse 2D Obstacle Scattering by Adaptive Iteration. *IEEE Transactions on Magnetics (accepted for publication)*, 1997.

4 Hanke, M. *Conjugate Gradient Type Methods for Ill-Posed Problems.* Pitman Research Notes in Mathematics Series. Longman Scientific & Technical, Harlow Essex, 1995.

5 Hansen, P. C. Numerical tools for analysis and solution of Fredholm integral equations of the first kind. *Inverse Problems*, **8**: 849–872, 1992.

6 Lehner, G. *Elektromagnetische Feldtheorie für Ingenieure und Physiker.* Springer, Berlin Heidelberg New York, 1994.

7 Maponi, P., Recchioni, M. C. and Zirilli, F. Three-Dimensional Time Harmonic Electromagnetic Inverse Scattering: The Reconstruction of the Shape and the Impedance of an Obstacle. *Computers Math. Applic.*, **31**: 1–7, 1996.

8 Rao, S. M., Wilton, D. R. and Glisson, A. W. Electromagnetic Scattering by Surfaces of Arbitrary Shape. *IEEE Transactions on Antennas and Propagation*, **30**: 409–418, 1982.

9 Tanabe, K. Conjugate-Gradient Method for Computing the Moore-Penrose Inverse and Rank of a Matrix. *Journal of Optimization Theory and Applications.*, **22**: 1–23, 1977.

An Algorithm for 3D Ultrasound Tomography

Frank Natterer

Institut für Numerische und instrumentelle Mathematik
Westfälische Wilhelms-Universität
Einsteinstrasse 62, D-48149 Münster, Germany
e-mail: frank.natterer@math.uni-muenster.de

Abstract: *Ultrasound tomography is modeled by the inverse problem of a 3D Helmholtz equation at fixed frequency with plane wave irradiation. It is assumed that the field is measured outside the support of the unknown potential f for finitely many incident waves. Starting out from an initial guess f^0 for f we propagate the measured field through the object f^0 to yield a computed field whose difference to the measurements is in turn backpropagated. The backpropagated field is used to update f^0. The propagation as well as the backpropagation are done by a finite difference marching scheme. The whole process is carried out in a single step fashion, i.e. the updating is done immediately after backpropagating a single wave. It is very similar to the well known ART method in X-ray tomography, with the projection and backprojection step replaced by propagation and backpropagation. Numerical experiments with a 3D breast phantom on a 65×65×65 grid are presented.*

1 Introduction

We consider the inverse problem for the 3D Helmholtz equation

$$\begin{aligned} \Delta u^j + k^2(1+f)u^j &= 0 \\ u^j &= u_i^j + v^j , \end{aligned} \tag{1.1}$$

where u_i^j, $j = 1, \ldots, p$ are the incoming waves, v^j satisfies the Sommerfeld radiation condition and the function f vanishes outside the ball of radius ρ. We want to recover f numerically from knowing $u^j = g^j$ on the sphere of radius ρ for $j = 1, \ldots, p$ and a fixed frequency k.

This is a model for ultrasonic tomography [6]. However we point out that in a real ultrasonic scanner the irradiating waves are no longer plane waves but standing waves in a finite container.

We start with a short survey on the extensive literature on numerical methods. With the exception of methods which use the Born or Rytov approximation [4], it seems that the only method which actually has been tested numerically in 3D is the Newton method combined with a finite Fourier expansion of f [5]. The other methods have been used in 2D only, even though an extension to 3D is possible in principle. The dual space method [2] reduces the problem from the whole space $\mathbb{R}^3$ with the far field (which we do not use) as data to an overposed boundary value problem in a finite volume which in turn is solved by optimization.

Of course we can always try to compute the Born series [3]. For this purpose we write (1.1) as an integral equation

$$u^j(x) = u_i^j(x) - k^2 \int G(x,y)f(y)u^j(y)dy \,. \tag{1.2}$$

Here, G is the Green's function of $\Delta + k^2$ with the radiation condition at ∞. The Born series (f_ℓ) is now obtained by solving

$$g^j(x) = u_i^j(x) - k^2 \int G(x,y)f_\ell(y)u_\ell^j(y)dy \,, \quad |x| = \rho \,, \quad j = 1, \ldots, p$$

for f_ℓ, where $u_0^j = u_i^j$ and

$$\Delta u_{\ell+1}^j + k^2(1 + f_\ell)u_{\ell+1}^j = 0$$

with $u_\ell^j = u_i^j + a$ function satisfying the radiation condition.

The generalized SOR-method of [7] also starts out from the integral equation (1.2). Writing for the integral operator in (1.2) simply G, this method minimizes the functional

$$\sum_{j=1}^{p} \left\{ \|g^j + u_i^j + k^2 G f u^j\|^2_{L_2(|x|=\rho)} + \gamma \|u^j - u_i^j + k^2 G f u^j\|^2_{L_2(|x|<\rho)} \right\} \tag{1.3}$$

with some weight factor γ. The minimization is done by

$$u_{\ell+1}^j = u_\ell^j + \alpha_\ell r_\ell^j \quad , \quad f_{\ell+1} = f_\ell + \beta_\ell d_\ell$$

where r_ℓ^j, d_ℓ are update directions and α_ℓ, β_ℓ are chosen so as to minimize (1.3).

A non iterative method has been suggested in [15]. With G_f the Green's function of $\Delta + k^2(1 + f)$, we can rewrite (1.1) as

$$v^j(x) = -k^2 \int G_f(x,y)f(x)u_i^j(y)dy \,. \tag{1.4}$$

Now form a linear combination of the incoming waves such that the resulting field peaks at z, i.e.

$$\sum_{j=1}^{p} \alpha_j(z)u_i^j(x) \sim \delta(x - z) \,.$$

Then, from (1.4),

$$-k^2 G_f(x,z)f(z) \sim \sum_{j=1}^{p} \alpha_j(z)v^j(x) \tag{1.5}$$

is approximately known on $|x| = \rho$. From the identity

$$G_f(x,y) - G(x,y) = -k^2 \int\limits_{|z|<\rho} G_f(x,z)f(z)G(z,y)dy$$

we get an approximation to $G_f(x,y)$ for $|x| = \rho$ which together with (1.5) determines f. The method still has to be tested.

A very efficient code has been given in [6]. It is very similar to ours in that it uses initial value techniques and ignores in the Jacobian entries which correspond to different incoming waves. The initial value technique is based on factoring the Helmholtz equation into a product of two first order differential operators.

The purpose of this note is to extend the method of [14] to 3D and to conduct 3D numerical experiments. We illuminate the object by plane waves $u_i^j = e^{ikx\cdot\theta_j}$ where the unit vectors θ_j are lying in a plane. We use an initial value technique as in [6], but we do not rely on parabolic approximations [8] of the Helmholtz equation.

2 The initial value method

Let Ω_j be the cube circumscribed to the reconstruction region $|x| \leq \rho$ with edges parallel to the coordinate axes and aligned with the direction $\theta_j = (\cos\varphi_j, \sin\varphi_j, 0)$ of the j-th incoming wave. Let $\Gamma_j^{\pm}$ be the face lying in the plane $x \cdot \theta_j = \pm\rho$, and let Γ_j be the other faces of Ω_j. Rather than working with the scattered fields v^j we use the scaled scattered fields $w^j = e^{-ikx\cdot\theta_j}v^j$ which satisfy

$$\Delta w^j + 2ik\theta_j \cdot \nabla w^j + k^2(1 + w^j)f = 0 \quad \text{in} \quad \Omega_j \,. \tag{2.1}$$

Note that we do not make the parabolic approximation [8], i.e. we do not assume that the second derivative of w^j in direction θ_j is small compared to $k\theta_j \cdot \nabla w^j$.

In [11] we haved analysed the stability of the initial value problem

$$w^j = g^j \quad \text{on} \quad \Gamma_j \cup \Gamma_j^- \,, \qquad \frac{\partial w^j}{\partial \nu} = h^j \qquad \text{on} \quad \Gamma_j^- \tag{2.2}$$

for this elliptic differential equation. Let

$$\hat{w}^j(\xi, s) = \frac{1}{2\pi} \int\limits_{x\cdot\theta_j = s} e^{-ix\cdot\xi} w^j(x)dx \quad , \quad \xi \perp \theta_j$$

be the 2D Fourier transform of w in the plane $x \cdot \theta_j = s$. We found that $\hat{w}^j(\xi, s)$ depends in a perfectly stable way on the initial values for w^j on Γ_j^-

for all frequencies $|\xi| < \kappa$ where κ is some number depending essentially on k and, to a minor extent, on f. More precisely we have the following result [11]:

Theorem 2.1 *Let $f \in C^1(\mathbf{R}^3)$ and $f = f_1 + \frac{i}{k} f_2$ with f_1, f_2 real, and let m_2, M_2 be constants such that*

$$-1 < m_1 \le f_1 \le M_1 \ , \quad |\frac{\partial f}{\partial x_2}| \le M_1 \ , \quad |f_2| \le M_2 \ .$$

Let w be the solution of the initial value problem

$$\Delta w + 2ik\theta \cdot \nabla w + k^2(1+w)f = 0$$

$$w = g \ , \qquad \frac{\partial w}{\partial \nu} = h \qquad on \quad x \cdot \theta = 0 \ .$$

Then, for $\kappa < k\sqrt{1+m_1}$, we have on $x \cdot \theta = s$

$$\|w_\kappa\|^2 \le e^{\alpha s}(\|h\|^2 + k^2(1+M_1)\|g\|^2 + 2k^4\|f(w - w_\kappa)\|^2)$$

with $\|\cdot\|$ the L_2-norm in $\mathbf{R}^2$, w_κ the low-pass filtered version of w, i.e.

$$w_\kappa(\xi) = \begin{cases} \hat{w}(\xi) \ , & |\xi| < \kappa \ , \\ 0 \ , & |\xi| \ge \kappa \ , \end{cases}$$

and

$$\alpha = 1 + \frac{2M_2}{\vartheta} + \frac{M_1}{\vartheta^2} \ , \quad \vartheta = \sqrt{1 + m_1 - (\frac{\kappa}{k})^2} \ .$$

Practically the theorem means that we have stability for the initial value problem provided that κ is slightly smaller than k. In our examples we used values of κ between $0.90k$ and $0.99k$.

Thus we may define a nonlinear map $R_j : L_2(|x| < \rho) \to L_2(\Gamma_j^+)$ by putting

$$R_j(f) = w_\kappa^j|_{\Gamma_j^+} \ . \tag{2.3}$$

The inverse scattering problem now calls for the solution of the nonlinear system

$$R_j(f) = g_j \quad , \quad g_j = g^j|_{\Gamma_j^+} \ , \qquad j = 1, \ldots, p \ . \tag{2.4}$$

As in [12] this is done by an ART-type procedure. Starting out from an initial approximation f^0, we put $f_0 = f^0$ and for $j = 1, \ldots, p$

$$f_j = f_{j-1} - \omega R_j'(f_{j-1})^* C_j^{-1}(R_j(f_{j-1}) - g_j) \ .$$

The first approximation f^1 is then defined to be f_p. For C_j we simply take the operator γI where γ is chosen such that, in the limit $k \to \infty$, $C_j \sim$

$R'_j(0)R'_j(0)^*$, i.e. $\gamma = k^2/\rho$ [10]. The evaluation of $R'_j(f)^*g$ for some $g \in L_2(\Gamma_j^+)$ can be done as follows: Solve the initial value problem

$$\begin{aligned} &\Delta z + 2ik\theta_j \cdot \nabla z - k^2 \overline{f} z = 0 \quad \text{in} \quad \Omega_j \\ &z = 0 \quad \text{on} \quad \Gamma_j \cup \Gamma_j^+ , \quad \frac{\partial z}{\partial \nu} = g \quad \text{on} \quad \Gamma_j^+ . \end{aligned} \tag{2.5}$$

Then,

$$R_j^*(f)g = k^2(1 + \overline{w}^j)z$$

where w^j is the solution of (2.1) - (2.2). Of course the stability properties of (2.5) are exactly as discussed above.

3 The finite difference method

The numerical solution of the initial value problems (2.1) - (2.2) and (2.5) can efficiently and conveniently be done by a finite difference method. In view of our stability result (see previous section) this suggestions itsself, but it has been used already in [9]. We simply use the usual five point discretization on the grid $\Omega_j^h = \{h\ell\theta_j + hm\theta_j^\perp + hne_3 : \ell, m, n = -q, \ldots, q\}$ where $\theta_j^\perp = (-\sin\varphi_j, \cos\varphi_j, 0)$ and $e_3 = (0,0,1)$ and $h = \rho/q$. Then, the finite difference approximation to (2.1) - (2.2) reads (the superscript j is omitted)

$$\begin{aligned} &w_{\ell+1,m.n} + w_{\ell-1,m,n} + w_{\ell,m+1,n} + w_{\ell,m-1,n} + w_{\ell,m,n+1} + w_{\ell,m,n-1} \\ &-6w_{\ell,m,n} + i\varepsilon(w_{\ell+1,m,n} - w_{\ell-1,m,n}) + \varepsilon^2(1 + w_{\ell,m,n})f_{\ell,m,n} = 0 , \\ &\qquad \ell, m, n = -q+1, \ldots, q-1 , \\ &w_{\ell,m,n} = g_{\ell,m,n} \quad \text{for} \quad |m| = q \ \text{or} \ |n| = q \ \text{and for} \ \ell = -q, -q+1 . \end{aligned} \tag{3.1}$$

The boundary conditions in (3.1) assume that the field can be measured everywhere on each of the faces of Ω_j. The values of $g_{\ell,m,n} = f^j(h\ell\theta_j^1 + hn\theta_j^2)$ for $\ell = q-1$ have to be computed from the field in $|x| \geq \rho$ by numerical differentiation. (3.1) is solved in a recursive way, i.e. if w in known on the levels ℓ, $\ell-1$ we compute it for $\ell+1$ by (3.1). In order to preserve stability we have to filter out the frequencies greater than κ at each stage of this process. This can be done by doing a 2D FFT on each matrix $(w_{\ell,m,n})_{m,n=-q,\ldots,q}$ as soon as it is computed, zeroing all the entries with $\sqrt{m^2+n^2} > 2\rho\kappa/\pi$, followed by a 2D inverse FFT. The total operator count for solving (2.1) - (2.2) once is $O(q^3 \log q)$. For details see [16]. (2.5) is solved exactly in the same way.

4 Numerical experiments

In a first test we checked the initial value method for accuracy. We solved the forward problem (1.1) for the function

$$f(x) = \begin{cases} 0.2 \, , \, |x| \leq 0.5 \\ 0 \quad , \text{ otherwise} \end{cases} \tag{4.1}$$

and $k = 50$ analytically and compared the exact solution with the approximate solution of the initial value method for $h = 32$. We found satisfactory agreement.

In a second test we used the exact data for (4.1) and $k = 50$ as input to our reconstruction method with $p = 25$, $q = 32$. As initial approximation we chose (4.1) with 0.2 replaced by 0.1. After 3 sweeps of our algorithm we obtained a reconstruction very similar to a low pass filtered version of f with cut off 50.

Finally we created a 3D breastphantom, see Fig. 1. It consists of fat, glandular tissue, a tumor and a cyst. The breast is suspended in a cube of sidelength 12 cm which is filled with water. Fig. 1 shows on the left hand side three vertical cross sections, taken at distances 3m apart. f is given by

$$f = \frac{c_0^2}{c^2} - 1 - i\frac{2\alpha c_0}{kc} \tag{4.2}$$

with $c_0 = 1500$ m sec^{-1} the speed of sound in water. The values of c and α (at 1MHz) are (k in units of m^{-1})

tissue	$c[\frac{m}{sec}]$	$\alpha[\frac{db}{m}]$	Ref	Imf
fat	1458	41	0.058	$-9.4/k$
glandular tissue	1519	80	−0.025	$-18.4/k$
tumor	1564	118	−0.080	$-27.2/k$
cyst	1568	10	−0.084	$-2.3/k$

Since the top face of the cube is not accessible, we have to modify the finite difference method (3.1). We stipulate the boundary condition $\partial w/\partial \nu = 0$ on the top face of Ω_j, i.e. we let n run from $-q$ to q and put $w_{\ell,m,q+1} = w_{\ell,m,q-1}$. The boundary value problem (2.5) for z has to be changed accordingly. Of course this procedure is questionable, but at the present state of our work we just don't know anything better.

We generated data for $p = 32$ equally spaced directions in $[0, 2\pi]$ using our initial value method with $q = 32$, i.e. $h = 6\text{cm}/32 = 1.875\text{mm}$. For 4 directions, the projections are displayed on the right hand side of Fig.1. The frequency of the iradiating waves was chosen to be 250 KHz, i.e. $k = 10.47\text{cm}^{-1}$. This corresponds to a wavelength of 6mm. The smallest tumor is at the resolution limit, i.e. his diameter coincides with the wavelength. After three sweeps of our algorithm we obtained a reconstruction in which

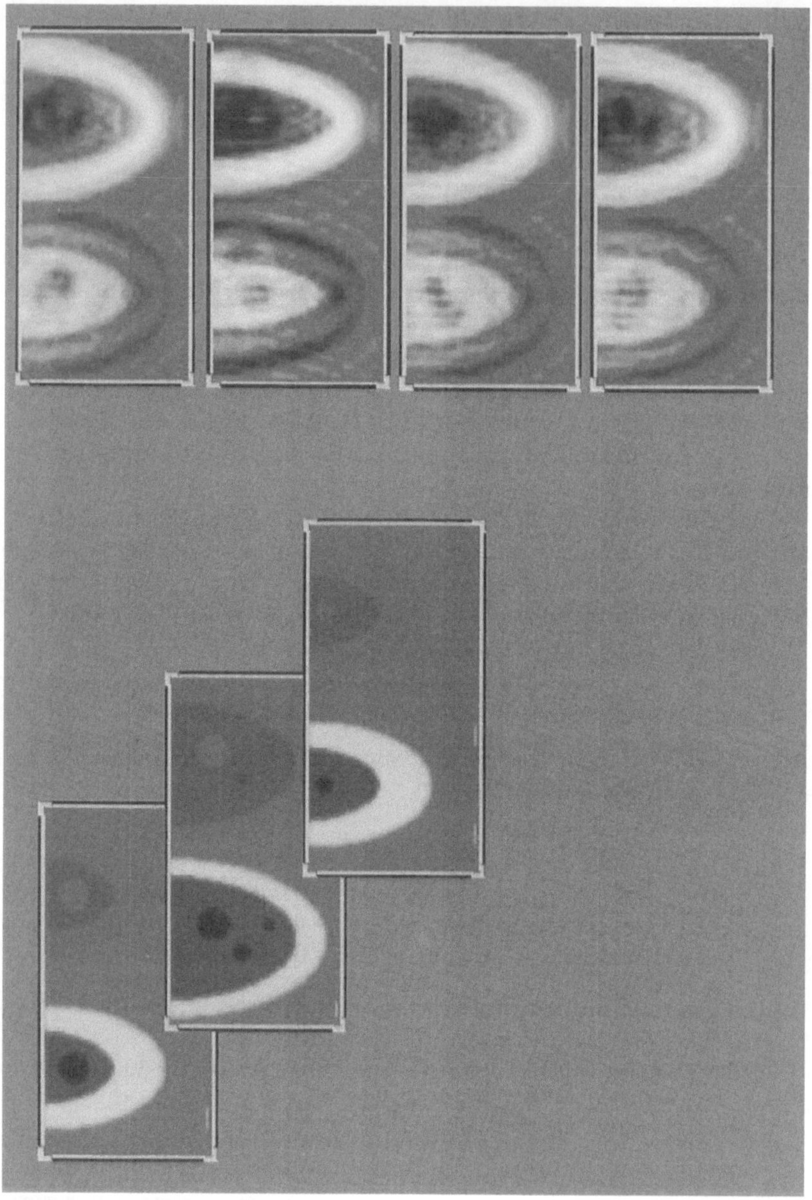

Fig 1: Breast phantom and data. Left we see 3 vertical cross sections through the suspended breast. Right we see the projections from 4 directions.

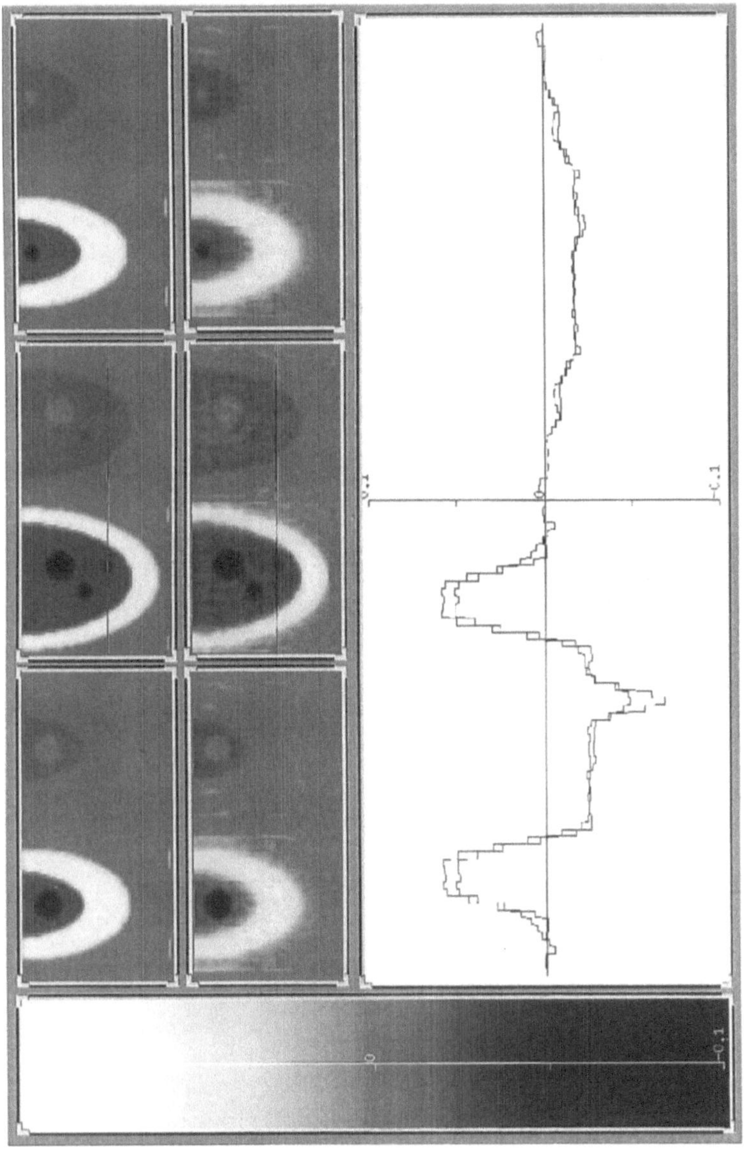

Fig 2: Breast phantom and reconstruction.

Top row: Original (same as in Fig. 1).

Middle row: Reconstructions

Bottom: Line plot of original and reconstruction along a horizontal line hitting the smallest tumor in the central cross section

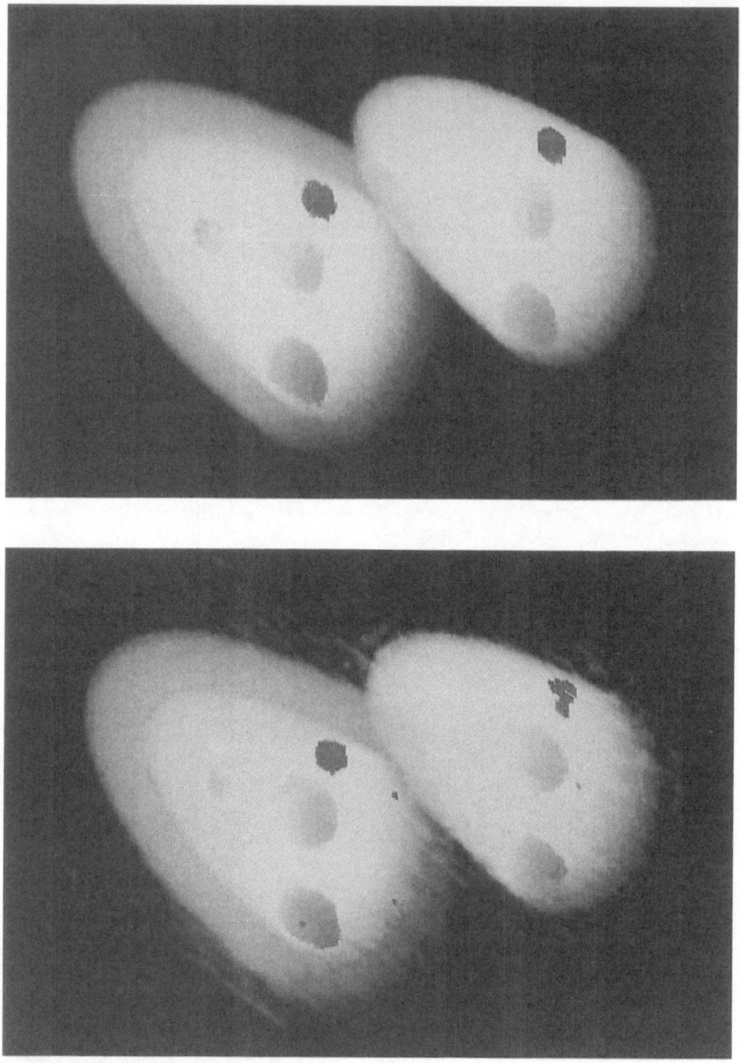

Fig 3: 3D semitransparent visualization of original (top) and reconstruction (bottom). Real part left, imaginary part right. The two little tumors show up clearly in the real part but are invisible in the imaginary part. This distinguishes them from the three cysts which can be seen both in the real and the imaginary part.

the cyst and the tumor where clearly visible and distinguishable. The results are shown in Fig. 2. In the top row we see the cross sections of Fig. 1. In the middle row we see the reconstructions of these cross sections. The bottom row shows a line plot of the original and the reconstruction for the horizontal line in the central cross section hitting the smallest tumor. The computing time per sweep on a SPARC 20 was 10 minutes.

References

Borup, D.T. - Johnson, S.A. - Kim, W.W. - Berggren, M.J.: Nonperturbative diffraction tomography via Gauss-Newton iteration applied to the scattering integral equation, *Ultrasonic Imaging* **14**, 69-85 (1992).

Colton, D. - Monk, P.: A modified dual space method for solving the electromagnetic inverse scattering problem for an infinite cylinder, *Inverse Problems* **10**, 87-108 (1994).

Chew, W.C. - Wang, Y.M.: Reconstruction of Two-Dimensional Permittivity Distribution Using the Distorted Born Iteration Method, *IEEE Transactions on Medical Imaging* **9**, 218-225 (1990).

Devaney, A.J.: A filtered backpropagation algorithm for diffraction tomography, *Ultrasonic Imaging* **4**, 336-350 (1982).

Gutman, S. - Klibanov, M.: Three-Dimensional inhomogeneous media imaging, *Inverse Problems* **10**, 239-246 (1994).

Johnson, S.A.: Progress in applied and inverse scattering for ground penetrating radar, sonar, and medical ultrasound. *Lecture on the IMA Workshop. Inverse Problems in Wave Propagation, March 6-17, 1995.*

Kleinman, R.E. - van den Berg, P.M.: A modified gradient method for two-dimensional problems in tomography, *J. Comp. Appl. Math.* **42**, 17-35 (1992).

Keller, J.B. - Papadakis, J.S.: Wave Propagation and Underwater Acoustics. *Lecture Notes in Physics*, **vol. 70**, Springer 1977.

Knightly, G.H. - Mary, D.F.St.: Stable marching schemes based on elliptic models of wave propagation, *J. Acoust. Soc. Am.* **93**, 1866-1872 (1993).

Natterer, F.: Finite difference methods for inverse problems and Applications to Geophysics, Industry, Medicine and Technology, Proceedings of the International Workshop on Inverse Problems, January 17-19, 1995, HoChiMinh City. *Publications of The HoChiMinh City Mathematical Society* **2**, 1995.

Natterer, F.: An initial value approach to the inverse Helmholtz problem at fixed frequency. Proceedings of the Conference "Inverse Problems in Medical Imaging and Nondestructive Testing", Oberwolfach, February 4-10, 1996. To appear in Springer 1996.

Natterer, F.: Numerical solution of bilinear inverse Problems, *Technical Report 19/96-N*, Fachbereich Mathematik der Universität Münster.

Natterer, F. - Wübbeling, F.: A finite difference method for the inverse scattering problem at fixed frequency, Lecture Notes in Physics, Vol. **422**, p. 157-166, Springer 1993.

Natterer, F. - Wübbeling, F.: A propagation - backpropagation method for ultrasound tomography, *Inverse Problems* **11**, 1225-1232 (1995).

Stenger, F. - O'Reilly, M.: A new approach to sonic inversion, to appear in *Inverse Problems.*

Wübbeling, F.: Direktes und inverses Streuproblem bei fester Frequenz, Thesis, Fachbereich Mathematik, Universität Münster (1994).

Regularity of an Inverse Problem in Wave Propagation

Gang Bao[1] and William W. Symes[2]

[1] Department of Mathematics, University of Florida, Gainesville, FL 32611, U.S.A.
[2] Department of Computational and Applied Mathematics, Rice University, Houston, TX 77251-1892, U.S.A.

Abstract. This work is devoted to determine appropriate domain and range of the *forward map* from the coefficients to the solutions of the multi-dimensional wave equation for which the forward map attains certain regularity properties. The first result concerns an explicit upper bound of its linearization or formal derivative and the properties of the coefficients on which the bound depends. In view of results for the smooth coefficient case, the estimate is optimal. We then present recent results on continuity and differentiability of the forward map as well as continuity of the linearized forward map. Information concerning regularity properties of the forward map is indispensable in the study of the inverse problem *via* smooth optimization methods. The usefulness of our results and some directions for future research are also discussed.

1 Introduction

The linear acoustic wave equation governs many physical processes such as seismic and acoustic wave propagation

$$\left(\frac{1}{c^2}\frac{\partial^2}{\partial t^2} - \Delta - \nabla\sigma\cdot\nabla\right)u = f\,.$$

Here $\sigma = \sigma(x)$ is the logarithm of the density, $c = c(x)$ is the sound speed of the medium, and $f = f(x,t)$ is the source term which introduces the energy to the problem. If σ, c and f are given along with appropriate side conditions, the forward (or direct) problem is to determine $u = u(x,t)$, the excess pressure. For appropriate choices of σ, c, and f, u is determined uniquely by standard linear hyperbolic theory of partial differential equations (*p.d.e.*). Thus the problem stated above defines a map from the coefficients to the solution of the wave equation. We are concerned with some aspects of the *regularity* of this map, and especially of its composition with the trace on a time-like hypersurface.

Throughout, we shall restrict ourselves to the special case of constant velocity c. An extension of the ideas to more general problems will be briefly discussed in Section 3.

To fix ideas, write $x \in \mathbf{R}^n$ as (x', x_n), where $x' \in \mathbf{R}^{n-1}$, $x_n \in \mathbf{R}$. We assume that the problem is set in the whole space $\mathbf{R}^n$ and $u = 0$ in the past $(t < 0)$. Take $f(x,t) = \delta(x,t)$ as an ideal point source. Thus the excess pressure u is the retarded fundamental solution:

$$\begin{aligned} \Box u - \nabla\sigma \cdot \nabla u &= \delta(x,t) \quad (x,t) \in \mathbf{R}^n \times \mathbf{R}\,, \\ u &= 0 \quad t < 0\,, \end{aligned}$$

where $\Box$ is defined to be $\partial_t^2 - \Delta$, and Δ is the Laplacian.

It is somewhat easier to understand the sensitivity of the solution to distant perturbations of the coefficients. Thus we will assume that the density σ and its perturbations are supported in the half space $\{x_n > 0\}$ and study the solution near the boundary $\{x_n = 0\}$.

Define the *forward map* F as:

$$F : \sigma \to (\phi u)\mid_{x_n=0}\,,$$

where $\phi \in C_0^\infty(\mathbf{R}^{n+1})$ is supported inside the conoid $\{t > |x|\}$ and near $\{x_n = 0\}$.

F is nonlinear even though the direct problem is linear. As a first step toward understanding the regularity of F, we study the formal linearization (or formal derivative) DF, with respect to the reference state $(\sigma_0,\, u_0)$. The first order perturbation theory gives, for a small change $\delta\sigma$, the following problem for the resulting change δu in u:

$$\begin{aligned} \Box\delta u - \nabla\sigma_0 \cdot \nabla\delta u &= \nabla\delta\sigma \cdot \nabla u_0\,, \\ \delta u &= 0 \quad t < 0\,. \end{aligned}$$

The formal derivative $DF(\sigma_0)$ with respect to the reference density σ_0 is defined by

$$DF(\sigma_0)\delta\sigma = (\phi\delta u)\mid_{x_n=0}\,. \tag{1}$$

Our first goal is to determine appropriate spaces of the domain and range of F for which

the formal derivative DF is bounded.

We are also interested in establishing continuity and differentiability estimates of F and continuity estimates of DF.

The study of the forward map is motivated by the *inverse problem* which arises in reflection seismology, oil exploration, ground-penetrating radar, etc. A highly over simplified version of the inverse problem is to determine the coefficient σ by knowing additional boundary value conditions of u. Since the inverse problem is just to invert the functional relation F, we are naturally interested in all the properties of this forward map.

To understand the problem, let us look at a simple exploration seismology experiment: Near the surface of the earth, a seismic source is fired at some point energy source. The seismic waves propagate into the earth. Since the earth's structure varies (as do its physical properties) part of the energy of the wave will be reflected back to the surface and can be measured. The inverse problem is then to deduce the interior properties of the earth from the recorded data.

A simple model of this reflection seismic inverse problem in this context is: given data $F_{data}(x', t)$, find a coefficient $\sigma(x)$ so that

$$F(\sigma) = F_{data}$$

or perhaps minimizing the error $(F_{data} - F(\sigma))$ in some suitable norm. An overview of various issues on formulating and solving the optimization problem may be found in Santosa and Symes (1989). Two questions arise immediately:

- Since the forward map F is nonlinear, one naturally considers its linearization. So far, most progress has been made through the study of the linearization. However, it seems that very little work has been done to justify this commonly used procedure. Does the linearization of F provide useful information in recovering the density?

- The large size of the typical data set demands fast means of solving the minimization problem. A natural candidate would be some Gauss-Newton like method. Under what conditions can one formulate such an algorithm?

To answer either one of the above questions requires the understanding of regularity of the forward map. In addition, local properties of the inverse problem may be obtained by examining the differentiability of the forward map. Also, continuity properties are crucial in the study of linearized forward map with respect to a nonsmooth reference density.

Numerical solution of this inverse problem by means of Newton's method and its relatives, such as the quasi-Newton, conjugate gradient, and variable metric methods, requires a choice of Banach space structure in the space of models σ and in the space of data $F(\sigma)$ in such a way that F is regular. This fact accounts for our reliance on the L^2-based Sobolev spaces. Here, we study the regularity properties of F: boundedness of DF, continuity and differentiability of F, and continuity of DF. We believe that the ideas will also allow investigation of coercive properties of DF, *i.e.*, the stability of linearized forward map, as is required by the theory of optimization.

When the spatial dimension is one or c and σ depend only on x_n (layered problem) there is a large literature available. For a similar problem in which the medium was assumed to be excited by an impulsive load on the surface $\{x_n = 0\}$ instead of point sources, the properties of the forward map have been studied fairly satisfactorily by Symes and others (see Symes (1986a) for references). It was shown by Symes that, in the constant wave speed case, the forward map defines a C^1*–diffeomorphism* between open sets in certain Hilbert spaces by applying the method of geometrical optics together with energy estimates.

When the spatial dimension $n > 1$ and c, σ depend on all space variables (nonlayered problem), very little is known in mathematics. See Symes (1983b), Symes (1986b), Sacks and Symes (1985), and Rakesh (1988), Bao and Symes (1996), and Bao (1996) for some partial results. The difficulties are essentially due to the ill-posed nature of the timelike hyperbolic Cauchy problem and the presence of nonsmooth coefficients. For the one dimensional wave equation, both coordinate directions are spacelike, which indicates that the problem is hyperbolic with respect to both directions. Apparently, this is not the case when the spatial dimension is larger than one. Recently, in the study of this class of inverse problems and other close related problems Bao and Symes (1993, 1995, 1996), we have employed nonsmooth microlocal analysis techniques of Beals and Reed (1982) to obtain the optimal timelike trace regularity under weaker hypotheses on the coefficients. Using these microlocal analysis techniques, we establish new estimates on the regularity of the forward map and the continuity of its linearization.

Rakesh (1988) studied a related linearized velocity inversion problem with constant density and point sources. Assuming smooth background velocity, he obtained both upper and lower bounds for the linearized forward map. The essential observation in Rakesh's work is that DF is a Fourier integral operator. The calculus of Fourier integral operators employed in Rakesh's work is not applicable to the nonsmooth reference velocity case since the linearized forward map is a Fourier integral operator only when the reference velocity is smooth. The approach does not lead to any regularity result for the forward map F. Nonetheless, for integer $l + (n-1)/2$, the regularity estimate for DF in Theorem 2.1 (loss of $(n-1)/2$ derivatives) is exactly the same as that proved in Rakesh (1988), and is optimal.

Symes (1983a) gave a pair of examples, based on the geometric optics construction, which show that both $DF(1)$ and $DF(1)^{-1}$ are unbounded for a slightly different problem. As the examples show, within the Sobolev scales no strengthening or weakening of topologies of the domain and range can make both DF and DF^{-1} bounded. This fact also implies a strategy of regularization: Change the topology in the domain so that DF becomes bounded, then ask for optimal regularization of DF^{-1} in the sense of best possible lower bound estimate for DF. In both examples of Symes, the unboundedness was caused by rapid oscillation of σ in the x'-direction or the tangential directions, hence the problem is actually "partially well-posed", i.e., only more smoothness of the coefficients in tangential directions (essentially grazing ray directions) will be required to cure the difficulty. For this reason, the results of Sacks and Symes (1985) were formulated using the anisotropic Sobolev spaces.

In Theorem 4.1 of Sacks and Symes (1985), they showed by using the method of sideways energy estimates that for a linearized density determination problem with constant velocity and plane wave sources, DF is bounded from $H^{1,1}$ to H^1, provided that the reference coefficient is in $H^{1,s}$ for some

$s > n+2$. They also proved the injectivity of DF. An extension of their reasoning shows that DF is bounded from $H^{l,1}$ to H^{l} provided that σ is in $H^{l,s}$ for $s > n+2$. Since $H^{l,s} \subset H^{l+s}$ and $H^{l,s} \not\subset H^{q}$ for $q < l+s$, the regularity condition on σ_0 in Theorem 2.1 is compatible with that of Sacks and Symes (1985). The bounds on DF are compatible as well, allowing for the difference between plane wave and point sources. We point out that in this setting several regularity results of F were also established in Symes (1983b), Symes (1986b). Our method is completely different from theirs. In particular, we believe that our method could be extended to study the velocity inversion problem, *i.e.*, to determine $c(x)$ when the density is fixed.

2 Results

For a real number α, denote $[\alpha]$ the smallest integer such that $\alpha \le [\alpha]$.

We obtain the following up-bound estimate for DF.

Theorem 2.1 *Assume $[l+(n-1)/2] > 1+n/2$, and that $s > [l+(n-1)/2]+[(n-1)/2]+n/2+2$. Then for σ_0, $\delta\sigma \in C_0^\infty(\{x_n > 0\})$,*

$$||DF(\sigma_0)\delta\sigma||_l \le C||\delta\sigma||_{[l+\frac{n-1}{2}]}\,, \tag{2}$$

where the constant C depends on the $||\sigma_0||_s$ and the support of ϕ, but is independent of $\delta\sigma$.

The proof is given in Bao and Symes (1996). Our proof is based on the method of nonsmooth microlocal analysis. Here we sketch the general procedure of our method:

First, our time-like trace regularity theorem in Bao and Symes (1993) indicates that $\delta u|_{x_n=0}$ can be as regular as δu itself, provided that microlocal restrictions against the tangential oscillations of the coefficient σ_0. In addition, an explicit estimate is available via the method of microlocal energy estimates. Thus the problem may be reduced to estimating $||\phi\delta u||_{[l+(n-1)/2]}$. A dual problem may be introduced next, which is a time-reversed wave equation with a smooth right hand side compactly supported near $\{x_n = 0\}$ and inside the characteristic "cone". This right hand side can be used as a test function to estimate the local norm of δu. Observe that the original differential equation for δu has a singular right side since u_0 is the fundamental solution.

The crucial part is to analyze the smoothness of u_0 and the solution to the dual problem microlocally. A microlocal cut-off technique is used to decompose the problem into three parts and analyze each part separately. Near a null bicharacteristic, the solution of the dual problem can be analyzed by the propagation of singularity theorem with an estimate. In the region inside the light cone, the problem requires the regularity of the fundamental solution u_0. To serve this purpose, the method of progressing wave expansions or the method of geometrical optics is employed. An important step is to analyze the solution regularity through a regularity study of the resulting transport equations. Finally, conormal properties of the wave operator are developed to estimate the remaining part.

We next examine the continuity and differentiability properties of the forward map.

Let σ_1 and σ_0 be the densities corresponding to the excess pressure u_1 and u_0 respectively, we have from the model equation that

$$\begin{aligned} (\Box - \nabla\sigma_1 \cdot \nabla)\tilde{u} &= \nabla\delta\sigma \cdot \nabla u_0 \,, \\ \tilde{u} &= 0 \quad t < 0 \,, \end{aligned} \tag{3}$$

where $\Box = \partial_t^2 - \Delta$, $\tilde{u} = u_1 - u_0$, and $\delta\sigma = \sigma_1 - \sigma_0$. Moreover,

$$(\phi\tilde{u})|_{x_n=0} = (\phi u_1)|_{x_n=0} - (\phi u_0)|_{x_n=0} \,,$$

where $\phi \in C_0^\infty$ supported inside $\{t > |x|\}$ and near $\{x_n = 0\}$.

In the following statements of theorems, we always assume that

$$[l + (n-1)/2] > 1 + n/2 \text{ and } \tau > [l + (n-1)/2] + [(n-1)/2] + n/2 + 2 \,.$$

Let $M_\tau > 0$ and define

$$\mathcal{M}_\tau = \{\sigma \in C_0^\infty\{x_n > 0\},\ ||\sigma||_\tau < M_\tau\}.$$

We also assume that the density and its perturbations are supported in the half space $\{x_n > 0\}$.

Theorem 2.2 *There exists a constant C depending on M_τ and the support of ϕ so that for σ_1 and $\sigma_0 \in \mathcal{M}_\tau$,*

$$||F(\sigma_1) - F(\sigma_0)||_l \le C||\sigma_1 - \sigma_0||_{[l+\frac{n-1}{2}]} \,.$$

The following results concern the differentiability of the forward map. The formal linearization DF of the forward map F, with respect to the reference state (σ_0, u_0), is defined by the linearized problem

$$\begin{aligned} (\Box - \nabla\sigma_0 \cdot \nabla)\delta u &= \nabla\delta\sigma \cdot \nabla u_0 \,, \\ \delta u &= 0 \quad t < 0 \end{aligned}$$

and

$$DF(\sigma_0)\delta\sigma = (\phi\delta u)|_{x_n=0} \,.$$

Recall that $\tilde{u}$ solves (3) and

$$F(\sigma_1) - F(\sigma_0) = (\phi\tilde{u})|_{x_n=0} \,,$$

then

$$F(\sigma_1) - F(\sigma_0) - DF(\sigma_0)\delta\sigma = (\phi(\tilde{u} - \delta u))|_{x_n=0} \,,$$

where

$$\begin{aligned} (\Box - \nabla\sigma_1 \cdot \nabla)(\tilde{u} - \delta u) &= \nabla\delta\sigma \cdot \nabla\delta u \,, \\ \tilde{u} - \delta u &= 0 \quad t < 0 \,. \end{aligned}$$

Theorem 2.3 *There exists a constant C depending on M_τ and the support of ϕ so that for σ_1 and $\sigma_0 \in \mathcal{M}_\tau$,*

$$||F(\sigma_1) - F(\sigma_0) - DF(\sigma_0)\delta\sigma||_l \leq C||\delta\sigma||_{\tau+1}||\delta\sigma||_{[l+\frac{n-1}{2}]}\,.$$

Theorem 2.4 *There exists a constant C depending on $M_{\tau+1}$ and the support of ϕ so that for σ_1 and $\sigma_0 \in \mathcal{M}_{\tau+1}$,*

$$||F(\sigma_1) - F(\sigma_0) - DF(\sigma_0)\delta\sigma||_l \leq C||\delta\sigma||_\tau||\delta\sigma||_{[l+\frac{n-1}{2}]}\,.$$

Note that the results of Theorem 2.3 and Theorem 2.4 do not imply that of Theorem 2.2 since the estimates depend on higher norms of the coefficients.

Finally, we present a continuity estimate for the linearized forward map. We assume that

$$[l+(n-1)/2] > 1+n/2 \text{ and } \tau > [l+(n-1)/2]+[(n-1)/2]+n/2+3\,.$$

Let $M > 0$ and define

$$\mathcal{M}_\tau = \{\sigma \in C_0^\infty\{x_n > 0\},\ ||\sigma||_\tau < M\}.$$

We once again assume that the density and its perturbations are supported in the half space $\{x_n > 0\}$.

Theorem 2.5 *(Bao (1996)) There exists a constant C depending on M and the support of ϕ so that for σ_1 and $\sigma_0 \in \mathcal{M}_\tau$. Then for $\eta \in C_0^\infty\{x_n > 0\}$*

$$||DF(\sigma_1)\eta - DF(\sigma_0)\eta||_l \leq C[||\sigma_1-\sigma_0||_{[l+\frac{n-1}{2}]}\,||\eta||_\tau + ||\eta||_{[l+\frac{n-1}{2}]}\,||\sigma_1-\sigma_0||_\tau]\,.$$

In particular, DF extends to a Lipschitz continuous map:

$$\sigma \to DF(\sigma)$$

$$H^\tau_{comp}(\mathbf{R}^{n-1}\times[0,\infty)) \to \mathcal{L}[H^\tau_{comp}(\mathbf{R}^{n-1}\times[0,\infty)),\ H^l(\mathbf{R}^{n-1}\times[0,\infty)].$$

The proofs of the above regularity results follow essentially the general ideas of Bao and Symes (1996) with some necessary modifications. Particularly, the regularity study of transport equations becomes more complicated and technically involved.

3 Discussions

It is known that the method of nonsmooth microlocal analysis can only deal with relatively weak singularities and the coefficients should be at least continuous. However, an earlier result of Bamberger et al (1979) on the one dimensional inverse problem with regular source terms allows the coefficients to be discontinuous or even bounded measurable. A similar multi-dimensional inverse problem with rough coefficients was recently studied by Fernandez et al (1993). Their general approach may be viewed as variational. The method in the several dimensional case was based on elliptic type energy estimates together with compactness arguments. However, no geometric property of the wave operator was used due to the rough coefficients. This presents a clear contrast to our model. Our model involves a singular right hand side but relatively regular coefficients (see the assumptions of the above theorems). Also, our approach relies heavily on the geometric properties of the wave operator. Therefore, a natural question arises: What happens when the coefficients are in the regime in between those of the two models? It remains to see whether the inverse problem in the regime could be studied by combining these two different methods – perhaps an interpolation technique.

Up to now, we only study the density determination problem. A more interesting problem is to study the dependence of the boundary values of the pressure field on the velocity c. At present, no regularity result for the multi-dimensional velocity inversion problem is available.

Denote $d_0(x) = c_0^{-2}(x)$, where $c_0(x)$ is the reference velocity. Further we assume that $0 < C_1 < d_0(x) < C_2$ with fixed constants C_1 and C_2. Consider the linearization of the model with respect to the reference state (d_0, u_0):

$$d_0(x)\partial_t^2 \delta u(x,t) - \Delta \delta u(x,t) = -\delta d(x) \partial_t^2 u_0(x,t) \;,$$
$$\delta u = 0 \qquad t < 0 \;.$$

With a smooth reference velocity, Rakesh (1988) showed that for $\delta u \in \mathbf{R}^n_\epsilon = \mathcal{E}'\{x \in \mathbf{R}^n;\; x_n > \epsilon\}$,

$$DF(d_0) : H^{s+(n-1)/2}_{comp} \to H^s_{loc} \text{ is bounded }, \tag{4}$$

and further, under certain geometric conditions, there exist a properly supported pseudo-differential operator Q of order 0 and a function $\phi \in C_0^\infty(\mathbf{R}^n_\epsilon)$, such that

$$||Q\delta d||_s^2 \leq C\{||\phi DF(d_0)\delta d||^2_{s-(n-1)/2} + ||\delta d||_l\} \tag{5}$$

for all $\delta d \in C_0^\infty(K)$, where K is a compact set of $\mathbf{R}^n_\epsilon$, $||\cdot||_j$ denotes the norm of the Sobolev space H^j, and l, s are real numbers. His proof was based on an important observation:

$$DF(d_0) \;\text{ is a Fourier Integral Operator (FIO) for smooth }\; d_0 \;,$$

together with the full machinery of calculus of FIO. Unfortunately, the technique is no longer available with the appearance of the nonsmooth reference velocity since in this case the linearized forward map is not a FIO. The difficulties seem clear: nonsmooth principal symbols, more complex ray geometry, and possible appearance of caustics. A challenging open problem is to determine the amount of smoothness of d_0 for which the results (4) and (5) remain valid. On this problem, we have made some progress recently. The time like trace regularity result, the calculus of nonsmooth symbols, and our theorem on propagation of singularity for pseudo-differential operator equations with nonsmooth principal parts in Bao and Symes (1995) are expected to be useful for solving the velocity inversion problem. Note that our symbol calculus generalizes the one of Beals and Reed (1984), in the sense that it is more suitable for the study of linear partial differential equations.

The reader is referred to Lewis and Symes (1991) for regularity results of the velocity coefficients to solution map and its linearization in the one dimensional case.

Acknowledgement

The work of W. W. Symes was partially supported by the National Science Foundation under grant DMS 86-03614 and DMS 89-05878, by the Office of Naval Research under contracts N00014-K-85-0725 and N00014-J-89-1115, by AFOSR 89-0363, by State of Texas, and by the Rice Inversion Project. The work of G. Bao was supported in part by the NSF grant DMS 95-01099 and a Research Development Award (University of Florida).

References

Bamberger A., Chavent G., and Laily P. (1979): About the stability of the inverse problem in 1-d wave equations– application to the interpretation of seismic profiles. Appl. Math. Optim. **5**, 1–47

Bao G. (1996): Smoothness between coefficients and boundary values for the wave equation. Preprint

Bao G. and Symes W. (1993): Trace regularity result for a second order hyperbolic equation with nonsmooth coefficients. J. Math. Anal. Appl. **174**, 370–389

Bao G. and Symes W. (1995): Time like trace regularity of the wave equation with a nonsmooth principal part. SIAM J. Math. Anal. **26**, 129–146

Bao G. and Symes W. (1996): On the sensitivity of solutions of hyperbolic equation to the coefficients. Comm. in P.D.E. **21**, 395–422

Beals M. and Reed M. (1982): Propagation of singularities for hyperbolic pseudodifferential operators and applications to nonlinear problems. Comm. Pure Appl. Math. **35**, 169–184

Beals M. and Reed M. (1984): Microlocal regularity theorems for nonsmooth pseudodifferential operators and applications to nonlinear problems. Trans. Amer. Math. Soc. **285**, 159-184

Fernandez E. M., Gauzellino P., Santos J. E., and Sheen D. (1993): Parameter estimation in Multidimensional acoustic media. Preprint

Lewis R.M. and Symes W. (1991): On the relation between the velocity coefficient and boundary value for solutions of the one-dimensional wave equation. Inverse Problems **7**, 597–631

Rakesh (1988): A linearized inverse problem for the wave equation. Comm. in P.D.E. **13**, 573–601

Sacks P. and Symes W. (1985): Uniqueness and continuous dependence for a multidimensional hyperbolic inverse problems. Comm. in P.D.E. **10**, 635–676

Santosa F. and Symes W. (1989): *An Analysis of Least-Squares Velocity Inversion* (Geophysical Monograph No. 4, Society of Exploration Geophysicists, Tulsa)

Symes W. (1983a): A trace theorem for solutions of the wave equation, and the remote determination of acoustic sources. Math. Meth. in the Appl. Sci. **5**, 131–152

Symes W. (1983b): Some aspects of inverse problems in several dimensional wave propagation. Proc. Conference on Inverse Problems, SIAM-AMS Proceedings 14, ed. D. W. McLaughlin, Amer. Math. Soc., Providence, R.I.

Symes W. (1986a): On the relation between coefficient and boundary values for solutions of Webster's horn equation. SIAM J. Math. Anal. **17**, 1400–1420

Symes W. (1986b): Linearization stability for an inverse problem in several dimensional wave propagation. SIAM J. Math. Anal. **17**, 132–151

Developments in Numerical Methods for Transient Scattering Problems

Patrick Joly[1]

INRIA, Domaine de Voluceau, Rocquencourt, 78153, Le Chesnay Cedex, France

1 Introduction

In recent years, solving time dependent problems of scattering by an obstacle has received considerable attention. Some facts are now commonly admitted:

- The time discretization must lead to schemes which are explicit. Indeed, although one has now at our disposal efficient iterative methods for the solution of linear systems, the inversion of a matrix at each time step must be prohibited, in particular because of the very huge size of the problems one generally has to deal with, especially in dimension 3.
- For a lot of applications, the usual second order methods are considered as insufficently accurate because of the numerical dispersion they induce.

In this talk, we shall discuss some methods for the space discretization of the equations of the problem which lead to explicit and possibly higher order methods. Among the various techniques that have been used and studied in the past, the finite difference method is one of the most attractive. This method uses a regular grid and hence is very efficient from the computational point of view. However, its great disadvantage is that it creates **numerical diffraction** when the obstacle boundary does not fit the grid mesh (see Fig. 1), which will necessary be the case as soon as the obtacle has a complex geometry (i.e. as soon as it is not a reunion of rectangles in 2D).

A possible solution to this drawback is the use of a finite element method. The finite element mesh can follow precisely the boundary of the object (see Fig. 2).

Nevertheless, some drawbacks are introduced. In particular, to obtain an explicit scheme, it appears **necessary to use mass lumping** which is still **difficult** to do in the case of **higher order** finite element methods, especially for **Maxwell's equations**. The objective of this presentation is to give an overview of two researches devoted to the solution of the difficulties related to each of the two approaches :

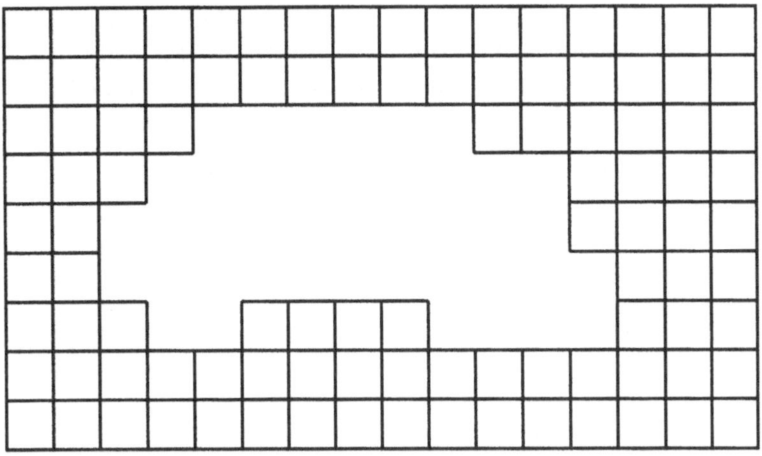

Fig. 1. Staircase approximation of the obstacle

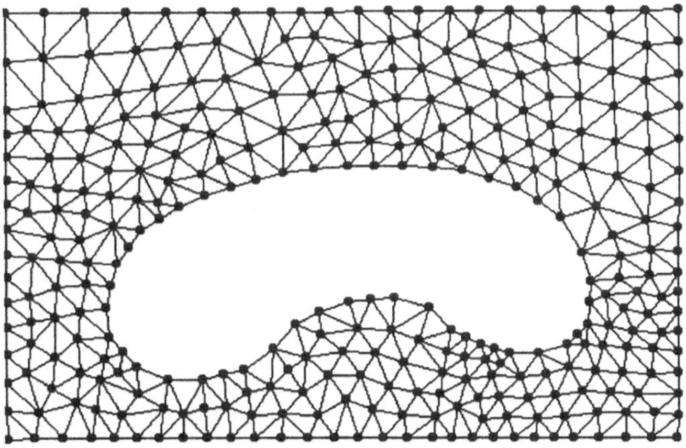

Fig. 2. Example of the conforming finite element mesh in 2D

1 **New higher order finite element spaces for mass lumping.**(Section 2) The case of Lagrange elements has been the object of a joint work with G. Cohen and N. Tordjman (Tordjman (1995)). We shall emphasize here the case of edge elements, which is a joint work with A. Elmkies (Elmkies and Joly (1996)).

2 **Fictitious domain methods for time dependent wave propagation problems.**(Section 3) Such methods intend to preserve the main

advantages of finite difference methods while ensuring a better respect of the geometry. The one I present here is the result of a joint work with F. Collino, S. Garces and F. Millot (Collino and Joly and Millot (1996)).

2 Edge finite elements and mass lumping for Maxwell's equations

2.1 Introduction

Edge finite elements such as they were introduced by Nédélec (1980) (see also Nédélec (1986)) are well known for providing natural methods for solving Maxwell's equations. Indeed, from a mathematical point of vue, they give spaces which are conforming approximations of the space $H(rot, \Omega)$ which naturally appears in the variational formulation of these equations. On the other hand, from a physical and practical point of vue, they allow to take in account boundary conditions as well as discontinuities of the electromagnetic fields at material interfaces. Moreover, one can then model complex geometries with the help of triangular (2D) or tetrahedric (3D) meshes. However, for such methods, the problem of mass lumping has not received, for the moment, a satisfactory solution. The aim of this work is to construct new edge finite element spaces which will solve this problem, including the case of anisotropic media. For simplicity we shall restrict ourselves to the 2D case. Let us consider Maxwell's equations in a homogeneous bidimensional medium Ω written as a second order system:

$$\frac{\partial^2 \mathbf{u}}{\partial t^2} + \mathbf{rot}(\mathrm{rot}\mathbf{u}) = 0 \quad x \in \Omega, t > 0 \tag{1}$$

If V_h is a finite element space, The space discretization of (1) by a finite element method in some space $V_h \subset H(rot, \Omega)$ results in a second order differential system:

$$M_h \frac{d^2 U_h}{dt^2} + A_h U_h = 0 \tag{2}$$

where M_h is the so called mass matrix whose entries are the $L^2(\Omega)$ inner products between basis functions of V_h. Mass lumping consists in **approximating M_h by a diagonal matrix** in a suitable basis of V_h. In such a case, the numerical scheme obtained after time discretization is fully explicit, which ensures the efficiency of the method, at least if the approximation of the mass matrix is **sufficiently accurate** to preserve the order of the method. In the case of the approximation of the scalar wave equation by Lagrange elements, this is obtained by calculating the integrals in the terms of M_h by a suitable **quadrature formula**. The quadrature points must coincide with the degrees of freedom of the finite element (cf. Tordjman (1995)). In the case of the edge finite elements built on squares, the specificity of the basis functions (which are actually tensor products) allow to use one dimensional quadrature rules and the method can be generalized to higher order elements (cf. Cohen and Monk (1995)). Nevertheless, in the case of triangular edge elements, such a strategy is unsuccessfull even for the first order element. Indeed, if K is a triangle, the space considered is $\mathcal{R}_1 = \{(\alpha_1, \alpha_2)^t + \beta(x_2, -x_1)^t, \ \alpha_1, \alpha_2, \beta \in \mathbb{R}\}$. To each edge is associated one degree of freedom which is the constant value of the tangential component of the vector fields along this edge (cf. figure 3). If $\mathbf{u}$ and $\mathbf{v}$ are two basis functions, $\int_K \mathbf{u}.\mathbf{v}dx$ is computed exactly using the

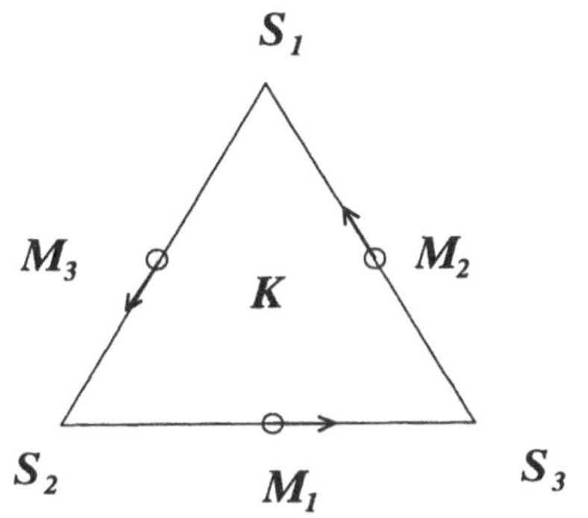

Fig. 3. Reference edge finite element of first order.

following quadrature rule:

$$\int_K \mathbf{u}.\mathbf{v}dx = \frac{mes(K)}{3} \sum_{i=1}^{3} \mathbf{u}(M_i).\mathbf{v}(M_i) \tag{3}$$

Let us make appear the tangential and the normal components of $\mathbf{u}$ et $\mathbf{v}$ on each edge:

$$\int_K \mathbf{u}.\mathbf{v}dx = \frac{mes(K)}{3} \sum_{i=1}^{3} u_\tau(M_i)v_\tau(M_i) + \frac{mes(K)}{3} \sum_{i=1}^{3} u_\nu(M_i)v_\nu(M_i) \tag{4}$$

If $\mathbf{u} \neq \mathbf{v}$, $u_\tau(M_i)v_\tau(M_i) = 0$ because of the definition of the degrees of freedom but in general, $u_\nu(M_i)v_\nu(M_i) \neq 0$. Thus, the failure of mass lumping can be attributed to the fact that only the tangential components of the vector field are degrees of freedom.

2.2 New triangular finite element

The problem of mass lumping for edge elements has been recently approached by Haugazeau and Lacoste (Lacoste (1994)). However, their approach is not completely satisactory since their applicability depends on the nature of the mesh and is restricted to lower order elements. In this work, we choose a completely different procedure based on the ideas of Tordjman (1995) for Lagrange elements. It consists in **incorporating the normal components** in the set of the degrees of freedom. We are led to enrich the $\mathcal{R}_1$ space and to introduce $\widetilde{\mathcal{R}}_1 = \mathcal{R}_1 \oplus [\mathbf{w}_1, \mathbf{w}_2, \mathbf{w}_3]$ where the $\mathbf{w}_i$ are P_2 vector fields whose tangential component vanishes on ∂K, in order to keep the conformity in $H(rot)$ (cf. figure 4). More precisely, if (i, j, k) is a permutation of $(1, 2, 3)$,

$$\mathbf{w}_i = \lambda_j \lambda_k \nabla \lambda_i$$

That is also for this purpose that, for the construction of the new finite element space, **we do not enforce the continuity of the normal component** of the field across the edges and we define the new approximation space:

$$\widetilde{V}_h = \left\{ \mathbf{v}_h \in H(\text{rot}, \Omega) / \ \forall K \in \mathcal{T}_h, \mathbf{v}_{h|K} \in \widetilde{\mathcal{R}}_1(K) \right\}$$

Therefore we have three degrees of freedom per edge, one tangential com-

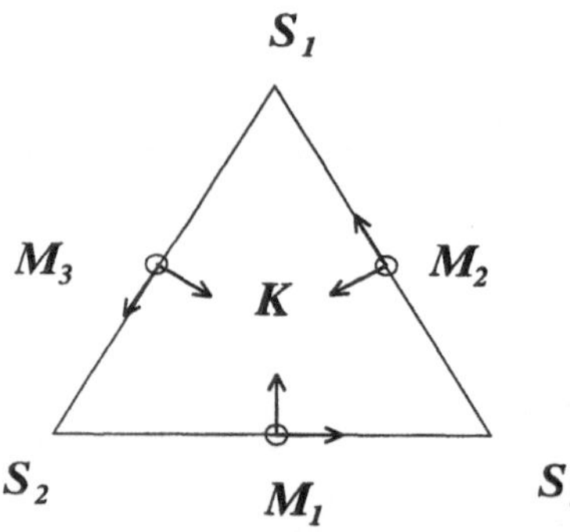

Fig. 4. 2D new finite element (first order).

ponent and two normal components (see Fig. 5). Let us emphasize the fact that, in an anisotropic medium, mass lumping is "almost" realized since one is reduced to the inversion of local 3×3 linear systems.The procedure can be generalized to higher order elements. The degrees of freedom which are moments on edges or triangles must be replaced by points values at some specific locations which will coincide with quadrature points. The new difficulty is thus to find a quadrature formula so that one does not lose any accuracy. In order to get this property , we can play on the weights of the quadrature fomula but also on the location of the interpolation points. Let us describe

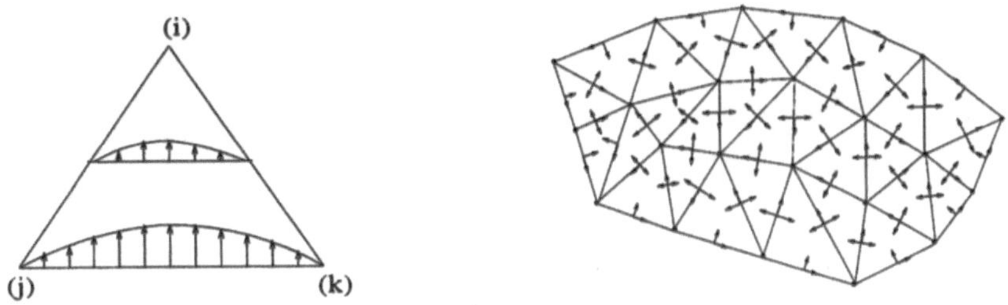

Fig. 5. basis function $\mathbf{w}_i$ (left) and triangular mesh for $\widetilde{\mathcal{R}}_1$ (right).

how we proceed for second order and third order elements (see also Figure 6).

Second order element: the usual Nédélec space $\mathcal{R}_2 \subset (P_2)^2$ is increased by 6 cubic vector fields and the location of the new interpolation points are given by

$$\begin{aligned} &M_{ij} = \alpha S_i + (1-\alpha) S_j \text{ with } \alpha = \frac{1}{2} - \frac{\sqrt{33}}{22} \\ &G \text{ barycenter of } K. \end{aligned} \tag{5}$$

and we use the following quadrature formula, exact on P_3:

$$\int_K f dx \simeq \operatorname{mes}(K) \{ \omega_m \sum_{i,j} f(M_{i,j}) + \omega_g f(G) \}$$

with $\omega_m = \dfrac{11}{240}$ and $\omega_g = \dfrac{9}{40}$.

Third order element: the usual Nédélec space $\mathcal{R}_3 \subset (P_3)^2$ is increased by 9 quadric vector fields and the location of the new interpolation points are given by

$$\begin{aligned} &M_{ij} = \alpha S_i + (1-\alpha) S_j \text{ and } \alpha = \frac{1}{2} - \frac{\sqrt{1785 + 168\sqrt{7}}}{126}. \\ &G_i = \beta S_i + \frac{1-\beta}{2} S_j + \frac{1-\beta}{2} S_k \text{ with } \beta = \frac{1}{3} + \frac{2\sqrt{7}}{21}. \end{aligned} \tag{6}$$

and we use the following quadrature formula, exact on P_5:

$$\int_K f dx \simeq \operatorname{mes}(K) \{ \omega_m \sum_{i,j} f(M_{i,j}) + \omega_a \sum_i f(M_i) + \omega_g \sum_i f(G_i) \}$$

with $\omega_m = \dfrac{7(1246 - 197\sqrt{7})}{361440}$, $\omega_a = \dfrac{4(49\sqrt{7} - 50)}{11295}$ and $\omega_g = \dfrac{7(14 - \sqrt{7})}{720}$.

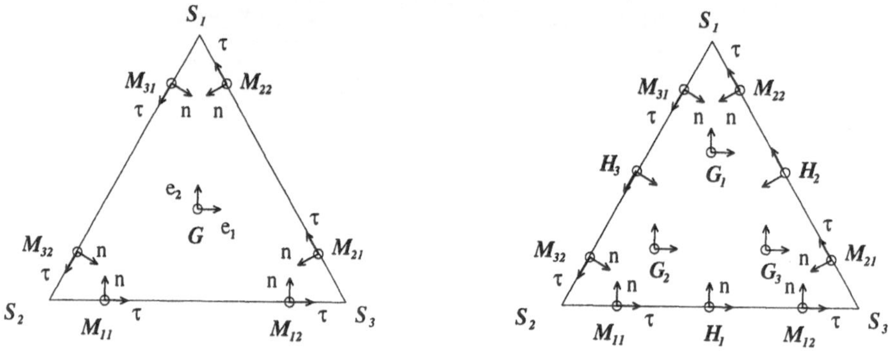

Fig. 6. New 2D triangular finite elements: second order (left) and third order (right).

Let us also mention that, with the same method, we can construct new edge finite elements on quadrangular meshes which are an alternative solution to the mass lumping problem to those of Cohen and Monk (cf. Cohen and Monk (1995)), with the advantage of succeeding to lumping the mass matrix in the case of anisotropic media.

2.3 Dispersion analysis

As we said previously, one of the main important features in the analysis of numerical methods for linear wave propagation is the study of their numerical dispersion on a regular mesh. We have considered below a uniform mesh made of rectangle triangles. On figure 7, we plot the variations of the adimensional phase velocity of a numerical plane wave (i.e. the ratio between the numerical phase velocity and the exact one) as a function of the inverse ot the number of meshpoints per wavelength. The different curves on each picture correspond to various propagation direction: this illustrates the anisotropy of the schemes. It turns out that in the case of the triangular finite elements of first and second order, we get a dispersion error which is $O(h^2)$ and $O(h^4)$ respectively. For the first order quadrangular element, we get a dispersion error which is $O(h^2)$ and appears to be smaller than the one obtained with the Cohen-Monk's method.

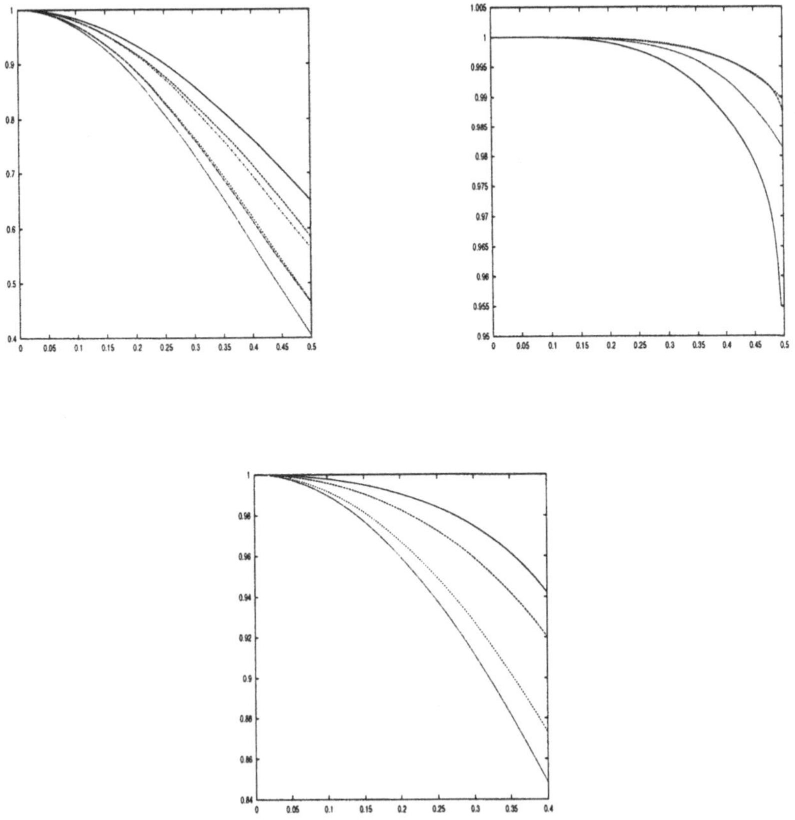

Fig. 7. Dispersion curves - Triangles: first order (top-left), second order (top-right), squares: first order (bottom)

3 A fictitious domain method for time dependent scattering problems

3.1 Introduction

Even when one can lump the mass matrix, one may prefer to use finite difference type meshes for various reasons (ease of implementation, bad stability condition induced by small elements in unstructured meshes,...). The fictitious domain method appears as an alternative method to finite element methods for solving time-dependent scattering problems. Such methods have been shown recently to have interesting potential for solving complicated problems (Astrakmantev (1978), Atamian and Glowinski and Periaux and Steve and Terrason (1989), Atamian and Joly (1993), Kuznetsov and Marchuk and Matsokin (1986), Finogenov and Kuznetsov (1988) Glowinski and Pan and Periaux (1994)), particularly in the stationary case.

The fictitious domain method, also called domain embedding method, consists in extending artificially the solution inside the obstacle so that the new domain of computation has a very simple shape (typically a rectangle in 2D). This extension requires the introduction of a **new variable defined only at the boundary of the obstacle**. This auxiliary variable allows one to **take into account the boundary condition**. It can be related to a singularity across the boundary of the obstacle of the extended function. This idea will be developed in section 3.2. The main point is that **the mesh for the solution** on the enlarged domain can be **chosen independently of the geometry of the obstacle**. In particular, the use of regular grids or structured meshes allows for simple and efficient computations.

Of course, we have to pay for this advantage in terms of some additional computational cost due to the determination of the new boundary unknown. However, the final numerical scheme appears to be a slight perturbation of the scheme for the problem without obstacle so that **this cost may be considered as marginal**. From the theoretical point of view, the convergence of the method is linked to the obtention of a uniform inf-sup condition which leads to a compatibility condition between the boundary mesh and the uniform mesh (see Girault and Glowinski (1994)). Theoretically, it implies that the two mesh grids can not be chosen completely independently, but in practice this is not a real constraint. Another important point is that the stability condition of the resulting scheme is the same as the one of the finite difference scheme. For simplicity, we have chosen to reduce our presentation to the scalar wave equation but the extension to Maxwell's equations is straightforward.

3.2 Presentation of the method

We consider first the scattering of a wave by an obstacle $\mathcal{O}$, $\mathcal{O} \subset \mathbf{R}^d$ with $d = 2$ or $d = 3$. The solution is governed by the wave equation in D, the open complement of the obstacle with a Dirichlet condition on the boundary (see Fig. 8):

$$\begin{cases} \dfrac{\partial^2 u}{\partial t^2} - \Delta u = 0 \text{ in } D \\ u = 0 \qquad\qquad\quad \text{on } \gamma = \partial D. \end{cases} \tag{7}$$

The incident wave is generated by initial conditions at time $t = 0$ given by

$$u(x,0) = u_0(x) \in H^1(D), \qquad \frac{\partial u}{\partial t}(x,0) = u_1(x,0) \in L^2(D). \tag{8}$$

For the sake of simplicity, a Dirichlet condition is assumed on the exterior boundary as well. For our purpose, we choose the geometry of the external boundary to be rectangular. We denote by Ω this bounded domain and by C the rectangle $\Omega \bigcup \mathcal{O}$ (see Fig. 8). We want to solve the simple problem

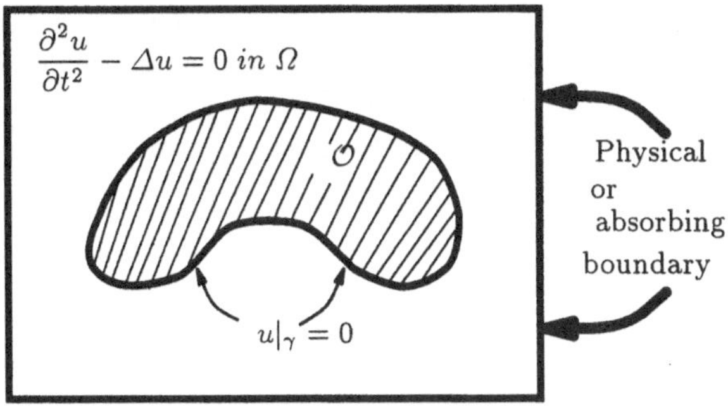

Fig. 8. Geometry of the problem

described by equation (7) by the fictitious domain method. Note that this method can be used also for more complicated problems as the scattering of an acoustic or electromagnetic wave in a heterogeneous medium.

The main idea of the fictitious domain method is to extend u from Ω to the enlarged domain C to a function (still denoted by u for simplicity) with $H^1(C)$ regularity. More precisely, we look for u in the space

$$u \in \tilde{V} = \left\{ v \in H^1(C);\ v = 0 \quad on \quad \gamma \right\}, \tag{9}$$

and we define u as the first argument of (u, λ) the solution of the following variational evolution problem

$$\begin{cases} \dfrac{d^2}{dt^2}(u,v) + a(u,v) = b(v,\lambda) \ \forall v \in X \\ b(u,\mu) = 0 \qquad\qquad\qquad\qquad\quad \forall \mu \in M, \end{cases} \tag{10}$$

where $X = H_0^1(C)$, $M = H^{-1/2}(\gamma)$, $H = L^2(C)$ and:

$$(u,v) = \int_C uv \, dx \tag{11}$$

$$a(u,v) = \int_C \nabla v \nabla u \, dx \tag{12}$$

The bilinear form $b(u, \mu)$ denotes the duality pairing between $H^{-1/2}(\gamma)$ and $H^{1/2}(\gamma)$ and is equal to

$$b(u,\mu) = \langle \mu, u \rangle_\gamma . \tag{13}$$

In its principle, the fictitious domain method consists in extending the solution in the enlarged computational domain and to introduce a new unknown at the boundary. The main difference with a standard conforming finite element approach lies in the fact that the Dirichlet condition is taken into account in a weak sense. This method has also some relationship with the integral equation method in the sense that the additional unknown λ is nothing but that the jump of the normal derivative of u across γ.

To understand (10), we can for instance consider the time t as a parameter and the function $f = -\dfrac{\partial^2 u}{\partial t^2}$ as a data. We have to solve now the following problem

$$\begin{cases} -\Delta u = f \text{ in } \Omega \\ u = 0 \qquad \text{on } \gamma. \end{cases} \tag{14}$$

It is equivalent to minimizing the functional

$$J(v) = \int_C (\frac{1}{2}|\nabla v|^2 - fv)\, dx$$

over the space V of functions of $H^1(\Omega)$ satisfying the constraint $v = 0$ on γ, which can be seen as the restrictions to Ω of functions of $\tilde{V}$. It is natural to consider the enlarged minimization problem defined by

$$\min_{\tilde{v} \in \tilde{V}} J(\tilde{v}) = \int_C (\frac{1}{2}|\nabla \tilde{v}|^2 - f\tilde{v})\, dx \tag{15}$$

where f has been extended to C. It is easy to verify that the restriction of the solution of problem (15) to Ω is exactly the solution of the problem (10). Problem (15) is a minimization problem with an equality constraint. Its solution is the first argument of the saddle point of the Lagrangian functional defined by $L(v, \mu) = J(v) - b(v, \mu)$. Writing that the derivative of this Lagrangian is equal to zero at the optimum (u, λ), we obtain:

$$\begin{cases} a(u, v) = b(v, \lambda) + (f, v) \; \forall v \in X \\ \\ b(u, \mu) = 0 \qquad\qquad \forall \mu \in M \end{cases} \tag{16}$$

which gives exactly the equations of (10) if we have written $f = -\dfrac{\partial^2 u}{\partial t^2}$.

3.3 Finite element approximation

Let X_h (resp M_h) be a finite dimensional subspace of X (resp M). We approximate the variational problem (10) by

$$\begin{cases} \text{Find } u_h \in X_h, \quad \lambda_h \in M_h \quad \text{such that} \\ \dfrac{d^2}{dt^2}(u_h, v_h) + a(u_h, v_h) = b(v_h, \lambda_h) \; \forall v_h \in X_h \\ \\ b(u_h, \mu_h) = 0 \qquad\qquad \forall \mu_h \in M_h \end{cases} \tag{17}$$

More precisely, X_h will be a finite element space based on a regular mesh in C (for example squares in 2D). On the other hand, M_h is directly related to the geometry of γ which can be, for instance, discretized into segments for 2D problems (see Fig 4). For instance we can take Q_1 elements for constructing X_h and piecewise constant functions for M_h. Let us introduce $\{v_j,\ 1 \leq j \leq p = \dim X_h\}$ and $\{w_\ell, 1 \leq \ell \leq q = \dim M_h\}$ two bases for the spaces X_h and M_h respectively. Indeed , we shall have respectively, if h denotes the step size of the meshes

$$\begin{cases} p = O(\frac{1}{h^2}) \quad \text{and } q = O(\frac{1}{h}) \quad \text{if d} = 2 \\ p = O(\frac{1}{h^3}) \quad \text{and } q = O(\frac{1}{h^2}) \text{ if d} = 3 \end{cases} \tag{18}$$

Let us define

- $M_h \in \mathcal{L}(X_h, X_h)$ = the $p \times p$ matrix associated to (u_h, v_h)
- $A_h \in \mathcal{L}(X_h, X_h)$ = the $p \times p$ matrix associated to $a(u_h, v_h)$
- $B_h \in \mathcal{L}(M_h, X_h)$= the $p \times q$ matrix associated to $b(u_h, \mu_h)$

If U_h (resp Λ_h) is the vector of the coordinates of u_h (resp. λ_h) in the basis $\{v_j\}$ (resp. $\{w_\ell\}$), we have

$$\begin{cases} M_h \dfrac{d^2 U_h}{dt^2} + A_h U_h = B_h \Lambda_h \\ B_h^t U_h = 0 \end{cases} \tag{19}$$

where B_h^t is the transpose of B_h. If M_h and A_h can be interpreted respectively as approximations of the identity and Laplace operators, B_h^t can be seen as a discrete trace operator from X_h to M_h. Note that problem (19) appears as a system of ordinary differential equations with an algebraic constraint. This establishes an analogy with problems of fluid dynamics in the incompressible case where the free divergence is the constrain. Of course, we apply mass lumping (which is very simple in quadrangular meshes) so that M_h is diagonal.

3.4 Time discretization

For time discretization, we consider a time step Δt and use a three time step finite difference explicit scheme, which leads to:

$$\begin{cases} U_h^{n+1} - 2U_h^n + U_h^{n-1} = -\Delta t^2 M_h^{-1} A_h U_h^n + \Delta t^2 M_h^{-1} B_h \Lambda_h^n \ (1) \\ B_h^t U_h^n = 0. \qquad\qquad\qquad\qquad\qquad\qquad\qquad\qquad (2) \end{cases} \tag{20}$$

To compute the solution explicitly, an apparent difficulty appears with the condition $B^t U_h^n = 0$. In fact, for practical computations, this condition is replaced by an equivalent equation which results from multiplying the first equation by B_h^t. More precisely, (20) can be shown to be equivalent to:

$$\begin{cases} U_h^{n+1} = 2U_h^n - U_h^{n-1} - (\Delta t)^2 M_h^{-1} A_h U_h^n + (\Delta t)^2 M_h^{-1} B_h \Lambda_h^n \\ B_h^t M_h^{-1} B_h \Lambda_h^n = B_h^t M_h^{-1} A_h U_h^n . \end{cases} \tag{21}$$

Finally, let us assume (U_h^{n-1}, U_h^n) to be known, U_h^{n+1} is computed by the following procedure:

- solve $B_h^t M_h^{-1} B \Lambda_h^n = B_h^t M_h^{-1} A_h U_h^n$ to compute Λ_h^n,
- compute U_h^{n+1} via ((21)-(1)).

Therefore, our method appears as a slight modification of the explicit finite difference scheme one would solve in the absence of the obstacle (which corresponds to the second step above). The additional cost is due to the computation of the Lagrange multiplier, for which we must invert the matrix $Q = B_h^t M_h^{-1} B_h$ which obviously satisfies:

- Q is symmetric and positive.
- The size of $Q,(q \times q)$ is very small compared to the size A_h since $q << p$.
- Q is a sparse matrix with narrow bandwidth.

Thus, if Q^{-1} exists, the inversion of Q can be performed by a Cholesky factorization or by a conjugate gradient algorithm. There remains the crucial question of the existence of this inverse which is linked to the uniform discrete inf-sup condition,

$$\exists C, \text{ independent of } h \text{ such that} \qquad \inf_{(\lambda \in M_h)} \sup_{(v \in X_h)} \frac{b(v,\lambda)}{\|\lambda\|_M \|v\|_X} = C > 0 \ . \tag{22}$$

This condition requires a compatibility relation between the two meshes. It imposes a condition between the dimensions of the two spaces X_h and M_h. Such a condition can be found in Girault and Glowinski (1994). In practice the space step used for the boundary mesh must be larger than the one used for the mesh of Ω. Finally, let us that one can show that our procedure preserves the conservation of some discrete energy. An important consequence is that our scheme is stable under the same stability condition than the usual finite difference scheme.

Actual numerical computations of scattering experiments show the real superiority in term of accuracy of the fictitious domain method with respect to a staircase approximation of the obstacle. This gain of precision can also be analyzed on a simple 1D problem. We refer the reader to Collino and Joly and Millot (1996) for more details.

Acknowlegments

The author would like to thank F. Collino and A. Elmkies for their help in the redaction of this article.

References

Astrakmantev, G.P. (1978): Methods of fictitious domains for a second order elliptic equation with natural boundary conditions. U.S.S.R Computational Math.and Math. Phys. **17**, 114–221

Atamian, C., Glowinski, R., Periaux, J., Steve, H., Terrason, G. (1989): Control approach to fictitious domain in electromagnetism, Conférence sur l'approximation et les méthodes numériques pour la résolution des équations de Maxwell, Hotel Pullmann, Paris.

Atamian, C., Joly, P. (1993): Une analyse de la methode des domaines fictifs pour le probleme de Helmholtz exterieur M2AN **27**, 251–288

Cohen, G., Monk, P. (1995): Efficient edge finite element schemes in computational electromagnetism, Proc. of the 3rd Int. Conf. on Mathematical and Numerical Aspects of Wave Propagation Phenomena, SIAM, April 1995.

Collino, F., Joly, P., Millot, F. (1996): A fictitious domain method for time dependent wave propagation problems, Rapport Interne INRIA n° 2963, Août 1996, submitted to J. Comp. Physics

Elmkies, A., Joly, P. (1996): Elements finis et condensation de masse pour les equations de Maxwell : le cas 2D. Rapport Interne INRIA n° 3035, Novembre 1996.

Finogenov, S. A., Kuznetsov, Y. A. (1988): Two stage fictitious components methods for solving the Dirichlet boundary value problem. Sov. J. Num. Anal. Math. Modelling **3**, 301–323

Girault, V., Glowinski, R. (1994): Error analysis of a fictitious domain method applied to a Dirichlet problem, preprint Université Paris 6.

Glowinski, R., Pan, T. W., Periaux, J. (1994): A fictitious domain method for Dirichlet problem and applications. Comp. Meth. in Appl. Mech. and Eng. Modelling , 283–303

Kuznetsov, Y. A., Marchuk, G. I., Matsokin, A. M. (1986): Fictitious domain and domain decomposition methods. Sov. J. Num. Anal. Math. Modelling **1**, 3–35

Lacoste, P. (1994): Les éléments finis des équations de Maxwell dans le code PALAS. Eléments finis nouveaux pour le cadre axisymétrique. La condensation des matrices masses, Thèse, Université de Bordeaux 1.

Nedelec, J. C. (1980): Mixed finite element in $\mathbb{R}^3$. Numerische Mathematik **35**, 315–341

Nedelec, J. C. (1986): A new family of mixed finite elements in $\mathbb{R}^3$. Numerische Mathematik **50**, 57–81

Tordjman, N. (1995): Eléments finis d'ordre élevé avec condensation de masse pour l'équation des ondes, Thèse, Université de Paris 9.

A Level-Set Approach for Eddy Current Imaging of Defects in a Conductive Half-Space

A. Litman[1], D. Lesselier[1] and F. Santosa[2]

[1] Laboratoire des Signaux et Systèmes, CNRS-Supélec, Plateau de Moulon, 91192 Gif-sur-Yvette cedex, France

[2] School of Mathematics, University of Minnesota, Minneapolis, MN 55455, USA

Abstract. The retrieval of the shape of a cylindrical defect of low conductivity buried in a conductive half-space is investigated from aspect-limited, frequency-diverse data. The sources of the interrogative fields and the receivers of the scattered (anomalous) fields are both placed on the same side of a particular interface. The defect is embedded on the other side. We derive an iterative process based on level-set methods. This level-set approach has been shown to be effective in treating problems with propagating fronts and is based on the ideas developed by Osher and Sethian. An iterative process is implemented: at each iteration, the boundary of the defect is moving with a speed term which minimizes the residual in the data fit. The resulting equation of motion is solved by employing entropy-satisfying upwind finite-differences schemes.

Introduction

In the well-known problem of electromagnetic inversion of objects in stratified environments, one is interested in reconstructing unkown hidden obstacles in known stratified media, see Lesselier and Duchêne (1996) for more references. Here we are concerned with a nondestructive evaluation of cracks in metallic structures from aspect-limited frequency-diverse data. A typical application is the eddy current probing of an air void or of an inclusion of low conductivity in a metal block of high conductivity, but other materials and frequency bands can be considered likewise.

The defects are modeled as infinitely long inhomogeneities with bounded cross-section embedded in a semi-infinite, isotropic metal block of known conductivity, with air above. The probing of the defect by a known interrogative field gives rise to an anomalous field. Data are values of this field which is radiated by Huygens-type sources whose support is the defect only. In the Born approximation, a first-kind integral equation links this field to the object function, here the contrast between the conductivity at a given point and the conductivity of the embedding. This object function is real-valued.

As source and receivers are both located in air, only reflection mode data are available. This *aspect-limited* configuration enhances the ill-posedness of the problem. To compensate, several frequencies are used. Furthermore, the defect is assumed to be homogeneous. The contrast of conductivity takes only two values: 0 outside the defect, and a prescribed value inside. The inverse problem simplifies: only the shape and location of the defect are sought. One major difficulty remains the lack of *a priori* topological information on the defects. Thus, the level-set modeling technique seems appropriate for such problems.

Indeed, since its introduction by Osher and Sethian (1988), the level-set approach has been widely used when moving interfaces are dealt with: crystal growth (Sethian and Strain 1992), shape modeling (Malladi et al. 1995) (Caselles et al. 1993). Santosa (1996) has shown the feasibility of such an approach in the case of inverse problems involving obstacles such as the reconstruction of a diffraction screen.

What we aim at is to start from an initial contour and find a suitable deformation which moves the contour closer to the actual shape of the defect. Kass et al. (1988) have proposed a modeling technique for active contours using a Lagrangian representation of the front. But this *snake* model does not permit, due to the parametrization of the curves, the treatment of several contours simultaneously and also suffers from instability (Caselles et al. 1993).

The modeling technique of Osher and Sethian uses an Eulerian representation of the front. The $(N-1)$-dimensional surface is embedded in a level-set function of N space dimensions. The contour of the defect is defined as the level-set 0. The hypersurface is then made to flow along its normal direction. This evolution is described by the evolution equation. Such a representation will handle naturally the splitting of the fronts. Furthermore, Osher and Sethian have derived a stable numerical scheme borrowed from hyperbolic conservation laws to solve the Hamilton-Jacobi type equation. The speed of the moving front is synthetized from the minimization of the residual in the data fit. The procedure stops when the front is close enough to the actual shape of the defect.

1 Wavefield formulation

1.1 Model

The model is the following (de Oliveira Bohbot et al. 1996): a z-orientated cylindrical object is embedded in a two half-space configuration, as depicted in Fig. 1. The upper half-space D_1, $(x \leq 0)$, is lossless and homogeneous, with permittivity ϵ_1 and permeability μ_1. The lower half-space is lossy with the same permittivity ϵ_1 and permeability μ_1, and a non-zero conductivity σ_2. The defect located inside D_2 is of limited cross-section D in the (x, y) plane. It is penetrable with permittivity ϵ_1 and permeability μ_1, and a constant conductivity σ_D.

A source at low angular frequency ω (time-dependence $\exp(-j\omega t)$) placed in D_1 radiates a known interrogative wave taken for simplicity as a E-polarized plane wave (the field has a unique non-zero component along the

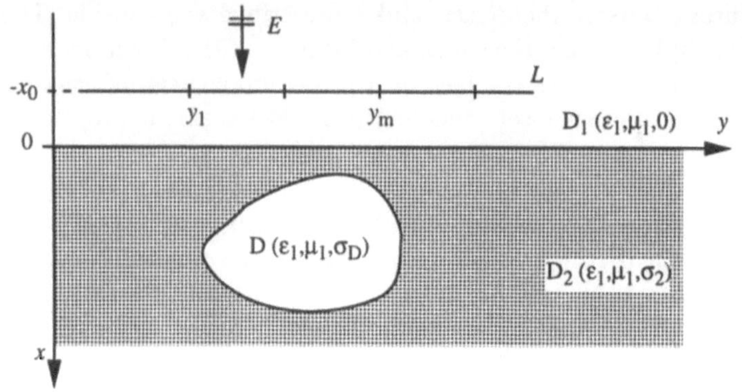

Fig. 1. Configuration

z axis) with normal incidence upon the air/metal interface. In the absence of defect D, the interrogative wave creates a known *incident* field E_0. In its presence, the anomalous field E_A is measured on a probing line L situated in D_1 at height $\mid x_0 \mid$ at different frequencies ω_f, $f = 1, \cdots, Freq$ and receivers y_m, $m = 1, \cdots, Mes$.

In the Born approximation framework, the two integral equations obtained by application of the Green's theorem to the Helmholtz equations with appropriate boundary conditions, are reduced to a single *observation* equation. This integral equation relates the anomalous field E_A and the object function of the defect, i.e. the contrast of conductivity $\chi(x,y) = \sigma(x,y)/\sigma_2 - 1$, defined in D_2, null outside D and valued to $\sigma_D/\sigma_2 - 1$ inside:

$$\frac{2j}{\delta^2} \int_D G_{12}(x_0, x, y_m, y, \omega_f)\, E_0(x, y, \omega_f)\, \chi(x,y)\, dxdy = E_A(x_0, y_m, \omega_f) \quad (1)$$

$$m = 1, \cdots, Mes, \quad f = 1, \cdots, Freq$$

$$G_{12}(x_0, x, y_m, y, \omega_f) = \frac{j}{2\pi} \int_{-\infty}^{+\infty} \frac{\exp\left[j(\beta_2 x - \beta_1 x_0)\right]}{\beta_1 + \beta_2} \exp\left[jK(y_m - y)\right]\, dK$$

$$\beta_h = \sqrt{k_h^2 - K^2},\ \Im(\beta_h) \geq 0,\ h = 1, 2.$$

The Green's function G_{12} represents the field observed at (x_0, y_m) in D_1 when a line source is located at (x, y) in D_2. k_1 (resp. k_2) is the wavenumber of D_1 (resp. D_2) and $\delta = \sqrt{2/\omega_f \mu_1 \sigma_2}$ corresponds to the skin depth.

A plane wave under normal incidence gives an *incident* wave of the following form $E_0(x, y, \omega_f) = T \exp(jk_2 x)$ in D_2, $T = 2k_1/(k_1 + k_2)$ is the Fresnel transmission coefficient.

1.2 Direct problem

We look for the anomalous field on the probing line induced by a known embedded defect illuminated by a known interrogative plane wave. Calculations are performed by a method of moments with pulse-basis functions and Dirac delta weight functions (de Oliveira Bohbot et al. 1996).

The defect D is divided into $Mx \times My$ square cells: $\Omega_{p,q} = \Lambda_p \times \Lambda_q$, $p = 1, \cdots, Mx$, $q = 1, \cdots, My$ with $\Lambda_p = [x_p - \lambda_x/2, x_p + \lambda_x/2]$ and $\Lambda_q = [y_q - \lambda_y/2, y_q + \lambda_y/2]$. In each cell, at a given frequency ω_f, the contrast and the field are assumed to be constant and take values $\chi(x_p, y_q)$ and $E_0(x_p, y_q, \omega_f)$. The anomalous field is then a simple sum of $Mx \times My$ terms:

$$E_A(x_0, y_m, \omega_f) = \sum_{p,q} \chi(x_p, y_q) E_0(x_p, y_q, \omega_f) \int_{\Omega_{p,q}} G_{12}(x_0, x, y_m, y, \omega_f) d\Omega \quad (2)$$

$$m = 1, \cdots, Mes, f = 1, \cdots, Freq$$

2 Definition of the cost function

In this specific inverse problem, the object function χ of the defect is constant but of unknown support. What we aim at is to find the shape and location of the defect, i.e. the object function χ, knowing the electrical properties as well as the incident field and the values on the probing line L of the anomalous field E_A at several frequencies. In the Born approximation, this problem is linear but still very ill-posed. Ill-posedness is taken into account by introducing in the inversion procedure *a priori* information about the defect we look for. For instance, the contrast of conductivity is purely real-valued. We separate (1) into two related real-valued integrals which are defined on D_2, the whole domain of definition of χ. In the following, the function $G(x, y_m, y, \omega_f)$ will correspond to the product $\frac{2j}{\delta^2}\, G_{12}(x_0, x, y_m, y, \omega_f)\, E_0(x_0, y_m, \omega_f)$, x_0 being implied. We have:

$$\int_D \Re G(x, y_m, y, \omega_f)\, \chi(x, y)\, dxdy = \Re E_A(y_m, \omega_f) \quad (3)$$

$$\int_D \Im G(x, y_m, y, \omega_f)\, \chi(x, y)\, dxdy = \Im E_A(y_m, \omega_f) \quad (4)$$

$$m = 1, \cdots, Mes, \quad f = 1, \cdots, Freq$$

Let us define the set $\mathcal{E} = \mathbb{R}^{2 \times Freq \times Mes}$. In the following, we keep the notation $\Re$ and $\Im$ for the real and imaginary part of vectors belonging to $\mathcal{E}$ such as E_A, and we denote by $A\chi$ the integral operator described in both (3) and (4). As the number of data differs from the number of unknowns, we look for a solution χ in the least-square sense, i.e. which minimizes the following cost function:

$$J(\chi) = \frac{1}{2} \parallel A\chi - E_A \parallel_{\mathcal{E}}^2 \quad (5)$$

The weighted scalar product on $\mathcal{E}$ is

$$< u, v >_{\mathcal{E}} = \sum_{f}^{Freq} \alpha_f \sum_{m}^{Mes} \Re u(y_m, \omega_f) \, \Re v(y_m, \omega_f) + \Im u(y_m, \omega_f) \, \Im v(y_m, \omega_f)$$

$$\frac{1}{\alpha_f} = \sum_{m}^{Mes} \mid E_A(y_m, \omega_f) \mid^2$$

3 Level-set formulation

We propose to solve the inverse problem by introducing a level-set description of the defect D and by following the evolution of this level-set along the iterations (Santosa 1996). This geometric description is an intrinsic one, i.e. the evolution of the curves does not depend on its particular parametrization. A fix grid can then be used along the iterations. Moreover, the changes such as expanding, shrinking, breaking of the fronts will be handled naturally by the hypersurface. This hypersurface $\phi(x, y, t)$ is defined everywhere on D_2:

$$\partial D_t = \{(x, y) : \phi(x, y, t) = 0\} \tag{6}$$

The inverse problem consists in following the evolution of ϕ, given a initial surface D_0 such that the surface D_t "tends" to the actual surface of the defect. This inverse problem becomes nonlinear. Indeed, there is a nonlinear dependence of $\chi(x, y, t)$ on $\phi(x, y, t)$:

$$\chi(x, y, t) = \begin{cases} \chi_{in} = \sigma_D/\sigma_2 - 1 & \{\phi(x, y, t) < 0\} = D_t \\ \chi_{out} = 0 & \{\phi(x, y, t) > 0\} = D_2 \backslash D_t \end{cases} \tag{7}$$

Once the initial surface D_0 is given, we have to find the evolution equation of the function ϕ. Two constraints are to be satisfied. First, each evolution should see the front getting closer to the actual defect boundaries, i.e. the cost function of (5) should decrease as time t increases. Second, this level-set function can only flow along its normal direction.

3.1 Derivation of the cost function

Let us look at the derivative according to t of the term of the cost function defined in (5) which only involves the real part of $(A\chi(t) - E_A)$. Let us denote this term by $\Re J(t)$ and by $\Im J(t)$ the other part of the cost function. If we assume that the functions under the integrals are regular enough, we have from Continuum Mechanics (Cea 1976):

$$\frac{d\Re J(t)}{dt} = \sum_{f}^{Freq} \alpha_f \sum_{m}^{Mes} \left[\Re A\chi(t) - \Re E_A\right](y_m, \omega_f)$$

$$\int_{\partial D_t} \chi_{in} \, \Re G(x, y_m, y, \omega_f) \, (\mathbf{v}_t \cdot \mathbf{n}_t) \, d\sigma$$

where $\mathbf{v}_t$ is the velocity of the points on the boundary ∂D_t and $\mathbf{n}_t$ the outward normal to this boundary at time t. Thus, the derivative of the cost function is such that:

$$\frac{dJ(t)}{dt} = \int_{\partial D_t} \chi_{in}\, A^* \left[A\chi(t) - E_A\right](x,y)\, (\mathbf{v}_t \cdot \mathbf{n}_t)\, d\sigma(x,y) \tag{8}$$

where the adjoint operator A^* is $A^*u(x,y) =< G(x,\cdot,y,\cdot), u >_{\mathcal{E}}$.

3.2 Evolution equation of ϕ

Following the idea of Osher and Sethian (1988), we look at the motion of the level-set $\partial D_t = \{\phi(x,y,t) = 0\}$. Let $(x(t), y(t))$ be the trajectory of a particle located on this level-set. As the points on the boundary of ∂D_t are assumed to be only moving along its normal direction, their velocity is $\mathbf{v}(x,y,t) = V(x,y,t)\ \mathbf{n}_t$. The normal vector $\mathbf{n}_t$ is equal to $\nabla\phi(x,y,t)/ \mid \nabla\phi \mid$. By the chain rule and substitution,

$$\frac{\partial\phi}{\partial t}(x,y,t) + \mathbf{v}(x,y,t)\cdot\nabla\phi(x,y,t) = 0$$

$$\frac{\partial\phi}{\partial t}(x,y,t) + V(x,y,t) \mid \nabla\phi \mid = 0 \tag{9}$$

This equation yields the motion of ∂D_t with normal velocity V on the level-set $\{\phi(x,y,t) = 0\}$. It is referred as a Hamilton-Jacobi type equation.

3.3 Choice of the speed

We have to define a speed function $V(x,y,t)$. This speed function must be such that the cost function decreases as time increases, i.e. such that the derivative given in (8) is negative. Santosa (1996) gave two possibilities for the speed. One is referred as the *evolution* approach with a velocity:

$$V(x,y,t) = -\chi_{in}\, A^* \left[A\chi(t) - E_A\right](x,y) \qquad (x,y) \in \partial D_t \tag{10}$$

The other is called the *approximation* approach and is inspired from the Gauss-Newton algorithm:

$$A^*A\, V(x,y,t) = -\chi_{in}\, A^* \left[A\chi(t) - E_A\right](x,y) \qquad (x,y) \in \partial D_t \tag{11}$$

But, the velocity is not defined for all the points of the domain D_2. For numerical reasons, we have to extend it to the whole domain. Following (10), we arrive at:

$$V(x,y,t) = -\chi_{in}\, A^* \left[A\chi(t) - E_A\right](x,y) \qquad \forall (x,y) \in D_2 \tag{12}$$

A similar extension can be made in the approximation scheme.

3.4 Hamilton-Jacobi equation and hyperbolic conservation laws

If the velocity $V(x, y, t)$ is assumed to be a constant V, independent of time, (9) is a first-order Hamilton-Jacobi equation (Kimia et al. 1995). Let us call a generalized solution, a solution which is locally Lipschitz and which satisfies this equation almost everywhere. There are many generalized solutions. The problem is to pick up the right one, called the viscosity solution. Barles (1985) has shown that the entropy solution of an hyperbolic conservation law equation is equivalent to the viscosity solution of the corresponding Hamilton-Jacobi equation. Sethian has expressed this entropy condition by: *Once a particle is burnt, it stays burnt*, if the boundary is viewed as a burning flame. This equivalence is the key of the numerical algorithm presented by Osher and Sethian and which chooses the correct viscosity solution.

The simplest numerical scheme is to replace the spatial derivatives by central differences and the time derivative by a forward difference. A selected part of the domain D_2 in which the defect is assumed to be found, is divided into $Nx \times Ny$ elementary cells $\Delta_{i,j}$, $i = 1, \cdots, Nx$, $j = 1, \cdots, Ny$ where the contrast, the field and the level-set function are constant. We denote by Δt the time step, by Δx and Δy the grid step in x and y. Unfortunately, this algorithm fails because it ignores the entropy condition. Osher and Sethian keep a forward difference for the time derivative:

$$\phi(x_i, y_j, t_{k+1}) = \phi(x_i, y_j, t_k) - \Delta t H(\phi)$$

but with a numerical Hamiltonian $H(\phi) = V(D\phi)^2$ such that: if $V \gtrless 0$

$$D\phi = {\max \atop \min}(D_x^- \phi, 0)^2 + {\min \atop \max}(D_x^+ \phi, 0)^2 + {\max \atop \min}(D_y^- \phi, 0)^2 + {\min \atop \max}(D_y^+ \phi, 0)^2$$

The standard definitions of the forward and backward difference operators are used:

$$D_x^- \phi(x_i, y_j, t_k) = \frac{\phi(x_i, y_j, t_k) - \phi(x_{i-1}, y_j, t_k)}{\Delta x}$$
$$D_x^+ \phi(x_i, y_j, t_k) = \frac{\phi(x_{i+1}, y_j, t_k) - \phi(x_i, y_j, t_k)}{\Delta x}$$

This conservative monotone scheme is an upwind method: the derivatives are calculated in the direction of the outward flowing normals. Thus, boundary conditions for the test domain do not flow backwards and do not create spurious solutions. This algorithm gives the correct entropy-satisfying weak solution to the moving boundary problem defined by (9) with constant velocity and for a given initial hypersurface $\phi(x, y, 0)$.

4 Algorithm

Once an initial shape D_0 is given, we initialize $\phi(x, y, 0)$ by the distance function $\pm dist((x, y), \partial D_0)$, where the minus (plus) sign is chosen if (x, y) is inside (outside) the initial boundary. Due to the discretization of the domain D_2, the integral operator A is considered in its discrete form:

$$A\chi(y_m, \omega_f) = \begin{bmatrix} \sum_{i,j}^{Nx,Ny} \chi(x_i, y_j, t) \int_{\Delta_{i,j}} \Re G(x, y_m, y, \omega_f)\, dxdy \\ \sum_{i,j}^{Nx,Ny} \chi(x_i, y_j, t) \int_{\Delta_{i,j}} \Im G(x, y_m, y, \omega_f)\, dxdy \end{bmatrix}. \quad (13)$$

Suppose now that we are at step $t = t_k$. We know the values of $\chi(x_i, y_j, t_k)$, $\phi(x_i, y_j, t_k)$ (as well as of the anomalous field induced by $\chi(x_i, y_j, t_k)$) and we want to update those functions. The algorithm is the following:

1. At each node (x_i, y_j), compute the extended speed function $V(x_i, y_j, t_k)$ from (12).
2. Solve (9) to find the update for ϕ. Use the numerical algorithm of Osher and Sethian as described in Sect. 3.4.
3. Find the new defect $D_{t_{k+1}}$ by constructing the level set $\phi = 0$. Deduce the values of the contrast at each grid node from (7).
4. Calculate the anomalous field from this defect following (2). Compare to the measured field. If the error is negligible, stop. Otherwise $k = k + 1$ and go back to Step 1.

In this algorithm, the velocity is evolving with time. This implies that, from one iteration to the next, the front can go back to positions it already occupied. This explains the ups and downs of the cost function along the iterations. No theoretical proof has been found concerning the convergence and the type of solution retrieved of such algorithm.

5 Numerical results

A material defect of known conductivity ($\sigma_D = 10\, \sigma_2$) affects a highly conductive metal half-space ($\sigma_2 = 10^7\ Sm^{-1}$). The upper half-space is air. A 20 mm long probing line is placed at $x_0 = 1.5$ mm above the metal with 64 samples taken every 0.4 mm. A limited number of probing frequencies is chosen on an almost log scale in between 10 and 500 kHz (12, 18, 26, 39, 58, 86, 127, 188, 278, 411 kHz). The test domain in which the defect is assumed to be found is a 2 mm-sided square centered at 1 mm depth from the interface. This square is divided into 20 × 20 square pixels. The defect is itself a 0.8 mm-sided square centered at 1 mm depth in the test domain, as shown in Fig. 2. The time step Δt is here equal to 0.001.

Three different velocities are considered. The first one corresponds to the evolution approach, as described in (10). The second one corresponds to the approximation approach, as described in (11). The last one is similar to the

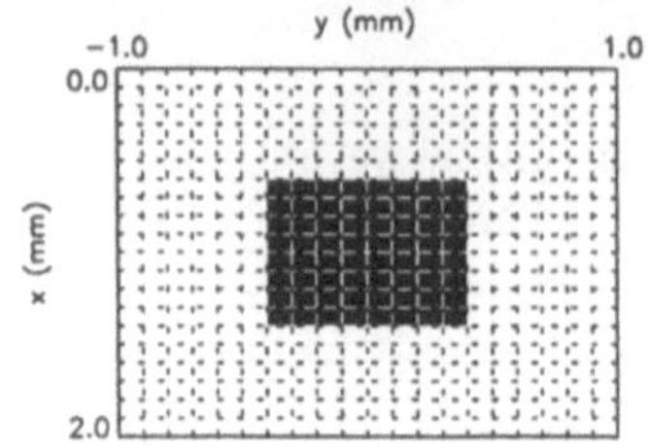

Fig. 2. Exact defect

approximation approach, but a weighting function is introduced in order to take into account the attenuation effect with depth (Litman et al. 1995); the velocity takes the form

$$V(x,y,t) = Q(x)V'(x,y,t) \qquad Q(x) = \frac{\beta^\beta}{\Gamma(\beta)}(\mu x)^{\beta-1}\exp(-\mu\beta x)$$

Q is a gamma distribution and we choose somewhat arbitrarily $\beta = 1.5$ and $\mu = 0.1$. Thus, the velocity is solution of:

$$AQ^*AQ\, V'(x,y,t) = -\chi_{in}\, AQ^*\left[A\chi(t) - E_A\right](x,y) \qquad (x,y) \in \partial D_t \quad (14)$$

where AQ is the operator defined in (13) multiplied by the Q function, and AQ^* its adjoint.

In each case, the initial contrast corrresponds to the "backpropagated" solution which has been reduced to binary values:

$$\chi(x,y,0) = \begin{cases} \chi_{in} & \text{if } A^*E_A(x,y) \geq 0 \\ \chi_{out} & \text{if } A^*E_A(x,y) < 0 \end{cases}$$

Figure 3 shows the resulting initial domain as well as the associated level-set function.

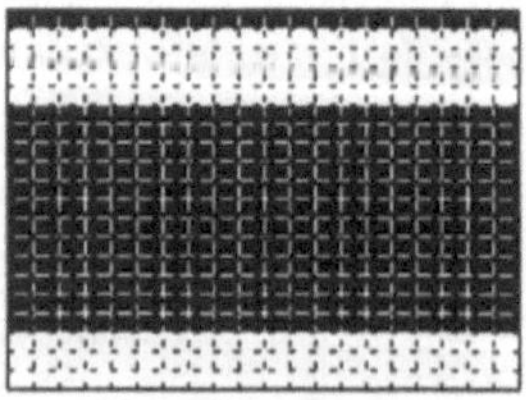

(a) $\chi(x,y)$, it=0

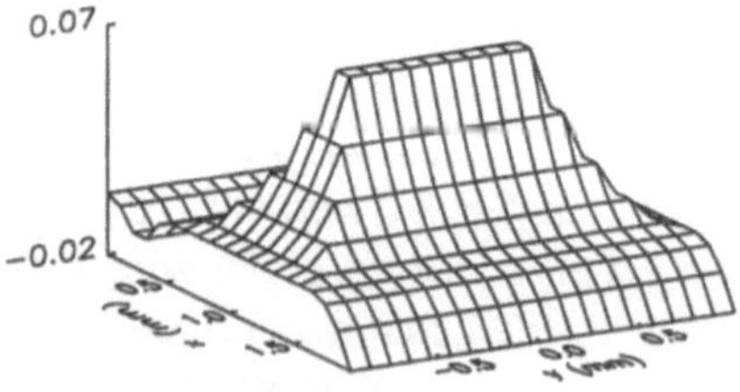

(b) $-\phi(x,y)$, it=0

Fig. 3. Initial estimates of the contrast and of the level-set

The algorithm stops when the residual is lower than a prescribed value, here 10^{-3}, or when the number of iterations reaches 500. The residuals in the data fit, as shown in Fig. 4, are not decreasing at each iteration as expected.

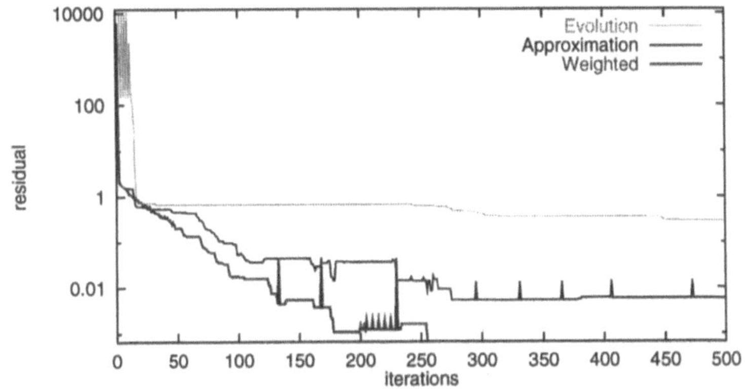

Fig. 4. Variation of the residual J along the iterations

We observe that the evolution method provides us with a high residual and a shape which is very different from the exact one (cf Fig.5.(a)). The residual obtained by the approximation method is 100 times lower than the previous residual. Still, the solution differs perceptibly from the exact one (cf Fig.5.(b)). Finally, the *weighted* approach gives us a very low residual and a shape very similar to the true one (cf Fig.5.(c)). Figure 6 gives an idea of the evolution of the level-set, the velocity and the map of the defect at various steps in the weighted approach.

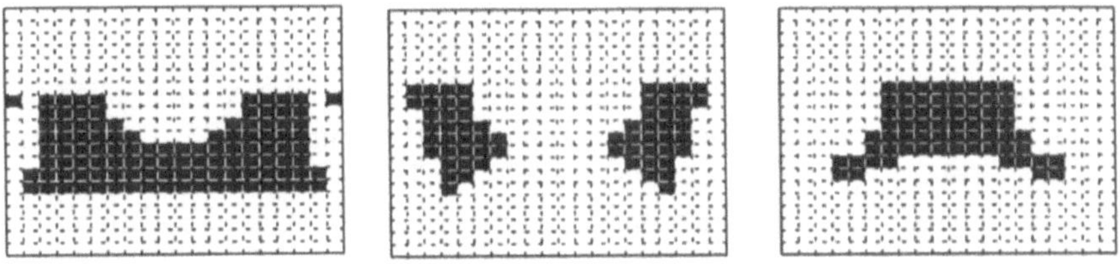

(a) Evolution it=500 (b) Approx. it=500 (c) Weighted it=255

Fig. 5. Defect found for the different velocities

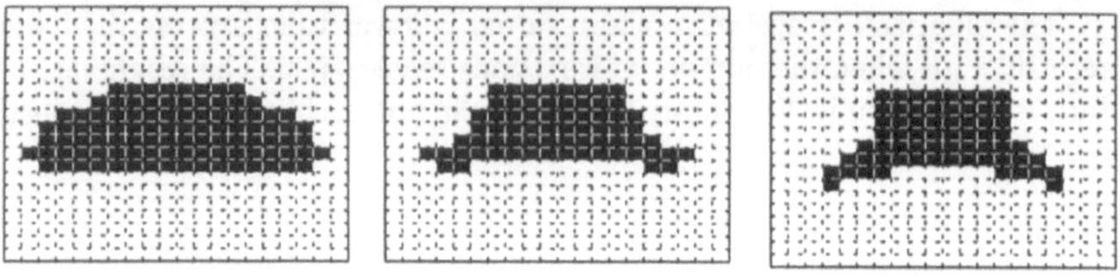

(a) $\chi(x,y)$, it=50 (b) $\chi(x,y)$, it=100 (c) $\chi(x,y)$, it=200

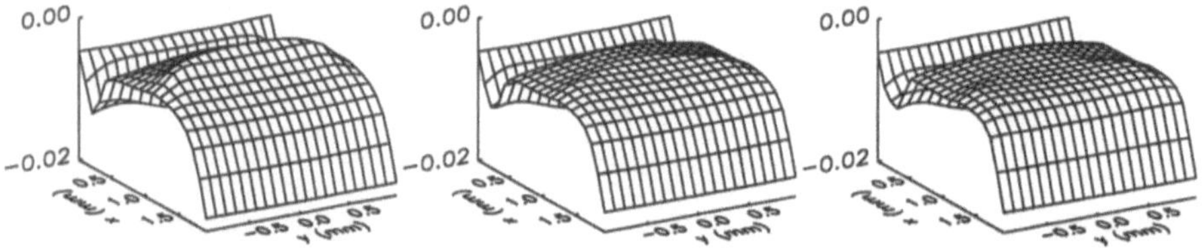

(d) $-\phi(x,y)$, it=50 (e) $-\phi(x,y)$, it=100 (f) $-\phi(x,y)$, it=200

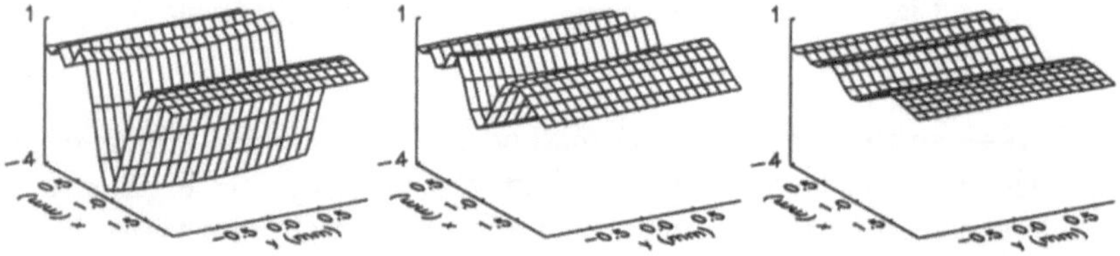

(g) $V(x,y)$, it=50 (h) $V(x,y)$, it=100 (i) $V(x,y)$, it=200

Fig. 6. Evolution of the contrast (top), level-set (median) and velocities (bottom) for the weighted approach

6 Conclusion

We have presented here preliminary results of a level-set approach applied to the eddy current nondestructive evaluation of defects in a stratified medium. Three computational schemes are considered. They only differ by the choice of the velocity for the level-set function. Santosa (1996) has successfully used the first two (evolution and approximation methods) for the deconvolution problem and the diffraction screen reconstruction problem, but they do not appear to work here. The third scheme takes into account the fact that the attenuation in the embedding space is very strong and shadows the deep parts of the defect. This scheme provides good results here. Notice that we have also implemented this level-set method in a similar acoustic case with no losses involved. In that case, the approximation method gives much better results than the evolution one, while the weighted scheme is obviously not useful anymore.

Thus, the choice of the velocity appears to be crucial to the success of the inversion algorithm. Also, the continuous extension of this velocity to the whole test domain is a key component of the algorithm. This should be one of the points to focus on in the future. There are also other pertinent questions: (i) The linearized Born approximation has been used here to reduce the complexity of the problem. It is necessary to consider the full (nonlinear) inverse problem in order to generalize the scope of the method. (ii) The value of the contrast of the defect with respect to its embedding is assumed to be known. In a more realistic case, this value is itself part of the problem. It should be seen whether a level-set description can yield both the shape and the contrast of the defect.

Acknowledgments

This work was supported under a CNRS-NSF International Cooperative Grant ♯INT-9415493, a NATO Collaborative Research Grant ♯CRG-940999. Santosa's research is partially supported by the National Science Foundation under grant ♯DMS-9503114, Department of Energy under grant ♯DE-FG02-94ER25225, the AirForce Office of Scientific Research URI-RIP grant ♯F49620-93-1-0500 and grant ♯F49620-95-1-0305.

References

Barles G. (1985): Remarks on a flame propagation model. Technical Report No. 464, INRIA Rapports de Recherche.

Caselles V., Catté F., Coll T., and Dibos F. (1993): A geometric model for active contours in image processing. Numer. Math. **66**, 1–31.

Cea J. (1976): Une méthode numérique pour la recherche d'un domaine optimal. In: *Publication IMAN*. Université de Nice.

Kass M., Witkin A., and Terzopoulos D. (1988): Snakes: active contour models. Int. J. Comput. Vision **1**, 321–331.

Kimia B.B., Tannenbaum A.R., and Zucker S.W. (1995): Shapes, shocks, and deformations I: the components of two-dimensional shape and the reaction-diffusive space. Int. J. Comput. Vision **15**, 189–224.

Lesselier D., and Duchêne B. (1996): Wavefield inversion of objects in stratified environments. From backpropagation schemes to full solutions. In: *Review of Radio Science 1993-1996* (Stone, ed). Oxford University Press, New York, 235–268.

Litman A., Lesselier D., and De Mol C. (1995): Mapping 2-D defects in a conductive half-space by eigenfunction expansions in K-space of Fourier-Laplace transforms. In: *Nondestructive Testing of Materials* (Collins et al., eds). IOS Press, Amsterdam, 175–183.

Malladi R., Sethian J.A., and Vemuri B.C. (1995): Shape modeling with front propagation: a level-set approach. IEEE Trans. Pattern Anal. Machine Intell. **17**, 158–175.

de Oliveira Bohbot R., Lesselier D., and Duchêne B. (1996): Mapping defects in a conductive half-space by simulated annealing with connectivity and size as constraints. J. Electromagn. Waves Applic. **10**, 983–1004.

Osher S., and Sethian J.A. (1988): Fronts propagating with curvature-dependent speed: Algorithms based on Hamilton-Jacobi formulations. J. Comput. Phys. **79**, 12–49.

Santosa F. (1996): A level-set approach for inverse problems involving obstacles. ESAIM: Cocv **1**, 17–33.

Sethian J.A., and Strain J. (1992): Crystal growth and dendritic solidification. J. Comput. Phys. **98**, 231–253.

An Inverse Problem for the Two-Dimensional Wave Equation in a Stratified Medium

Lorella Fatone[1], Pierluigi Maponi[1], Cristina Pignotti[1] and Francesco Zirilli[2]

[1] Dipartimento di Matematica e Fisica,
Università di Camerino, 62032 Camerino (MC), Italy
[2] Dipartimento di Matematica "G.Castelnuovo",
Università di Roma "La Sapienza", 00185 Roma, Italy

Abstract. Let $D = \{(x, y) \in \mathbb{R}^2 \,|\, x > 0,\ y \in \mathbb{R}\}$ and $u(x, y, t)$ be the solution of an initial-boundary value problem for the two-dimensional wave equation in the half plane D. The half plane D carries a velocity stratification given by flat layers parallel to the boundary of the half plane characterized by a thickness and a constant velocity. We consider the following inverse problem: given the initial data, from the knowledge of $u(0, 0, t)$, $t > 0$, reconstruct the stratification. We give an algorithm to solve this problem based on an explicit formula for $u(0, 0, t)$ and we report some numerical experience.

1 Introduction

Let $\mathbb{R}$ be the set of the real numbers, k be a positive integer and $\mathbb{R}^k$ be the k dimensional real euclidean space. Let $D = \{(x, y) \in \mathbb{R}^2 \,|x > 0\ ,\ y \in \mathbb{R}\} \subset \mathbb{R}^2$. Let n be a non negative integer, let $x_{j-1} \geq 0$, $c_j > 0$, $j = 1, 2, \ldots, n+1$, be given constants, such that $c_j \neq c_{j+1}$, $x_{j-1} < x_j$, $j = 1, 2, \ldots, n$ and $x_0 = 0$. For every $x \geq 0$ we consider the following piecewise constant function:

$$c(x) = \begin{cases} c_j, & x_{j-1} \leq x < x_j,\ j = 1, 2, \ldots, n, \\ c_{n+1}, & x \geq x_n, \end{cases} \tag{1}$$

that we call $n-$jump function. Let $\mathcal{E}_n$ be the set of the $n-$jump functions and $\mathcal{E} = \bigcup_{n \geq 0} \mathcal{E}_n$.

Let $n \geq 0$, $c \in \mathcal{E}_n$, we denote with u_n the unique solution of the following initial-boundary value problem:

$$\frac{\partial^2 u_n}{\partial t^2}(x, y, t) = div\Big(c(x) \nabla u_n(x, y, t)\Big)\ , \quad (x, y) \in D,\ t > 0, \tag{2}$$

$$\frac{\partial u_n}{\partial x}(0, y, t) = 0\ , \quad y \in \mathbb{R}.\ t > 0, \tag{3}$$

$$u_n(x, y, 0) = \delta(x, y)\ , \quad (x, y) \in D, \tag{4}$$

$$\frac{\partial u_n}{\partial t}(x, y, 0) = 0\ , \quad (x, y) \in D, \tag{5}$$

where *div* and ∇ denote respectively the divergence operator and the gradient operator respect to the variables x, y, moreover $\delta(x, y)$ is the Dirac delta and the solution of (2), (3), (4), (5) must be interpreted as the limit for $\epsilon \to 0^+$ of the solution of the same problem with $\delta(x - \epsilon, y)$ in (4) as initial condition.

In this paper we consider the following two problems:

Problem 1.1 (Direct problem) *Given* $n \geq 0$, $c \in \mathcal{E}_n$, *find the solution* $u_n(x, y, t)$ *for* $(x, y) \in D$, $t > 0$, *of the initial-boundary value problem (2), (3), (4), (5).*

Problem 1.2 (Inverse problem) *Given* $F(t)$, $t > 0$ *find* $n \geq 0$ *and* $c \in \mathcal{E}_n$ *such that* $F(t) = u_n(0, 0, t)$ *for* $t > 0$, *where* $u_n(x, y, t)$ *for* $(x, y) \in D$, $t > 0$ *is the solution of problem (2), (3), (4), (5).*

These problems can be regarded as model problems of some interesting questions in several application fields such as: geological prospecting, civil engineering, materials technology.

We note that Problem 1.2 for a general coefficient c is an ill posed problem, here it is "stabilized" with the a priori assumption $c \in \mathcal{E}_n$, for some n. Moreover Problem 1.2 with the assumption $c \in \mathcal{E}_n$, is ill conditioned and must be solved with a special algorithm. We present an algorithm to solve Problem 1.2 based on a formula that gives $u_n(0, 0, t)$, the solution of (2), (3), (4), (5) in the n-jump case evaluated at the origin, in terms of the parameters x_{j-1}, c_j, $j = 1, 2, \ldots, n+1$ that define the function c. That is $u_n(0, 0, t)$ is given by the following recurrence relation:

$$
\begin{aligned}
u_m(0,0,t) &= u_{m-1}(0,0,t) + \\
&+ \frac{4^m}{2\pi}\frac{\partial}{\partial t} \sum_{k_m^m=1}^{\infty} \sum_{k_{m-1}^m=1}^{\infty} \cdots \sum_{k_1^m=1}^{\infty} \left\{ \frac{\sqrt{c_m}\,\frac{t_{m,\underline{k}^m}}{\tau_m(k_m^m)} \prod_{l=2}^{m-1} c_l \frac{t_{l,\underline{k}^m}^2}{\tau_l^2(k_l^m)}}{\prod_{l=1}^{m-1} \left(\frac{\sqrt{c_l}\, t_{l,\underline{k}^m}}{\tau_l(k_l^m)} + \frac{\sqrt{c_{l+1}}\, t_{l+1,\underline{k}^m}}{\tau_{l+1}(k_{l+1}^m)} \right)^2} \beta_{\underline{k}^m} \cdot \right. \\
&\cdot \left(\frac{c_m t_{m,\underline{k}^m} - \sqrt{c_{m+1}[c_{m+1} t_{m,\underline{k}^m}^2 + (c_m - c_{m+1})\tau_m^2(k_m^m)]}}{c_m t_{m,\underline{k}^m} + \sqrt{c_{m+1}[c_{m+1} t_{m,\underline{k}^m}^2 + (c_m - c_{m+1})\tau_m^2(k_m^m)]}} \right)^{k_m^m} \cdot \\
&\left. \cdot \frac{H\left(t - \sum_{l=1}^{m} \tau_l(k_l^m)\right)}{\sum_{l=1}^{m} \frac{\sqrt{c_l}\,\tau_l(k_l^m)}{t_{l,\underline{k}^m}} \sqrt{t_{l,\underline{k}^m}^2 - \tau_l^2(k_l^m)}} \right\}, \\
&\qquad t > 0\,, \quad m = 1, 2, \ldots, n, \qquad (6)
\end{aligned}
$$

where

$$
u_0(0,0,t) = \frac{1}{\pi c_1} \frac{d}{dt} \frac{H(t)}{t} = -\frac{1}{\pi c_1} \frac{1}{t^2}, \qquad t > 0, \qquad (7)
$$

$H(\cdot)$ is the Heaviside function, $\underline{k}^m = (k_1^m, k_2^m, \ldots, k_m^m)$, $m = 1, 2, \ldots, n$, is a multi-index,

$$\beta_{\underline{k}^1} = \frac{\tau_1^2(k_1^1)}{c_1 t_{1,\underline{k}^1}^2}, \qquad k_1^1 = 1, 2, \ldots, \tag{8}$$

$$\beta_{\underline{k}^m} = \sum_{j_{m-1}=1}^{j(m-1,m)} \sum_{j_{m-2}=1}^{j(m-2,m)} \cdots \sum_{j_1=1}^{j(1,m)} (-1)^{j_1+k_1^m+\cdots+j_{m-1}+k_{m-1}^m} \cdot$$

$$\cdot \prod_{l=1}^{m-1} \binom{k_{l+1}^m + k_l^m - j_l}{k_{l+1}^m} \binom{k_{l+1}^m - 1}{j_l - 1} \left(\frac{\frac{\sqrt{c_{l+1}} t_{l+1,\underline{k}^m}}{\tau_{l+1}(k_{l+1}^m)} - \frac{\sqrt{c_l} t_{l,\underline{k}^m}}{\tau_l(k_l^m)}}{\frac{\sqrt{c_{l+1}} t_{l+1,\underline{k}^m}}{\tau_{l+1}(k_{l+1}^m)} + \frac{\sqrt{c_l} t_{l,\underline{k}^m}}{\tau_l(k_l^m)}} \right)^{k_{l+1}^m + k_l^m - 2j_l}$$

$$k_1^m, k_2^m, \ldots, k_m^m = 1, 2, \ldots, \quad m = 2, 3, \ldots, n, \tag{9}$$

$$j(l, m) = \min\{k_l^m,\ k_{l+1}^m\}, \quad l = 1, 2, \ldots, m-1, \quad m = 1, 2, \ldots, n, \tag{10}$$

$$\tau_l(k_l^m) = 2k_l^m \frac{x_l - x_{l-1}}{\sqrt{c_l}}, \quad l = 1, 2, \ldots, m, \quad m = 1, 2, \ldots, n. \tag{11}$$

Moreover the functions $t_{l,\underline{k}^m} = t_{l,\underline{k}^m}(t)$, $k_1^m, k_2^m, \ldots, k_m^m = 1, 2, \ldots$, $l = 1, 2, \ldots, m$, $m = 1, 2, \ldots, n$, for $t \geq \sum_{l=1}^{m} \tau_l(k_l^m)$, are defined by the following equalities-inequalities system:

$$t_{l,\underline{k}^m} > \tau_l(k_l^m)\ ,\ l = 1, 2, \ldots, m,\ m = 1, 2, \ldots, n, \tag{12}$$

$$\frac{1}{c_1}\left(\frac{t_{1,\underline{k}^m}^2}{\tau_1^2(k_1^m)} - 1\right) = \frac{1}{c_2}\left(\frac{t_{2,\underline{k}^m}^2}{\tau_2^2(k_2^m)} - 1\right) = \cdots = \frac{1}{c_m}\left(\frac{t_{m,\underline{k}^m}^2}{\tau_m^2(k_m^m)} - 1\right) \tag{13}$$

$$t_{1,\underline{k}^m} + t_{2,\underline{k}^m} + \cdots + t_{m,\underline{k}^m} = t. \tag{14}$$

The functions $t_{l,\underline{k}^m}(t)$, $k_1^m, k_2^m, \ldots, k_m^m = 1, 2, \ldots$, $l = 1, 2, \ldots, m$, $m = 1, 2, \ldots, n$ for $t < \sum_{l=1}^{m} \tau_l(k_l^m)$, can be defined as arbitrary constants since they are multiplied by zero in (6). Finally in formulas (6), (9) and in the following when in a product such as $\prod_{i=\cdot}^{\cdot}$ the lower index is greater than the upper one the product must be understood equal to one. Formula (6) must be interpreted in distribution sense and has been obtained in Maponi et al. using the spectral theory of self-adjoint operators on Hilbert spaces and some explicit formulas for the spectral measure associated to the right hand side of (2) with the boundary condition (3). Inverse problems analogous to the one considered here for the one dimensional wave and diffusion equations have been considered in Bartoloni et al. and Giordana et al.. The case of two dimensional diffusion equations has been considered in Mochi et al..

In section 2 we analyse the structure of the signal $u_n(0, 0, t)$ given by (6) and we exploit this structure to develop an algorithm to solve Problem 1.2.

In section 3 we report some numerical experience obtained with synthetic data using the algorithm of section 2.

2 An algorithm to solve the inverse problem

Let us analyse the structure of $u_n(0,0,t)$ given by (6).

We note that for $n = 0$ we have the well known solution of the two-dimensional wave equation in an homogeneous half plane. In order to fix the ideas we take $n = 1$, we can rewrite (6) as follows:

$$u_1(0,0,t) = S_0(t) + R_0(t) + \sum_{k_1^1=1}^{\infty} \left[S_{\underline{k}^1}(t) + R_{\underline{k}^1}(t) \right], \quad t > 0, \tag{15}$$

where:

$$S_0(t) = \frac{1}{\pi c_1} \frac{\delta(t)}{t} = 0 \,, \quad t > 0, \tag{16}$$

$$R_0(t) = -\frac{1}{\pi c_1} \frac{H(t)}{t^2} = -\frac{1}{\pi c_1} \frac{1}{t^2} \,, \quad t > 0, \tag{17}$$

$$S_{\underline{k}^1}(t) = \frac{2}{\pi c_1} \frac{\delta\left(t - \tau_1(k_1^1)\right)}{\sqrt{t_{1,\underline{k}^1}^2 - \tau_1^2(k_1^1)}} \left(\frac{c_1 t_{1,\underline{k}^1} - \sqrt{c_2[c_2 t_{1,\underline{k}^1}^2 + (c_1 - c_2)\tau_1^2(k_1^1)]}}{c_1 t_{1,\underline{k}^1} + \sqrt{c_2[c_2 t_{1,\underline{k}^1}^2 + (c_1 - c_2)\tau_1^2(k_1^1)]}} \right)^{k_1^1},$$
$$t > 0, \ k_1^1 = 1, 2, \ldots, \tag{18}$$

$$R_{\underline{k}^1}(t) = \frac{2}{\pi c_1} H\left(t - \tau_1(k_1^1)\right) \cdot$$
$$\cdot \frac{\partial}{\partial t} \left[\frac{1}{\sqrt{t_{1,\underline{k}^1}^2 - \tau_1^2(k_1^1)}} \left(\frac{c_1 t_{1,\underline{k}^1} - \sqrt{c_2[c_2 t_{1,\underline{k}^1}^2 + (c_1 - c_2)\tau_1^2(k_1^1)]}}{c_1 t_{1,\underline{k}^1} + \sqrt{c_2[c_2 t_{1,\underline{k}^1}^2 + (c_1 - c_2)\tau_1^2(k_1^1)]}} \right)^{k_1^1} \right],$$
$$t > 0, \ k_1^1 = 1, 2, \ldots, \tag{19}$$

where $\tau_1(k_1^1) = 2k_1^1 \dfrac{x_1}{\sqrt{c_1}}$, and $t_{1,\underline{k}^1}(t) = t$ for $t \geq \tau_1(k_1^1)$.

In (15) we have used formulas such as:

$$\frac{\partial}{\partial t} \frac{H(t-a)}{\sqrt{|t-a|}} = \frac{\delta(t-a)}{\sqrt{|t-a|}} + H(t-a) \frac{\partial}{\partial t} \frac{1}{\sqrt{|t-a|}} \,, \quad a > 0 \,, \quad t > 0. \tag{20}$$

We note that the left hand side of (20) is a legitimate distribution while the two addenda on the right hand side are only formal expressions.

The signal $u_1(0,0,t)$ corresponding to $c \in \mathcal{E}_1$ is singular for $t = 0$ and for $t = k_1^1 T_1$ where $T_1 = 2\dfrac{x_1}{\sqrt{c_1}}$, and $k_1^1 = 1, 2, \ldots$. We note that $T_1 = 2\dfrac{x_1}{\sqrt{c_1}}$ is the first travel time of our medium, that is the time necessary to the initial pulse travelling at speed $\sqrt{c_1}$ to reach the discontinuity of the medium located at $x = x_1$ and to come back to the origin. When the pulse, leaving the origin at $t = 0$, reaches the discontinuity at $x = x_1$ it is splitted in two waves: the first one is transmited through the discontinuity of the medium in the second layer and the second one is reflected by the discontinuity. This second wave comes back toward the origin and when arrives at the origin at time $t = T_1$ we have a singularity in $u_1(0,0,t)$. The reflected wave that comes back toward the origin is reflected by the boundary $x = 0$ and goes back toward the discontinuity located at $x = x_1$, where the phenomenon described above takes place again and so on. That is the term $S_{\underline{k}^1}(t) + R_{\underline{k}^1}(t)$ at $t = k_1^1 T_1$, $k_1^1 = 1, 2, \ldots$ can be interpreted as the $\underline{k}^1$-echo coming from the first discontinuity in the medium, see Figure 1.

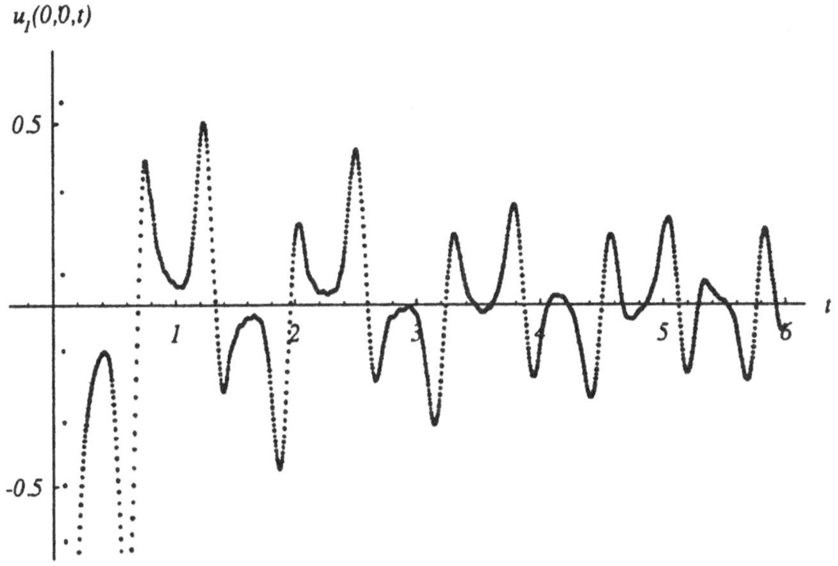

Fig. 1. Finite difference approximation of $u_1(0,0,t)$ when $c_1 = 9$, $c_2 = 25$, $x_1 = 1$

Morever $u_1(0,0,t)$ for $t \neq k_1^1 T_1$, $k_1^1 = 0, 1, \ldots$, is a smooth function of t, and for $k_1^1 = 1, 2, \ldots$ the following relations hold:

$$\frac{1}{c_1} = \lim_{t \to 0+} \left(-\pi t^2 R_0(t)\right), \tag{21}$$

$$\left(\frac{\sqrt{c_1} - \sqrt{c_2}}{\sqrt{c_1} + \sqrt{c_2}}\right)^{k_1^1} = \lim_{t \to \tau_1(k_1^1)+} -\frac{\pi c_1}{2} \frac{\left(t_{1,\underline{k}^1}^2 - \tau_1^2(k_1^1)\right)^{\frac{3}{2}}}{t} R_{\underline{k}^1}(t). \tag{22}$$

From (21), (22) and the fact that $x_1 = \dfrac{\sqrt{c_1}T_1}{2}$ it is easy to reconstruct the velocity profile $c \in \mathcal{E}_1$, that is x_1, c_1, c_2. Finally we note that $R_0(t) = \mathcal{O}\left(\frac{1}{t^2}\right)$, $R_{\underline{k}^1}(t) = \mathcal{O}\left(\frac{1}{t^2}\right)$, as $t \to +\infty$ for $k_1^1 = 1, 2, \ldots$.

The analysis of $u_n(0,0,t)$, for $n > 1$, needs a preliminary investigation of the solutions $t_{l,\underline{k}^m}(t)$, for $t > \tau_1(k_1^m) + \cdots + \tau_m(k_m^m)$, $l = 1, 2, \ldots, m$, $m = 1, 2, \ldots, n$ of the equalities-inequalities system (12), (13), (14).

Lemma 2.1 *Let m be an integer, with $1 \le m \le n$, given a multi index $\underline{k}^m = (k_1^m, k_2^m, \ldots, k_m^m)$, let $\tau_1(k_1^m), \tau_2(k_2^m), \ldots, \tau_m(k_m^m) > 0$ be defined by (11), then for every $t > \sum_{j=1}^{m} \tau_j(k_j^m)$ there exists a unique solution $t_{l,\underline{k}^m} = t_{l,\underline{k}^m}(t)$, $l = 1, 2, \ldots, m$, of the system (12), (13), (14). Moreover, let $t'_{l,\underline{k}^m}(t) = \dfrac{dt_{l,\underline{k}^m}}{dt}(t)$, $l = 1, 2, \ldots, m$, then we have:*

$$t'_{l,\underline{k}^m}(t) = \frac{c_l \dfrac{\tau_l^2(k_l^m)}{t_{l,\underline{k}^m}(t)}}{\displaystyle\sum_{j=1}^{m} c_j \frac{\tau_j(k_j^m)^2}{t_{j,\underline{k}^m}(t)}}, \quad t > \sum_{j=1}^{m} \tau_j(k_j^m), \quad l = 1, 2, \ldots, m. \tag{23}$$

Proof Introducing the parameter $s \ge 0$ equations (13) can be rewritten as

$$s = \frac{1}{c_l}\left(\frac{t_{l,\underline{k}^m}^2}{\tau_l^2(k_l^m)} - 1\right), \quad l = 1, 2, \ldots, m. \tag{24}$$

We note that when $s = 0$ we have $t_{l,\underline{k}^m} = \tau_l(k_l^m)$, $l = 1, 2, \ldots, m$ and $t = \sum_{l=1}^{m} t_{l,\underline{k}^m} = \sum_{l=1}^{m} \tau_l(k_l^m)$ and when $s > 0$ we have $t_{l,\underline{k}^m} > \tau_l(k_l^m)$ for $l = 1, 2, \ldots, m$, and $t > \sum_{l=1}^{m} \tau_l(k_l^m)$. Let $t_{l,\underline{k}^m}(s)$, $s \ge 0$, $l = 1, 2, \ldots, m$, be the non negative functions defined implicitely by (24).

Let:

$$T(s) = t_{1,\underline{k}^m}(s) + t_{2,\underline{k}^m}(s) + \cdots + t_{m,\underline{k}^m}(s) = \sum_{l=1}^{m} \tau_l(k_l^m)\sqrt{c_l s + 1}, \quad s \ge 0, \tag{25}$$

then the equation:

$$t = T(s), \tag{26}$$

defines implicitely a unique function $s = s(t)$ for every $t \ge \sum_{j=1}^{m} \tau_j(k_j^m)$, in fact we have: $T(0) = \sum_{j=1}^{m} \tau_j(k_j^m) \le t$, $\lim_{s\to\infty} T(s) = +\infty$ and $\dfrac{dT}{ds} > 0$ for $s \ge 0$.

Thus with abuse of notation we denote with $t_{l,\underline{k}^m}(t) = t_{l,\underline{k}^m}(s(t))$, $l = 1, 2, \ldots, m$ the unique solutions of (12), (13), (14). Moreover we have:

$$t'_{l,\underline{k}^m}(t) = \frac{\tau_l(k_l^m)}{2\sqrt{c_l s(t)+1}}\frac{ds}{dt} > 0\,, \qquad l = 1, 2, \ldots, m. \tag{27}$$

Finally, from (13), (14) we obtain:

$$\sum_{j=1}^{m} t'_{j,\underline{k}^m}(t) = 1 \ \text{ for } \ t > \sum_{j=1}^{m} \tau_j(k_j^m), \tag{28}$$

$$\frac{1}{c_1}\frac{t_{1,\underline{k}^m}(t)t'_{1,\underline{k}^m}(t)}{\tau_1^2(k_1^m)} = \frac{1}{c_2}\frac{t_{2,\underline{k}^m}(t)t'_{2,\underline{k}^m}(t)}{\tau_2^2(k_2^m)} = \cdots = \frac{1}{c_m}\frac{t_{m,\underline{k}^m}(t)t'_{m,\underline{k}^m}(t)}{\tau_m^2(k_m^m)}. \tag{29}$$

From (27), (28), (29) we have (23). This concludes the proof.

When $n > 1$ we can rewrite $u_n(0, 0, t)$ as follows:

$$u_n(0,0,t) = S_0(t) + R_0(t) + \sum_{l=1}^{n}\left\{\sum_{k_l^l=1}^{\infty}\cdots\sum_{k_2^l=1}^{\infty}\sum_{k_1^l=1}^{\infty} S_{\underline{k}^l}(t) + R_{\underline{k}^l}(t)\right\},\ t > 0, \tag{30}$$

where S_0, R_0 are given by formulas (16), (17), $\underline{k}^l = (k_1^l, k_2^l, \ldots, k_l^l)$ and $S_{\underline{k}^l}$, $R_{\underline{k}^l}$ for $k_1^l, k_2^l, \ldots, k_l^l = 1, 2, \ldots$, $l = 1, 2, \ldots, n$, can be easily identified from (6), in analogy with the previously introduced $S_{\underline{k}^1}(t)$, $R_{\underline{k}^1}(t)$.

We note that the structure of the signal $u_n(0, 0, t)$ is considerably more complicated than the structure of $u_1(0, 0, t)$. This is due to the fact that every discontinuity between two layers splits a wave propagating in the medium that meets the discontinuity in two different parts: the trasmitted wave and the reflected wave. Thus $u_n(0, 0, t)$ is the result of the action on the initial pulse of the n discontinuities in the medium, that is given l and $\underline{k}^l = (k_1^l, \ldots, k_l^l)$, the terms $S_{\underline{k}^l}$, $R_{\underline{k}^l}$ in (30) represent the contribution to $u_n(0, 0, t)$ coming from the medium made of the first $l + 1$ layers of $c_n(x)$ from the part of the initial pulse that has traveled k_i^l times back and forth in the i-th layer of the medium, $i = 1, 2, \ldots, l$. We note that $\sum_{i=1}^{l}\tau_i(k_i^l)$, $l = 1, 2, \ldots, n$ is the travel time for such a path. The signal $u_n(0, 0, t)$ is singular for $t = \tau_1(k_1^l) + \cdots + \tau_l(k_l^l)$ and the term $S_{\underline{k}^l}(t) + R_{\underline{k}^l}(t)$, $t \geq \tau_1(k_1^l) + \cdots + \tau_l(k_l^l)$ is named the $\underline{k}^l$-echo, see Figure 2.

From Lemma 2.1 we have:

$$\left(\frac{\sqrt{c_l} - \sqrt{c_{l+1}}}{\sqrt{c_l} + \sqrt{c_{l+1}}}\right)^{k_l^l} = \lim_{t\to(\tau_1(k_1^l)+\cdots+\tau_l(k_l^l))^+} -\frac{2\pi}{4^l}\frac{\prod_{s=1}^{l-1}\left(\sqrt{c_s} + \sqrt{c_{s+1}}\right)^2}{\sqrt{c_l}\prod_{s=2}^{l-1} c_s}\cdot$$

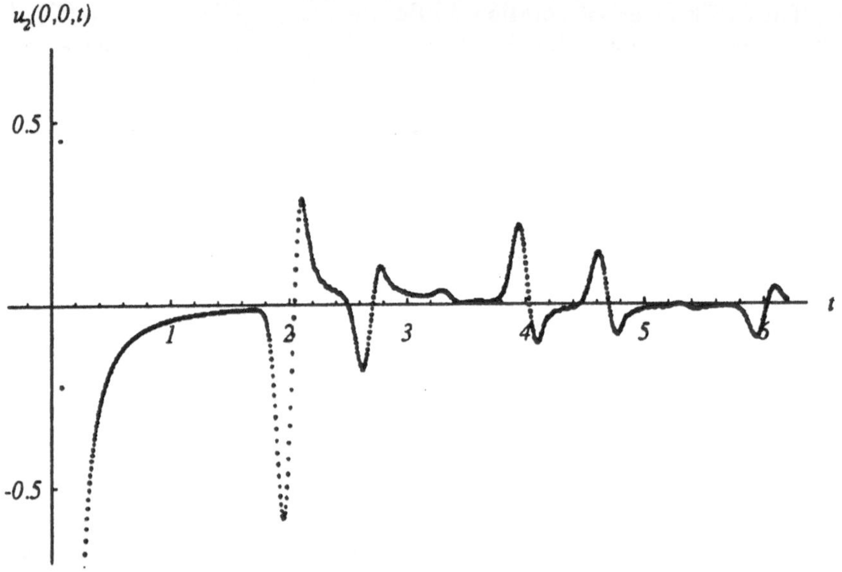

Fig. 2. Finite difference approximation of $u_2(0,0,t)$ when $c_1 = 4$, $c_2 = 9$, $c_3 = 16$, $x_1 = 2$, $x_2 = 3$

$$\cdot \frac{\left(\sum_{s=1}^{l} c_s \tau_s(k_s^l)\right)^2}{\left(\sum_{s=1}^{l} c_s \tau_s^2(k_s^l)\right)^{\frac{3}{2}}} \frac{\left(\sum_{s=1}^{l} t_{s,\underline{k}^l}^2 - \sum_{s=1}^{l} \tau_s^2(k_s^l)\right)^{\frac{3}{2}}}{\gamma_{\underline{k}^l}} R_{\underline{k}^l}(t),$$

$$k_1^l, k_2^l, \ldots, k_l^l = 1, 2, \ldots, \quad l = 1, \ldots, n, \qquad (31)$$

where:

$$\gamma_{\underline{k}^1} = \frac{1}{c_1}, \quad k_1^1 = 1, 2, \ldots, \qquad (32)$$

$$\gamma_{\underline{k}^l} = \sum_{j_{l-1}=1}^{j(l-1,l)} \sum_{j_{l-2}=1}^{j(l-2,l)} \cdots \sum_{j_1=1}^{j(1,l)} (-1)^{j_1+k_1^l+\cdots+j_{l-1}+k_{l-1}^l} \cdot$$

$$\prod_{s=1}^{l-1} \binom{k_{s+1}^l + k_s^l - j_s}{k_{s+1}^l} \binom{k_{s+1}^l - 1}{j_s - 1} \left(\frac{\sqrt{c_{s+1}} - \sqrt{c_s}}{\sqrt{c_{s+1}} + \sqrt{c_s}}\right)^{k_{s+1}^l + k_s^l - 2j_s},$$

$$k_1^l, k_2^l, \ldots, k_l^l = 1, 2, \ldots, \quad l = 1, \ldots, n, \qquad (33)$$

and finally $R_{\underline{k}^l}(t) = \mathcal{O}\left(\frac{1}{t^2}\right)$, when $t \to +\infty$, for $k_1^l, k_2^l, \ldots, k_l^l = 1, 2, \ldots$, $l = 1, 2, \ldots, n$.

The following algorithm to solve Problem 1.2 is based on the previous analysis of the structure of $u_n(0,0,t)$, $t \geq 0$. The purpose of this algorithm is to find the function c with the smallest number of jumps that fits the data $F(t)$, $t > 0$ to a given accuracy.

Algorithm 2.2 *Given $F(t)$ for $0 < t < T_F$, given three real positive parameters Δ_E, Tol_E, F_T, perform the following steps:*

1. *Compute $\tilde{c}_1$ (see formula (21)) as follows:*

$$\tilde{c}_1 = -\lim_{t \to 0+} \frac{1}{\pi t^2 F(t)}. \tag{34}$$

2. *Set: $n = 0$, $\tilde{x}_0 = 0$, $\tilde{c}(x) = \tilde{c}_1$, $x > \tilde{x}_0$, $T_0 = \Delta_E$.*
3. *Inspect the signal $F(t)$, for $T_0 < t < T_F$, to find T_E, the first echo in the interval (T_0, T_F) that does not come from multiple reflections due to the structure $\tilde{c}(x)$ already reconstructed. We recognize T_E as an echo if the following condition holds:*

$$|F(t_1)| > |F(t_2)| > F_T \text{ for every } t_1, t_2 \text{ such that } T_E \leq t_1 < t_2 \leq T_E + \Delta_E. \tag{35}$$

4. *If $F(t)$, for $T_0 \leq t < T_F$, has no echoes of the type described in step 3 then go to step 7.*
5. *Let T_E be the echo found in step 3, increase the counter n by 1 and define (see formulas (31), (32), (33)):*

$$\tilde{x}_n = \tilde{x}_{n-1} + \begin{cases} \dfrac{\sqrt{\tilde{c}_1}}{2} T_E & , \quad n = 1, \\[2ex] \dfrac{\sqrt{\tilde{c}_n}}{2} \left(T_E - \displaystyle\sum_{j=1}^{n-1} 2 \frac{\tilde{x}_j - \tilde{x}_{j-1}}{\sqrt{\tilde{c}_j}} \right) & , \quad n > 1, \end{cases} \tag{36}$$

$$\tilde{c}_{n+1} = \tilde{c}_n \left(\frac{1 - L_E}{1 + L_E} \right)^2, \tag{37}$$

$$\tilde{c}(x) = \begin{cases} \tilde{c}_j, & \tilde{x}_{j-1} \leq x < \tilde{x}_j, \ j = 1, 2, \ldots, n, \\ \tilde{c}_{n+1}, & x \geq \tilde{x}_n, \end{cases} \tag{38}$$

where

$$L_E = \lim_{t \to (\tilde{\tau}_1(1) + \cdots + \tilde{\tau}_n(1))+} -\frac{2\pi}{4^n} \frac{\displaystyle\prod_{l=1}^{n-1} \left(\sqrt{\tilde{c}_l} + \sqrt{\tilde{c}_{l+1}} \right)^2}{\sqrt{\tilde{c}_n} \displaystyle\prod_{l=2}^{n-1} \tilde{c}_l} \frac{\left(\displaystyle\sum_{l=1}^{n} \tilde{c}_l \tilde{\tau}_l(1) \right)^2}{\left(\displaystyle\sum_{l=1}^{n} \tilde{c}_l \tilde{\tau}_l^2(1) \right)^{\frac{3}{2}}} \cdot$$

$$\cdot \frac{\left(\displaystyle\sum_{l=1}^{n} \tilde{t}^2_{l,\underline{1}^n} - \sum_{l=1}^{n} \tilde{\tau}_l^2(1) \right)^{\frac{3}{2}}}{\tilde{\gamma}_{\underline{1}^n}} F(t), \tag{39}$$

$$\tilde{\gamma}_l(1) = 2\frac{\tilde{x}_l - \tilde{x}_{l-1}}{\sqrt{\tilde{c}_l}}, \quad l = 1, 2, \ldots, n, \tag{40}$$

$$\tilde{\gamma}_{\underline{1}^1} = \frac{1}{\tilde{c}_1}, \tag{41}$$

$$\tilde{\gamma}_{\underline{1}^n} = 1, \tag{42}$$

where $\tilde{t}_{l,\underline{1}^n}(t)$, $l = 1, 2, \ldots, n$ *are the solutions of (12), (13), (14) when* $k_l^n = 1$, $l = 1, 2, \ldots, n$, *that is* $\underline{1}^n = (1, 1, \ldots, 1) \in \mathbb{R}^n$.

6. *Set* $T_0 = T_E + Tol_E$ *and go to step 3.*
7. *Stop.*

We note that Algorithm 2.2 reconstructs the function c calculating the pairs c_j, x_{j-1}, $j = 1, 2, \ldots, n+1$, one at the time. This is useful to deal with ill conditioned character of Problem 1.2. The reconstruction is done analysing the echoes in the signal $F(t)$. The parameters Δ_E, Tol_E and F_T play important rules in Algorithm 2.2. In particular Δ_E and F_T control the recognition of the echoes in the signal $F(t)$ and Tol_E is the tolerance in the recognition of echoes in the signal $F(t)$ that comes from new structure and echoes that comes from the coefficient function $\tilde{c}$ reconstructed from the signal $F(t)$, $t \leq T_0$. Finally in Algorithm 2.2 we have not considered the degenerate case, where the first echo in $F(t)$ related to new structure in c is superimposed with an echo in $F(t)$ due to multiple reflections on the structure already reconstructed. In such cases some extra care must be used in the reconstruction.

3 The numerical experience

In Table 1 we present some numerical results obtained with an implementation of Algorithm 2.2 in a FORTRAN 77 code.

We have computed $F(t)$ solving Problem 1.1 by a finite difference scheme on a rectangle R with side lengths $L_x, 2L_y$, that is $R = (0, L_x) \times (-L_y, L_y)$. The finite difference scheme uses h_s as spatial mesh size and h_t as temporal mesh size. We have considered a simple form of absorbing boundary conditions, see Reynolds, to simulate the halfplane D. Finally, the Dirac delta appearing in condition (4) is approximated by the following function:

$$\tilde{\delta}(x, y; h_s) = \frac{1}{2\pi h_s} e^{-\frac{(x-h_s)^2 + y^2}{2h_s}}, \quad (x, y) \in R. \tag{43}$$

The data $F(t)$, $0 < t < T_F$ are obtained as follows: given a positive integer J_F let $T_F = J_F h_t$, $t_j = jh_t$, $j = 1, 2, \ldots, J_F$, $F(t_j) = F_j$, $j = 1, 2, \ldots, J_F$ where F_j is the finite difference approximation of $u_n(0, 0, t_j)$, $j = 1, 2, \ldots, J_F$. In Example 3.1 we have chosen $L_x = 5$, $L_y = 2.5$, $h_s \approx 1.11 \cdot 10^{-2}$, $h_t \approx 8.73 \cdot 10^{-4}$ and values of the same order of magnitude have been chosen for these parameters in the remaining examples.

Table 1. Numerical experience

	Originals	Reconstructions
Example 3.1	$c_1 = 81$	$\tilde{c}_1 = 81.19$
	$c_2 = 16$	$\tilde{c}_2 = 16.27$
	$x_1 = 3.5$	$\tilde{x}_1 = 3.48$
Example 3.2	$c_1 = 9$	$\tilde{c}_1 = 10.25$
	$c_2 = 25$	$\tilde{c}_2 = 19.9$
	$x_1 = 1$	$\tilde{x}_1 = 0.98$
Example 3.3	$c_1 = 4$	$\tilde{c}_1 = 3.84$
	$c_2 = 9$	$\tilde{c}_2 = 9.22$
	$c_3 = 16$	$\tilde{c}_3 = 18.32$
	$x_1 = 2$	$\tilde{x}_1 = 1.92$
	$x_2 = 3$	$\tilde{x}_2 = 2.94$
Example 3.4	$c_1 = 6$	$\tilde{c}_1 = 6.12$
	$c_2 = 3$	$\tilde{c}_2 = 2.25$
	$c_3 = 8$	$\tilde{c}_3 = 6.67$
	$x_1 = 2$	$\tilde{x}_1 = 1.98$
	$x_2 = 3$	$\tilde{x}_2 = 2.84$
Example 3.5	$c_1 = 4$	$\tilde{c}_1 = 3.99$
	$c_2 = 9$	$\tilde{c}_2 = 9.57$
	$c_3 = 4$	$\tilde{c}_3 = 3.52$
	$c_4 = 9$	$\tilde{c}_4 = 9.92$
	$x_1 = 2$	$\tilde{x}_1 = 1.95$
	$x_2 = 3$	$\tilde{x}_2 = 2.98$
	$x_3 = 4$	$\tilde{x}_3 = 3.91$
Example 3.6	$c_1 = 4$	$\tilde{c}_1 = 4.1$
	$c_2 = 8$	$\tilde{c}_2 = 7.82$
	$c_3 = 24$	$\tilde{c}_3 = 23.95$
	$c_4 = 30$	$\tilde{c}_4 = 38.66$
	$x_1 = 2$	$\tilde{x}_1 = 1.97$
	$x_2 = 3$	$\tilde{x}_2 = 2.96$
	$x_3 = 4$	$\tilde{x}_3 = 3.99$

The results reported in Table 1 are obtained using Algorithm 2.2 on the synthetic data described above. We note that the computation of the limits (34), (39) in Algorithm 2.2 must be handled with care since only a signal $F(t)$ of low quality can be expected near $t = T_E$, a travel time.

The results given in Table 1 are obtained with the following values of the parameters of Algorithm 2.2: $Tol_E = 0.2$, $\Delta_E = 0.05$, $F_T = 0.005$.

These results are relative to one, two and three jumps functions. Table 1 shows accurate reconstructions. Similar results can be obtained with slight more complicated structures, but reconstructions of coefficient function c with many more jumps needs more sophisticated algorithms than Algorithm 2.2.

In fact the echoes coming from complicated structures are misunderstood by the simple recognition procedure of Algorithm 2.2.

References

Maponi P., Zirilli F.: Inverse problem for a class of two-dimensional wave equations with piecewise constant coefficients, preprint.

Bartoloni A., Lodovici C., Zirilli F. (1993): Inverse problem for a class of one-dimensional wave equations with piecewise constant coefficients, J. of Optim. Theory and Appl., **76**, 13–32.

Giordana C., Mochi M., Zirilli F. (1992): The numerical solution of an inverse problem for a class of one-dimensional diffusion equations with piecewise constant coefficients, SIAM J. on Appl. Math., **52**, 1992, 428–441.

Mochi M., Pacelli G., Recchioni M. C., Zirilli F.: An inverse problem for a class of two dimensional diffusion equations with piecewise constant coefficients, preprint.

Reynolds A. C. (1978): Boundary conditions for the numerical solution of the wave propagation problems, Geophysics, **43**, 1099–1110.

On the Inverse Seismic Problem for Horizontally Layered Media: Subsidiary Study

Michel Cuer[1] and Jean Louis Petit[2]

[1] Laboratoire de Physique Mathématique, Université de Montpellier II, cc 050, Place Eugène Bataillon, 34095 Montpellier, Cedex, France
[2] Institut Français du Pétrole (IFP), Géophysique et Instrumentation 1 et 4 Avenue de Bois-Préau, BP 311, 92852 Rueil Malmaison Cedex, France

Abstract. The starting point of this work is the inversion of vertical seismic profiling (VSP) data. The usual processing of VSP data by inverse techniques is restricted to $1D$ propagation model. In this case, the parameters to identify are the acoustic impedance as function of travel time and the seismic source so that we have as unknowns two functions of one variable and as data a function of two variables, the time and the depth positions of geophones. The problem is thus largely overdetermined and an elementary mathematical analysis can be made. The source is modelled as a boundary condition at the top of the geophones zone. So this boundary condition replaces the true source function and the medium parameters above the geophones zone. The question asked by V. Richard from IFP was the "management" of this unknown source when $3D$ propagation effects are taken into account in horizontally layered medium where the propagation equations are parametrized by the k parameter of the Hankel transform. Now we think that the answer is that it is impossible to work round the fact that there are at least two unknown functions, the source and the medium parameters above the geophones zone. During this study, we have searched for some non local boundary conditions and this was the opportunity to obtain some results on exact transparent conditions for $3D$ propagation in $1D$ media (preliminary communication was made by Petit and Cuer (1994)) and on the discretization of such conditions in the acoustic case (preliminary communication was made by Cuer and Petit (1995)). This is the mathematical substance of this work in which the Poisson summation formula is used to prove the stability of a discrete non local boundary condition.

1 Introduction

During recent years, there has been increasing interest in full-waveform model-based inversion algorithms for multioffset seismic data. Without pretence to completeness, we refer to the papers of Jurado et al. (1995a,b) for some bibliographical references. The first of these articles contains a numerical method for solving the direct elastodynamic problem for $3D$ propagation in $1D$ media. This method is based on the Candel algorithm (Candel 1981) for the numerical computation of Hankel transforms and on a finite difference scheme on a staggered grid in depth and time as used by Virieux (1986) and Madariaga (1976). The choice of these algorithms results only from simplicity considerations and other more common techniques such as reflectivity or characteristics methods or more sophisticated schemes of high orders or spectral methods should be included in an ideal computing library. The second article contains numerical experiments on the inverse problem based on the least squares approach, the gradient of the misfit functional being computed by the adjoint state technique (see Lions 1968, Chavent 1974). When the misfit functional is displayed as function of the P-wave velocity v_P or as function of the bottom depth of a particular layer in a layer-cake model, we observe local minima which can prevent descent methods to work. When the earth model is parametrized with the travel time $\tau(z) = \int_0^z \frac{d\xi}{v_P(\xi)}$ to measure the depth (here z is the vertical coordinate), the behaviour of the misfit functional is improved. More precisely, as function of the velocity expressed in travel-time depth, the misfit functional tends to a quasi-quadratic form. As function of the depth interface the problem is the same but the picking is more easy since then the depths of interfaces are directly accessible on the seismograms. Thus an inversion procedure with manual picking of interfaces is possible and an automatic one probably requires a combination of some kind of layer stripping as the ones proposed by Yagle and Levy (1985) or Carazzone (1986) with the usual least-squares formulation. In the hope to perform processing of field data that are of immediate interest in practice, we have then studied an inverse problem for borehole data.

The VSP technique is a borehole seismic survey technique in which the seismic source is near the surface of the earth and the geophones are located at various depths in the well. This technique gives more accurate depth calibration than that is obtained by using velocities deduced from surface seismic data. Besides the signal bandwidth is closer to the ordinary seismic data than do the sonic logs so that the comparison with seismic surface data is more easy. With such data, the wave field along the well is measured and the distinction between downgoing and upgoing waves and the estimation of seismic velocities are not very difficult. The basic steps of a conventional processing are traces editing, separation of the downgoing waves from the upgoing waves (with, for example, an f-k dip filter), static correction, deconvolution and filtering and stacking. This gives a "corridor stacking" that is an alternative to zero offset "synthetic seismogram" derived from the sonic log: it can be compared with the CMP stack of surface seismic data at the well location (see Yilmaz (1987) and Mari and Coppens (1989) and the cited references for more information).

Inverse techniques have been proposed by Macé and Lailly (1986) and used for example by Lefebvre (1985) to obtain the acoustic impedance (or reflection coefficients) from the VSP data (the conventional processing does not produce this information). In these works the propagation model is $1D$ and formulated in the travel-time variable $\tau(z) = \int_{z_{min}}^z \frac{d\xi}{c(\xi)}$ ($c(\xi) = v_P(\xi)$ for longitudinal waves). The problem solved by least squares minimization, the gradient being computed by the adjoint-state technique, is the following: determine the acoustic impedance $\sigma(\tau)$ ($= \rho(z(\tau))c(z(\tau))$ where ρ is the density and c the sound velocity) as function of the travel time τ and the seismic pulse $g(t)$ from the data $\partial_t u(\tau,t)$ for $0 \le \tau \le \tau^{max}, 0 \le t \le t^{max}$, the vertical displacement u being solution of the $1D$ wave equation:

$$\sigma(\tau)\partial_{tt}u(z,t) - \partial_\tau(\sigma(\tau)\partial_\tau u(\tau,t)) = 0 \tag{1}$$

for $0 \le \tau \le \tau^{max}, 0 \le t \le t^{max}$,

$$\sigma(0)\partial_\tau u(0,t) = g(t) \tag{2}$$

$$\partial_t u(\tau^{max},t) + \partial_\tau u(\tau^{max},t) = 0 \tag{3}$$

for $0 \le t \le t^{max}$,

$$u(\tau,0) = \partial_t u(\tau,0) = 0 \tag{4}$$

for $0 \le \tau \le \tau^{\max}$. The second boundary condition (3) for $\tau = \tau^{max}$ (which is an exact transparent condition) is included here only for convenience: nothing essential is lost if $\tau^{max} = \infty$ (without boundary condition at ∞). In this inverse problem, if a couple (σ, g) is a solution, then $(\alpha\sigma, \alpha g)$ is also a solution for any constant α. In practice this is not a drawback because $\sigma(0)$ can always be fixed. Macé and Lailly (1986) have reported an excellent reliability of the inversion even when $\partial_t u(\tau,t)$ is given only for $0 \le \tau \le \tau_0^{max} < \tau^{max}$. **Note that the unknown function $g(t)$ is not necessarily the true source since the source can be located in the region $\tau << 0$: $g(t)$ is the trace of $\sigma\partial_\tau u$ at the first depth of investigation.** This is an advantage of the technique: the zone to invert can be chosen. Note also that on the elementary mathematical side, these results are "natural" since the velocity $v(z,t) = \partial_t u(z,t)$ in the original depth variable z being solution of:

$$\rho(z)\partial_{tt}v(z,t) - \partial_z(\rho(z)c(z)^2\partial_z v(z,t)) = 0 \tag{5}$$

writing:

$$\partial_z(\rho(z)c(z)^2\partial_z v(z,t)) =$$
$$\partial_z(\rho(z)c(z))c(z)\partial_z v(z,t) + \rho(z)c(z)\partial_z(c(z)\partial_z v(z,t)) \tag{6}$$

we obtain, if $c(z)$ is known, the formula giving the acoutic impedance (if $\rho(z^{min})c(z^{min})$ is known):

$$\rho(z)c(z) =$$
$$\rho(z^{min})c(z^{min})\exp(\int_{z^{min}}^{z}(\frac{\partial_{tt}v(\zeta,t) - c(\zeta)^2\partial_{\zeta\zeta}v(\zeta,t))}{c(\zeta)^2\partial_\zeta v(\zeta,t)} - \frac{\partial_\zeta c(\zeta)}{c(\zeta)})d\zeta)\ . \tag{7}$$

The deduction of $c(z)$ from the data, a work made by the conventional processing, is also an elementary mathematical result since writing that the time derivative of the expression under the integral sign in (7) is zero, we obtain:

$$c(z)^2 = \frac{(\partial_z v)\partial_{ttt}v - (\partial_{tt}v)\partial_{tz}v}{(\partial_z v)\partial_{zzt}v - (\partial_{zz}v)\partial_{tz}v}\ . \tag{8}$$

The fact that when $\partial_t u(\tau, t)$ is given only for $0 \leq \tau \leq \tau_0^{max} < \tau^{max}$, $\sigma(\tau)$ can be also identified follows from the ordinary results on the $1D$ inverse problem (see for example Bamberger et al. (1979) and Bube and Burridge (1983)).

The starting point of this study is thus the extension of this inversion technique when $3D$ propagation effects are taken into account, the characteristic of the media being only z dependent. This contribution is now divided into three sections. In the next section 2, some problems found in formulating the inversion of VSP data with $3D$ propagation effects are presented. The section 3 is devoted to a related question of exact transparent boundary conditions. The section 4 is devoted to the stability of the discretization of this boundary condition in the acoustic case.

2 Formulating the Problem of Inversion of VSP Data with 3*D* Propagation Effects in Layered Media

Let us consider the axisymmetric acoustic case in which the equations of motion in cylindrical coordinates (r, θ, z) are:

$$\rho(z)\partial_t v^r(r, z, t) = \partial_r p(r, z, t) \tag{9}$$

$$\rho(z)\partial_t v^z(r, z, t) = \partial_z p(r, z, t) \tag{10}$$

$$\frac{1}{\rho(z)c(z)^2}\partial_t p(r, z, t) =$$

$$\partial_r v^r(r, z, t) + \frac{v^r(r, z, t)}{r} + \partial_z v^z(r, z, t) + \frac{\delta(r)\delta(z - z_s)g(t)}{2\pi r\rho(z)c(z)^2} \tag{11}$$

with null initial condition and the boundary conditions ((12) is a "natural" boundary condition, (13) results from the axisymmetry and (14) is an absorbing condition introduced for convenience):

$$p(r, z = 0, t) = 0 \tag{12}$$

$$v^r(r = 0, z, t) = \partial_r v^z(r = 0, z, t) = \partial_r p(r = 0, z, t) = 0 \tag{13}$$

$$(p + \rho c v^z)(r, z^{max}, t) = 0 \ . \tag{14}$$

Here $\rho(z)$ and $c(z)$ are respectively the density and the sound velocity, v^r and v^z are respectively the radial and vertical components of the velocity, p is the pressure with a change of sign to keep the elasticity convention (opposed to the fluid mechanic one) and $g(t)$ is the source function ($g(t) = 0$ for $t < 0$). Clearly the terms $\partial_r v^r(r, z, t) + \frac{v^r(r,z,t)}{r}$ prevent the analysis of the $1D$ case. However results of Romanov presented in a book of Lavrent'ev, Reznitskaya and Yakhno (1986) (pages 52 – 58) show that if ρ is constant and c is known for $0 \leq z \leq z_g$ (where $z_s < z_g$), then when the source function $g(t)$ is known, the measurement of $v^z(0, z_g, t)$ for $t \geq 0$ suffices to obtain $c(z)$ for $z > z_g$. The proof exploits the support properties of Green functions of general acoustic media, well known properties of Volterra integral equation and some non trivial geometrical properties of wave fronts (in the geometrical optic sense). We do not have made a more detailled analysis of this point: note only that when $z >> z_s$ the plane wave approximation works so that $\partial_r v^r(r,z,t) + \frac{v^r(r,z,t)}{r} \simeq_{z>>zs} 0$ and the $1D$ analysis is probably a good approximation.

In an attempt to find a boundary condition at $z = z^{min}$ with $z_s < z^{min}$ (the measurement being made at depths $z \geq z^{min}$), we have considered the second order wave propagation equation satisfied by the pressure p (with $\mathbf{x}_s = (0,0,z_s)$):

$$\frac{1}{\rho(z)c(z)^2}\partial_{tt}p(\mathbf{x},t) - \mathrm{div}(\frac{1}{\rho(z)}\ \mathbf{grad}\ p(\mathbf{x},t)) = \frac{\delta(\mathbf{x}-\mathbf{x}_s)g'(t)}{\rho(z)c(z)^2} \qquad (15)$$

and the equation satisfied by the Green's function $G(\mathbf{x},t;\mathbf{y})$ corresponding to a density function $\tilde{\rho}(\mathbf{x})$ and a sound velocity $\tilde{c}(\mathbf{x})$ (with the usual initial and boundary conditions):

$$\frac{1}{\tilde{\rho}(\mathbf{x})\tilde{c}(\mathbf{x})^2}\partial_{tt}G(\mathbf{x},t;\mathbf{y}) - \mathrm{div}(\frac{1}{\tilde{\rho}(\mathbf{x})}\ \mathbf{grad}\ G(\mathbf{x},t;\mathbf{y})) = \delta(\mathbf{x}-\mathbf{y})\delta(t) \qquad (16)$$

all the spacial derivative being made with respect to $\mathbf{x} = (x,y,z) = (r,\theta,z)$. Using a convolution in t and the Green formula in $\Omega = \{\mathbf{x} = (x,y,z) : 0 < z < z^{min}\}$, it follows from (15) and (16) that if $\tilde{\rho} = \rho$ and $\tilde{c} = c$ in Ω, then for $\mathbf{y} \in \Omega$:

$$p(\mathbf{y},t) = \frac{\int_0^t G(\mathbf{x}_s,t-\tau;\mathbf{y})g'(\tau)d\tau}{\rho(z_s)c(z_s)^2} +$$

$$\int_0^t (\int_{\partial\Omega} \frac{1}{\rho(z)}(G(\mathbf{x},t-\tau;\mathbf{y})\frac{\partial p(\mathbf{x},\tau)}{\partial n} - p(\mathbf{x},\tau)\frac{\partial G(\mathbf{x},t-\tau;\mathbf{y})}{\partial n})d\sigma(\mathbf{x}))d\tau \quad (17)$$

where $\frac{\partial}{\partial n} = \mathbf{n}^T\mathbf{grad}$ is the derivative along the unit normal exterior to $\partial\Omega$ and $d\sigma(\mathbf{y})$ is the area element on $\partial\Omega$. **Taking the limit $\mathbf{y} \to \partial\Omega$ we obtain a possible way to find a boundary condition at $z = z^{min}$.** But this formula depends on the unknown function G and even in the simple case where ρ and c are constant in Ω so that we can choose as G the Green function of the homogeneous half space $z > 0$:

$$G(\mathbf{x},t;\mathbf{y}) = \frac{\rho}{4\pi}(\frac{\delta(t- \parallel \mathbf{x}-\mathbf{y} \parallel /c)}{\parallel \mathbf{x}-\mathbf{y} \parallel} - \frac{\delta(t- \parallel \mathbf{x}-\mathbf{y}' \parallel /c)}{\parallel \mathbf{x}-\mathbf{y}' \parallel}) \qquad (18)$$

(where $\mathbf{y}'$ is obtained from $\mathbf{y}$ by orthogonal symmetry across the plane $z = 0$) one finds the formula, Γ_{min} being the plane $z = z^{min}$:

$$p(\mathbf{y},t) = \frac{1}{2\pi c^2}(\frac{g'(t- \parallel \mathbf{x}_s-\mathbf{y} \parallel /c)}{\parallel \mathbf{x}_s-\mathbf{y} \parallel} - \frac{g'(t- \parallel \mathbf{x}_s-\mathbf{y}' \parallel /c)}{\parallel \mathbf{x}_s-\mathbf{y}' \parallel})$$

$$+\frac{1}{2\pi}\int_{\Gamma_{\min}} \frac{1}{\parallel \mathbf{x}-\mathbf{y} \parallel}(\partial_z p(\mathbf{x},t- \parallel \mathbf{u}-\mathbf{y} \parallel /c))_{\mathbf{u}=\mathbf{x}}d\sigma(\mathbf{x})$$

$$-\frac{1}{2\pi}\int_{\Gamma_{\min}}\Big(\frac{1}{\| \mathbf{x}-\mathbf{y}' \|}\partial_z p(\mathbf{x},t-\| \mathbf{u}-\mathbf{y}' \| /c)$$

$$+\frac{1}{c}\partial_z(\ln(\| \mathbf{x}-\mathbf{y}' \|))\partial_t p(\mathbf{x},t-\| \mathbf{u}-\mathbf{y}' \| /c)$$

$$-\partial_z\Big(\frac{1}{\| \mathbf{x}-\mathbf{y}' \|}\Big)p(\mathbf{x},t-\| \mathbf{u}-\mathbf{y}' \| /c)\Big)_{\mathbf{u}=\mathbf{x}}d\sigma(\mathbf{x}) \tag{19}$$

which after the Hankel transform is (we denote by $\hat{\varphi}_\nu(k)=\int_0^\infty J_\nu(kr)\varphi(r)rdr$ the direct Hankel transform of order ν of a function φ where J_ν is the Bessel function of order ν):

$$\hat{p}_0(k,z^{min},t)=$$

$$\frac{1}{2\pi c}\Big(Y\big(t-\frac{z^{min}-z_s}{c}\big)\int_{\frac{z^{min}-z_s}{c}}^{t}J_0\Big(kc\sqrt{s^2-\big(\frac{z^{min}-z_s}{c}\big)^2}\Big)g'(t-s)ds$$

$$-Y\big(t-\frac{z^{min}+z_s}{c}\big)\int_{\frac{z^{min}+z_s}{c}}^{t}J_0\Big(kc\sqrt{s^2-\big(\frac{z^{min}+z_s}{c}\big)^2}\Big)g'(t-s)ds\Big)$$

$$+\rho c\hat{v}_0^z(k,z^{min},t)-k\rho c^2\int_0^t J_1(kcs))\hat{v}_0^z(k,z^{min},t-s)ds$$

$$-Y\big(t-\frac{z^{min}+z_s}{c}\big)\Big(\rho c\hat{v}_0^z\big(k,z^{min},t-\frac{2z^{min}}{c}\big)$$

$$+\rho c\int_{\frac{2z^{min}}{c}}^{t}\partial_s J_0\Big(kc\sqrt{s^2-\big(\frac{2z^{min}}{c}\big)^2}\Big)\Big)\hat{v}_0^z(k,z^{min},t-s)ds$$

$$+\int_{\frac{2z^{min}}{c}}^{t}J_0\Big(kc\sqrt{s^2-\big(\frac{2z^{min}}{c}\big)^2}\Big)\Big)\Big(\frac{2z^{min}}{cs}\partial_t\hat{p}_0(k,z^{min},t-s)$$

$$+\frac{2z^{min}}{cs^2}\hat{p}_0(k,z^{min},t-s)\Big)ds\Big) \tag{20}$$

where Y is the Heaviside function. The formula (20) can be deduced from (19) by the technique of the next section. From the complexity of this partial result, we conclude that when $3D$ propagation effects are taken into account, there is no simple way to introduce a boundary condition allowing a choice of the zone to invert. Practically it is necessary to discretize all the borehole, partially with macro layers if necessary (thus carrying out some kind of homogeneization if necessary). Besides the results of Romanov is probably a convenient starting point for a theoretical analysis of the inverse problem of VSP data.

3 Exact Transparent Boundary Conditions for the Axisymmetric Problem after Hankel Transform

The computation made for (17) can be modified to write the exact transparent condition for $z = z^{max}$ if ρ and c are constant for $z \geq z^{max}$. More precisely if one chooses $G(\mathbf{x}, t; \mathbf{y}) = \frac{\rho}{4\pi} \frac{\delta(t - \|\mathbf{x}-\mathbf{y}\|/c)}{\|\mathbf{x}-\mathbf{y}\|}$, the Green function in $\mathbb{R}^3$ in the homogeneous case, and $\Omega = \{\mathbf{x} = (x, y, z) : z > z^{max}\}$, then using a convolution in t and the Green formula in Ω, it follows from (15) and (16) that for $\mathbf{y} \in \Omega$:

$$
\begin{aligned}
p(\mathbf{y}, t) &= \int_0^t \left(\int_{\partial\Omega} \frac{1}{\rho(z)} (G(\mathbf{x}, t-\tau; \mathbf{y}) \frac{\partial p(\mathbf{x}, \tau)}{\partial n} - p(\mathbf{x}, \tau) \frac{\partial G(\mathbf{x}, t-\tau; \mathbf{y})}{\partial n}) d\sigma(\mathbf{x}) \right) d\tau \\
&= \frac{1}{4\pi} \int_{\partial\Omega} \Big(\frac{1}{\| \mathbf{x} - \mathbf{y} \|} \frac{\partial p}{\partial n}(\mathbf{x}, t - \| \mathbf{u} - \mathbf{y} \| / c) \\
&\quad + \frac{1}{c} \frac{\partial}{\partial n} (\ln \| \mathbf{x} - \mathbf{y} \|) \frac{\partial p}{\partial t}(\mathbf{x}, t - \| \mathbf{u} - \mathbf{y} \| / c) \\
&\quad - \frac{\partial}{\partial n} \left(\frac{1}{\| \mathbf{x} - \mathbf{y} \|} \right) p(\mathbf{x}, t - \| \mathbf{u} - \mathbf{y} \| / c) \Big)_{\mathbf{u}=\mathbf{x}} d\sigma(\mathbf{x}) \qquad (21)
\end{aligned}
$$

which is the Kirchhoff formula (which can also be obtained from a particular property of the Laplacian in $\mathbb{R}^3$: see Smirnov (1964)).

Taking the limit $\mathbf{y} \to \partial\Omega$, the exact transparent boundary condition is obtained. In our case $\partial\Omega$ is the plane $z = z^{max}$ and the result is that for $\mathbf{y} = (x, y, z^{max})$:

$$
p(\mathbf{y}, t) = \frac{1}{2\pi} \int_{\partial\Omega} \left(\frac{1}{\| \mathbf{x} - \mathbf{y} \|} \frac{\partial p}{\partial n}(\mathbf{y}, t - \| \mathbf{u} - \mathbf{y} \| / c) \right)_{\mathbf{u}=\mathbf{x}} d\sigma(\mathbf{x}) \ . \qquad (22)
$$

Computing the integral in (22) in polar coordinates centered at $\mathbf{y}$ and using the axisymmetry of the problem and the causal nature of p, we first obtain:

$$
p(r, z^{max}, t) =
$$

$$
-\frac{1}{2\pi} \int_0^{ct} \tau d\tau \int_0^{2\pi} \left(\frac{1}{\tau} \partial_z p((r^2 + \tau^2 + 2r\tau \cos\varphi)^{1/2}, z^{max}, t - \tau/c) \right) d\varphi \qquad (23)
$$

for $r \geq 0, t \geq 0$. Then using the inverse Hankel transform:

$$
\partial_z p((r^2 + \tau^2 + 2r\tau \cos\varphi)^{1/2}, z^{max}, t - \tau/c) =
$$

$$
\int_0^\infty J_0(k(r^2 + \tau^2 + 2r\tau \cos\varphi)^{1/2}) \partial_z \hat{p}_0(k, z^{max}, t - \tau/c) k dk \qquad (24)
$$

and an addition formula of Bessel functions (Watson (1922), p 358–361):

$$
J_0(k(r^2 + \tau^2 + 2r\tau \cos\varphi)^{1/2}) = \sum_{m \in \mathbb{Z}} (-1)^m J_m(kr) J_m(k\tau) e^{im\varphi} \qquad (25)
$$

we find:

$$\hat{p}_0(k, z^{max}, t) = -\int_0^{ct} \partial_z \hat{p}_0(k, z^{max}, t-\tau/c) J_0(k\tau) d\tau \tag{26}$$

for $k \geq 0$, $t \geq 0$ that is to say, since $\partial_z \hat{p}_0(k,z,t) = \rho(z)\partial_t \hat{v}_0^z(k,z,t)$ and after an integration by parts in τ ($J_0'(x) = -J_1(x)$, $J_0(0) = 1$ are used and in (27), ρ and c are taken for $z = z^{max}$):

$$\hat{p}_0(k, z^{max}, t) + \rho c \hat{v}_0^z(k, z^{max}, t)$$

$$-k\rho c^2 \int_0^t J_1(kc(t-s)) \hat{v}_0^z(k, z^{max}, s) ds = 0 \ . \tag{27}$$

This is the exact transparent boundary condition after the Hankel transform in the acoustic case adapted to the following equations of propagation written as first order hyperbolic systems:

$$\rho(z)\partial_t \hat{v}_1^r(k,z,t) = -k\hat{p}_0(k,z,t) \tag{28}$$

$$\rho(z)\partial_t \hat{v}_0^z(k,z,t) = \partial_z \hat{p}_0(k,z,t) \tag{29}$$

$$\frac{1}{\rho(z)c^2(z)}\partial_t \hat{p}_0(k,z,t) = k\hat{v}_1^r(k,z,t) + \partial_z \hat{v}_0^z(k,z,t) + \frac{\delta(z-z_s)}{2\pi\rho(z)c^2(z)} s(t) \tag{30}$$

$$\hat{p}_0(k, z=0, t) = 0 \tag{31}$$

$$\hat{p}_0(k,z,0) = \hat{v}_1^r(k,z,0) = \hat{v}_0^z(k,z,0) = 0 \ . \tag{32}$$

In the isotropic elastodynamic case, the wave operator is replaced by the operator $\mathbf{u} \longrightarrow \rho\partial_{tt}u_j - \partial_l(\delta_{lj}\lambda\partial_k u_k + \mu(\partial_j u_l + \partial_l u_j))$ and the fundamental solution, for ρ, λ, μ constant (with the right hand side $\delta(\mathbf{x})\delta(t)\delta_{ij}$) is the matrix $\mathbf{G}$ with coefficients:

$$G_{ij}(\mathbf{x}, t) =$$

$$\frac{1}{4\pi\rho}\frac{3\gamma_i\gamma_j - \delta_{ij}}{\|\mathbf{x}\|^3} t\left(Y\left(t - \frac{\|\mathbf{x}\|}{\alpha}\right) - Y\left(t - \frac{\|\mathbf{x}\|}{\beta}\right)\right)$$

$$+\frac{1}{4\pi\rho\alpha^2}\frac{\gamma_i\gamma_j}{\|\mathbf{x}\|}\delta\left(t - \frac{\|\mathbf{x}\|}{\alpha}\right) - \frac{1}{4\pi\rho\beta^2}\frac{\gamma_i\gamma_j - \delta_{ij}}{\|\mathbf{x}\|}\delta\left(t - \frac{\|\mathbf{x}\|}{\beta}\right) \tag{33}$$

where Y is the Heaviside function, $\alpha = \sqrt{\frac{\lambda+2\mu}{\rho}} = v_P$, $\beta = \sqrt{\frac{\mu}{\rho}} = v_S$ and $\gamma_i = \frac{x_i}{\|\mathbf{x}\|}$. Similar computations as the previous ones where the Green formula is replaced by the Betti formula, allow to obtain exact transparent condition in the elastic case. Using polar coordinates as in (23), using (25) and:

$$\begin{pmatrix} \cos\vartheta \\ \sin\vartheta \end{pmatrix} J_1(k(r^2 + \tau^2 + 2\tau r\cos\varphi)^{1/2}) =$$

$$\sum_{m\in\mathbb{Z}} J_{m+1}(kr) J_m(k\tau) \begin{pmatrix} (-1)^m \cos m\varphi \\ (-1)^{m+1} \sin m\varphi \end{pmatrix} \tag{34}$$

where $\cos\vartheta = \frac{r+\tau\cos\varphi}{\sqrt{r^2+\tau^2+2\tau r\cos\varphi}}$, $\sin\vartheta = \frac{\tau\sin\varphi}{\sqrt{r^2+\tau^2+2\tau r\cos\varphi}}$, the obtained exact transparent conditions are:

$$\hat{\sigma}_1^{rz}(k, z^{max}, t) + \rho\beta\hat{v}_1^r(k, z^{max}, t) - k\beta \int_0^t (J_1(k\alpha(t-s))$$

$$+ \int_{k\beta(t-s)}^{k\alpha(t-s)} \frac{J_1(\xi)}{\xi} d\xi)\hat{\sigma}_1^{rz}(k, z^{max}, s)ds - k\lambda\frac{\beta}{\alpha}\int_0^t J_0(k\alpha(t-s))\hat{v}_0^z(k, z^{max}, s)ds$$

$$+2k^2\mu\beta \int_0^t ((t-s)\int_{k\beta(t-s)}^{k\alpha(t-s)} \frac{J_0(\xi)}{\xi} d\xi)\hat{v}_0^z(k, z^{max}, s)ds = 0 \qquad (35)$$

$$\hat{\sigma}_0^{zz}(k, z^{max}, t) + \rho\alpha\hat{v}_0^z(k, z^{max}, t) - k\alpha \int_0^t (J_1(k\beta(t-s)) -$$

$$\int_{k\beta(t-s)}^{k\alpha(t-s)} \frac{J_1(\xi)}{\xi} d\xi)\hat{\sigma}_0^{zz}(k, z^{max}, s)ds$$

$$+\rho\alpha \int_0^t (\frac{\beta^2}{\alpha^2}\frac{J_1(k\alpha(t-s))}{t-s} + \frac{3k\beta^2}{\alpha}J_1'(k\alpha(t-s)) - 2k\beta J_1'(k\beta(t-s))$$

$$-3k^2\beta^2(t-s)\int_{k\beta(t-s)}^{k\alpha(t-s)} \frac{J_1''(\xi)}{\xi} d\xi)\hat{v}_1^r(k, z^{max}, s)ds = 0 \qquad (36)$$

for the longitudinal (or $P-SV$) waves, and:

$$\hat{\sigma}_1^{\theta z}(k, z^{max}, t) + \rho\beta\hat{v}_1^{\theta}(k, z^{max}, t)$$

$$-k\beta \int_0^t J_1(k\beta(t-s))\hat{\sigma}_1^{\theta z}(k, z^{max}, s)ds = 0 \qquad (37)$$

for the torsional (or SH) waves.

Remark 1. This process giving (21) can be generalized: for example, if the original problem is in the half space $z > 0$ with the natural boundary condition $p = 0$ for $z = 0$ and if Ω is the exterior of the half sphere $\{\mathbf{x} = (x, y, z) : x^2 + y^2 + z^2 > \text{Const}, z > 0\}$, the limit of the first formula in (20) when $\mathbf{y} \to \partial\Omega$ gives the exact transparent condition on $\Gamma_a = \{\mathbf{x} = (x, y, z) : x^2 + y^2 + z^2 = \text{Const}, z > 0\}$ if (18) is chosen for $G(\mathbf{x}, t; \mathbf{y})$. This could be applied to the so called corner problem. In elastodynamics such a process is complicated because the Green function of the half space is non trivial (the Rayleigh surface wave must be taken into account).

Remark 2. To construct exact transparent conditions, instead of the limit of (21), we can use the formula obtained by applying $\frac{\partial}{\partial \mathbf{n}_y}$ to the two sides of (21). Using results on layer potentials which can be found in the book of Colton and Kress (1983), the obtained result is that for $\mathbf{y} \in \partial\Omega$, if $G(\mathbf{x}, t; \mathbf{y}) = \frac{\rho}{4\pi}\frac{\delta(t-\|\mathbf{x}-\mathbf{y}\|/c)}{\|\mathbf{x}-\mathbf{y}\|}$ can be chosen:

$$\frac{\partial p}{\partial n_y}(\mathbf{y},t) = \frac{1}{2\pi}\int_{\partial\Omega}(\frac{\partial}{\partial n_y}(\frac{1}{\| \mathbf{x}-\mathbf{y} \|})\frac{\partial p}{\partial n_x}(\mathbf{x}, t- \| \mathbf{u}-\mathbf{y} \| /c)$$

$$+(\frac{\partial}{\partial n_x}(\frac{1}{\| \mathbf{x}-\mathbf{y} \|})\mathbf{n}_y - (\mathbf{n}_y^T \mathbf{n}_x)\mathbf{grad}_y(\frac{1}{\| \mathbf{x}-\mathbf{y} \|}))^T(\mathbf{Grad}_x p(\mathbf{x}, t- \| \mathbf{u}-\mathbf{y} \| /c)$$

$$+\frac{\| \mathbf{x}-\mathbf{y} \|}{c}\mathbf{Grad}_x(\frac{\partial p}{\partial t}(\mathbf{x}, t- \| \mathbf{u}-\mathbf{y} \| /c))$$

$$-\frac{\| \mathbf{x}-\mathbf{y} \|}{c^2}\mathbf{Grad}_y(\| \mathbf{x}-\mathbf{y} \|)\frac{\partial^2 p}{\partial t^2}(\mathbf{x}, t- \| \mathbf{u}-\mathbf{y} \| /c))$$

$$-\frac{1}{\| \mathbf{x}-\mathbf{y} \|}\frac{\partial \| \mathbf{x}-\mathbf{y} \|}{\partial n_y}(\frac{1}{c}\frac{\partial^2 p}{\partial t \partial n_x}(\mathbf{x}, t- \| \mathbf{u}-\mathbf{y} \| /c)$$

$$+\frac{\partial \| \mathbf{x}-\mathbf{y} \|}{\partial n_x}\frac{1}{c^2}\frac{\partial^2 p}{\partial t^2}(\mathbf{x}, t- \| \mathbf{u}-\mathbf{y} \| /c)))_{\mathbf{u}=\mathbf{x}}d\sigma(\mathbf{x}) \qquad (38)$$

where **grad** is the $3D$-gradient and **Grad** the surfacic gradient.

Remark 3. Some very simple approximations of (22) and (38) give the Engquist-Majda formula (1977, 1979) of first and second orders. More precisely, if Ω is the half space $x > 0$ (the computational domain is thus $x < 0$) and if p is independent of the variable z (two dimensional problem) then (22) can be written:

$$p(x=0,y,t) = -\frac{1}{2\pi}\int_0^{2\pi} d\theta \int_0^{ct} \frac{\partial p}{\partial x}(0, y+\rho\cos\theta, t-\rho/c)d\rho \qquad (39)$$

and the approximation $\frac{\partial p}{\partial x}(0, y+\rho\cos\theta, t-\rho/c) \cong \frac{\partial p}{\partial x}(0, y, t-\rho/c)$ gives the first order condition:

$$\frac{1}{c}\frac{\partial p}{\partial t}(0,y,t) = -\frac{\partial p}{\partial x}(0,y,t) \ . \qquad (40)$$

In the same way (38) can be written:

$$\frac{\partial p}{\partial x}(0,y,t) = \frac{1}{2\pi}\int_0^{2\pi} d\theta \int_0^{ct} (\frac{\cos\theta}{\rho}\frac{\partial p}{\partial y}(0, y+\rho\cos\theta, t-\rho/c)$$

$$+\cos\theta\frac{1}{c}\frac{\partial^2 p}{\partial t \partial y}(0, y+\rho\cos\theta, t-\rho/c) - \frac{1}{c^2}\frac{\partial^2 p}{\partial t^2}(0, y+\rho\cos\theta, t-\rho/c))d\rho =$$

$$\frac{1}{2\pi}\int_0^{2\pi} d\theta \int_0^{ct} (\sin^2\theta\frac{\partial^2 p}{\partial y^2}(0, y+\rho\cos\theta, t-\rho/c)+$$

$$\cos\theta\frac{1}{c}\frac{\partial^2 p}{\partial t\partial y}(0,y+\rho\cos\theta,t-\rho/c)-\frac{1}{c^2}\frac{\partial^2 p}{\partial t^2}(0,y+\rho\cos\theta,t-\rho/c))d\rho \quad (41)$$

and the same approximation ($\varphi(0,y+\rho\cos\theta,t-\rho/c)\cong\varphi(0,y,t-\rho/c)$) gives the second order condition:

$$\frac{1}{c}\frac{\partial^2 p}{\partial t\partial x}(0,y,t)=\frac{1}{2}\frac{\partial^2 p}{\partial y^2}(0,y,t)-\frac{1}{c^2}\frac{\partial^2 p}{\partial t^2}(0,y,t)\ . \quad (42)$$

Note also that, for $c=1$, (41) is an other form of the initial equation used by Engquist-Majda:

$$\frac{\partial p}{\partial x}(0,y,t)=-\int_{\mathbb{R}^2} i\xi(1-\frac{\omega^2}{\xi^2})^{1/2}e^{i\xi t+i\omega y}\hat{p}(0,\xi,\omega)d\xi d\omega \quad (43)$$

where the chosen square root is $(1-\frac{\omega^2}{\xi^2})^{1/2}=\mid 1-\frac{\omega^2}{\xi^2}\mid^{1/2}(-i\ \mathrm{sgn}(\xi))$ for $\omega^2>\xi^2$ and $\hat{p}$ is the (y,t)-Fourier transform of p:

$$\hat{p}(x,\xi,\omega)=\frac{1}{(2\pi)^2}\int_{\mathbb{R}^2} e^{-i\xi t-i\omega y}p(x,y,t)dydt\ . \quad (44)$$

Writing also (39) in this way, we find the identities (which can also be obtained by direct calculus — in the distributions sense):

$$\frac{1}{\pi}\int_0^\pi d\theta\int_0^\infty(\sin^2\theta(i\omega)^2+\cos\theta(i\xi)(i\omega)-(i\xi)^2)e^{-i\xi\rho+i\omega\rho\cos\theta}d\rho=$$

$$-\mid\omega^2-\xi^2\mid^{1/2}, \quad (45)$$

$$-\frac{1}{\pi}\int_0^\pi d\theta\int_0^\infty e^{-i\xi\rho+i\omega\rho\cos\theta}d\rho=-\frac{1}{i\xi(1-\omega^2/\xi^2)^{1/2}}\ . \quad (46)$$

Note finally that when Ω is the exterior of a vertical cylinder of radius r^{max} and p is independent of the angular variable θ in cylindrical coordinates, we can also obtained the first order condition:

$$\frac{1}{c}\frac{\partial p}{\partial t}(r^{max},z,t)+\frac{\partial p}{\partial r}(r^{max},z,t)+\frac{p(r^{max},z,t)}{2r^{max}}=0 \quad (47)$$

and the second order one:

$$\frac{1}{c}\frac{\partial^2 p}{\partial t\partial r}(r^{max},z,t)+\frac{1}{c^2}\frac{\partial^2 p}{\partial t^2}(r^{max},z,t)-\frac{1}{2}\frac{\partial^2 p}{\partial z^2}(r^{max},z,t)$$

$$-\frac{1}{2cr^{max}}\frac{\partial p}{\partial t}(r^{max},z,t)-\frac{p(r^{max},z,t)}{8(r^{max})^2}=0\ . \quad (48)$$

Remark 4. The right hand side of (27) can be interpreted as a "upgoing wave". This kind of waves splitting appears in numerous studies on inverse problems of wave propagation in $1D$ media and is used in the work of He (1991) and He and Karlsson (1993) as an alternative of invariant imbedding technique used for example by He and Ström (1992).

4 Stability of the Discretizations of the Exact Transparent Boundary Condition (for the Axisymmetric Problem after Hankel Transform) in the Acoustic Case

The problem (28)–(32),(27) is clearly well-posed since it is equivalent to a well posed problem in the half-space $z > 0$. But this result can be proved directly using the properties of the Fourier transforms of Bessel functions. Briefly, in the case where the source terms in (30) is replaced by a function $\frac{\xi(k,z,t)}{\rho(z)c^2(z)}$ such that $\int_0^\infty (\int_0^{z^{max}} \frac{|\xi(k,z,t)|^2}{\rho(z)c^2(z)} dz)^{1/2} dt < \infty$, making the product of (28), (29), (30) by $\hat{v}_1^r$, $\hat{v}_0^z$, $\hat{p}_0$ respectively, we find by summation and integration in z:

$$\frac{dE(k,t)}{dt} + a(k,t) \leq \sqrt{2E(k,t)}S(k,t) \tag{49}$$

where:

$$E(k,t) = \frac{1}{2}\int_0^{z^{max}} (\frac{|\hat{p}_0(k,z,t)|^2}{\rho(z)c^2(z)} + \rho(z) |\hat{v}_1^r(k,z,t)|^2 + \rho(z) |\hat{v}_0^z(k,z,t)|^2) dz \tag{50}$$

is the acoustic energy,

$$S(k,t) = (\int_0^{z^{max}} \frac{|\xi(k,z,t)|^2}{\rho(z)c^2(z)} dz)^{1/2} \tag{51}$$

and:

$$a(k,t) = \rho(z^{max})c(z^{max}) |\hat{v}_0^z(k,z^{max},t)|^2 - k\rho(z^{max})c^2(z^{max})\hat{v}_0^z(k,z^{max},t) \int_0^t J_1(kc(t-t'))\hat{v}_0^z(k,z^{max},t')dt' \ . \tag{52}$$

(We think that when $s(t)$ in (30) is sufficiently regular this study allows to obtain convergence results because error terms in finite differences discretisation are function of the same kind as ξ).

When the exact transparent condition (27) is replaced by the absorbing one (approximation of first order):

$$\hat{p}_0(k,z^{max},t) + \rho c \hat{v}_0^z(k,z^{max},t) = 0 \tag{53}$$

the integral in (52) disappears and from $a(k,t) \geq 0$ it follows the a priori inequality (giving the wellposedness):

$$\sqrt{E(k,t)} \leq \frac{1}{\sqrt{2}} \int_0^t S(k,\tau) d\tau \ . \tag{54}$$

In the case of the boundary condition (27) the same inequality can be deduced from the fact that:

$$X(t) \equiv \int_0^t a(k,\tau) d\tau \geq 0 \tag{55}$$

(since then $E(k,t) \leq \int_0^t \sqrt{2}S(k,\tau)\sqrt{E(k,\tau)}d\tau = W(k,t)$ so that $W'(k,t) \leq \sqrt{2}S(k,t)\sqrt{W(k,t)}$ and $\sqrt{E(k,t)} \leq \sqrt{W(k,t)} \leq \frac{1}{\sqrt{2}}\int_0^t S(k,\tau)d\tau$ QED).

Introducing the Fourier transform:

$$\hat{\varphi}_t(\omega) = \frac{1}{\sqrt{2\pi}}\int_0^t e^{-i\omega s}\hat{v}_0^z(k, z^{max}, s)ds \tag{56}$$

and using (with the square root determination $\sqrt{k^2c^2+s^2} \cong_{|s|\to\infty} s$):

$$\frac{1}{\sqrt{2\pi}}\int_0^\infty e^{-i\omega t}J_1(kct)dt = \frac{1}{kc\sqrt{2\pi}}(1 - \frac{i\omega}{\sqrt{k^2c^2-\omega^2}}) \tag{57}$$

we find:

$$X(t) = \rho c \int_{|\omega|\geq kc} \frac{|\omega|^2}{\sqrt{\omega^2-k^2c^2}} \mid \hat{\varphi}_t(\omega) \mid^2 d\omega \tag{58}$$

and (55) and (54) follows.

We now obtain an analog result for the discretization of (28)–(32),(27), by elementary finite differences. The unknowns $\hat{p}_0$, $\hat{v}_0^z$, $\hat{v}_1^r$ are sampled at nodes $(j+1/2)\Delta z$ or $j\Delta z$, $0 \leq j \leq J-1$ (with $J\Delta z = z^{max}$) and at times $n\Delta t$ or $(n+1/2)\Delta$t. Here j and n are integers, and Δz and Δt are the steps in space and time, respectively. So, we introduce as unknowns of the (direct) problem the three sequences of vectors:

$$\mathbf{p}^n = (\hat{p}^n_{0,1/2}, \ldots, \hat{p}^n_{0,J-1/2})^T, \mathbf{vz}^{n+1/2} = (\hat{v}^{z,n+1}_{0,0}, \ldots, \hat{v}^{z,n+1/2}_{0,J})^T$$

$$\mathbf{vr}^{n+1/2} = (\hat{v}^{r,n+1}_{1,1/2}, \ldots, \hat{v}^{r,n+1/2}_{1,J-1/2})^T \tag{59}$$

where $\hat{p}^n_{0,j+1/2} = \hat{p}_0((j+1/2)\Delta z, n\Delta t)$ for example.

The discrete form of (28)–(32),(27) are then the following recursive equations (the term $\mathbf{t}a^n(\mathbf{vz})$ which follows from the transparent boundary condition will be defined later in (65) and (66)):

$$\mathbf{M}_3\frac{\mathbf{p}^{n+1}-\mathbf{p}^n}{\Delta t} = k\mathbf{vr}^{n+1/2} - \mathbf{K}^T\mathbf{vz}^{n+1/2} + \mathbf{s}^{n+1/2} \tag{60}$$

$$\mathbf{M}_1\frac{\mathbf{vr}^{n+3/2}-\mathbf{vr}^{n+1/2}}{\Delta t} = -k\mathbf{p}^{n+1} \tag{61}$$

$$\mathbf{M}_2\frac{\mathbf{vz}^{n+3/2}-\mathbf{vz}^{n+1/2}}{\Delta t} = \mathbf{K}\mathbf{p}^{n+1} - \frac{1}{2}\mathbf{A}(\mathbf{vz}^{n+3/2}+\mathbf{vz}^{n+1/2}) + \mathbf{t}a^n(\mathbf{vz}) \tag{62}$$

for $n \geq 0$, with $\mathbf{p}^0 = \mathbf{0}$, $\mathbf{vr}^{1/2} = \mathbf{0}$, $\mathbf{vz}^{1/2} = \mathbf{0}$. The discretization of the source term is $\mathbf{s}$. The matrices $\mathbf{M}_1$, $\mathbf{M}_2$, $\mathbf{M}_3$, $\mathbf{K}$ and $\mathbf{A}$ of order $J\times J$, $(J+1)\times(J+1)$,

$J \times J$, $(J+1) \times J$ and $(J+1) \times (J+1)$ respectively are easy to construct from (28)–(32),(27), for example:

$$\mathbf{M}_2 = \begin{pmatrix} 1/2\rho_0 & & & & \\ & \rho_1 & & & \\ & & \ddots & & \\ & & & \rho_{J-1} & \\ & & & & 1/2\rho_J \end{pmatrix}, \mathbf{K} = \frac{1}{\Delta z} \begin{pmatrix} 1 & & & & \\ -1 & 1 & & & \\ & -1 & \ddots & & \\ & & \ddots & 1 & \\ & & & -1 & 1 \\ & & & & -1 \end{pmatrix},$$

$$\mathbf{A} = \frac{\rho(z^{max})c(z^{max})}{\Delta z} \begin{pmatrix} & & 0 \\ & 0 & . \\ & & 0 \\ 0 & . & 0 \; 1 \end{pmatrix} \tag{63}$$

where $\rho_j = \rho(j\Delta z)$.

When the boundary condition is (53), the term $\mathbf{t}a^n(\mathbf{vz})$ in (62) is zero and it is easy to show that the stability condition is of the form (this is the CFL condition in the homogeneous case; for the general case see for example Jurado et al. (1995a)):

$$\Delta t < \frac{\Delta z}{v_P\sqrt{1+(k^{max})^2\Delta z^2/4}} \; . \tag{64}$$

In the following we are interested in the two "natural" discrete forms of the convolution term of the exact transparent boundary condition (27):

$$\mathbf{t}a^n(\mathbf{vz}) = kc\Delta t \sum_{l=0}^{n} J_1(kc(n+1/2-l)\Delta t)\mathbf{A}\mathbf{vz}^{l+1/2} \tag{65}$$

(corresponding to the midpoint rule) and:

$$\mathbf{t}a^n(\mathbf{vz}) = kc\Delta t \sum_{l=0}^{n} J_1(kc(n+1-l)\Delta t)\mathbf{A}(\mathbf{vz}^{l+1/2} + \mathbf{vz}^{l-1/2}) \tag{66}$$

which corresponds to the trapezoidal formula. Multipying (60) by $(\mathbf{p}^{n+1} + \mathbf{p}^n)^T$, (61) by $(\mathbf{vr}^{n+3/2} - \mathbf{vr}^{n+1/2})^T$ and (62) by $(\mathbf{vz}^{n+3/2} - \mathbf{vz}^{n+1/2})^T$ we find that the discrete analogue of X in (55) is with $u^{n+1/2} = \mathbf{vz}_J^{n+1/2}$:

$$X = \rho c\Delta t \sum_{n=0}^{N+1} (\frac{1}{4}(u^{n+1/2} + u^{n-1/2})^2$$

$$-\frac{kc\Delta t}{2}(u^{n+1/2} + u^{n-1/2}) \sum_{l=0}^{n} J_1(kc(n-l+1/2)\Delta t)u^{l-1/2}) \tag{67}$$

in the midpoint case (65) and:

$$X = \rho c \Delta t \sum_{n=0}^{N+1} (\frac{1}{4}(u^{n+1/2} + u^{n-1/2})^2$$

$$-\frac{kc\Delta t}{2}(u^{n+1/2} + u^{n-1/2}) \sum_{l=0}^{n} J_1(kc(n-l)\Delta t)(u^{l-1/2} + u^{l+1/2})/2) \qquad (68)$$

in the trapezoidal case (66). Introducing $\hat{U}(z) = \frac{\Delta t}{\sqrt{2\pi}} \sum_{n=0}^{N+1} u^{n-1/2} z^n$, $b = u^{N+3/2}$, it follows that in the midpoint case (65) we have:

$$X = \frac{\rho c}{\Delta t} \int_{-\pi}^{\pi} ((\cos^2(\theta/2) - \cos(\theta/2) Re(\sqrt{2\pi} kce^{i\theta/2} \hat{J}_a(e^{i\theta}))) \mid \hat{U}(e^{i\theta}) \mid^2$$

$$+Re(e^{-i(N+1)\theta}(1 - \sqrt{2\pi} kc\hat{J}_a(e^{i\theta}))\hat{U}(e^{i\theta})b) + b^2)d\theta \qquad (69)$$

with:

$$\hat{J}_a(z) = \frac{\Delta t}{\sqrt{2\pi}} z^{-1/2} \sum_{n=0}^{\infty} J_1(kc(n+1/2)\Delta t) z^{n+1/2} \ . \qquad (70)$$

Similarly, if $\hat{W}(z) = \frac{\Delta t}{\sqrt{2\pi}} \sum_{n=0}^{N+1} (u^{n-1/2} + u^{n+1/2})/2z^n$, in the trapezoidal case (66) we have:

$$X = \frac{\rho c}{\Delta t} \int_{-\pi}^{\pi} (1 - \sqrt{2\pi} kc\hat{J}_b(e^{i\theta})) \mid \hat{W}(e^{i\theta}) \mid^2 d\theta \qquad (71)$$

with:

$$\hat{J}_b(z) = \frac{\Delta t}{\sqrt{2\pi}} \sum_{n=0}^{\infty} J_1(kcn\Delta t) z^n \ . \qquad (72)$$

Using finally the following particular case of the Poisson summation formula (see for example Henrici (1977)) for $Re(s) \geq 0$:

$$kc\Delta t \sum_{n \in \mathbb{Z}: nkc\Delta t + \tau \geq 0} e^{-s(nkc\Delta t + \tau)} J_1(nkc\Delta t + \tau) =$$

$$\sum_{n=-\infty}^{\infty} f(s + \frac{2i\pi n}{kc\Delta t}) \exp(\frac{2i\pi n\tau}{kc\Delta t}) \qquad (73)$$

with:

$$f(s) = \int_0^{\infty} e^{-st} J_1(kct) dt = \frac{1}{kc}(1 - \frac{s}{\sqrt{k^2c^2 + s^2}}) \qquad (74)$$

we find that in the midpoint case (65) (here $\tau = 1/2$ in (73)):

$$X = \frac{\rho c}{\Delta t} \int_{-\pi}^{\pi} ((\cos^2(\theta/2) - \cos(\theta/2)$$

$$(1+\sum_{n=-\infty,n\neq 0}^{\infty}\frac{(-1)^{n+1}}{\sqrt{(\frac{2\pi n-\theta}{kc\Delta t})^2-1}(\sqrt{(\frac{2\pi n-\theta}{kc\Delta t})^2-1}+1)}))$$

$$|\,\hat{U}(e^{i\theta})\,|^2 + \text{terms in } b \text{ and } b^2)d\theta \qquad (75)$$

which can be < 0 for $b = 0$, and in the trapezoidal case (66) (here $\tau = 0$ in (73)):

$$X =$$

$$\frac{\rho c}{\Delta t}\int_{-\pi}^{\pi}\sum_{n=-\infty,n\neq 0}^{\infty}\frac{1}{\sqrt{(\frac{2\pi n-\theta}{kc\Delta t})^2-1}(\sqrt{(\frac{2\pi n-\theta}{kc\Delta t})^2-1}+1)}\,|\,\hat{W}(e^{i\theta})\,|^2\,d\theta \quad (76)$$

which is always ≥ 0 as in the continuous case. So the trapezoidal rule is stable. For the midpoint rule we cannot conclude by the present analysis, but an experiment made by Petit and reported in Fig. 1 shows that good results can be obtained. The computation is made with an homogeneous model of depth $z^{max} = 500m$, with $\rho = 1g/cm^3$, $v_P = 1500m/s$, the source is located at $z_s = 400m$, the source wavelet being $s(t) = \frac{d^2}{dt^2}e^{-\alpha(t-t_0)^2}$, with $\alpha = \pi^2 f^2$ and $f = 30Hz$, $t_0 = 53.39ms$. The depth of geophones is $350m$ and the intertrace $100m$. The parameters of the numerical computation are $k^{max} = 1.86m^{-1}$, $\Delta k = 0.0018m^{-1}$, $\Delta z = 1.25m$, $\Delta t = 0.5ms$.

5 Conclusions and Acknowledgements

We are aware of that we have only sticked pins into the true inverse seismic problem in the framework of horizontally layered media. However we can conclude into two points.

The first is that the usual least-squares formulation of the inverse seismic problem for classic acquisition data (surface or marine data) is insufficient, although the choice of the P-wave travel time to measure the depth, improves the situation in a manual picking procedure. An automatic procedure probably requires a combination of some kind of layer stripping with these least-squares techniques. Note also that the results of Petit in an other work show that the usual least-squares work for VSP data.

The second conclusion is that the $3D$ formulation of the inversion of VSP data is a true inverse problem. A starting point of its theoretical analysis could be the work by Romanov. As subsidiary study to this problem, we have obtained, using essentially the Green formula and the summation formula of Bessel functions, the exact transparent boundary conditions in the time domain adapted to $1D$ medium with $3D$ propagation. Using the Poisson summation formula we have then shown the stability of a discretization of one of these conditions.

We thank D. Macé, F. Jurado, D. Lebrun and V. Richard from the Institut Français du Pétrole and the organizers of the congress in particular G. Chavent and P.C. Sabatier.

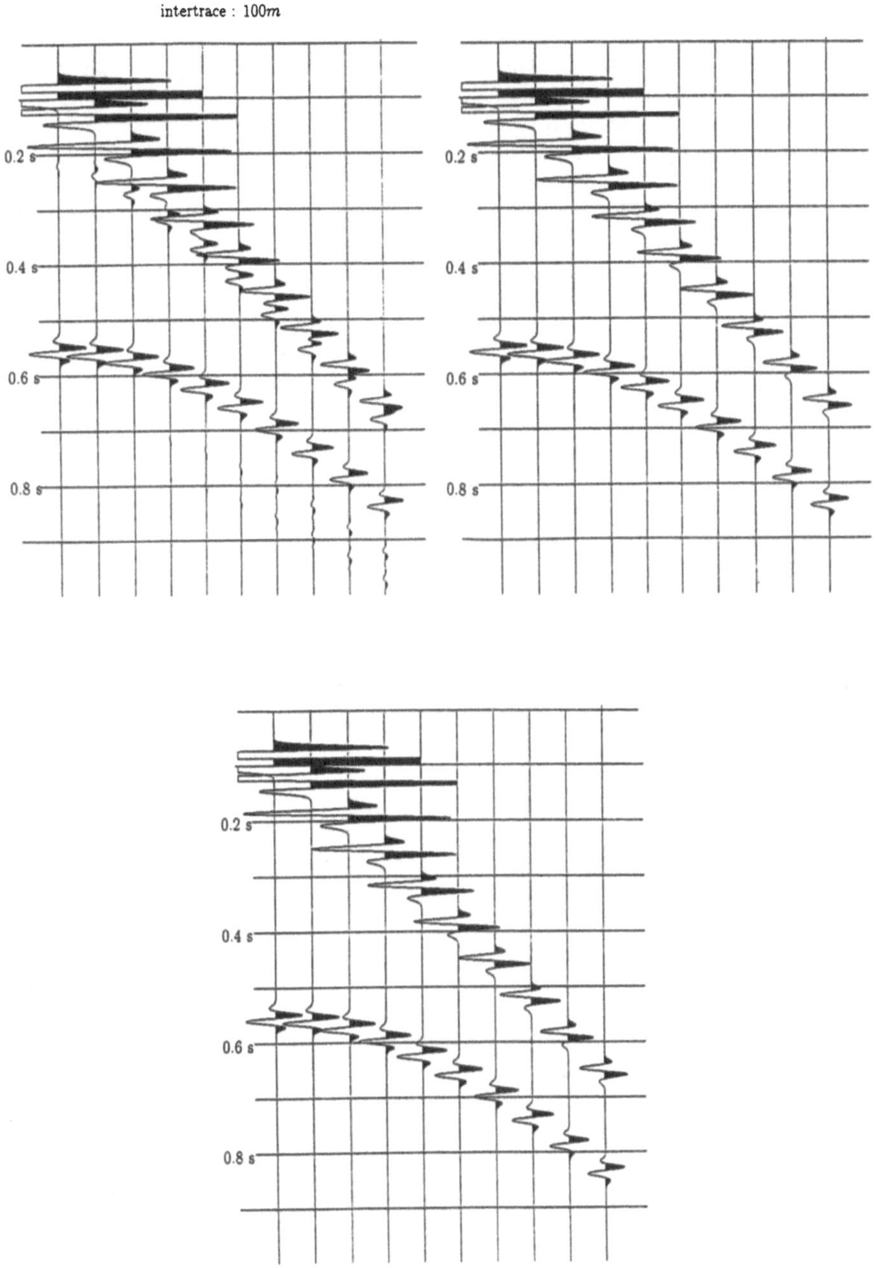

Fig. 1. This is a result obtained in an homogeneous domain with the first order absorbing condition (top left), with the discretized exact transparent condition (65) (top right) and using a sufficiently large domain (bottom)

References

Bamberger A., Chavent G., Lailly P. (1979): About the stability of the inverse problem in 1-D wave equations: Application to the interpretation of seismic profiles. Appl. Math. Optim., **5**, 1–47.

Bube K.P., Burridge R. (1983): The one-dimensional inverse problem of reflection seismology. Siam Rev., **25**, 497–559.

Candel S.M. (1981): Simultaneous calculation of Fourier-Bessel transforms up to order N. J. of Comp. Phys., **44**, 243–261.

Carazzone J.J. (1986): Inversion of $P-SV$ seismic data. Geophysics, **51**(5), 1056–1068.

Chavent G. (1974): Identification of functional parameters in partial differential equation: in "Identification of parameter distributed systems", Goodson, R.E. and Polis, Eds.. NY, ASME.

Colton D., Kress R. (1983): Integral equation methods in scattering theory. John Wiley & Sons, N.Y.

Cuer M., Petit J.L. (1995): Schémas de discrétisation de conditions transparentes exactes pour la propagation $3D$ des ondes sismiques en milieu stratifié $1D$ cas acoustique). Communication au 27 ième Congrès National d'Analyse Numérique, 179–180.

Engquist B., Majda A. (1977): Absorbing boundary conditions for the numerical simulation of waves. Math. Comp., **31**(139), 629–651.

Engquist B., Majda A. (1979): Radiation boundary conditions for acoustic and elastic wave calculations. Comm. Pure and Appl. Math., **32**, 313–357.

He S. (1991): Factorization of a dissipative wave equation and the Green functions technique for axially symmetric fields in a stratified slab. J. Math. Phys. **33**(3), 953–966.

He S., Karlsson A. (1993): Time domain Green function technique for a point source over a dissipative stratified half-space. Radio Science, **28**(4), 513–526.

He S., Ström A. (1992): The electromagnetic inverse problem in the time domain for a dissipative slab and a point source using invariant imbedding: Reconstruction of the permittivity and conductivity: J. Comput. Appl. Math., **42**, 137–155.

Henrici P. (1977): Applied and computational complex analysis, Vol 2, 270–276. John Wiley & Sons, N.Y.

Jurado F., Cuer M., Richard V. (1995a): $1-D$ layered media: Part $1, 3-D$ elastic modeling. Geophysics, **40**(6), 1843–1856.

Jurado F., Cuer M., Richard V. (1995b): $1-D$ layered media: Part 2, Layer-based waveform inversion. Geophysics, **40**(6), 1857–1869.

Lavrent'ev M.M., Reznitskaya K.G., Yakhno V.G. (1986): One-dimensional inverse problems of mathematical physics. American Mathematical Society Translations, Series 2, Volume 130, Providence, Rhode Island.

Lefebvre D. (1985): Inversion of VSP data. 55th Ann. Internat. Mtg., Soc. Expl. Geophys., Expanded Abstract, 69–72.

Lions J.L. (1968): Contrôle de système gouvernés par des équations aux dérivées partielles. Dunod, Paris.

Macé D., Lailly P. (1986): Solution of the VSP one-dimensional inverse problem. Geophysical Prospecting, **34**, 1002–1021.

Madariaga R., (1976): Dynamics of an expanding circular fault. Bull. Seis. Soc. Am., **66**, 639–666.

Mari J.L., Coppens F. (1989): La sismique de puits. Editions Technip, Paris.

Petit J.L., Cuer M. (1994): Conditions transparentes exactes pour la propagation $3D$ des ondes sismiques en milieu stratifié. Communication au 26 ième Congrès National d'Analyse Numérique, 207–209.

Smirnov V.I. (1964): A course of higher mathematics, Vol II: Pergamon Press, Oxford.

Virieux J. (1986): $P-SV$ wave propagation in heterogeneous media: velocity-stress finite-difference method. Geophysics, **51**, 889–901.

Yagle A.E., Levy B.C. (1985): A layer-stripping solution of the inverse problem for a one dimensional elastic medium. Geophysics, **50**(3), 425–433.

Yilmaz O. (1987): Seismic data processing. Investigations in Geophysics no 2, Society of Exploration Geophysicists, Tulsa, OK.

Watson, G.N. (1922): A treatise on the theory of Bessel functions. Cambridge at the University Press, London.

Scattering of Guided Waves in Laterally Varying Layered Media

Fabian Ernst and Gérard Herman

Department of Applied Mathematics, Delft University of Technology, Mekelweg 4, 2628 CD Delft, The Netherlands

Abstract. A major part of the unwanted signals in seismic data consists of waves which are scattered at heterogeneities near the surface of the earth. In the present paper we present a method to remove these waves. We formulate the scattering process within the distorted Born approximation in a background medium consisting of thin, laterally varying, layers. It is vital to be able to compute the propagation in the background medium in an efficient way. To this extent a modal approach is used, where the Green's functions consist of vertical modes and horizontal rays. The formulation for the scattered field then consists of factors accounting for lateral propagation, and a factor representing mode-to-mode scattering.

1 Introduction

In geophysical exploration, one tries to obtain an image of the structure of the deeper parts of the Earth in order to assess the probability of hydrocarbons being present. For this purpose seismic reflection methods are being used, where a source at the surface generates a wavefield which is scattered by the Earth and recorded by receivers located at the Earth's surface. Unfortunately, the strongest signals in these data are often caused by scattering of the wavefield at heterogeneities in the near-surface region. This part of the signal contains no information on the deeper layers and can therefore be considered as "coherent noise". In order to remove this type of noise, methods like "statics" or "f-k-filtering" (Yilmaz, 1987) are currently being employed. These methods have been very successful in some areas, however, they fail in regions with a rather complex near-surface structure.

Recently a theory for removing this type of scattering in a laterally invariant medium has been developed which was based on wave theory (Blonk and Herman, 1994). This method has been applied succesfully to seismic field data (Blonk et al., 1995). In the present paper, this method is extended to the case where the assumption of lateral invariance is no longer valid. We assume that lateral variations in the background medium occur on a length scale which is large compared to the dominant wavelength. In the vertical direction we assume that the near surface consists of thin layers, i.e., the thickness of the layers is of the same order of magnitude as the dominant wavelength.

The method we use consists of four basic steps (Ernst and Herman, 1995):

1. Estimate certain characteristics of the source wavelet and the background medium.
2. Estimate a scatterer distribution in the near surface which is *consistent* with the data.
3. Compute the scattered field arising from this near surface scatterer distribution.
4. Remove the computed scattered field from the seismic data.

In each of the first two steps we have to solve an inverse problem. Especially the inverse problem in step 2 is ill-posed, in the sense that solutions are non-unique. However, we are not interested in recovering the *actual* distribution, but merely in finding a distribution which is *consistent* with the data, in order to be able to compute the scattered field in step 3 and remove it from our data in step 4.

We formulate the inverse problem within the distorted Born approximation. In order to be able to compute the scattered field it is vital to be able to compute the relevant Green's functions of the background medium in an efficient manner. To this extent, a formulation is used for the Green's function in laterally varying media that is based on a modal expansion. The accuracy of this approximation is investigated by means of reciprocity. Finally, the full expression for the scattered field is derived, consisting of factors accounting for lateral propagation and mode-to-mode scattering.

2 Formulation of the Scattering Problem

2.1 The Forward Problem

The key step in our method is to find a scatterer distribution which can explain (the near-surface scattered part of) our data. To this extent, we use a data fitting approach which uses all data available. The method is derived for the acoustic case, however, the derivation for the elastodynamic case is analoguous.

We suppose that the acoustic velocity in the medium can be split into a long-wavelength (background) part c_0 and a short-wavelength part δc. The total pressure field p_{tot} can then be written as the sum of an incident field p_{inc}, arising from the action of the point source on the background medium, and a scattered field p_{sc} due to the interaction with δc. Both incident and total pressure field are solutions of the Helmholtz equation in a halfspace:

$$\Delta p + \frac{\omega^2}{c^2} p = -W(\omega)\delta(\mathbf{x} - \mathbf{x_s}), \tag{1}$$

where $c = c_0$ for the incident field p_{inc} and $c = c_0 + \delta c$ for the total field p_{tot}, $\delta(\mathbf{x})$ denotes the Dirac delta function, and $W(\omega)$ is the frequency-domain representation of the source waveform. Together with appropriate boundary conditions at the pressure-free surface and at infinity, and with continuity conditions of pressure and velocity at interfaces, we have a unique solution p.

2.2 The Inverse Problem

Denoting $P_{sc}^{est}(\mathbf{r}, t, \mathbf{s})$ as the scattered field at time t arising from the action of a source located at $\mathbf{s}$ acting on the near-surface scatterer distribution ξ^{est} and recorded at receiver $\mathbf{r}$, and $P_{tot}^{act}(\mathbf{r}, t, \mathbf{s})$ as our field data, our aim is to minimize the error functional:

$$\min_{[\xi^{est}]} \sum_{\mathbf{s}} \sum_{\mathbf{r}} \int_t dt \parallel H(t)(P_{tot}^{act}(\mathbf{r}, t, \mathbf{s}) - P_{sc}^{est}(\mathbf{r}, t, \mathbf{s})) \parallel_2^2, \tag{2}$$

where $H(t)$ is a weight function which can be used to suppress parts of the data which contain no near-surface scattered field and/or as a preconditioner.

In the frequency domain equation (2) can be written as follows:

$$\min_{[\xi^{est}]} \sum_{\mathbf{s}} \sum_{\mathbf{r}} \int_\omega d\omega \parallel h(\omega) * (p_{tot}^{act}(\mathbf{r}, \omega, \mathbf{s}) - p_{sc}^{est}(\mathbf{r}, \omega, \mathbf{s})) \parallel_2^2, \tag{3}$$

where the $*$ denotes convolution and the lowercase variables are the Fourier transforms of their uppercase equivalents.

2.3 An Integral Representation for the Scattered Field

An integral representation for the scattered field $p_{sc} = p_{tot} - p_{inc}$ has been given by many authors (see for instance Clayton and Stolt (1981)) and is given by:

$$p_{sc}(\mathbf{r}, \omega, \mathbf{s}) = -\omega^2 \int_D dx G_0(\mathbf{r}, \omega, \mathbf{s}) p_{tot}(\mathbf{x}, \omega, \mathbf{s}) \delta m(\mathbf{x}), \tag{4}$$

where

$$\delta m = \frac{1}{(c_0 + \delta c)^2} - \frac{1}{c_0^2}. \tag{5}$$

In (4), D is the (finite) domain containing the scatterers, and G_0 is the Green's function of the layered background medium.

On the assumption that the scatterers are limited in strength, we can linearize the total field around the incident field. Retaining only the zeroth-order term leads to the distorted Born approximation:

$$p_{sc}(\mathbf{r}, \omega, \mathbf{s}) = -\omega^2 W(\omega) \int_D d\mathbf{x} G_0(\mathbf{r}, \omega, \mathbf{x}) G_0(\mathbf{x}, \omega, \mathbf{s}) \delta m(\mathbf{x}), \tag{6}$$

where we have written $p_{inc}(\mathbf{r}, \omega, \mathbf{s}) = W(\omega) G_0(\mathbf{r}, \omega, \mathbf{s})$. The integral representation for the scattered field is now reduced to a readily solvable integral, provided we can compute the Green's function of the background medium.

The remainder of this paper is devoted to an efficient way of computing this Green's function of a laterally varying background medium, consisting of thin layers.

3 Computation of the Green's Function

In this section we derive an expression for the Green's function of a laterally varying medium, where the length scales of the lateral variations are large compared to the dominant wavelength.

3.1 The Modal Approach

Our derivation is based on a modal approach, and can as such be considered similar to the approaches which have been used in integrated optics (Marcuse, 1974), underwater acoustics (Weinberg and Burridge, 1974), and global seismology (Woodhouse, 1974) for many years.

The essence of the modal approach is a splitting of the wave field in lateral and vertical components. Therefore we will make the following "ansatz" for the wavefield away from the source region:

$$G_0(\mathbf{x}, \omega, \mathbf{s}) = \sum_m L_m(\mathbf{x_h}, \omega, \mathbf{s_h}) \phi_m(x_z, \omega, \mathbf{x_h}). \tag{7}$$

This formulation is the concept of "local modes" (see also (Marcuse, 1974)), i.e., modes which satisfy a local eigenproblem, but are no solution of the total Helmholtz equation. The subscript h denotes the two lateral components of the vector, and the subscript z denotes the vertical component. Effects of the source are accounted for in initial values for L_m; this is discussed later in this paper. Note that we have here a summation over all existing modes.

3.2 High-frequency Approximation

Under the assumption that the characteristic length scale L of the lateral variations is small compared to the dominant wavelength, we may assume that the lateral part of the Green's function L_m can be written in its high-frequency asymptotic form (see also Blok (1979)):

$$L_m(\mathbf{x_h}, \omega, \mathbf{s_h}) = \exp(-i\frac{\omega L}{c^0}\theta_m(\mathbf{x_h}, \omega, \mathbf{s_h})) \sum_p \left(\frac{\omega L}{c^0}\right)^p A_{mp}(\mathbf{x_h}, \omega, \mathbf{s_h}), \tag{8}$$

where θ_m is the phase function, A_{mp} are the p-th order amplitudes, and c^0 is a reference velocity of the same order of magnitude as the velocities occurring in the actual medium. We assume that the θ_m and A_{mp} vary slowly in lateral direction, i.e., $\nabla_h \theta_m = O(\frac{c^0}{\omega L})$ and $\nabla_h A_{mp} = O(\frac{c^0}{\omega L})$.

We can now derive expressions for the various parts of the far-field Green's function by substituting expressions (7) and (8) into (1).

3.3 Local Normal Modes

Consider first the solutions $\phi_m(x_z, \omega, \mathbf{x_h})$ of the local eigenvalue problem:

$$\frac{d^2\phi_m}{dz^2} + \left\{\frac{\omega^2}{c_0^2} - \kappa_m^2\right\}\phi_m = 0,\ \forall m, \tag{9}$$

together with appropriate boundary conditions at $z = 0$ and $z = \infty$. The eigenvalue κ_m can be interpreted as the horizontal wavenumber, and the eigenfunction $\phi_m(x_z, \omega, \mathbf{x_h})$ can be considered as the shape of the mode. It can be shown that the modes are orthogonal with respect to the inner product $\langle\phi_m, \phi_n\rangle = \int_0^\infty \phi_m(x_z, \omega, \mathbf{x_h})\phi_n(x_z, \omega, \mathbf{x_h})dx_z$, and they can be normalized to unit energy, so that we have

$$\int_0^\infty \phi_m(x_z, \omega, \mathbf{x_h})\phi_n(x_z, \omega, \mathbf{x_h})dx_z = \delta_{mn}. \tag{10}$$

3.4 Lateral Propagation

To find an expression for the lateral propagation L_m of the modes, we substitute (7) in the Helmholtz equation, and use the definition of the local normal modes (9). After multiplication by $\phi_n(x_z, \omega, \mathbf{x_h})$, integration over the x_z-coordinate and with the aid of the orthogonality relation (10), we arrive at

$$\begin{aligned} &\Delta_h L_n(\mathbf{x_h}, \omega, \mathbf{s_h}) + \kappa_n^2 L_n(\mathbf{x_h}, \omega, \mathbf{s_h}) = \\ &-\sum_m \int_0^\infty dx_z \{2\nabla_h L_m(\mathbf{x_h}, \omega, \mathbf{s_h}) \cdot \phi_n(x_z, \omega, \mathbf{x_h})\nabla_h\phi_m(x_z, \omega, \mathbf{x_h}) \\ &+ L_m(\mathbf{x_h}, \omega, \mathbf{s_h})\phi_n(x_z, \omega, \mathbf{x_h})\Delta_h\phi_m(x_z, \omega, \mathbf{x_h})\}. \end{aligned} \tag{11}$$

The left hand side of this equation has the standard form for propagation of a single mode in a medium, whereas the right hand side accounts for the coupling of the modes. The equations for the phase θ_n and the zeroth-order amplitude A_{n0} can now be derived by substituting equation (8) into (11) and collecting terms of lowest order in $\frac{c^0}{\omega L}$. For terms of $O(1)$ this leads an eikonal equation for the phase θ_n:

$$\kappa_n^2 - (\frac{\omega L}{c^0}\nabla_\mathbf{h}\theta_n)^2 = 0, \tag{12}$$

so we see that the coupling of the modes has no influence on the phase of the modes, and the eigenvalue κ_n can be interpreted as the horizontal wavenumber for the n'th mode.

Collecting terms of $O(\frac{c^0}{\omega L})$ leads to an amplitude equation for the zeroth-order amplitude A_{n0}:

$$2\nabla_{\mathbf{h}} A_{n0} \cdot \nabla_{\mathbf{h}} \theta_n + A_{n0} \Delta_h \theta_n = -2 \int_0^\infty dx_z \phi_n(x_z, \omega, \mathbf{x_h}) \times \tag{13}$$
$$\sum_m \left(\nabla_{\mathbf{h}} \theta_m \cdot \nabla_{\mathbf{h}} \phi_m(x_z, \omega, \mathbf{x_h}) \exp(-\frac{i\omega L}{c^0}(\theta_m - \theta_n)) A_{m0} \right).$$

If we neglect the effects of mode coupling, which will be investigated later on, and note that the right-hand side vanishes for $m = n$ due to our choice of normalization, equation (13) reduces to the conventional ray-tracing amplitude equation

$$2\nabla_{\mathbf{h}} A_{n0} \cdot \nabla_{\mathbf{h}} \theta_n + A_{n0} \Delta_h \theta_n = 0. \tag{14}$$

The resulting first order approximation for the Green's function of a laterally varying medium can be written as:

$$G_0 = \sum_m A_{m0} \exp(-i\theta_m) \phi_m, \tag{15}$$

which implies that we have normalized modes in vertical direction, and an amplitude and phase factor which can be computed by conventional ray tracing methods (see for example (Cervený et al., 1988)) in lateral directions.

3.5 Accounting for the Source Strength

Now that we have the three basic equations (9), (12) and (14) we can compute the Green's function. However, for a complete expression we need initial values for the amplitude and the phase. These initial values can be derived from the presence of the source by means of the method of matched asymptotic expansions (Kevorkian and Cole, 1981). The expressions to be matched are the *outer expansion* based on equation (15) and the expression for the field due to a point source in a homogeneous medium; this *inner expansion* has the following standard form (Aki and Richards, 1980), (Ernst and Herman, 1995):

$$G_0^{\text{inner}} = \sum_m i H_0^{(2)}(\kappa_m r) \kappa_m \phi_m(x_z, \omega, \mathbf{x_h}) \phi_m(s_z, \omega, \mathbf{s_h}) \left(\frac{dD}{d\kappa} \right)^{-1}_{\kappa=\kappa_m} \tag{16}$$
$$+ \frac{1}{\pi i} \int_B \kappa H_0^{(2)}(\kappa r) g(s_z, x_z, \kappa) d\kappa.$$

From the zeroth-order and first-order matching conditions we find the initial values for the phase function θ_m and zeroth-order amplitude A_{m0}:

$$\theta_m(\mathbf{s_h}, \omega, \mathbf{s_h}) = 0, \tag{17}$$

$$\lim_{\mathbf{x_h} \to \mathbf{s_h}} \sqrt{\| \mathbf{x_h} - \mathbf{s_h} \|} A_{m0}(\mathbf{x_h}, \omega, \mathbf{s_h}) = a_{m0} \phi_m(s_z, \omega, \mathbf{s_h}), \tag{18}$$

where a_{m0} is given by

$$a_{m0} = \sqrt{\frac{\kappa_m}{2\pi}} \exp(\frac{3\pi}{4} i) \left(\frac{dD}{d\kappa}\right)^{-1}\Bigg|_{\kappa=\kappa_m} . \tag{19}$$

3.6 Example

In order to verify the assumption concerning mode coupling underlying the step from equation (13) to (14) we have carried out some numerical tests. This verification is done by means of reciprocity. For a given structure of the background model, we have the reciprocity relation

$$p(\mathbf{r}, \omega, \mathbf{s}) = p(\mathbf{s}, \omega, \mathbf{r}) \tag{20}$$

for the total pressure field, i.e., summed over all modes and without neglecting mode-coupling. We have computed the pressure field for three cases with a homogeneous layer over an homogeneous halfspace. We put our source and receiver at $x_A = (0, 0, 2)$ and $x_B = (1000, 0, 4)$.

The choice for the parameters in this background model are:
Model A: Layer thickness h varying linearly from 40 m at the source position to 20 m at the receiver position. Acoustic velocities in upper and lower layer are constant at 750 m/s and 1600 m/s, respectively.
Model B: Constant layer thickness of $h = 40\ m$. The acoustic velocity in the upper layer varies linearly from 750 m/s at the source to 1200 m/s at the receiver. In the lower layer the acoustic velocity is constant at 1600 m/s.
Model C: Constant layer thickness of $h = 40\ m$. The acoustic velocity in the upper layer is constant at 750 m/s, whereas in the lower layer it varies linearly from 1600 m/s at the source to 900 m/s at the receiver.
These variations have a characteristic order of magnitude for the problems we are interested in.

The source wavelet is a bandlimited wavelet with unit strength between 10 and 25 Hz, and tapered smoothly towards zero at the edges by means of a $\cos^2$-function.

In figure 1 the responses for model A, B, and C are shown. The solid line denotes the response of the receiver at x_B for the source at x_A, and the dashed line denotes the response for the opposite case. From these figures it becomes clear that our formulation (without mode coupling) satisfies the reciprocity relation quite well.

4 Computation of the Scattered Field

The total expression for the scattered field in the Born approximation now follows from expressions (6) and (15):

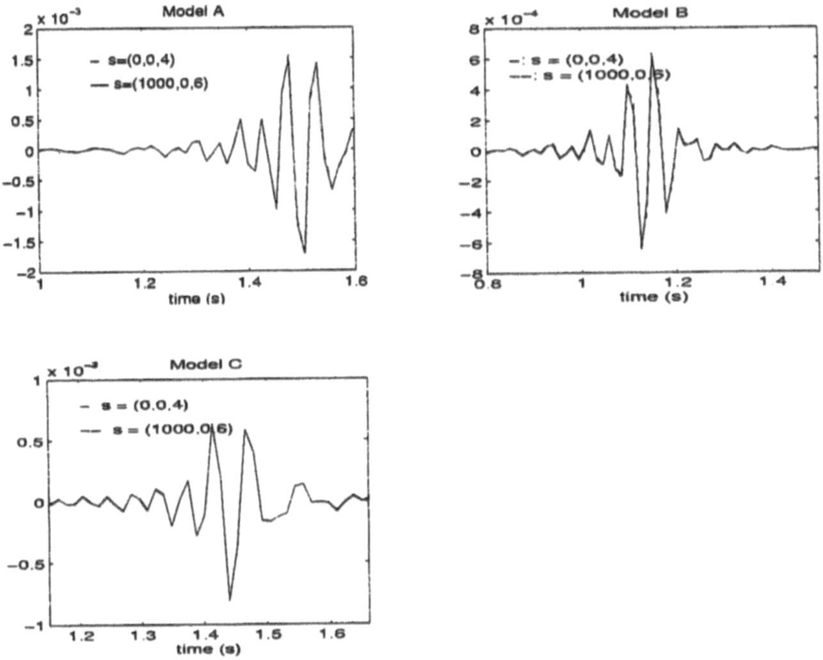

Fig. 1. Reciprocity tests for three laterally varying models. In model A we have variation of layer thickness, in model B variation of acoustic velocity in the upper layer and in model C variation of acoustic velocity in the lower layer. It is seen that our first order approximation without mode coupling is accurate enough for the models of interest.

$$
\begin{aligned}
p_{sc}(\mathbf{r},\omega,\mathbf{s}) = & -\omega^2 W(\omega)\int_D d\mathbf{x}\delta m(\mathbf{x}) \\
& \times \sum_m \phi_m(x_z,\omega,\mathbf{x_h})A_{m0}(\mathbf{x_h},\omega,\mathbf{s_h})e^{-i\omega\tau_m(\mathbf{x_h},\omega,\mathbf{s_h})}\phi_m(s_z,\omega,\mathbf{s_h}) \\
& \times \sum_n \phi_n(r_z,\omega,\mathbf{r_h})A_{n0}(\mathbf{r_h},\omega,\mathbf{x_h})e^{-i\omega\tau_n(\mathbf{r_h},\omega,\mathbf{x_h})}\phi_n(x_z,\omega,\mathbf{x_h})
\end{aligned}
\tag{21}
$$

For numerical computations, we need to discretize the scatterer distribution. Assume therefore that we have

$$
\delta m(\mathbf{x}) = \delta m_i, \quad x \in D_{h,i} \times D_{x_z,i} \tag{22}
$$

with

$$
D = \bigcup_i (D_{h,i} \times D_{x_z,i}), \tag{23}
$$

Substituting this expression in (22) we arrive at

$$p_{sc}(\mathbf{r},\omega,\mathbf{s}) = -\omega^2 W(\omega)\sum_i \int_{D_{h,i}} d\mathbf{x_h}\delta m_i \tag{24}$$
$$\times \sum_m A_{m0}(\mathbf{x_h},\omega,\mathbf{s_h})e^{-i\omega\tau_m(\mathbf{x_h},\omega,\mathbf{s_h})}\phi_m(s_z,\omega,\mathbf{s_h})$$
$$\times \sum_n A_{n0}(\mathbf{r_h},\omega,\mathbf{x_h})e^{-i\omega\tau_n(\mathbf{r_h},\omega,\mathbf{x_h})}\phi_n(r_z,\omega,\mathbf{r_h})$$
$$\times \qquad M_{mn,i}(\omega,\mathbf{x_h}),$$

where

$$M_{mn,i}(\omega,\mathbf{x_h}) = \int_{D_{x_z,i}} \phi_m(x_z,\omega,\mathbf{x_h})\phi_n(x_z,\omega,\mathbf{x_h})dx_z \tag{25}$$

is the scattering pattern from mode m to mode n due to a unit scatterer at lateral position $\mathbf{x_h}$, which has the form of an inproduct of modes weighed by the vertical extension of the scatterer domain.

Introducing the matrices M_i, G^s and G^r, and the vectors $\mathbf{S}$ and $\mathbf{R}$ as follows:

$$M_{mn,i} = M_{mn,i}(\omega,\mathbf{x_h}) \tag{26}$$
$$G^s_{mp} = A_{m0}(\mathbf{x_h},\omega,\mathbf{s_h})\exp(-i\omega\tau_m(\mathbf{x_h},\omega,\mathbf{s_h})\delta_{mp} \tag{27}$$
$$G^r_{nq} = A_{n0}(\mathbf{r_h},\omega,\mathbf{x_h})\exp(-i\omega\tau_n(\mathbf{r_h},\omega,\mathbf{x_h})\delta_{nq} \tag{28}$$
$$\mathbf{S}_m = \phi_m(s_z,\omega,\mathbf{s_h}) \tag{29}$$
$$\mathbf{R}_n = \phi_m(r_z,\omega,\mathbf{r_h}) \tag{30}$$

we can write the scattered field in compact notation as:

$$p_{sc}(\mathbf{r},\omega,\mathbf{s}) = -\omega^2 W(\omega)\sum_i \delta m_i \int_{D_{h,i}} \mathbf{S}^T\left(G^s\right)^T \cdot M_i \cdot G^r \mathbf{R} d\mathbf{x_h}, \tag{31}$$

where T denotes the transpose of a matrix.

5 Application to Simulated Data

We now illustrate our method by applying it to simulated data resembling real seismic data. The model considered consists of a layer with a thickness of 30 meters overlying a homogeneous halfspace with an acoustic velocity of 1600 m/s. The acoustic velocity in is 600 m/s in the southwestern corner and increases linearly to 900 m/s in the northwestern corner.

The acquisition geometry consists of a source line in east-west direction and a receiver line in north-south direction. The sources are located 300 m and 750 m respectively from the middle of the receiver line. The spacing between the receiver line is 15 m.

The scattering area is 500 m by 500 m, and is discretized into blocks of 20 m by 20 m. The distance from the scattering area is 200 m to both the source line and the receiver line. A top view of this geometry is depicted in figure 2.

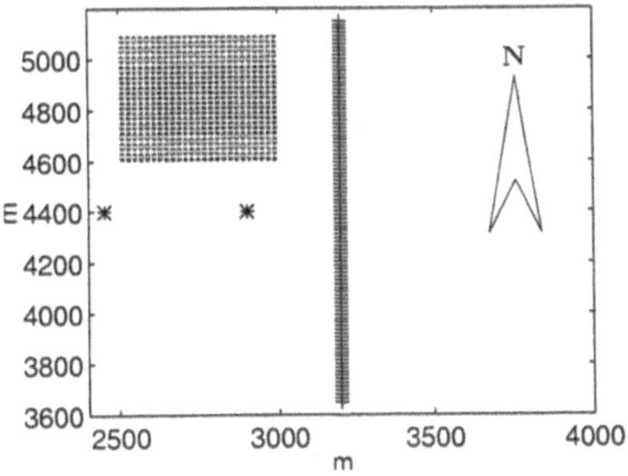

Fig. 2. A top view of the acquisition geometry. The scatterers are located in the shaded area in the northwestern part of the region. A * denotes a source; a + a receiver.

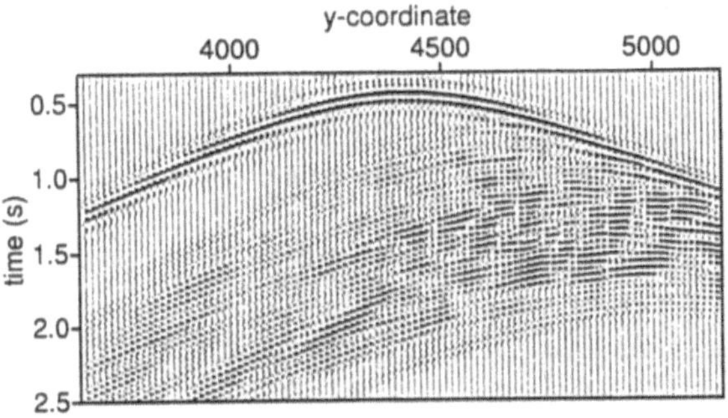

Fig. 3. Input data, shot 2. Note the dispersive character of the incoming as well as of the scattered waves. Reflection from deeper layers, if present, would not be visible in the right-hand part of the figure due to the strong scattered waves.

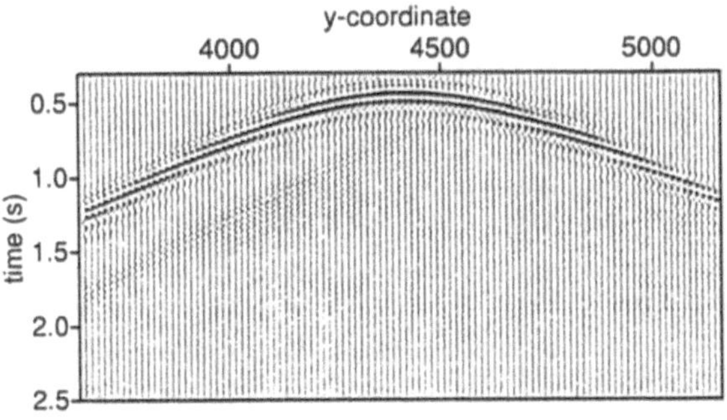

Fig. 4. Resulting data after application of our method. Although the direct wave was present in the inversion step, it has hardly been affected, whereas the scattered waves have been severely attenuated.

The simulated data are depicted in figure 3. Note the dispersive character of the incident and scattered wave, as well as the complex interference pattern. After estimating an optimal scatterer distribution and subtracting the scattered waves, we arrive at figure 4. Although the direct wave was still present in the inversion step, we see that the scattered waves have been strongly attenuated without affecting the direct wave.

6 Discussion

In this paper a method is discussed for the removal of scattered guided waves from seismic data by means of a data fitting procedure.

An important aspect of this method is an accurate and efficient computation of Green's function for laterally varying media. This computation is performed by means of an expansion in local normal modes, where propagation in lateral directions can be described by high-frequency asymptotics, i.e., a standard ray-tracing approach. Suitable initial conditions can be derived from the presence of the point source. By numerically verifying a reciprocity relation, we have found that neglecting mode coupling is accurate enough for our purposes.

This formulation for the forward problem gives then rise to a formulation for the scattered field in the distorted Born approximation which consists of factors accounting for lateral propagation in the background medium and a factor accounting for mode-to-mode scattering.

Acknowledgement

This research was carried out with financial support from the Dutch Technology Foundation STW and Shell.

References

Aki, K. and Richards, P. (1980). *Quantitative Seismology, Theory and Methods.* W.H. Freeman, San Francisco.

Blok, H. (1979). Ray theory of electromagnetic surface-wave modes in integrated optical systems. Radio Science, **14**, 333–339.

Blonk, B. and Herman, G. (1994). Inverse scattering of surface waves: A new look at surface consistency. Geophysics, **59**, 963–972.

Blonk, B., Herman, G., and Drijkoningen, G. (1995). An elastodynamic inverse scattering method for removing scattered surface waves from field data. Geophysics, **60**, 1897–1905.

Cervený, V., Klimeš, L., and Pšencik, I. (1988). Complete seismic-ray tracing in three-dimensional structures. In Doornbos, D., editor, *Seismological Algorithms*, chapter II-1, pages 89–168. Academic, London.

Clayton, R. and Stolt, R. (1981). A Born-WKBJ inversion method for acoustic reflection data. Geophysics, **46**, 1559–1567.

Ernst, F. and Herman, G. (1995). Computation of Green's function of laterally varying media by means of a modal expansion. In *65th Ann. Int. Mtg. Soc. Expl. Geoph., Expanded Abstracts*, pages 623–626. SEG.

Kevorkian, J. and Cole, J. (1981). *Perturbation methods in applied mathematics.* Springer, New York.

Marcuse, D. (1974). *Theory of Dielectric Optical Waveguides.* Academic, New York.

Weinberg, H. and Burridge, R. (1974). Horizontal ray theory for ocean acoustics. J. Acoust. Soc. Am., **55**, 63–79.

Woodhouse, J. (1974). Surface waves in a laterally varying layered structure. Geophys. J. R. Astr. Soc., **37**, 461–490.

Yilmaz, O. (1987). *Seismic Data Processing.* SEG, Tulsa.

Born Inversion in Realistic Backgrounds by Means of Recursive Green's Functions

T.J. Moser [1,2], M. Biryulina[1], and G. Ryzhikov[1]

[1] Institute of Solid Earth Physics, University of Bergen, Norway
[2] now Alexander-von-Humboldt fellow at: Geophysikalisches Institut, Karlsruhe, Germany

Abstract. The commonly applied methods for seismic inversion are based on some drastic assumptions regarding the known background or macro-velocity model and the data acquisition, that limit their applicability in geologies of realistic complexity or to realistic, noisy and incomplete, data sets. The background is usually assumed smooth, often to such a degree that the wave field can be described by simple ray theory without caustics or multipathing; the data are assumed to be complete and noise-free. Correspondingly, the algorithms for ray-based Green's functions are, until now, mostly developed for smooth media. To be able to image in realistic backgrounds and with realistic data sets, the assumptions have to be weakened. This must be done on two fronts: the imaging formula and the Green's functions. A new, generalized, imaging formula has been developed that takes into account that real data are incomplete, noisy and have a limited frequency band. A new approach for Green's functions allows the backgrounds in the inversion to be non-smooth, and accounts for reflected and transmitted ray fields by organizing the ray tracing recursively. Combined, the two approaches allow a systematical target-oriented inversion, in which upper parts of the Earth model are assumed known and fixed, and the attention is concentrated on important details below. The new imaging formula, together with the realistic Green's functions, has been successfully applied on the imaging of a complicated horst structure from the North sea.

Keywords. Born inversion, raytracing, Green's functions

1 Introduction

A seismic inversion in a realistic background remains a challenge even for 2D problems, because of the absence of any a priori symmetry. An advanced inversion formula, reducing inversion to a sort of migration, has been suggested by Beylkin ([2], [3]), and applied with relative success ([7], and others). Nevertheless, this formula has a limited applicability in areas with a realistic complexity, or for realistic seismic data acquisitions. The inversion is based on the generalized Radon transform, which supposes that for a given background and a given source/receiver acquisition, there exists a one-to-one correspondence between a record sample and the parameters of an isochron crossing a background point. Even in a layered background this supposition fails due to existence of up- and downgoing rays, crossing the background point. Besides, the inversion formula is not valid for incomplete data (for instance, reflection data from a narrow aperture acquisition), also it is hard to adapt the formula for data from a few source/receiver pairs, noisy data, narrow frequency band signal, etc. Also, its validity depends strictly on the validity of ray theory, which, despite the computational speed of ray tracing, is rather limited. That is why it is preferable to deal with an inversion based on a general wave field evaluation, rather than explicitly on ray theory. The inversion formula suggested by Ryzhikov and Troyan ([12], [13], [14]) allows to perform "wave field + Born" -, or "wave field + linearized response"- inversion. A detailed derivation is included in the appendix.

Based on this more generalized inversion scheme, this paper presents an attempt to image a complicated horst structure in the North Sea. The background, or the macro-velocity model, is allowed to be non-smooth in this numerical experiment. The Green's functions are provided by the recursive seismic ray modeling scheme ([9], [10], [11]), which is summarized below. To show the viability of the proposed imaging scheme, the inversion of a synthetic data set is presented in this paper; real data are inverted by the same scheme in [10].

2 Inversion

Let us try to summarize why so elegant a concept on the interpretation of dynamical seismic data as the generalized Radon transform fails, when it is applied to raw real data. Mathematically, this is caused first of all by the background's Green's function, appearing in the Lippmann-Schwinger equation. To deal with data in terms of the first Born approximation ('linear'/'single-scattering'/'single-diffracting'), we have to start from the background's Green's function being far from the ray-theoretical one, otherwise we have to include many, or even an infinite number, of terms in the Born series (e.g. a perturbation like a full layer can be represented only by an infinite number of terms). If we want to maintain the concept of an inversion of the generalized Radon transform, then the 2×2 Jacobian of the mapping from local parameters describing an isochron to a generic source/receiver location should exist. Apart from that, there should be a one-to-one correspondence between the pair 'sample of a source-receiver pair record' - 'location + orientation of isochron' (for a 3D problem these are 3 parameters of a plane, tangent to an isochron). These requirements have quite drastic consequences for the inversion. Namely, they imply that:

a) the reference medium should be fairly simple: a signal propagation just in terms of rays (every sample has its own 'travel time'; every unit volume element of the medium provides a piece of information that, for a given source-receiver pair record, is associated just with the pair of rays 'source - medium point - receiver'), no caustics, no shadow zones, locally plane isochrons, etc. (which is not valid even for a stratified reference medium or a medium with a constant velocity gradient). Note that the representation of a (generally curved) isochron by a plane is only locally valid, in a vicinity of the touching point of the tangent plane to the isochron (e.g. for a small object in a far field zone of a near to homogeneous reference medium);

b) the effect of the reference medium perturbation should be linear, which corresponds to a weak and single scattering/diffraction;
c) for a given source the spatial parameters of the receiver should run over the full sphere surrounding the target of the imaging. This may be realizable in technical devices (e.g. tomograph), but it is unfeasible for lots of remote sensing, geophysical, and technical problems, that naturally suffer from a finite aperture. Any attempt to complete the data by an interpolation/extrapolation in data space indirectly induces an a priori representation of reconstructed medium parameters (e.g. a smoothing of data is associated with a corresponding smoothing of unknown media parameters - hard to analyze).

Mathematical problems arise already when the ray-theoretical representation is fairly good, but apart from that, we have to:
1. introduce a non-smooth reference model (e.g. layered medium);
2. use a sounding signal of a finite frequency band, i.e. a signal of finite time duration, not a delta impulse;
3. deal with data from acquisition involving a few sources: it is hard to avoid overlapping of data;
4. interprete data from a realistic experiment design.

Besides, as a rule real data are incomplete: they are not given in continuous-, but in digital form, and the set of receiver positions is finite, sparse or even random. Also the data are noisy, due to variety of known and unknown sources; this makes regularization difficult. But, of course, the main problems are generated by wave phenomena, which depend upon medium supposed to be known a priori and upon the ratios 'dominant wave length/perturbation size/source-object-receiver distances'. Very often such phenomena can not be described properly in terms of the ray theory; the latter is the base to treat dynamical data in terms of integrals over isochrons.

Therefore we follow another way, which is not so strongly associated with the kinematics of first arrivals. We start from realistic assumptions about the data, taking into account their strong incompleteness, i.e., their inadequacy to determine the unknown medium parameters. For example, for seismic exploration the data are represented by unstacked digitized finite-band noisy seismograms, recorded by a sparse and probably random net of sources and receivers. The strategy of our approach is tied to dealing with such segments of data, which can be decoded properly. It means that these segments should be described by a linearized forward model with an accuracy induced by the value of data noise, - the latter is caused by the physical registration channel and unidentified sources. The proper input for the inversion are data residuals: observed data reduced with the wave field evaluated for unperturbed background. Note here, that such a reduced data set can be easily obtained directly in a problem of reconstruction of inclusions in a laterally uniform background or in a problem of seismic monitoring: these are just differential data. The inverse problem is posed as an optimization one: to find the perturbation of a background, that provides the best fitting of corresponding synthetic (linear) responses with input data for all source-receiver pairs. The linear response, caused by a volume unit perturbation consists of: evaluation of the wave field in a background generated by a source (incoming wave field); application of an operator of interaction - the result can be interpreted as a secondary source; evaluation of the wave field (in the background), generated by the secondary source (outgoing wave field). The operator of interaction consists of a Fréchet-derivative of a wave operator with respect to medium parameters; e.g. for the acoustic wave equation this operator is the well-known second time derivative, other examples of Fréchet derivatives/operators of interaction for an arbitrary background are given in [12]. The actual inversion formula is the result of a simplified solution of the optimization problem, and provides the local inversion, just as Beylkin's formula does: to get the result of inversion in a given volume, we need to collect just linear responses of data, induced by unit perturbation in this volume only. It allows to parallelize the inverse problem very efficiently. The formula can be interpreted as follows: to reconstruct the value of perturbation in a given volume, it is necessary to evaluate a correlation coefficient between data residuals for all (relevant) source-receiver pairs and all linear responses induced by a unit perturbation in the given volume element. This correlation coefficient contains a factor, being a ratio of an average amplitude of data residuals and average amplitude of linear responses.

A short summary of our approach is following (a detailed derivation is given in the appendix):

1. No need to deal with ray/isochron representation: the inversion is expressed in terms of in- and outgoing wave fields in a reference medium, no matter how they are evaluated (e.g. by finite differences). This allows an essential reduction of errors, that is to be accounted for by a proper forward model, being a base for inversion. When the approximation ray + Born is adequate to real phenomena, our forward model is the same as the conventional one, except that we take into account directly the effect of registration channel (to be realized a channel should have a finite frequency band).
2. The algebraization/discretization of the problem is straightforward - it corresponds to finite number of records with digital data.
3. The problem is then reduced to an optimization one, that allows to treat ill-posed problems properly.

It is evident that strong perturbations generally can not be expressed in terms of the Born/generalized Born finite series. For example, if the perturbation of a stratified reference medium is represented by a thick layer, it leads to a phase shift with respect to the propagation in the reference medium. This shift can not be described properly by the first term of the Born series, nor by any finite number of terms. However, there are data segments that can be represented well as linear response. For example, when an overburden is known and our problem is to reconstruct an unknown interface of a bedrock. Although the linear response is not valid for the transmitted data, it is then still a good approximation for the short-offset reflected data. In our numerical experiments we supposed that these data can be simulated with the Born-diffractors located in a vicinity of the interface [4]. The solution of a more general 3D nonlinear inverse problem was shown in [14].

3 Recursive construction of Green's functions

Like the one presented in this paper, most inversion schemes heavily rely on the possibility of a computationally fast forward modeling of the wave propagation, or the availability of Green's functions for the wave equation, in a presumed known Earth model (i.e. reference- or background model, see appendix equations [A4] and [A5]). Although the proposed inversion procedure is valid for any type of forward modeling, ray theory still provides by far the fastest algorithms. The ray-theoretical representation of the Green's function is $G(\boldsymbol{x}, \boldsymbol{y}, \omega) = \sum_k A_k(\boldsymbol{x}, \boldsymbol{y}) \exp i\omega T_k(\boldsymbol{x}, \boldsymbol{y})$, where $T_k(\boldsymbol{x}, \boldsymbol{y})$ is the traveltime between two locations $\boldsymbol{x}$ and $\boldsymbol{y}$, $A_k(\boldsymbol{x}, \boldsymbol{y})$ the corresponding (possibly complex-valued) amplitude and the summation over k accounts for multipathing. In recent years, schemes have been developed for ray-based isochron- or wavefront tracing in smooth media without reflecting interfaces ([5], [6], [15], [16], [17]). They construct the Green's function for one fixed point $\boldsymbol{y}$ (for instance a source or receiver point) and for $\boldsymbol{x}$ running over the model, by propagating a set of rays from $\boldsymbol{y}$ and resample the rays at each computed new isochron, in order to achieve a uniform coverage of the model by rays. In such a way receiver captures are easily detected and each time the wavefront passes a receiver point the ray quantities necessary for a local evaluation of the wave field (arrival time, amplitude, phase shift) can be easily assessed by a simple interpolation from neighboring rays. By allowing the wave front to fold over itself at caustics, the algorithms properly take multivaluedness into account. The two-point raytracing problem is thereby eliminated, and the whole model is filled with (possibly multivalued) ray arrivals.

The Born approximation for scattering by an obstacle is valid under the condition that the magnitude of the scattered wave, measured in some norm, is much smaller than that of the incoming- or reference wave. This implies that it is a weak-scattering- and low-frequency approximation (refer to appendix equation [A10]), contrary to the ray-based Green's functions, which are high-frequent approximate. A successful application of Green's functions in Born inversion therefore depends on how small the scatterer is, in other words, the allowed degree of complexity of the background; the more realistic the background, the smaller is the unknown scatterer and the better is the image. Ray-based Green's functions in smooth media impose severe restrictions in this respect, for a variety of reasons. When a smooth background is to be improved in order to approach a non-smooth reality (according to criteria not discussed in this paper), the ray fields become more and more multivalued. As a result, this implies that the computation time, for the Green's functions as well as for the imaging, increases accordingly. Also, along the caustics the ray-amplitude is infinite, so that the image quality is distorted (often only thanks to numerical inaccuracies a complete break-down is prevented). Next, for a sufficient accuracy the integration step-size along the rays has to decrease, increasing again the CPU. Finally, ray theory is not valid any more in smooth, but rapidly varying media, and does not represent any more the actual (finite-frequency) wave field. The effects of such fields on the image quality is shown in the section 4.

All these considerations motivate the introduction of discontinuities in the background and modify the smooth medium's Green's functions algorithms to include reflections and transmissions. One disadvantage of non-smoothness is that, unless diffractions are taken into account (which is expensive), it can introduce shadow zones in the Green's functions. However, we think that, as long as the Green's functions from all data points together provide a sufficient coverage of the target region, imaging with shadow zones, i.e. zero amplitudes, is to be preferred to imaging with caustics, i.e. infinite amplitudes. Apart from diffractions, the ray- and wave field is already considerably more complicated in presence of discontinuities than in smooth models, due to multiple reflections, which themselves may be multivalued again, not only because of smooth model variations, but also because of possible curvature of the discontinuities. The organization and storage of all these different phases can cause formidable difficulties.

Table 1. Synthetic velocity model

Layer	Velocity (km/s)
1	1.5
2	2.2
3	2.9
4	2.4
5	2.9
6	3.4
7	4.2

Table 2. Brage velocity model

Layer	Name	Velocity (km/s)
1	water	1.478
2	upper sediments	1.900
3	Tertiary I	2.020
4	Tertiary II	1.815
5	Cretacious	2.680
6	Balder	2.350
7	Shetland	2.950
8	Draupne	2.900
9	Fensfjord	3.464
10	Brent	3.100
11	fault zone	2.800
12	Shetland	3.100
13	Statfjord	3.500

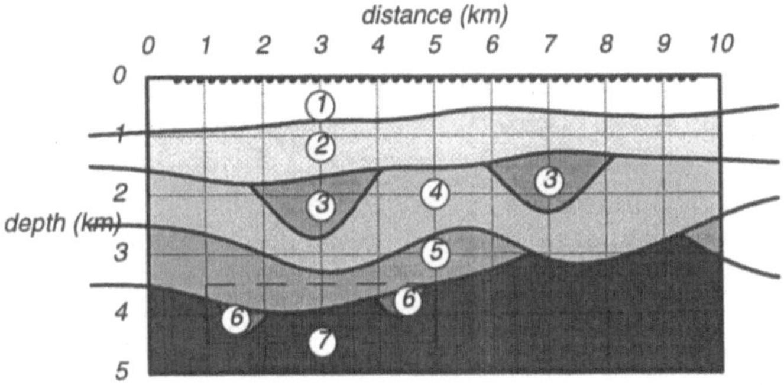

Fig. 1. Synthetic velocity model. The labels correspond to layers tabulated in Table 1. The gray scale represents velocities. The box bounds the target region. The dots along the surface denote the shot/receiver locations.

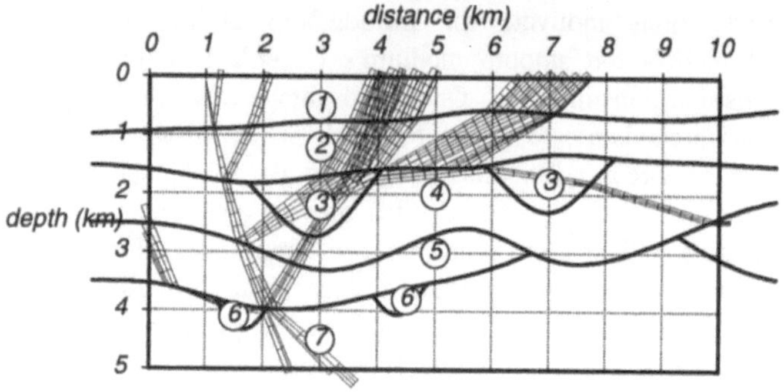

Fig. 2. Recursive raytracing of one ray cell in the synthetic model. Transmissions and first order reflections are shown.

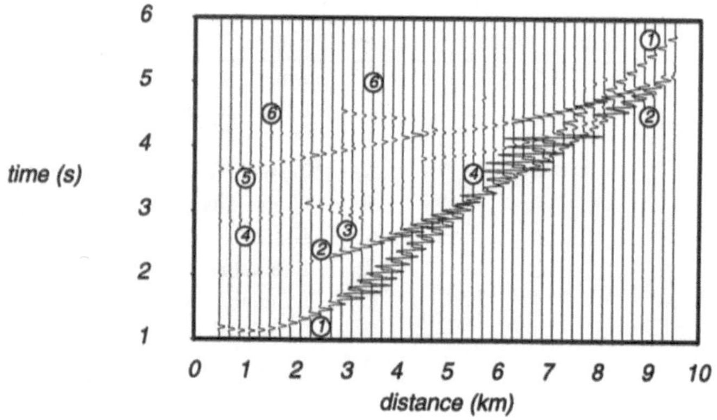

Fig. 3. Synthtic section for a shot at 1.0km in the synthetic model, computed by recursive raytracing. Only first order reflections are shown. The labels refer to the layer in which the reflection occurred. The arrivals labeled 5 and 6 are used for the images in Figures 4 to 7. Wavelet: 10 Hz Gabor.

A recursive treatment ([9], [10], [11]) solves these problems by reversing the order of the computations; instead of operating with wave fronts and isochrons, it works with only one ray cell at a time, defined by two (in 2D) and three (in 3D) neighboring rays and two successive isochrons. This ray cell is originated at a source point or at an initial surface, and propagated all the way through the medium until it meets some termination criterion. At interfaces, the ray cell splits into a reflected and a transmitted cell, that both continue their own way, independently from each other. Both cells possibly hit new interfaces, each time generating new offsprings, that behave similarly to the original cell. The tree structure of ray segments, that thus appears, is conveniently handled by recursion. The storage requirement for the tree is negligible, because it depends only logarithmically on the number of subdivisions in the family of one initial cell. One particular advantage of recursion is that all relevant phases on a seismic section are generated automatically, without human interaction, together with all information necessary to analyze or select them; this advantage has been exploited in phase identification in vertical seismic profiling [10]. A detailed description and evaluation of the recursion algorithm is given in [10], both in 2D and 3D. Also in [10], an approach to include edge diffractions in the recursion is presented; diffractions fill in shadow zones, but may be expensive.

4 Numerical tests

The numerical tests have been performed on two models, a synthetic model, to illustrate the statements in the previous sections, and a real model, the Brage oil field from the Northern part of the North Sea, which is in production by Norsk Hydro. In both cases synthetic data are used, in the synthetic model provided by raytracing, in order to isolate possible imaging artefacts, in the Brage model provided by a finite-difference solution of the wave equation, in order to test the viability of the imaging formula, without interference of typical problems of real data (like noise, multiples, 3D effects, elastic effects, or inaccurate information on the overburden). Real Brage data are inverted with the same imaging technique and with non-smooth medium's Green's functions algorithms in [10].

The synthetic model is shown in Figure 1 and its velocities (constant per layer) are tabulated in Table 1. Special attention is given to the 'channel' structures, labeled '6' is Figure 1. Figure 2 illustrates the recursive raytracing of one cell in the

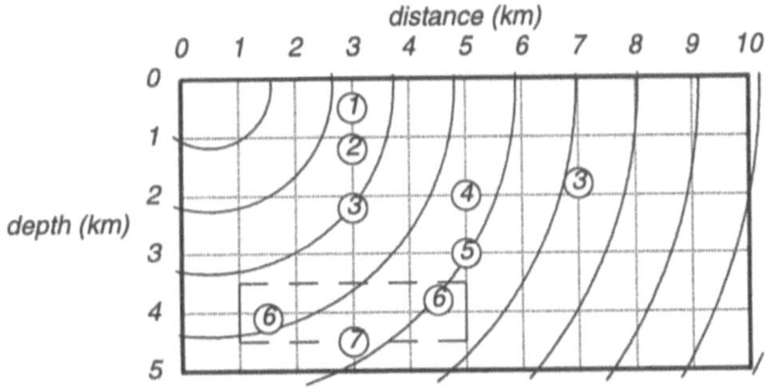

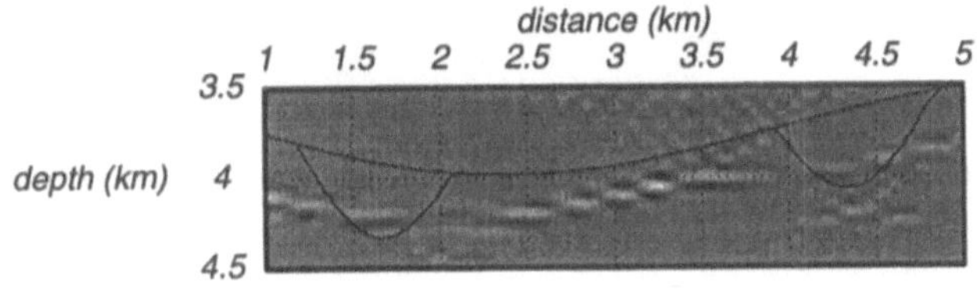

Fig. 4. Isochrons of the Green's function in a constant background (above). Rays are not shown. The reconstructed image in the target region is shown below (gray scale: perturbation of squared slowness on original model, thin lines: original interfaces).

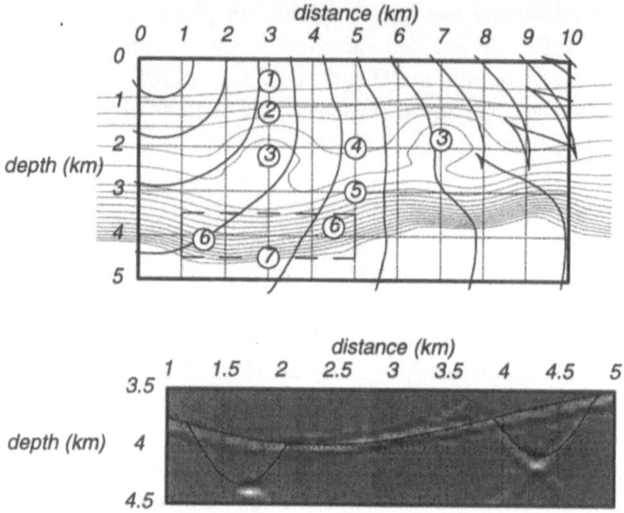

Fig. 5. Isochrons of the Green's function in a slowly varying background (above) and image belonging to it (below).

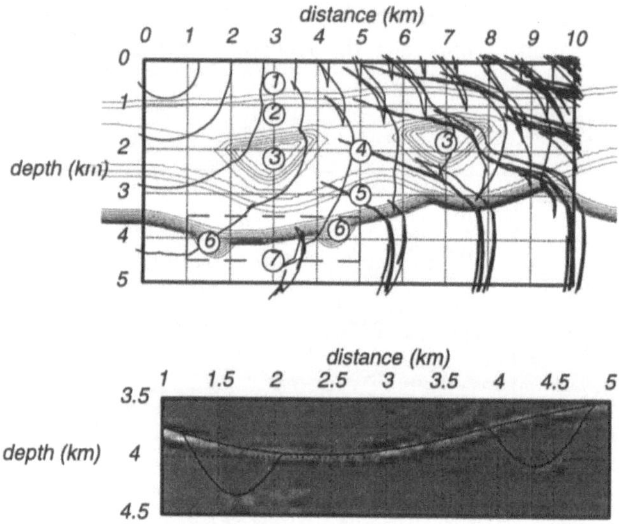

Fig. 6. Isochrons of the Green's function in a rapidly varying background (above) and image belonging to it (below).

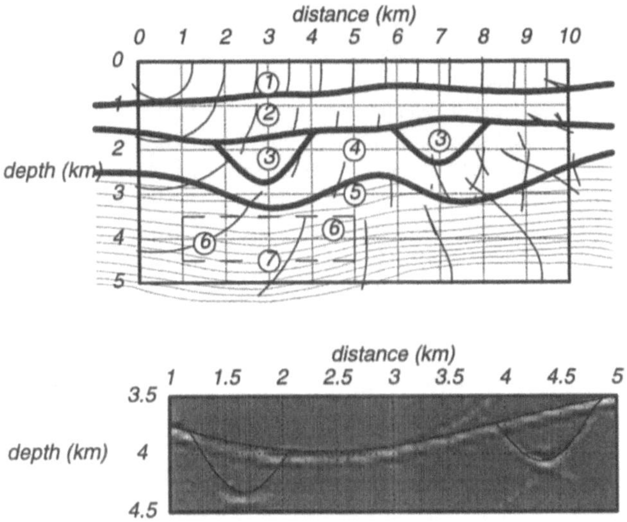

Fig. 7. Isochrons of the Green's function in a non-smooth background (above) and image belonging to it (below).

synthetic model. For a complete coverage of the model, and to obtain the shot record in Figure 3, ray cells are initiated in al directions from the source point. In Figure 2, one sees how, already from one initial cell, a quite complicated tree of ray segments, reflected and transmitted, emanates. Thanks to the recursive organization, it is possible to accumulate the ray history along the ray and to distinguish the phases in Figure 3 by the layer in which they were reflected. 46 locations act as sources and receivers in the synthetic model, from where Green's functions into the model are calculated. 32 receiver locations symmetrically around each source point are involved in the imaging. Figures 4 to 7 show the imaging with the formula [A19] (appendix), for a sequence of reference models that are increasingly complex and close to the actual model. In Figures 4 to 6, the reference models are smooth. It is clearly visible how the Green's functions become more and more multivalued, developing caustics and finally distorting the image quality (Figure 6). Figure 7 shows the imaging with a non-smooth reference field. Shadow zones appear, but the Green's functions are much less multivalued and therefore realistic, and the image is correspondingly better (and cheaper to compute).

An interpreted depth section of the Brage model is shown in Figure 8 and its velocities are tabulated in Table 2. It contains lots of diffracting edges, so that the ray field is disconnected and exhibits shadow zones (Figure 9, for one initial cell, and Figure 10 a shot record). Yet, a comparison with finite-differences, which is supposed to provide the complete wave field (at the expense of a much higher CPU) shows agreement to a high degree in arrival times, amplitudes and phase shifts (Figure 10 against 11). The data acquisition used for the imaging is tabulated in Table 3. The non-smooth reference model has been chosen equal to the velocity

Table 3. Data acquisition in Brage

Number of shots	400
Number of receivers/shot	120
First shot location (km)	-10.0
Shot location increment (km)	+0.0375
First receiver offset (km)	+0.1375
Receiver offset increment (km)	+0.025
Shot/receiver depth (km)	+0.008
Sampling rate (s)	+0.002

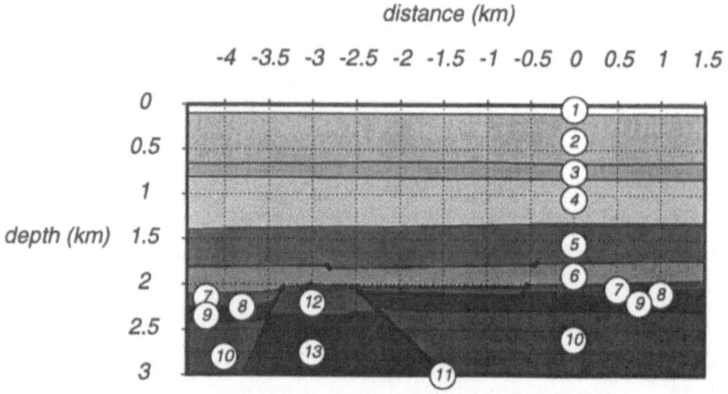

Fig. 8. The Brage velocity model. The labels correspond to the stratigraphic units tabulated in Table 2. The gray scale represents velocities. The black dots denote diffracting edges.

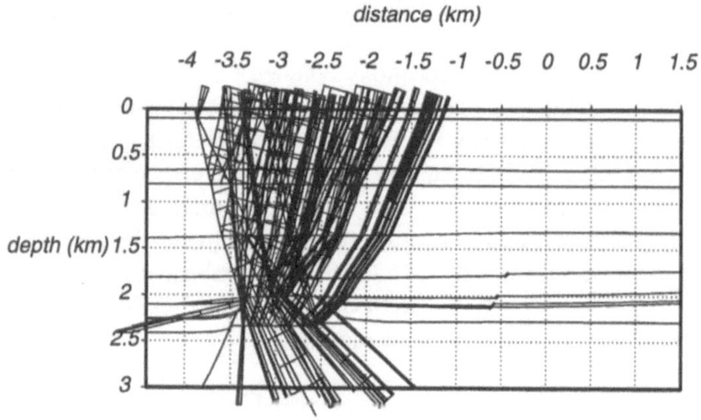

Fig. 9. Recursive raytracing of one ray cell in the Brage model. Transmissions and first order reflections.

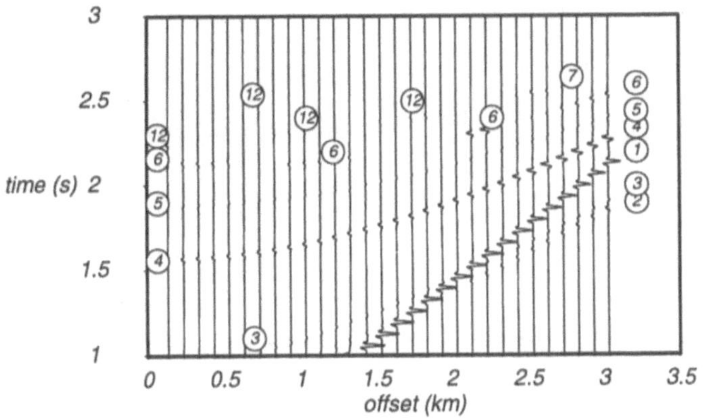

Fig. 10. Synthetic section for a shot in the Brage model at -3.9km, computed by recursive raytracing. The labels refer to reflections on the bottom of layers in Figure 8. Wavelet: 30 Hz Gabor.

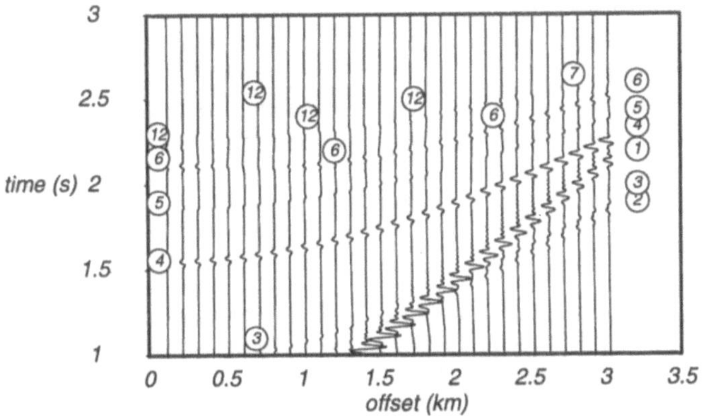

Fig. 11. Finite-difference section for a shot in the Brage model at -3.9km. The labels are taken from Figure 10. The arrivals labeled 12 are used in the inversion to obtain the image of Figure 13. Wavelet: 30 Hz zero-phase Ricker.

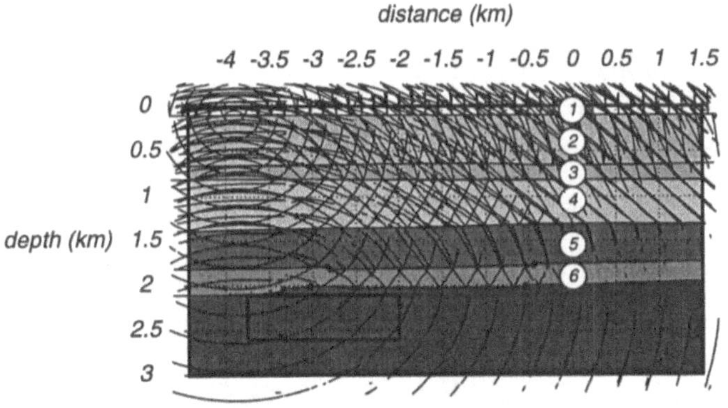

Fig. 12. Non-smooth reference model for inversion. The target is bounded by the rectangular box. The gray scale represents velocities. The thin lines denote the isochrons of the Green's function for a source at -3.9 km.

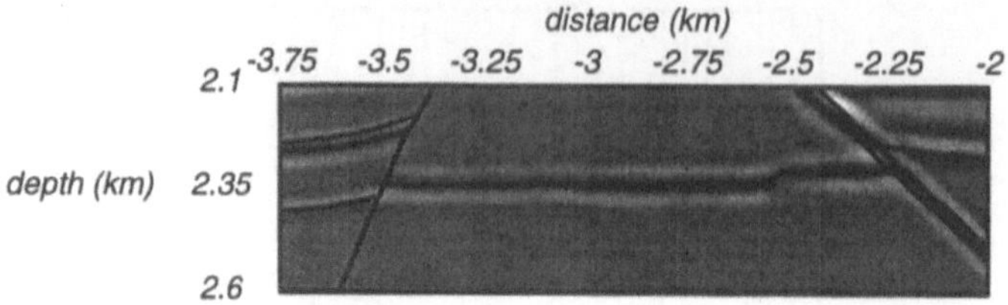

Fig. 13. Image of the horst interior. The grey scale denotes the reconstructed perturbation of the squared slowness on the reference model given in Figure 12. The thin lines represent the geometry of the Brage-model, in which the (finite-difference) synthetic data have been generated (see Figure 11).

model from the interpreted depth section (Figure 8), as far as the overburden is concerned (indicated by indices 1 to 6), and chosen equal to a constant representative velocity in the lower part of the model. The target zone is bounded by the rectangular box in Figure 12 and contains the most important, interior, part of the horst. Also shown are the isochrons of the Green's functions for an exemplary source point at -3.9km. The image derived from the finite-difference data is displayed in Figure 13. The geometry of the interfaces is very accurately reconstructed. Dynamical information is available as well, in the form of a reflectivity measure; the right fault has a sharp contrast, the fault on the left almost no contrast.

Acknowledgements

The authors acknowledge financial support from the Norwegian Scientific Council for the EU-project 3D Asymptotic Seismic Imaging, NRF No 30995. We are grateful to Dr J. Pajchel from Norsk Hydro for his invaluable contributions to the work and for running many numerical tests. The first author is grateful to the Alexander-von-Humboldt Foundation for financial support.

References

[1] Beydoun, W.B., and Mendes, M., 1989. Elastic ray-Born l_2-migration/inversion, *Geophysical Journal*, **97**, 151-160.

[2] Beylkin, G., 1985. Imaging of discontinuities in the inverse scattering problem by inversion of a causal generalized Radon transform, *J. Math. Phys.*, **26**, 99-108.

[3] Beylkin, G., and Burridge, R., 1990. Linearized inverse scattering problem in acoustics and elasticity, Wave Motion, **12**, 15-52.

[4] Biryulina, M., and Ryzhikov, G. 1995. Rytov-Born decomposition in 3D reflection seismics, 57th Annual EAGE Meeting Glasgow, Expanded Abstracts.

[5] Lambaré, G., Lucio, P.S., and Hanyga, A., 1994. 2D fast asymptotic Green's function, 56th Annual EAEG Meeting Vienna, Expanded Abstracts.

[6] Lucio, P.S., Lambaré, G., and Hanyga, A., 1995. 3D multivalued travel time and amplitude maps, 57th Annual EAEG Meeting Glasgow, Expanded Abstracts.

[7] Miller, D., Oristaglio, M., and Beylkin, G., 1987. A new slant on seismic imaging: Migration and integral geometry, *Geophysics*, **52**, 943-964.

[8] Moser, T.J., 1995. Inversion in a non-smooth background, 57th Annual EAEG Meeting Glasgow, Expanded Abstracts.

[9] Moser, T.J., and Pajchel, J., 1996. 3D seismic profiling by means of recursion, 58th Annual EAGE meeting Amsterdam, Expanded Abstracts.

[10] Moser, T.J., and Pajchel, J., 1996. Recursive seismic ray modeling: applications in inversion and VSP, submitted to *Geophysical Prospecting*.

[11] Pajchel, J., and Moser, T.J., 1995. Recursive cell raytracing, 51th Annual EAEG Meeting Glasgow, Expanded Abstracts.

[12] Ryzhikov, G. and Troyan, V., 1992. 3D diffraction tomography. Part 1: Construction and interpretation of tomography functionals, in *Expanded Abstracts, Russian - Norwegian Oil Exploration Workshop II*, Voss.

[13] Ryzhikov, G. and Troyan, V., 1992. 3D diffraction tomography. Part 2: Reconstruction algorithm with statistical regularization, in *Expanded Abstracts, Russian - Norwegian Oil Exploration Workshop II*, Voss.

[14] Ryzhikov, G., Biryulina, M., and Hanyga, A., 1995. 3D nonlinear inversion by entropy of image contrast optimization, *Nonlinear Processes in Geophysics*, **2**, 228-240.

[15] Sun, Y., 1993. Computation of 2-D multivalued arrival travel time fields by an interpolative shooting method, 63th Annual SEG Meeting, Expanded Abstracts.

[16] Vinje, V., Iversen, E., Gjøystdal, H., and Åstebøl, K., 1994. Estimation of multivalued arrivals in 3D models using wavefront construction, 56th Annual EAEG Meeting Vienna, Expanded Abstracts.

[17] Vinje, V., Iversen, E., and Gjøystdal, H., 1993. Traveltime and amplitude estimation using wavefront construction, *Geophysics*, **58**, 1157-1166.

Appendix - Imaging formula

Scattering phenomena for the scalar wave equation

$$\Delta\phi(\boldsymbol{x}, t) - u^2(\boldsymbol{x}) \frac{\partial^2 \phi(\boldsymbol{x}, t)}{\partial t^2} = S(\boldsymbol{x}, t) \qquad \text{[A1]}$$

can be analyzed by separating the squared slowness $u^2(\boldsymbol{x})$ in a reference field and a perturbation on it:

$$u^2(\boldsymbol{x}) = u_0^2(\boldsymbol{x}) + \varepsilon u_1^2(\boldsymbol{x}), \qquad \text{[A2]}$$

and expanding the wave field $\phi(\boldsymbol{x}, t)$ in a Born series

$$\phi(\boldsymbol{x}, t) = \sum_{n=0}^{\infty} \varepsilon^n \phi_n(\boldsymbol{x}, t). \qquad \text{[A3]}$$

Inserting [A2] and [A3] in [A1] and collecting the ε^0 terms gives the wave propagation in the reference model:

$$\Delta\phi_0(\boldsymbol{x}, t) - u_0^2(\boldsymbol{x}) \frac{\partial^2 \phi_0(\boldsymbol{x}, t)}{\partial t^2} = S(\boldsymbol{x}, t), \qquad \text{[A4]}$$

which is formally solved by the Green's function $G(\boldsymbol{x}, \boldsymbol{y}, t, \tau)$:

$$\phi_0(\boldsymbol{x}, t) = \int_M \int_{-\infty}^{+\infty} G(\boldsymbol{x}, \boldsymbol{y}, t, \tau) S(\boldsymbol{y}, \tau) \, d\boldsymbol{y} \, d\tau. \qquad \text{[A5]}$$

In [A5], M denotes the spatial support of the integrand; the time integration extends from $-\infty$ to $+\infty$, but in practice over the length of a seismic trace. In [A1], [A4], and [A5], $S(\boldsymbol{x}, t)$ denotes the source density. For a point source, located at $\boldsymbol{x}_s$,

$$S(\boldsymbol{x}, t) = \delta(\boldsymbol{x} - \boldsymbol{x}_s) s(t) \qquad \text{[A6]}$$

and

$$\phi_0(\boldsymbol{x}, \boldsymbol{x}_s, t) = \int_{-\infty}^{+\infty} G(\boldsymbol{x}, \boldsymbol{x}_s, t, \tau) s(\tau) \, d\tau. \qquad \text{[A7]}$$

Collecting the ε^1 terms results in the equation for the first-order scattering:

$$\Delta\phi_1(\boldsymbol{x}, t) - u_0^2(\boldsymbol{x}) \frac{\partial^2 \phi_1(\boldsymbol{x}, t)}{\partial t^2} = u_1^2(\boldsymbol{x}) \frac{\partial^2 \phi_0(\boldsymbol{x}, t)}{\partial t^2}, \qquad \text{[A8]}$$

which is again formally solved by the Green's function

$$\phi_1(\boldsymbol{x}, t) = \int_M \int_{-\infty}^{+\infty} G(\boldsymbol{x}, \boldsymbol{y}, t, \tau) u_1^2(\boldsymbol{y}) \frac{\partial^2 \phi_0(\boldsymbol{y}, \tau)}{\partial \tau^2} \, d\boldsymbol{y} \, d\tau. \qquad \text{[A9]}$$

The frequency domain expression for [A9] reads

$$\phi_1(\boldsymbol{x}, \omega) = -\omega^2 \int_M G(\boldsymbol{x}, \boldsymbol{y}, \omega) u_1^2(\boldsymbol{y}) \phi_0(\boldsymbol{x}, \omega)\, \mathrm{d}\boldsymbol{y} . \qquad \text{[A10]}$$

For a reference wave excited by a point source (see [A7])

$$\phi_1(\boldsymbol{x}, \boldsymbol{x}_s, t) = \int_M \int_{-\infty}^{+\infty} G(\boldsymbol{x}, \boldsymbol{y}, t, \tau) u_1^2(\boldsymbol{y}) \frac{\partial^2 \phi_0(\boldsymbol{y}, \boldsymbol{x}_s, \tau)}{\partial \tau^2}\, \mathrm{d}\boldsymbol{y}\, \mathrm{d}\tau . \qquad \text{[A11]}$$

The image is derived from the least-square misfit between the first-order scattered wave and the observed wave ϕ^{obs}:

$$F[u^2] = \frac{1}{2} \sum_{r,s} \int_{-\infty}^{+\infty} (\phi_0(\boldsymbol{x}_r, \boldsymbol{x}_s, t) + \phi_1(\boldsymbol{x}_r, \boldsymbol{x}_s, t)[u^2] - \phi^{obs}(\boldsymbol{x}_r, \boldsymbol{x}_s, t))^2\, \mathrm{d}t . \text{[A12]}$$

In [A12], the summation is over all sources $\boldsymbol{x}_s$ and receivers $\boldsymbol{x}_r$ available in the data acquisition. Here, and in what follows, we put $\varepsilon = 1$. $F[u^2]$ is a functional, that assigns a real number to each squared slowness distribution $u^2(\boldsymbol{x})$; the dependence on u^2 is denoted by square brackets. Its gradient with respect to u^2, $\nabla_{u^2} F[u^2]$, is a function of a spatial location $\boldsymbol{y}$. In the first Born approximation, the scattered wave ϕ_1 depends linearly on u^2, and $F[u^2]$ has one minimum, which follows from

$$\nabla_{u^2} F[u^2](\boldsymbol{y}) = 0 . \qquad \text{[A13]}$$

An expression for $\nabla_{u^2} F[u^2]$ is obtained by differentiating [A12]:

$$\nabla_{u^2} F[u^2](\boldsymbol{y}) = \sum_{r,s} \int_{-\infty}^{+\infty} (\phi_0(\boldsymbol{x}_r, \boldsymbol{x}_s, t) + \phi_1(\boldsymbol{x}_r, \boldsymbol{x}_s, t)[u^2] - \phi^{obs}(\boldsymbol{x}_r, \boldsymbol{x}_s, t)) \nabla_{u^2} \phi_1(\boldsymbol{x}_r, \boldsymbol{x}_s, t)[u^2](\boldsymbol{y})\, \mathrm{d}t . \qquad \text{[A14]}$$

In [A14], $\nabla_{u^2} \phi_1(\boldsymbol{x}_r, \boldsymbol{x}_s, t)[u^2](\boldsymbol{y})$ is a function denoting the gradient of the scattered wave with respect to the squared slowness perturbation. It can be expressed with help of [A11]:

$$I(\boldsymbol{x}_r, \boldsymbol{x}_s, \boldsymbol{y}, t) \equiv \nabla_{u^2} \phi_1(\boldsymbol{x}_r, \boldsymbol{x}_s, t)[u^2](\boldsymbol{y}) = \int_{-\infty}^{+\infty} G(\boldsymbol{x}_r, \boldsymbol{y}, t, \tau) \frac{\partial^2}{\partial \tau^2} \phi_0(\boldsymbol{y}, \boldsymbol{x}_s, \tau)\, \mathrm{d}\tau . \qquad \text{[A15]}$$

The solution to [A13] can be found by expanding $\nabla_{u^2} F[u^2]$ around the reference field u_0^2:

$$\nabla_{u^2} F[u^2](\boldsymbol{y}) = \nabla_{u^2} F[u_0^2](\boldsymbol{y}) + \int_M \nabla_{u^2} \nabla_{u^2} F[u_0^2](\boldsymbol{y}, \boldsymbol{z})(u^2(\boldsymbol{z}) - u_0^2(\boldsymbol{z}))\, \mathrm{d}\boldsymbol{z} . \qquad \text{[A16]}$$

The Hessian operator of $F[u^2]$ is represented by the integral kernel

$$\nabla_{u^2}\nabla_{u^2}F[u_0^2](y, z) = \sum_{r,s} \int_{-\infty}^{+\infty} I(\boldsymbol{x}_r, \boldsymbol{x}_s, y, t) I(\boldsymbol{x}_r, \boldsymbol{x}_s, z, t)\, \mathrm{d}t\,. \qquad [A17]$$

Following [1], we approximate the Hessian by replacing it by its diagonal:

$$\nabla_{u^2}\nabla_{u^2}F[u_0^2](y, z) \sim \sum_{r,s} \int_{-\infty}^{+\infty} I(\boldsymbol{x}_r, \boldsymbol{x}_s, y, t) I(\boldsymbol{x}_r, \boldsymbol{x}_s, z, t) \delta(y - z)\, \mathrm{d}t\,. \qquad [A18]$$

The image follows then from [A13] and [A16]:

$$u_1^2(y) = u^2(y) - u_0^2(y) =$$

$$-\frac{\sum_{r,s} \int_{-\infty}^{+\infty} (\phi_0(\boldsymbol{x}_r, \boldsymbol{x}_s, t) - \phi^{obs}(\boldsymbol{x}_r, \boldsymbol{x}_s, t)) I(\boldsymbol{x}_r, \boldsymbol{x}_s, y, t)\, \mathrm{d}t}{\sum_{r,s} \int_{-\infty}^{+\infty} I(\boldsymbol{x}_r, \boldsymbol{x}_s, y, t)^2\, \mathrm{d}t}\,. \qquad [A19]$$

The enumerator of [A19] is proportional to the data misfit and vanishes for a background equal to the real model. The denominator is referred to as information sensitivity and quantifies the possibility to reconstruct the image, given a certain data acquisition. The implementation of [A19] requires Green's functions to be calculated for all source and receiver locations towards the target region.

Nonlinear Inversion of Seismic Reflection Data by Simulated Annealing

Pascal Amand and Jean Virieux

UMR Géosciences Azur, Rue A. Einstein, 06560 Valbonne, France

Abstract. We present a non-linear inversion scheme in order to recover the shape of seismic reflectors using reflection traveltime data without need of traveltime picking. The use of fully non-linear approach such as simulated annealing technique reduces the importance of a-priori assumptions on model parameters but requires many computations of the forward problem. We have implemented a parallel ray tracing (PRT) code based on PVM message-passing library which speeds up 2D two-point ray tracing and runs on general purpose multi-computers or on massively parallel machines such as Cray T3D and Connection Machine CM-5. For 30 sources and 96 receivers per source, we are able for the investigation of the shape of an interface beneath an heterogeneous layer in approximatively 600 forward modelings and less than 4 hours on 8-node T3D partition. We find that simulated annealing method permits us to minimize the objective function based on semblance for synthetic data of complex Earth models.

1 Introduction

Truly non-linear multi-parameter optimization in seismology has not been widely developed in spite of pioneer work of Press (1968) because the computation of the associated forward problem turns out to be too time-consuming for most cases. Only specific problems as static estimation (Rothman (1985), Rothman (1986)) or velocity analysis (Landa et al. (1989), Jin and Madariaga (1993)) in seismic exploration, earthquake location (Sambridge and Gallagher (1993)), focal mechanism (Kobayashi and Nakanishi (1994)) or simple seismic waveform inversion (Sen and Stoffa (1991), Sen and Stoffa (1992), Sambridge and Drijkoningen (1992)) have been tackled using non-linear inversions. The main characteristic of these approaches is a relatively simple estimation of the forward problem used in the optimization.

Many geophysical inverse problems have misfit functions which are non-quadratic and which present local minima. We shall be interested in the seismic reflection inversion which is one of these non-linear problems. Traveltimes for reflected waves depend on ray paths which locate velocity anomalies of the different media as well as reflection points on the unknown interfaces. The solution of this problem is based on linearized approaches (Farra and Madariaga (1988), Ivansson (1986)) where one starts from an initial model and tries to determine locally a better model which makes the misfit function decrease. A good strategy to find an improved model is based on the gradient of the misfit function but a step procedure as the simplex method (Nelder and Mead (1965)) is an alternative which prevents the relatively unstable gradient estimation.

Global methods as grid search or Monte-Carlo approach will still be too time-consuming for such problem where the most important step is the ray tracing forward modeling in order to compute synthetic traveltimes. Intermediate methods named semi-global searches as simulated annealing algorithms or genetic algorithms are worth investigating alternatives to linearized approaches. We suggest such analysis in this article where we shall focus our attention in efficient ray tracing tool. A brief presentation of a parallel ray tracing strategy will be given. Using such fast formulation of the forward problem, we shall analyze inversion approaches which mix local features as well as semi-global attempts and we shall show that, indeed, semi-global search is feasible for the seismic reflection inversion.

2 Forward modeling of traveltimes

2.1 Ray tracing for seismic wave interpretation

The forward problem is the computation of traveltimes from a source point at each observer and, for such purpose, one may perform ray tracing between the source and the receiver. Tracing rays inside a medium is a powerful tool for extracting information, because the computed quantities (traveltime, slowness vector, polarization vectors and amplitude) are related to simple quantities in a seismogram and are perfectly associated with different features of the medium (Červený et al. (1977), Chapman (1985), Virieux and Farra (1991)). Ray tracing method is a too time-consuming algorithm which cannot be used as it is for semi-global inversion. An intrinsic property is very appealing for speeding up the forward computation: each ray can be traced independently from the others. This feature can be exploited in parallel computing where different rays are computed on different machines.

2.2 Parallel Ray Tracing method (PRT)

Whereas, in a sequential algorithm, rays are computed one after the other, in the parallel ray tracing, rays are computed on different processors at the same time leading to major speed up in term of computation time. This approach, namely ray distribution, is opposed to a data distribution strategy where the medium is split into sub-media among processors and where rays are passed from processor to processor during their propagation as they cross sub-medium boundaries. The ray distribution approach requires the knowledge of the entire medium at each processor. As soon as the technology has provided enough memory for each individual processor, this ray distribution approach has been a worth investigated strategy and turns out to be simpler than the data distribution approach.

3 Inversion

Our fast forward problem allows us to investigate non-linear strategy for inversion of reflection traveltime data to recover reflector shape. We define the model space and the semi-global strategy of inversion.

3.1 Parametrization

We will represent model parameter by a vector m given by

$$m = [m_1, m_2, \ldots, m_M]^T$$

while data d is a set of seismic traces for different source-receiver pairs

$$d = \{U_{ij}\}$$

where i denotes sources and j receivers. We are interested in seismic reflectors which delimit 2 layers of different acoustical impedances. Different parametrizations are possible to describe such an Earth model, but we choose in order to allow for complicated shapes to parametrize reflector by B-splines (Virieux and Farra (1991)). B-splines are parametric curves approximation of control point positions. For 2D geometry, each control point $C_i(x, z)$ has 2 components, the offset position x and the depth z. The index i denotes point position in the ordered list of control points. It is hard to solve both x and z at the same time because this can lead to unphysical solution for example with splines that intersect. We choose for all the control points to set the x coordinate and to look only for the z coordinate. Thus, each model parameter m_i is the vertical location of a control point which defines part of the searched reflector.

If we assume that each parameter takes its value in a finite dimensional discrete set of size P, the size of model parameter space for N parameters is then P^N (actually the total number of possible models). Increasing in size of the parameter space is exponential with number of parameters. If we link inversion problem size with parameter space size, we can state that complexity of the inversion grows in $O(N!)$ which denotes a NP-complete problem, that is a problem which cannot be solved in a time polynomial function of the number of parameters. We shall discuss later on how to solve NP-complete problems.

We see the importance to consider a restricted set of parameters to reduce inversion problem size. But there is a tradeoff between the number of parameters (number of control points) of the inversion and the complexity of the solution (complexity of the reflector shape). Another problem which prevents from taking into account only a small number of points is the fact that they are not completely independent. If they were, we could consider the inversion of one point at a time. Nevertheless, by definition, B-spline control points have a local control on the shape of reflector. Therefore it is possible to split the inversion of a long reflector into inversion of shorter parts with possible overlap, which must be found empirically.

Because of acquisition geometry, we cannot invert seismic reflection data for both velocity and reflector shape at the same time. Even if the velocity information is present in the large offset traces of reflection data, trying to recover both shape and depth of reflector and velocity above leads to non-unique solutions. Thus, let us assume the possibly heterogeneous velocity above the searched reflector to be known, determined by some velocity analysis techniques or any a priori information we can have.

3.2 Error function

For a given model m, the ray tracing forward modeling operator g permits to compute traveltimes τ which can be multivalued.

$$g(m) = \tau_{ijb}(m)$$

where b denotes different branches with different arrival times.

We define the error function $E(m)$ as the opposite of measure of trace coherency or semblance within a given time window from computed traveltime, thus following the notation of Landa et al. (1995) :

$$-E(m) = \sum_b \sum_i \sum_{k=0}^{K} \frac{\left\{ \sum_j U_{ij}\left[k\delta t + \tau_{ijb}(m)\right] \right\}^2}{\sum_j \left\{ U_{ij}[k\delta t + \tau_{ijb}(m)] \right\}^2}$$

were δt denotes the sample interval and $K\delta t$ the time window for semblance calculation.

3.3 Optimization by simulated annealing

Inversion process is recast into minimization of the error function. Because relation between data and model is non-linear we expect the error function to have several local minima and we need a fully non-linear approach to solve the problem. The NP-complete nature of the problem prevents from applying grid search enumeration method, thus some heuristic as to be employed. We shall now discuss the simulated annealing (SA) strategy.

SA is a stochastic method derived from Monte-Carlo. Pure Monte-Carlo is a random search in the model space that possibly spends time sampling regions of low interest where error function is high. SA and genetic algorithms (GA) permit to restrict investigation around regions of high interest where the error function is low. Let us describe one of the basics SA algorithm called Metropolis (Metropolis et al. (1953)) : given a current model m_i with error function $E(m_i)$ do a perturbation of m_i into m_{i+1} and compute the new error function $E(m_{i+1})$. Let ΔE be the difference in the error function between m_i and m_{i+1} i.e. :

$$\Delta E = E(m_{i+1}) - E(m_i)$$

If $\Delta E \leq 0$ replace m_i with m_{i+1}, if $\Delta E > 0$ do the replacement with the probability

$$P = exp(-\frac{\Delta E}{T})$$

where T is a parameter called temperature. The perturbation-acceptance test is repeated for a large number of iterations at a fixed temperature before temperature is lowered, following a cooling schedule. We shall address the following issues : choice of initial temperature, choice of cooling schedule, law of perturbation of the models, and stopping criterion.

3.4 Very fast simulated reannealing (VFSR)

The VFSR algorithm proposed by Ingber (1989) uses a probability distribution for model parameter generation such that a slow cooling of temperature is no longer required (Sen and Stoffa (1995)). Let m_i^k be the ith model parameter at iteration k such that

$$m_i^{min} \leq m_i^k \leq m_i^{max}$$

where m_i^{min} and m_i^{max} are minimum and maximum values of the model parameter m_i and m_i^{k+1} the model parameter at next iteration such that

$$m_i^{k+1} = m_i^k + y_i \left(m_i^{max} - m_i^{min}\right)$$

where y_i is a random number in the range $[0,1]$ from Ingber proposed distribution. A random number u drawn from uniform distribution $U[0,1]$ can be mapped into Ingber distribution using the following relation

$$y_i = sgn\left(u - \frac{1}{2}\right)\ T_i\ \left[\left(1 + \frac{1}{T_i}\right)^{|2u-1|} - 1\right]$$

where sgn() is a function that returns the sign of its argument and T_i is current temperature. From this distribution, global minimum can statistically be obtained by using the following cooling schedule

$$T_i\,(k) = T_0\ exp\left(-c_i\ k^{1/M}\right)$$

Where k is the annealing step, T_0 the initial temperature and c_i a decay factor that permits tuning the cooling schedule for specific problems, we choose $c_i = 1$. The acceptance test remains Metropolis. There is also possibility with VFSR to control each parameter with different temperature this can be interesting when dealing with parameters of different nature. A strategy for reannealing (increase of the temperature) is also proposed.

3.5 Uncertainty estimation

The best fit model parameters obtained during first step is one element of the solution of an inverse problem. A second step which shall permit to quantify uncertainties is necessary to deliver the complete solution. We propose to perform an importance sampling around the solution obtained by VFSR for different temperatures and apply a technique from multidimensional statistical description of numerical data.

From the total number of tested models n during this second step, we can build matrix of observations $\mathcal{M}$ such that

$$\mathcal{M} = \begin{bmatrix} m_1^1 & \ldots\ldots\ldots & m_M^1 \\ \vdots & \ddots & \\ \vdots & & \\ m_1^n & & m_M^n \end{bmatrix}$$

The jth model $m^j = [m_1^j, \ldots, m_M^j]$ has an error function $E(m^j)$ which mesure the fit of the model with data. From this error function, we express probability distribution for model m^j by

$$\mathcal{P}(m^j) = \frac{exp(-(E(m^j) + C))}{\sum_j exp(-(E(m^j) + C))}$$

where C is a constant which aim is to make the error function positive. We choose $C = -min(E(m^j))$. For a given parameter p and for every tested models j we represent function

$$x = m_p^j, \; y = \mathcal{P}(m^j)$$

(see Fig. 3).

We group these probabilities into a diagonal matrix

$$\mathcal{D} = \begin{bmatrix} \mathcal{P}(m^1) & & 0 \\ & \mathcal{P}(m^2) & \\ & & \ldots \\ 0 & & \mathcal{P}(m^n) \end{bmatrix}$$

We derive mean model

$$g = \mathcal{M}^T \mathcal{D} 1$$

where 1 is a vector from $\Re^n$ with all its components set to 1. And finally variance-covariance matrix

$$V = \mathcal{M}^T \mathcal{D} \mathcal{M} - g g^T$$

Standard deviation represents square root of diagonal elements from V.

4 Synthetic example

The aim is to investigate the inversion procedure to recover a given interface from synthetic reflection data. We shall discuss model parameters, misfit function for the inversion kernel, influence of noise and uncertainties in the shape of the interface.

The model we build is one reflector with complex shape which develops multivalued traveltimes (Fig. 1). In order to generate synthetic data we choose a geometry of acquisition which is typical in marine seismic exploration with 96 traces, 100 shots located at the surface, 25 m between traces, 300 m near offset, and 50 m between shots. An sample of the synthetic data is shown Fig. 2.

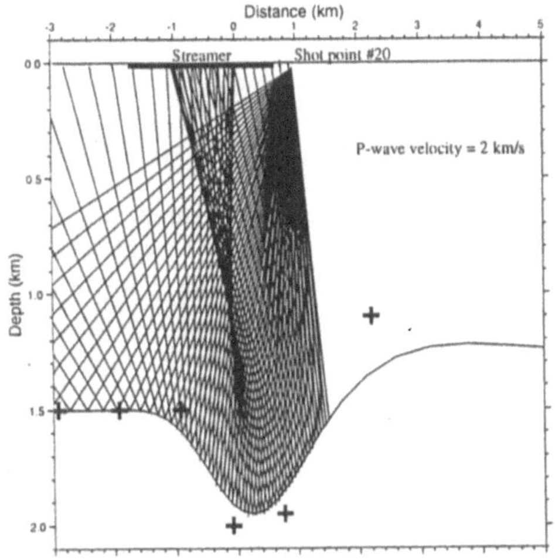

Fig. 1. Ray tracing from shot point number 20 in the subsurface model built for validation of inversion method by VFSR. Crosses denote B-spline control points defining the reflector shape. Notice the focusing/defocusing effect due to the syncline shape of the reflector.

Searched reflector is parametrized by a set of 24 control points 1 km spaced in x in the range $[-3, 20]$ km. On the edges of the model at position -3 km and 20 km three superimposed control points delimit the reflector extend, the effect of these hard bounds is to bias obtained solution on the edges (see Fig. 3). The search space for z coordinate is the interval $[0.8, 2.5]$ km. Model parameters are a subset of the 24 points which correspond to the points that are actually resolved by the data in the following way : we consider a set of shot points, the traced rays from these shots that reflect on the interface we want to image hit a piece of this interface at a given

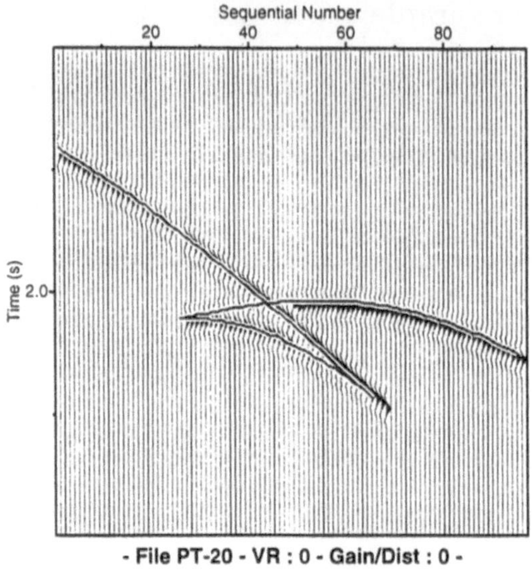

Fig. 2. Synthetic data for shot point number 20 located at 1 km offset at the surface. The thin line represents traveltimes obtained by forward modeling applied to the solution of the inversion by VFSR. Notice that the shape of the reflector gives rise to multivalued traveltime which are recovered.

location, it is easy from hit point to derive which control points influence area around hit point. These control points are added to the set of model parameters. Thus, model parameters are not determined prior but deduced from which data is used for inversion. If we invert only one shot, parameter set will be smaller than if we invert for 10 shots because rays from the 10 shots illuminate a larger portion of the reflector. Number of shots determines number of model parameters. If we take few shots, few model parameters are involved, thus dimension of model space is small and inversion can be fast, but it is possible that we do not have enough data to uniquely determine the solution. If we take a lot of shots, model parameter space dimension is larger and inversion longer and solution may become instable. The VFSR algorithm gives good results very fast (less than 3 hours on a Cray T3D) for a number of parameter around 10. We choose to consider inversion of 30 shot point data, given current geometry these 30 shots illuminate 7 control points. As we want to image all the reflector, after a VFSR result has been found, we shift the 30 shot points to recover next piece of reflector. Solution with 100 shots is shown Fig. 3.

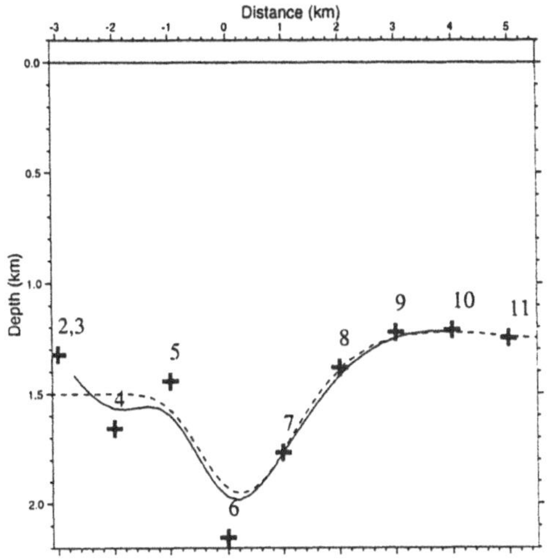

Fig. 3. Solid lines are the result of the inversion by VFSR of synthetic reflection data for 100 shots and 96 receivers for each shot. Dash line is the original reflector. This solution for 100 shots has been constructed from several pieces of solution from the inversion of 30 shot sliding windows. The retained control points in the final solution are those that have the maximum hit count. The solution is not good around point 2 and 3 due to low hit count in this area and to the fact that limit of this edge is fixed by 3 superimposed control points at the same position. It is not the case on the other edge where limit is not hard fixed. The gray area figures uncertainties calculated by Gibbs importance sampling 200 m around the result of the inversion by VFSR. The absolute value of these uncertainties has no meaning, the interest resides in the relativity of the measure when considering different areas of the reflector.

5 Conclusions and perspectives

Thanks to parallel formulation of forward problem we can trace rays very efficiently in an heterogeneous medium. Therefore, we are able to apply fully non-linear technique to solve traveltime inversion problem by stochastic exploration of model parameter space. Monte-Carlo techniques are powerfull for such exploration but are still to expensive. Sophisticated Monte-Carlo procedures such as VFSR permit to rapidly find a relative good solution avoiding to get trapped into local minima of error function. They also permit to characterize uncertainties, thus solving the inverse problem. Coherency function can be used in the framework of seismic reflection inversion in order not to have to perform traveltime picking. Plans are to invert real data set and to study behaviour in presence of many reflectors and noise.

6 Acknowledgements

This work has been partly founded by European Commission and Norwegian Research Council in the framework of the JOULE II program (Project "3-D Asymptotic Seismic Imaging"). Computational funds have been partly provided by the french national supercomputer Centres CNCPST and IDRIS. Contribution UMR Géosciences Azur N°104.

References

Červený V., Molotkov I.A., Pšenčik I. (1977): Ray method in seismology. Studia geoph. et geod., **27**, 9–18

Chapman C.H. (1985): Ray theory and its extension: WKBJ and maslov seismograms. Journal of Geophysics, **58**, 27–43

Farra V., Madariaga R. (1988): Nonlinear reflection tomography. Geophys. J., **95**, 135–147

Ingber L. (1989): Very fast simulated reannealing. Mathl. Comput. Modeling, **12(8)**, 967–993

Ivansson S. (1986): Some remarks concerning seismic reflection tomography and velocity analysis. Geophys. J. R. Astron. Soc., **87**, 539–557

Jin S., and Madariaga R. (1993): Background velocity inversion with a genetic algorithm. Geophys. Res. Lett., **20**, 93–96

Kobayashi R., and Nakanishi I. (1994): Application of genetic algorithms to focal mechanism determination. Geophys. Res. Lett., **21**, 729–732

Landa E., Beydoun W., Tarantola A. (1989): Reference velocity model estimation from prestack waveforms: coherency optimization by simulated annealing. Geophysics, **54**, 984–990

Landa E., Keydar S., Kravtsov A. (1995): Determination of a shallow velocity-depth model from seismic refraction data by coherence inversion. Geophysical Prospecting, **43**, 177–190

Metropolis N., Rosenbluth A., Rosenbluth M., Teller A., Teller E. (1953): Equation of state calculations by fast computing machines. J. Chem. Phys., **21**, 1087–1092

Nelder J.A., Mead R. (1965): A simplex method for function minimization. The Computer Journal, **7**, 308–313

Press F. (1968): Earth models obtained by monte carlo inversions. J. Geophys. Res., **73**, 5223–5234

Rothman D.H. (1985): Nonlinear inversion, statistical mechanics and residual statics estimation. Geophysics, **50**, 2784–2796

Rothman D.H. (1986): Automatic estimation of large residual statics corrections. Geophysics, **51**, 332–346

Sambridge M., Drijkoningen G. (1992): Genetic algorithms in seismic waveform inversion. Geophys. J. Int., **109**, 323–342

Sambridge M., Gallagher K. (1993): Earthquake hypocenter locations using genetic algorithms. Bull. Seismol. Soc. Am., **83**, 1467–1491

Sen M., Stoffa P.L. (1991): Nonlinear one dimensional seismic waveform inversion using simulated annealing. Geophysics, **56**, 1624–1638

Sen M., Stoffa P.L. (1992): Rapid sampling of model space using genetic algorithms: examples from seismic waveform inversion. Geophys. J. Int., **108**, 281–292

Sen M., Stoffa P.L. (1995): *Global optimization methods in geophysical inversion* Elsevier

Virieux J., Farra V. (1991) Ray tracing in 3-d complex isotropic media: an analysis of the problem. Geophysics, **56**, 2057–2069

Asymptotic Theory for Imaging the Attenuation Factors Q_P and Q_S

Alessandra Ribodetti and Jean Virieux

UMR Géosciences-Azur Sophia Antipolis, 250 Avenue A. Einstein, Bâtiment 3, F-06560 Valbonne, France

Abstract. Linearized inverse scattering problem in anelasticity is solved for perturbations in different parameters treating P-to-P, P-to-S,S-to-P and S-to-S data. Three steps are required for finding the material parameters of the medium, i.e. the density and the complex relaxation functions. In a given smooth reference medium, an high-frequency Green function is expressed as a function of traveltime, amplitude and attenuation factors. For a slightly different medium, the perturbation of the asymptotic Green function is expressed as a linear integral over the diffracting region containing the model perturbations using the first-order Born approximation. The inversion scheme is developed in the frequency domain where we were enable to set up an analytical kernel for the Born approximation of asymptotic anelastic solutions used for the forward problem and an approximate analytical kernel for the linearized inversion. Radiation patterns are analysed to show that the simultaneous multiparameter inversion is possible when one takes into account the parameters related to attenuation. The iterative asymptotic inversion might resolve the difference between the elastic parameters and the attenuation factors.

Introduction

Many practical problems of nondestructive testing, medical imaging, seismic exploration, consist in finding variations of material parameters.

Many workers have tried to elaborate algorithms for the reconstruction of elastic parameters (Bleistein (1987), Beydoun and Mendes (1989), Crase et al.(1990), Beylkin and Burridge (1990), Jin et al.(1991), Lambaré et al.(1992)) using different approximations and approaches. Few have tried to elaborate algorithms using the complete seismogram for recovering the attenuation (Tarantola (1988)). In this paper we present an asymptotic method for imaging the attenuation factors Q_P and Q_S for P and S waves, following the methods developed for the reconstruction of elastic parameters (Jin et al.(1991), Lambaré et al.(1992)).

Jin et al.(1991) have proposed a new procedure for finding the Earth's structure from seismic data. The inversion method follows the optimization approach but uses specific features of the selected asymptotic operator relating parameters and seismograms in order to express the inversion through analytical expressions. This approach is fast and allows different data acquisition geometries (Lambaré et al.(1992)). This method has been extended to diffusive electromagnetic phenomena by Virieux et al.(1994) and in viscoacoustic media by Ribodetti et al.(1995). In viscoelastic media description of wave propagation requires several different scattering modes, such as P-to-P, P-to-S, S-to-P and S-to-S, which are developed in this paper.

We present an high-frequency approximation with attenuation in a smooth reference medium. The attenuation feature of the propagation is represented by a complex Lamé coefficients $\hat{\lambda}$ and $\hat{\mu}$ while the density may also vary spatially in this smooth reference medium.

Linearization of the inverse scattering problem is achieved by considering the actual medium as a perturbation of (a spatially varying) background medium. Then by using the single scattering (Born) approximation, we obtain a linear relation between model perturbation parameters and seismogram perturbations. This linear relation is asymptotically inverted in the frequency domain using an iterative quasi-Newtonian inversion based on a least-squares criterion.

The radiation patterns analysis shows that the simultaneous inversion for the elastic and anelastic parameters is possible and the attenuation does not change the elastic radiation pattern components. A good description of the model parameters is studied. In this paper mathematical analysis of the linearized inverse scattering problem for anelastic medium is proposed.

1 Wave propagation and Born approximation

1.1 Modeling anelastic attenuation

Seismic attenuation is a potentially useful parameter for characterizing and monitoring hydrocarbon reservoirs in conjunction with seismic velocity. The most common measures of attenuation are the dimensionless quality factor Q and its inverse Q^{-1}. As an intrinsic property of rock, Q is a ratio of stored energy to dissipated energy. O'Connel and Budiansky (1978) discussed various definitions of Q and their relationships to the viscoelastic constitutive equations for a given material.

The amount of intrinsic attenuation present in seismic waves is not accurately known and the additional contribution of local scattering to the Q factor increases the complexity of the analysis (Hatzidimitriou (1995)). We assume that the Q factor, sometimes called apparent Q factor, incorporates effects which modify locally the amplitude of the wave during propagation. The attenuative properties of rocks can be estimated by a wide range of measures. For low-loss linear solids, a definition of Q may be found through extented stress-strain relations. In these anelastic media, the correspondence principle (White (1965), Eringen (1980), Ferry (1961)) allows one to replace in the frequency domain the elastics moduli λ and μ by a complex $\hat{\lambda} = \lambda_R + i\, \lambda_I$ and $\hat{\mu} = \mu_R + i\, \mu_I$. In this model, the parameters λ_R, μ_R and the density ρ controle wave velocity, and the imaginary moduli $i\lambda_I$ and $i\mu_I$ govern energy damping. A linear model for attenuation of waves is presented, with Q, or the portion of energy lost during each cycle or wavelength, exactly independent of frequency (Kjartansson (1979), Aki and Richards (1980)). If the attenuation is small, then the velocities and attenuations coefficients can be expressed convenientely (Mavko and Nur (1978)). For compressional waves,

$$c_P = \sqrt{\frac{\lambda_R + 2\mu_R}{\rho}}, \tag{1}$$

and

$$\frac{1}{Q_P} = \frac{\lambda_I + 2\mu_I}{\lambda_R + 2\mu_R}, \tag{2}$$

where $\lambda_R = \kappa_R - 2/3\mu_R$, κ_R is the real part of the complex bulk modulus. For shear waves

$$c_S = \sqrt{\frac{\mu_R}{\rho}}, \tag{3}$$

and

$$\frac{1}{Q_S} = \frac{\mu_I}{\mu_R}. \tag{4}$$

In the following, we will consider an anelastic model characterized by a density ρ and complex relaxation functions

$$\hat{\lambda}(\mathbf{x}) = \lambda(\mathbf{x}) + i\Lambda(\mathbf{x}), \tag{5}$$

$$\hat{\mu}(\mathbf{x}) = \mu(\mathbf{x}) + i\nu(\mathbf{x}), \tag{6}$$

where $\lambda(\mathbf{x})$ and $\mu(\mathbf{x})$ are the elastic parts and the factors $\Lambda(\mathbf{x})$ and $\nu(\mathbf{x})$ are the terms related to specific attenuation.

1.2 Ray theory

We study the propagation with attenuation and we shall introduce the corresponding high frequency approximation. For a source position $\mathbf{s} = (s_1, s_2, s_3)$, let $\mathbf{G}(\mathbf{s}, \mathbf{x}, \omega)$ be the displacement field at the point $\mathbf{x}$ which satisfies the elastodynamic equation of motion,

$$\rho\omega^2 G_{jl} + (g_{lmpq} G_{jp,q})_{,m} = -\delta_{jl}\delta(\mathbf{x} - \mathbf{s}), \tag{7}$$

where, for an isotropic medium, we have the following expression for the elastic parameters

$$g_{lmpq}(\mathbf{x}) = \lambda(\mathbf{x})\delta_{lm}\delta_{pq} + \mu(\mathbf{x})(\delta_{lp}\delta_{mq} + \delta_{lq}\delta_{mp}). \tag{8}$$

The solution $G_{jl}(\mathbf{s},\mathbf{x},\omega)$ is the jth component of displacement at the point $\mathbf{x}$ due to a point force in the l−direction applied at the source $\mathbf{s}$. The density is denoted by ρ and the krönecker's symbol δ_{jl}. For convenience in writing, we have omitted the dependence of G_{jl} on parameters $(\mathbf{s},\mathbf{x},\omega)$.

Through the correspondence principle, the attenuation is described by changing the Lamé constants λ and μ by their complex equivalent expression $\hat{\lambda}$ and $\hat{\mu}$ and the expression (9)

$$\rho\omega^2 G_{jl} + (\hat{g}_{lmpq} G_{jp,q})_{,m} = -\delta_{jl}\delta(\mathbf{x}-\mathbf{s}), \tag{9}$$

expresses the propagation with attenuation.

We change from component notation to intrinsic notation by letting $\mathbf{G}_j = (G_{j1}, G_{j2}, G_{j3})$, $j = 1,2,3$ and $\mathbf{I}_j$ the vector having 1 at the place j and 0 elsewhere . The equation (9) becomes

$$\rho\omega^2 \mathbf{G}_j + \nabla \cdot (\hat{g}(\nabla \mathbf{G}_j + \nabla \mathbf{G}_j{}^T)) = -\mathbf{I}_j \delta(\mathbf{x}-\mathbf{s}), \tag{10}$$

which can be modified into

$$\rho\omega^2 \mathbf{G}_j + \nabla \hat{g}(\nabla \mathbf{G}_j + \nabla \mathbf{G}_j{}^T) + \hat{g}[\nabla(\nabla \cdot \mathbf{G})_j + \Delta \mathbf{G}_j] = -\mathbf{I}_j \delta(\mathbf{x}-\mathbf{s}). \tag{11}$$

For a smooth heterogeneous medium, the asymptotic time-harmonic expression for the 3D Green function is

$$\mathbf{G}_0(\mathbf{s},\mathbf{x},\omega) = \mathbf{A}_0(\mathbf{s},\mathbf{x},\omega) e^{i\omega T(\mathbf{s},\mathbf{x})} \tag{12}$$

for a point source located at $\mathbf{s}$; $\mathbf{A}_0 = (A_{j01}, A_{j02}, A_{j03})$, $j = 1,2,3$. The traveltime T is the integration along the ray of the slowness

$$T^P(\mathbf{s},\mathbf{x}) = \int_{\sigma_0(\mathbf{s})}^{\sigma(\mathbf{x})} \sqrt{\rho(\xi)/(\lambda(\xi)+2\mu(\xi))}\, d\xi \tag{13}$$

and

$$T^S(\mathbf{s},\mathbf{x}) = \int_{\sigma_0(\mathbf{s})}^{\sigma(\mathbf{x})} \sqrt{\rho(\xi)/\mu(\xi)}\, d\xi \tag{14}$$

where σ is the arclength along the ray between the source $\mathbf{s}$ and the point $\mathbf{x}$; the superimposed P and S indicate P-wave and S-wave. The amplitude $\mathbf{A}_0$, the first term of the serie $\sum_{k=0}^{+\infty} \mathbf{A}_k(\mathbf{s},\mathbf{x},\omega)/(i\omega)^k$, has a frequency dependence related to our selected dependence of the attenuation. The amplitude follows from the transport equation obtained with some calculation following the same procedure as Cervený and Hron (1980) and as Caviglia et al.(1990):

$$\mathbf{A}_0^P(\mathbf{s},\mathbf{x},\omega) = \sqrt{\frac{(\rho c_P J)(\sigma_0(\mathbf{s}))}{(\rho c_P J)(\sigma(\mathbf{x}))}} \mid \mathbf{A}_0(\sigma_0(\mathbf{s})) \mid$$
$$e^{-\omega \int_{\sigma_0(S)}^{\sigma(X)} (\Lambda(\xi)+2\nu(\xi))/2c_P(\xi)(\lambda(\xi)+2\mu(\xi))\, d\xi}, \tag{15}$$

and

$$\mathbf{A}_0^S(\mathbf{s},\mathbf{x},\omega) = \sqrt{\frac{(\rho c_S J)(\sigma_0(\mathbf{s}))}{(\rho c_S J)(\sigma(\mathbf{x}))}} \mid \mathbf{A}_0(\sigma_0(\mathbf{s})) \mid e^{-\omega \int_{\sigma_0(\mathbf{S})}^{\sigma(\mathbf{X})} \nu(\xi)/2c_S(\xi)\mu(\xi)\, d\xi} \tag{16}$$

where $c_P = \sqrt{(\lambda + 2\mu)/\rho}$ for P-wave and $c_S = \sqrt{\mu/\rho}$ for S-wave are the phase velocities and J is the Jacobian of the mapping via rays. The first term is the geometrical spreading due to wavefront expansion and the initial value $\mathbf{A}_0(\sigma_0(\mathbf{s}))$ turns out to be $1/4\pi$ by matching the high-frequency solution and the complete solution for an homogeneous medium. Of course, we may deduce the expression for a non-attenuating medium by letting Λ and ν to be zero. The attenuation effect results in an exponential decay of the amplitude along the rays. We define

$$\alpha^P(\mathbf{s},\mathbf{x}) = e^{-\omega \int_{\sigma_0(\mathbf{S})}^{\sigma(\mathbf{X})} (\Lambda(\xi)+2\nu(\xi))/2c_P(\xi)(\lambda(\xi)+2\mu(\xi))\, d\xi}, \tag{17}$$

and

$$\alpha^S(\mathbf{s},\mathbf{x}) = e^{-\omega \int_{\sigma_0(\mathbf{S})}^{\sigma(\mathbf{X})} \nu(\xi)/2c_S(\xi)\mu(\xi)\, d\xi}. \tag{18}$$

These terms can be interpreted as the imaginary part of a complex traveltime $T + i\alpha$. Whatever is the selected notation, it represents the attenuation along the raypath between source at $\mathbf{s}$ and diffracting-point at $\mathbf{x}$.

The effects of attenuation on seismograms are analysed in Figure 1 a) in a medium with constant velocity field ($c_P = 2000$ m s^{-1}) and constant quality factor ($Q_P = 10$). The amplitude, traveltime and attenuation are obtained in Figure 2 using the algorithm presented by Lambaré et al.(1996). The attenuated seismogram are superimposed with non-attenuated seismogram. The attenuation effects decrease when the quality factor increases. In Figure 1 b) we have considered a medium with constant $Q_P = 100$.

1.3 Born approximation

Following the same procedure proposed by Beylkin and Burridge (1990) for acoustic and elastic media, we present an extension for an anelastic isotropic 3D medium. We assume that the medium can be separated in two parts: a smooth reference medium for which the Green function may be computed by ray theory and an unknown weak local perturbation of medium parameters. For developing wave equation theory perturbation, we choose the density $\rho(\mathbf{x})$ and the relaxations functions for P and S waves $\hat{\lambda}(\mathbf{x})$ and $\hat{\mu}(\mathbf{x})$, because it makes expressions simpler.

The parameters of the global medium can be written

$$\begin{aligned} \rho(\mathbf{x}) &= \rho_0(\mathbf{x}) + \delta\rho(\mathbf{x}) \\ \hat{\lambda}(\mathbf{x}) &= \hat{\lambda}_0(\mathbf{x}) + \delta\hat{\lambda}(\mathbf{x}) \\ \hat{\mu}(\mathbf{x}) &= \hat{\mu}_0(\mathbf{x}) + \delta\hat{\mu}(\mathbf{x}) \end{aligned} \tag{19}$$

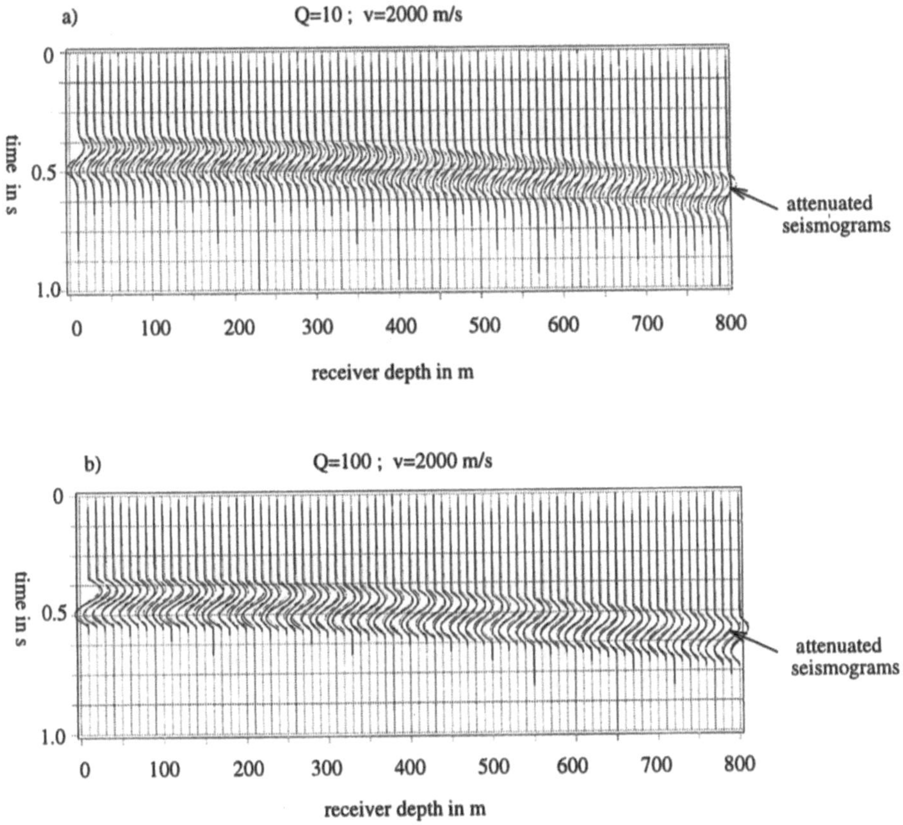

Fig. 1. Seismograms obtained from the traveltime, amplitude and attenuation given by ray tracing, in Figure 2. On the top anelastic-seismograms ($Q_P = 10$, $c_P = 2000$ m s^{-1}) are compared with elastic seismograms ($c_P = 2000$ m s^{-1}). On the bottom anelastic-seismograms ($Q_P = 100$, $c_P = 200$ m s^{-1}) are compared with elastic seismograms ($c_P = 2000$ m s^{-1}). The source signature is s(t)= sinc(t). The receivers are located along a vertical line; the first receiver is in $x = 700$ m from the origin and at $z = 10$ m and the receivers sampling step is 10 m.

where $\delta\rho$, $\delta\hat{\lambda}$ and $\delta\hat{\mu}$ are corresponding perturbations of the density and of the relaxation functions. The perturbation area, as for the reference medium, has an attenuation which is complex. Then from (6) we have

$$\hat{\lambda}(\mathbf{x}) = \lambda_0(\mathbf{x}) + \delta\lambda(\mathbf{x}) + i(\Lambda_0(\mathbf{x}) + \delta\Lambda(\mathbf{x})),$$
$$\hat{\mu}(\mathbf{x}) = \mu_0(\mathbf{x}) + \delta\mu(\mathbf{x}) + i(\nu_0(\mathbf{x}) + \delta\nu(\mathbf{x})), \tag{20}$$

in which we distinguish between the elastic part and the part related to dissipation.

The complete Green function $\mathbf{G}$ will be split into the known Green function $\mathbf{G}_0$ and the unknown perturbation $\delta\mathbf{G}$ due to the scattering from the

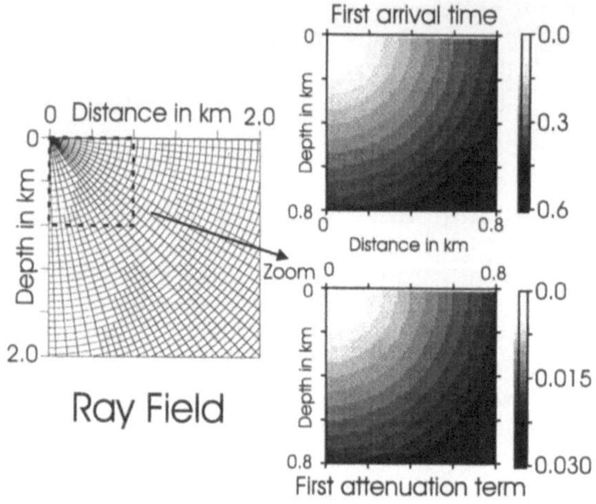

Fig. 2. Ray field and maps for time and attenuation by algorithm of Lambaré et al. (1996). The source is at origin.

perturbations of model parameters. The local equation at $\mathbf{r} = (r_1, r_2, r_3)$

$$\rho\omega^2 \mathbf{G}(\mathbf{s}, \mathbf{r}, \omega) + \nabla \cdot (\hat{g}(\mathbf{r})\nabla)\mathbf{G}(\mathbf{s}, \mathbf{r}, \omega) = -\delta(\mathbf{r} - \mathbf{s}), \tag{21}$$

for an arbitrary point $\mathbf{x}$ of the medium, can be expanded into the following form

$$\begin{aligned} &\rho_0(\mathbf{x})\,\omega^2 \delta\mathbf{G}(\mathbf{s}, \mathbf{x}, \omega) + \nabla \cdot (\hat{g}_0(\mathbf{x})\nabla)\delta\mathbf{G}(\mathbf{s}, \mathbf{x}, \omega) = \\ &- \quad (\omega^2 \delta\rho(\mathbf{x})\mathbf{G}(\mathbf{s}, \mathbf{x}, \omega) + \nabla \cdot (\delta\hat{g}(\mathbf{x})\nabla)\mathbf{G}(\mathbf{s}, \mathbf{x}, \omega)). \end{aligned} \tag{22}$$

The solution of equation (22) can be written as a convolution over the domain $\mathcal{M}$ of diffracting points of the Green function $\mathbf{G}_0(\mathbf{x}, \mathbf{r}, \omega)$, solution of the equation (21) for the reference medium, with the source term $\omega^2\delta\rho\mathbf{G} + \nabla \cdot (\delta\hat{g}\nabla)\mathbf{G}$ which yields

$$\begin{aligned} \delta\mathbf{G}(\mathbf{s}, \mathbf{r}, \omega) = \int_{\mathcal{M}} &\mathbf{G}_0(\mathbf{x}, \mathbf{r}, \omega)[\delta\rho(\mathbf{x})\omega^2 \mathbf{G}(\mathbf{s}, \mathbf{x}, \omega) \\ &+ \nabla \cdot (\delta\hat{g}(\mathbf{x})\nabla)\mathbf{G}(\mathbf{s}, \mathbf{x}, \omega)]d\mathbf{x}. \end{aligned} \tag{23}$$

The first-order Born approximation is obtained by replacing the total field $\mathbf{G}$ by the incident field $\mathbf{G}_0$ in the integral (23), leading to the following linear operator between $\delta\rho$, $\delta\hat{g}$ and $\delta\mathbf{G}$

$$\delta \mathbf{G}(\mathbf{s}, \mathbf{r}, \omega) = \int_{\mathcal{M}} \mathbf{G}_0(\mathbf{x}, \mathbf{r}, \omega)[\delta\rho(\mathbf{x})\omega^2 \mathbf{G}_0(\mathbf{s}, \mathbf{x}, \omega) + \nabla \cdot (\delta\hat{g}(\mathbf{x})\nabla)\mathbf{G}_0(\mathbf{s}, \mathbf{x}, \omega)]d\mathbf{x}. \quad (24)$$

Integrating the second term on the right hand by parts and noticing that the boundary term vanishes since perturbations $\delta\rho$ and $\delta\hat{g}$ are zero on the boundary $\partial\mathcal{M}$ of the domain $\mathcal{M}$, we obtain the linear relation

$$\delta \mathbf{G}(\mathbf{s}, \mathbf{r}, \omega) = \int_{\mathcal{M}} [\mathbf{G}_0(\mathbf{x}, \mathbf{r}, \omega)\mathbf{G}_0(\mathbf{s}, \mathbf{x}, \omega)\delta\rho(\mathbf{x})\omega^2 - (\delta\hat{g}(\mathbf{x})\nabla)\mathbf{G}_0(\mathbf{x}, \mathbf{r}, \omega) \cdot \nabla\mathbf{G}_0(\mathbf{s}, \mathbf{x}, \omega)]d\mathbf{x}. \quad (25)$$

Let us remark that the Born approximation requires the scattering zone to have a weak amplitude and a small extension (Wu (1989)). For an isotropic medium, the leading singular term of the Green functions in equation (25) can be written in the form

$$\mathbf{G}_0(\mathbf{s}, \mathbf{x}, \omega) = \mathbf{G}_0^P(\mathbf{s}, \mathbf{x}, \omega) + \mathbf{G}_0^S(\mathbf{s}, \mathbf{x}, \omega) \quad (26)$$

and

$$\mathbf{G}_0(\mathbf{x}, \mathbf{r}, \omega) = \mathbf{G}_0^P(\mathbf{x}, \mathbf{r}, \omega) + \mathbf{G}_0^S(\mathbf{x}, \mathbf{r}, \omega) \quad (27)$$

where

$$\mathbf{G}_0^P(\mathbf{s}, \mathbf{x}, \omega) = \mathbf{A}_0^P(\mathbf{s}, \mathbf{x})e^{i\omega T^P(\mathbf{s},\mathbf{x})}e^{-\omega\alpha^P(\mathbf{s},\mathbf{x})}, \quad (28)$$

$$\mathbf{G}_0^S(\mathbf{s}, \mathbf{x}, \omega) = \mathbf{A}_0^S(\mathbf{s}, \mathbf{x})e^{i\omega T^S(\mathbf{s},\mathbf{x})}e^{-\omega\alpha^S(\mathbf{s},\mathbf{x})}, \quad (29)$$

$$\mathbf{G}_0^P(\mathbf{x}, \mathbf{r}, \omega) = \mathbf{A}_0^P(\mathbf{x}, \mathbf{r})e^{i\omega T^P(\mathbf{x},\mathbf{r})}e^{-\omega\alpha^P(\mathbf{x},\mathbf{r})}, \quad (30)$$

$$\mathbf{G}_0^S(\mathbf{x}, \mathbf{r}, \omega) = \mathbf{A}_0^S(\mathbf{x}, \mathbf{r})e^{i\omega T^S(\mathbf{x},\mathbf{r})}e^{-\omega\alpha^S(\mathbf{x},\mathbf{r})}. \quad (31)$$

The leading singular terms of the spatial derivatives of the Green functions are as follows

$$\nabla\mathbf{G}_0^P(\mathbf{s}, \mathbf{x}, \omega) = i\omega(\nabla T^P(\mathbf{s}, \mathbf{x}) + i\nabla\alpha^P(\mathbf{s}, \mathbf{x}))\mathbf{A}_0^P(\mathbf{s}, \mathbf{x})e^{i\omega T^P(\mathbf{s},\mathbf{x})}e^{-\omega\alpha^P(\mathbf{s},\mathbf{x})}, \quad (32)$$

$$\nabla\mathbf{G}_0^S(\mathbf{s}, \mathbf{x}, \omega) = i\omega(\nabla T^S(\mathbf{s}, \mathbf{x}) + i\nabla\alpha^S(\mathbf{s}, \mathbf{x}))\mathbf{A}_0^S(\mathbf{s}, \mathbf{x})e^{i\omega T^S(\mathbf{s},\mathbf{x})}e^{-\omega\alpha^S(\mathbf{s},\mathbf{x})}, \quad (33)$$

$$\nabla\mathbf{G}_0^P(\mathbf{x}, \mathbf{r}, \omega) = i\omega(\nabla T^P(\mathbf{x}, \mathbf{r}) + i\nabla\alpha^P(\mathbf{x}, \mathbf{r}))\mathbf{A}_0^P(\mathbf{x}, \mathbf{r})e^{i\omega T^P(\mathbf{x},\mathbf{r})}e^{-\omega\alpha^P(\mathbf{x},\mathbf{r})}, \quad (34)$$

$$\nabla\mathbf{G}_0^S(\mathbf{x}, \mathbf{r}, \omega) = i\omega(\nabla T^S(\mathbf{x}, \mathbf{r}) + i\nabla\alpha^S(\mathbf{x}, \mathbf{r}))\mathbf{A}_0^S(\mathbf{x}, \mathbf{r})e^{i\omega T^S(\mathbf{x},\mathbf{r})}e^{-\omega\alpha^S(\mathbf{x},\mathbf{r})}. \quad (35)$$

The scattered field might be split into scattering modes, thanks to the asymptotic hypothesis We have the following scattering modes and therefore the scattered field are

$$\delta \mathbf{G}(\mathbf{s},\mathbf{r},\omega) = \delta \mathbf{G}^{PP}(\mathbf{s},\mathbf{r},\omega) + \delta \mathbf{G}^{PS}(\mathbf{s},\mathbf{r},\omega) + \delta \mathbf{G}^{SP}(\mathbf{s},\mathbf{r},\omega) + \delta \mathbf{G}^{SS}(\mathbf{s},\mathbf{r},\omega). \tag{36}$$

The SS notation includes $\delta \mathbf{G}^{SVSV}$ and $\delta \mathbf{G}^{SHSH}$ related to two independent polarizations of S wave, i.e. $SVSV$ and $SHSH$ modes of propagation. We define the total effect of the propagation with the matrix

$$\mathbf{E} = \begin{pmatrix} E^{PP} & E^{PS} & E^{SP} & E^{SVSV} & E^{SHSH} \end{pmatrix},$$

$$\begin{aligned}
\mathbf{E}^{PP} &= \omega^2 \mathbf{A}_0^{PP}(\mathbf{s},\mathbf{x}) \mathbf{A}_0^{PP}(\mathbf{x},\mathbf{r}) e^{i\omega(T^{PP}(\mathbf{s},\mathbf{x})+T^{PP}(\mathbf{x},\mathbf{r}))} e^{-\omega(\alpha^{PP}(\mathbf{s},\mathbf{x})+\alpha^{PP}(\mathbf{x},\mathbf{r}))},\\
\mathbf{E}^{PS} &= \omega^2 \mathbf{A}_0^{PS}(\mathbf{s},\mathbf{x}) \mathbf{A}_0^{PS}(\mathbf{x},\mathbf{r}) e^{i\omega(T^{PS}(\mathbf{s},\mathbf{x})+T^{PS}(\mathbf{x},\mathbf{r}))} e^{-\omega(\alpha^{PS}(\mathbf{s},\mathbf{x})+\alpha^{PS}(\mathbf{x},\mathbf{r}))},\\
\mathbf{E}^{SP} &= \omega^2 \mathbf{A}_0^{SP}(\mathbf{s},\mathbf{x}) \mathbf{A}_0^{SP}(\mathbf{x},\mathbf{r}) e^{i\omega(T^{SP}(\mathbf{s},\mathbf{x})+T^{SP}(\mathbf{x},\mathbf{r}))} e^{-\omega(\alpha^{SP}(\mathbf{s},\mathbf{x})+\alpha^{SP}(\mathbf{x},\mathbf{r}))},\\
\mathbf{E}^{S} &= \omega^2 \mathbf{A}_0^{S}(\mathbf{s},\mathbf{x}) \mathbf{A}_0^{S}(\mathbf{x},\mathbf{r}) e^{i\omega(T^{S}(\mathbf{s},\mathbf{x})+T^{S}(\mathbf{x},\mathbf{r}))} e^{-\omega(\alpha^{S}(\mathbf{s},\mathbf{x})+\alpha^{S}(\mathbf{x},\mathbf{r}))}.
\end{aligned}$$

In the last expression, S is respectively for the $SVSV$ and $SHSH$ modes of propagation.

Using asymptotic Green functions given by equation (12) and, by properties of the perturbation area, we found the most singular [1] term of the scattered integral

$$\delta \mathbf{G}(\mathbf{s},\mathbf{r},\omega) = \int_{\mathcal{M}} \mathbf{E}(\mathbf{s},\mathbf{x},\mathbf{r},\omega)\, \mathbf{W}(\mathbf{s},\mathbf{x},\mathbf{r})\, \mathbf{K}\, \mathbf{f}(\mathbf{x})\, d\mathbf{x}. \tag{37}$$

The diagonal matrix

$$\mathbf{K} = \text{ diag } \begin{pmatrix} 1, \; 1, \; 1, \; i, \; i \end{pmatrix},$$

is deduced from the specific rheology of the medium (the imaginary terms are related to complex relaxation functions we have assumed). The Ray-Born scattering matrix $\mathbf{W}$ can be written under a mode-to-mode conversion of waves

$$\begin{pmatrix}
W_\lambda^{PP} & 0 & 0 & 0 & 0 \\
W_\mu^{PP} & W_\mu^{PS} & W_\mu^{SP} & W_\mu^{SVSV} & W_\mu^{SHSH} \\
W_\rho^{PP} & W_\rho^{PS} & W_\rho^{SP} & W_\rho^{SVSV} & W_\rho^{SHSH} \\
W_\Lambda^{PP} & 0 & 0 & 0 & 0 \\
W_\nu^{PP} & W_\nu^{PS} & W_\nu^{SP} & W_\nu^{SVSV} & W_\nu^{SHSH}
\end{pmatrix},$$

and $\mathbf{f}(\mathbf{x}) = (\delta\lambda(\mathbf{x}), \delta\mu(\mathbf{x}), \delta\rho(\mathbf{x}), \delta\Lambda(\mathbf{x}), \delta\nu(\mathbf{x}))$ is the perturbation parameters vector. The complex terms of this matrix are described in the Appendix. In

[1] We shall consider the most singular term of the integral representations of the single scattered fields. Following the same procedure of Beylkin and Burridge (1990), we used the most singular part of the Green's functions and its derivatives (in the high frequency approximation).

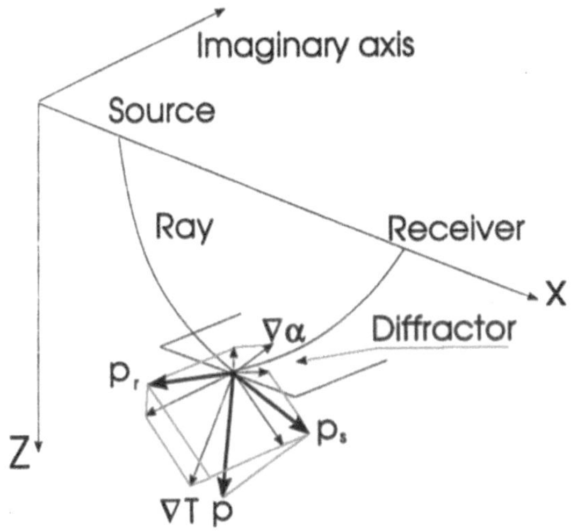

Fig. 3. Diffraction geometry where the source and the receiver are connected through a diffracting point with specific hitting angles of rays which control the recovered amplitude of the anomaly. The vector ∇T is the gradient of the two-way traveltime and is the sum of the local traveltime of these two rays. The vector $\nabla\alpha$, related to dissipation, is the sum of the vector in the direction of maximal spatial attenuation between the source and the diffracting point and the vector in the direction of maximum spatial attenuation between the diffracting point and the receiver. The slowness vector $\mathbf{p}$ is the sum of ∇T and $\nabla\alpha$.

the equation (37), we obtained integral equations relating the singly scattered field **linearly** to the unknown parameters.

The physical interpretation of equation (37) can be described in the following way. A diffracting point $\mathbf{x}$ reacts at the wave arriving from the source and emits a diffracted wave. This diffracted field recorded at the receiver (Figure 3) is proportional to the amplitude of perturbations of the medium parameters through the equation (37).

The scattering matrix, for a given position $\mathbf{x}_0$ of the scattering point depends only on the geometry between incident and scattered ray, as shown in Figure 3. The element $\nabla T(\mathbf{s}, \mathbf{x}_0, \mathbf{r})$ is in the direction of phase propagation, i.e., it is perpendicular to the planes (or surfaces) of constant phase $T = const$; $\nabla\alpha(\mathbf{s}, \mathbf{x}_0, \mathbf{r})$ is in the direction of maximum spatial attenuation, i.e., it is perpendicular to the planes (or surfaces) of constant amplitude (Hearn and Krebes (1990b)). For an attenuating medium the total complex slowness vector has the form $\mathbf{p}(\mathbf{s}, \mathbf{x}_0, \mathbf{r}) = \nabla T(\mathbf{s}, \mathbf{x}_0, \mathbf{r}) + i\nabla\alpha(\mathbf{s}, \mathbf{x}_0, \mathbf{r})$. The initial value of the attenuation angle can be determined by Fermat's principle (Hearn and Krebes (1990a)).

The scattering matrix $\mathbf{W}$ is analysed for an anelastic medium having $Q_{0_S} = 20$, $c_{0_S} = 800$ m s^{-1}, $Q_{0_P} = 2.25 \times Q_{0_S}$, $c_{0_P} = \sqrt{3}c_{0_S}$; $\theta_1 = 0.52$ rad, $\gamma_1 = 1.05$, $\gamma_2 = 0.52$ rad. The elements of the matrix $\mathbf{W}$ are plotted in Figure 4 and Figure 5.

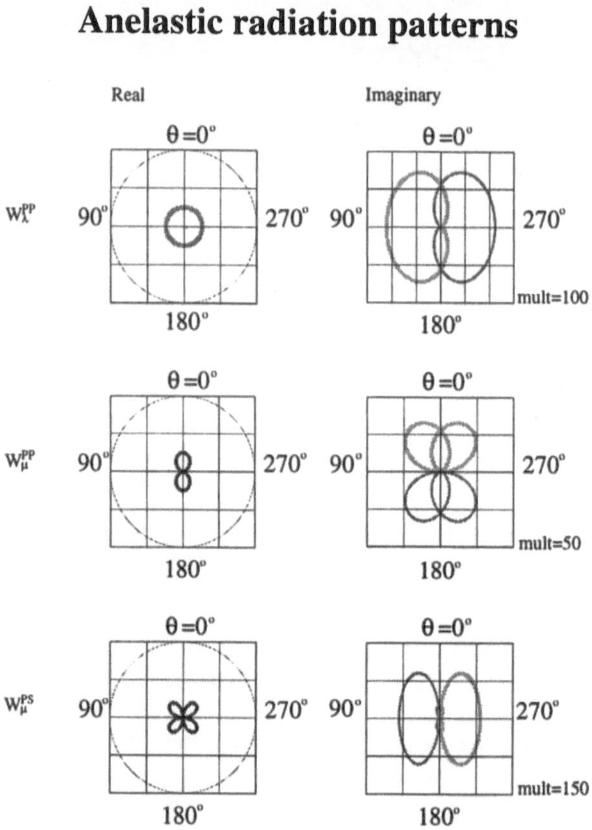

Fig. 4. The radiation patterns for $\delta\lambda$ and $\delta\mu$ are presented for each mode of wave propagation. On the left of the panel the real part of the scattering matrix are compared with imaginary part for each perturbation parameters. On the right, a zoom of the imaginary parts are represented. The real part of W_λ^{PP} is plotted a), and the imaginary part of W_λ^{PP} using a multiplier factor of 100; the real part of W_μ^{PP} is displayed b), and the imaginary part of W_μ^{PP} using a multiplier factor of 50; c) on the left the real part of W_μ^{PS} is plotted, on the right the imaginary part of W_μ^{PS} using a multiplier factor of 150.

We represent in Figure 6 the radiation patterns for the pure elastic parameters inversion; the same results for the elastic case are presented by Forgues (1996).

The attenuation does not change the scattering diagrams of the elastic perturbation parameters. The real part of the radiation-anelastic-pattern of $\delta\lambda$, plotted in green color in Figure 4 a) on the top-left, is identical to radiation-elastic-pattern plotted in Figure 6 a). The real part of the radiation-anelastic-patterns of $\delta\mu$, in red color (Figure 4 and Figure 5) are identical to pure elastic-patterns of $\delta\mu$ (Figure 6 in red color). The imaginary parts represented in Figure 4 and in Figure 5 are more small that the correspondent

Anelastic radiation patterns

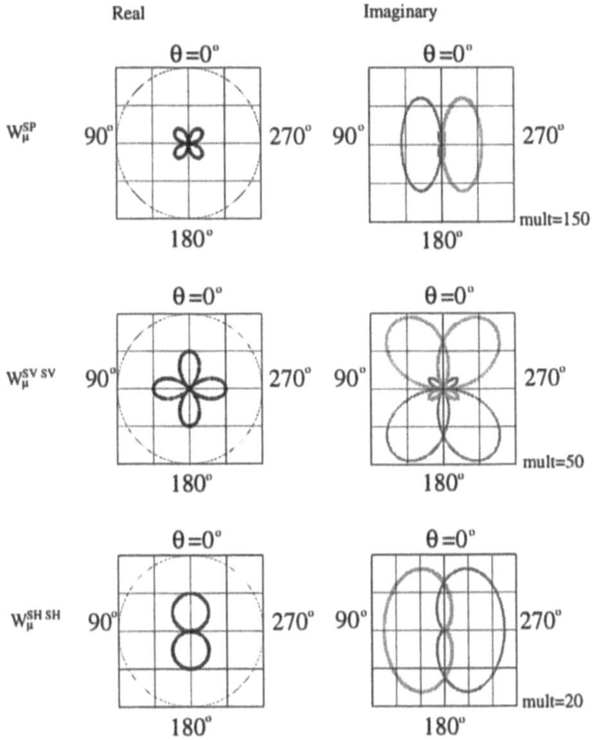

Fig. 5. a) on the left the real part of W_μ^{SP} is plotted, on the right the imaginary part of W_μ^{SP} using a multiplier factor of 150; b) on the left the real part of W_μ^{SVSV} is plotted, on the right the imaginary part of W_μ^{SVSV} using a multiplier factor of 50; c) on the left the real part of W_μ^{SHSH} is plotted, on the right the imaginary part of W_μ^{SHSH} using a multiplier factor of 20.

real parts. For the representation we use a specific multiplier factor. This important result shows that it is possible to extract more informations by the complete seismogram when the attenuation effects are taken into account and the other parameter are not effected. We remark a similar behaviour between the imaginary diagrams of $\delta\lambda$ for PP mode of propagation and the imaginary part of $\delta\mu$ for $SHSH$ mode of propagation. The same behaviour is evident for the imaginary parts for PS and SP mode of propagation.

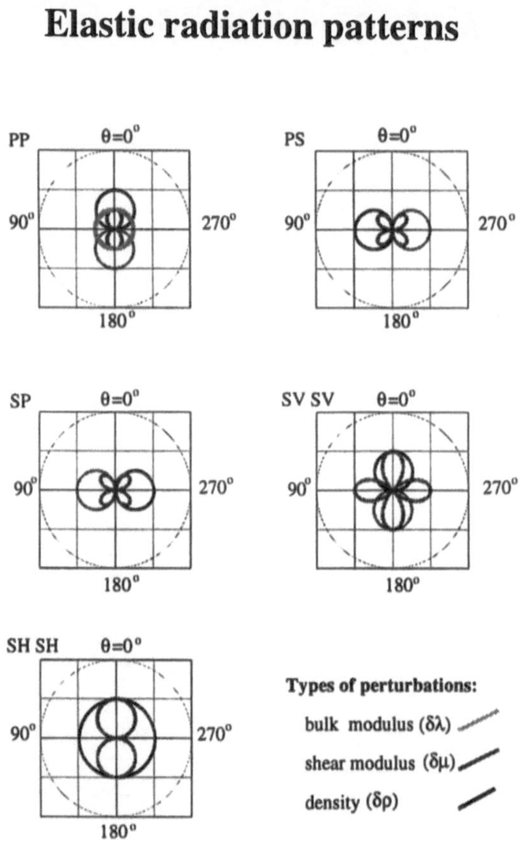

Fig. 6. The radiation patterns for elastic parameters $\delta\lambda$,$\delta\mu$ and $\delta\rho$ are plotted. a) The elements of the scattering matrix associated to $\delta\lambda$ in green, to $\delta\mu$ in red and to $\delta\rho$ in blue for PP diffraction are displayed; b) the radiation pattern related to $\delta\mu$ in red and $\delta\rho$ in blue for PS diffraction are represented; c) the diagrams for $\delta\mu$ and $\delta\rho$ for SP diffraction are shown; d) the elements associated to $\delta\mu$ and $\delta\rho$ for SV SV diffraction are plotted; c) the diagrams for $\delta\mu$ and $\delta\mu$ are shown for SH SH mode of wave propagation.

2 Asymptotic inversion theory

2.1 General approach to inversion

Let us define both model and data spaces with associated metrics and operators between these two spaces. The model space $\mathcal{M}$ is the space of all possible perturbations of the density $\rho(\mathbf{x})$, the complex relaxation functions $\hat{\lambda}(\mathbf{x})$ and $\hat{\mu}(\mathbf{x})$ in which we distinguish between the elastic parts $\lambda(\mathbf{x})$ and $\mu(\mathbf{x})$ and the factors related to attenuation $\Lambda(\mathbf{x})$ and $\nu(\mathbf{x})$. It will be denoted by the vector $\mathbf{f} = (\delta\lambda, \delta\mu, \delta\rho, \delta\Lambda, \delta\nu)$ depending on the position $\mathbf{x}$.

The data space $\mathcal{D}$ consists of the entire set of perturbed seismograms $\delta \mathbf{G}$ which will be denoted $\mathbf{u}$ recorded at the free surface. The data acquisition system depends on source and receiver positions at $\mathbf{s}$ and at $\mathbf{r}$, respectively, and on the angular frequency ω.

We may write the linear forward problem (37) in the compact operator form

$$\mathbf{u}_{\mathcal{D}} = \mathcal{G}\, \mathbf{f}_{\mathcal{M}}, \tag{38}$$

where $\mathcal{G} : \mathcal{M} \to \mathcal{D}$ is the integral operator on the right hand side of (37). The solution of the linearized inverse problem consists in finding the inverse operator of $\mathcal{G}$ applied on seismograms $\mathbf{u}$ in order to recover the model $\mathbf{f}$.

2.2 Inversion by the least-squares method

The inverse solution of the equation (37) is obtained through the optimization method for which a misfit function between observed and calculated seismograms has to be minimized. We adopt the least-squares norm $\mathcal{L}^2$ of the difference between observed and predicted seismograms.

Following the same approach as Jin et al.(1991), we introduce the following definition of the inner product in the data space

$$\begin{aligned}\langle \mathbf{u} \mid \mathbf{v} \rangle_D = \sum_{\mathbf{s},\mathbf{r},l} \int_{\Omega} d\omega \mathbf{u}_l^{\dagger}(\mathbf{s},\mathbf{r},\omega) \mathcal{Q}_{lk}^{\dagger}(\mathbf{s},\mathbf{x}_0,\mathbf{r},\omega) \\ \times\, \mathcal{Q}_{kl}(\mathbf{s},\mathbf{x}_0,\mathbf{r},\omega) \mathbf{v}_l(\mathbf{s},\mathbf{r},\omega)\end{aligned} \tag{39}$$

where $\mathbf{u}$ and $\mathbf{v}$ are two sets of seismograms and $\dagger$ denotes the complex conjugate. The index l from 1 to 5 denotes the scattering modes PP, PS, SP, $SVSV$, $SHSH$, $l = 1$ indicates the PP mode, $l = 2$ the PS mode, etc.. The sum in the equation (39) extends over the data space of seismograms. The matrix $\mathcal{Q}$ is the covariance matrix of the stochastic inversion (Tarantola (1986)) and will be designed such that the first iteration of the inversion provides an approximate inverse solution. As proposed by Jin et al.(1991), at the point $\mathbf{x}_0$ of the diffracted domain $\mathcal{M}$ when we want to recover model values, the matrix $\mathcal{Q}$ has the form

$$\mathcal{Q} = \begin{pmatrix} \mathcal{Q}_{\lambda_0\lambda_0} & 0 & 0 & 0 & 0 \\ 0 & \mathcal{Q}_{\mu_0\mu_0} & 0 & 0 & 0 \\ 0 & 0 & \mathcal{Q}_{\rho_0\rho_0} & 0 & 0 \\ 0 & 0 & 0 & \mathcal{Q}_{\Lambda_0\Lambda_0} & 0 \\ 0 & 0 & 0 & 0 & \mathcal{Q}_{\nu_0\nu_0} \end{pmatrix} \tag{40}$$

where the terms are defined with ray quantities by

$$\mathcal{Q}^l_{\lambda_0\lambda_0}(\mathbf{s},\mathbf{x}_0,\mathbf{r},\omega) = \frac{|\nabla T^l(\mathbf{s},\mathbf{x}_0,\mathbf{r}) + i\nabla\alpha^l(\mathbf{s},\mathbf{x}_0,\mathbf{r})|}{2\pi\sqrt{2\pi}\omega\mathbf{A}_{0_l}(\mathbf{s},\mathbf{x}_0,\mathbf{r})e^{-2\alpha^l(\mathbf{s},\mathbf{x}_0,\mathbf{r},\omega)}},$$

$$\mathcal{Q}^l_{\Lambda_0\Lambda_0}(\mathbf{s},\mathbf{x}_0,\mathbf{r},\omega) = \frac{|\nabla T^l(\mathbf{s},\mathbf{x}_0,\mathbf{r}) + i\nabla\alpha^l(\mathbf{s},\mathbf{x}_0,\mathbf{r})|}{2\pi\sqrt{2\pi}\, i\,\omega\mathbf{A}_{0_l}(\mathbf{s},\mathbf{x}_0,\mathbf{r})e^{-2\alpha^l(\mathbf{s},\mathbf{x}_0,\mathbf{r},\omega)}}, \tag{41}$$

where $|\nabla T^l(\mathbf{s},\mathbf{x}_0,\mathbf{r}) + i\nabla\alpha^l(\mathbf{s},\mathbf{x}_0,\mathbf{r})|$ is the complex module of the slowness vector; $\mathcal{Q}^l_{\mu_0\mu_0} = \mathcal{Q}^l_{\rho_0\rho_0} = \mathcal{Q}^l_{\lambda_0\lambda_0}$ and $\mathcal{Q}^l_{\nu_0\nu_0} = \mathcal{Q}^l_{\Lambda_0\Lambda_0}$

We must underline that the preconditioning associated to the matrix $\mathcal{Q}$ varies with the current position $\mathbf{x}_0$ of the diffracted domain. The particular form of the covariance matrix $\mathcal{Q}$ corrects for geometrical spreading, for obliquities of rays, as well as for the spectral contents of the Green function. We remark that $Q^l_{\Lambda_0\Lambda_0}$ and $Q^l_{\nu_0\nu_0}$ are designed to correct also for the complex dependence of the medium rheology.

For the inverse problem, we also need a definition of the inner product between any two functions $\mathbf{f}(\mathbf{x})$ and $\mathbf{l}(\mathbf{x})$ in the model space $\mathcal{M}$:

$$\langle \mathbf{f}(\mathbf{x}) \mid \mathbf{l}(\mathbf{x})\rangle_{\mathcal{M}} = \int_M \sum_k f_k^\dagger(\mathbf{x}) l_k(\mathbf{x})\, d\mathbf{x}, \tag{42}$$

where k stands for a discrete summation over components in the model space while a continuous integration is performed over the diffracted domain. From the equation (39), we obtain the $\mathcal{L}^2$ misfit function

$$S(\mathbf{f}) = 1/2\langle \mathbf{u} - \mathcal{G}\,\mathbf{f} \mid \mathbf{u} - \mathcal{G}\,\mathbf{f}\rangle_{\mathcal{D}} \tag{43}$$

where $\mathbf{u} = \delta\mathbf{G}^{obs}$ are observed data and $\mathcal{G}\,\mathbf{f}$ are synthetic seismograms estimated through the equation (37). With these definitions, we formulate the inversion problem such as

$$\text{find} \quad \mathbf{f} \quad \text{which} \quad \text{minimizes} \quad S(\mathbf{f}). \tag{44}$$

This formulation leads to the classical "system" of normal equations written as

$$\mathcal{G}^\dagger\mathcal{G}\,\mathbf{f} = \mathcal{G}^\dagger\mathbf{u} \tag{45}$$

where $\mathcal{G}^\dagger$ is the adjoint function of the Green function $\mathcal{G}$. This adjoint operator is defined by the classical relationship

$$\langle \mathbf{u} \mid \mathcal{G}\,\mathbf{f}\rangle_{\mathcal{D}} = \langle \mathcal{G}^\dagger\mathbf{u} \mid \mathbf{f}\rangle_{\mathcal{M}}, \tag{46}$$

which enables us to construct the kernel $\mathcal{K}$ of this adjoint operator $\mathcal{G}^\dagger$ through the integral definition

$$\mathcal{G}^\dagger\mathbf{u} = -\sum_{\mathbf{s},\mathbf{r},l} \int_\Omega d\omega \mathcal{K}^\dagger_{il}(\mathbf{s},\mathbf{x},\mathbf{r},\omega)\mathbf{u}_l(\mathbf{s},\mathbf{r},\omega), \tag{47}$$

where

$$\mathcal{K}^{\dagger}_{il}(\mathbf{s},\mathbf{x},\mathbf{r},\omega) = W^{\dagger}_{ij}(\mathbf{x})\mathcal{Q}^{\dagger}_{pj}(\mathbf{s},\mathbf{x}_0,\mathbf{r},\omega)\mathcal{Q}_{qp}(\mathbf{s},\mathbf{x}_0,\mathbf{r},\omega)K^{\dagger}_{ql} \times \mathbf{A}_0(\mathbf{s},\mathbf{x},\mathbf{r})\ \omega^2 e^{-i\omega T^l(\mathbf{s},\mathbf{x},\mathbf{r})} e^{-\omega\alpha^l(\mathbf{s},\mathbf{x},\mathbf{r})}. \tag{48}$$

The matrix $\mathbf{W}^{\dagger}$ and $\mathbf{K}^{\dagger}$ are respectively the transposed of $\mathbf{W}$ and $\mathbf{K}$.

This definition of operators is the last definition needed for the gradient definition and the Hessian reconstruction associated to the linear system (45).

3 Gradient estimation and Hessian approximation

Following the same approach proposed by Jin et al.(1991), we found that the formal inverse at the diffracting point $\mathbf{x}_0$ can be written

$$\mathbf{f}(\mathbf{x}_0) = \begin{bmatrix} \delta\lambda(\mathbf{x}_0) \\ \delta\mu(\mathbf{x}_0) \\ \delta\rho(\mathbf{x}_0) \\ \delta\Lambda(\mathbf{x}_0) \\ \delta\nu(\mathbf{x}_0) \end{bmatrix} = \mathbf{H}^{-1}(\mathbf{x}_0,\mathbf{x})\gamma^0(\mathbf{x}), \tag{49}$$

with an explicit expression of the gradient at the point x when one wants the image at the point $\mathbf{x}_0$

$$\begin{bmatrix} \gamma^0_{\lambda_0}(\mathbf{x}) \\ \gamma^0_{\mu_0}(\mathbf{x}) \\ \gamma^0_{\rho_0}(\mathbf{x}) \\ \gamma^0_{\Lambda_0}(\mathbf{x}) \\ \gamma^0_{\nu_0}(\mathbf{x}) \end{bmatrix} = -\sum_{\mathbf{s},\mathbf{r},l} \frac{1}{8\pi^3} \int_{\Omega} d\omega \frac{\mathbf{A}_{0_l}(\mathbf{s},\mathbf{x},\mathbf{r})}{\mathbf{A}^2_{0_l}(\mathbf{s},\mathbf{x}_0,\mathbf{r})} \times \mid \nabla T^l(\mathbf{s},\mathbf{x}_0,\mathbf{r}) + i\nabla\alpha^l(\mathbf{s},\mathbf{x}_0,\mathbf{r}) \mid^2 e^{-i\omega T^l(\mathbf{s},\mathbf{x},\mathbf{r})} e^{\omega\alpha^l(\mathbf{s},\mathbf{x},\mathbf{r})} \times \underline{\mathbf{K}}^{\dagger}_{kl} \times W^{\dagger}_{lk} \times \mathbf{u}_l(\mathbf{s},\mathbf{r},\omega), \tag{50}$$

where $\underline{\mathbf{K}}^{\dagger} = \operatorname{diag}\left(1,\ 1,\ 1,\ -i,\ -i\right)$. The operator $\mathbf{H}^{-1}$ in the equation (49) is the formal inverse of

$$\mathbf{H} = \mathcal{G}^{\dagger}\mathcal{G} = \frac{\partial^2 S}{\partial \mathbf{f}^2}, \tag{51}$$

where the function S is the misfit function over frequencies. The inverse of the Hessian cannot be calculated analytically. Observing that the diagonal terms are dominant, we obtain an Hessian approximated by the following relationship

$$\mathbf{H}(\mathbf{x},\mathbf{x}_0) \sim M_{\mathbf{sr}}\ \mathbf{W}^{\dagger} \times \mathbf{W} \cdot \delta(\mathbf{x}-\mathbf{x}_0) \tag{52}$$

where, for a discrete distribution of sources and receivers, Jin et al.(1991) and

Lambaré et al.(1992) found that equation (51) must be normalized by $M_{\mathbf{sr}} = 0.5 \cdot N_{\mathbf{s}}/\Delta\mathbf{r}$, where $N_{\mathbf{s}}$ is the number of sources and $\Delta\mathbf{r}$ is the interval between receivers when they are regularly spaced. Because the Hessian expression in equation (52) is only an approximation, we proceed iterativaly to obtain the best solution using a quasi-Newtonian approach.

4 Pertinent choice of adapted parameters for the model description

For a good description of the medium rheology, it is possible to recover the global velocities c_P and c_S as well as the global quality factors Q_P Q_S according to

$$\mathfrak{f}_2(\mathbf{x}_0) = \begin{bmatrix} c_P(\mathbf{x}_0) \\ c_S(\mathbf{x}_0) \\ Q_P(\mathbf{x}_0) \\ Q_S(\mathbf{x}_0) \end{bmatrix} = \begin{bmatrix} c_{0_P}(\mathbf{x}_0) + \delta c_P(\mathbf{x}_0) \\ c_{0_S}(\mathbf{x}_0) + \delta c_S(\mathbf{x}_0) \\ Q_{0_P}(\mathbf{x}_0) + \delta Q_P(\mathbf{x}_0) \\ Q_{0_S}(\mathbf{x}_0) + \delta Q_S(\mathbf{x}_0) \end{bmatrix} \tag{53}$$

where

$$\begin{aligned} c_{0_P}(\mathbf{x}_0) &= \sqrt{(\lambda_0(\mathbf{x}_0) + 2\mu_0(\mathbf{x}_0))/\rho_0(\mathbf{x}_0)} \\ c_{0_S}(\mathbf{x}_0) &= \sqrt{\mu_0(\mathbf{x}_0)/\rho_0(\mathbf{x}_0)} \\ Q_{0_P}(\mathbf{x}_0) &= (\lambda_0(\mathbf{x}_0) + 2\mu_0(\mathbf{x}_0))/(\Lambda_0(\mathbf{x}_0) + 2\nu_0(\mathbf{x}_0)) \\ Q_{0_S}(\mathbf{x}_0) &= \mu_0(\mathbf{x}_0)/\nu_0(\mathbf{x}_0) \end{aligned} \tag{54}$$

are the background quantities and δc and δQ indicate the (first order) perturbations of the velocities and the quality factors following the relationship:

$$\begin{aligned} \delta c_P(\mathbf{x}_0) &= \frac{1}{2}\sqrt{\frac{\rho_0(\mathbf{x}_0)}{(\lambda_0(\mathbf{x}_0) + 2\mu_0(\mathbf{x}_0))}} \left[\frac{\delta\lambda(\mathbf{x}_0) + 2\delta\mu(\mathbf{x}_0)}{\rho_0(\mathbf{x}_0)} + \right. \\ &\quad \left. - \frac{(\lambda(\mathbf{x}_0) + 2\mu_0(\mathbf{x}_0))\delta\rho(\mathbf{x}_0)}{\rho_0^2(\mathbf{x}_0)} \right] \\ \delta c_S(\mathbf{x}_0) &= \frac{1}{2}\sqrt{\frac{\rho_0(\mathbf{x}_0)}{\mu_0(\mathbf{x}_0)}} \left[\frac{\delta\mu(\mathbf{x}_0)}{\rho_0(\mathbf{x}_0)} - \frac{\mu_0(\mathbf{x}_0)\delta\rho(\mathbf{x}_0)}{\rho_0^2(\mathbf{x}_0)} \right] \\ \delta Q_P(\mathbf{x}_0) &= \frac{\delta\lambda(\mathbf{x}_0) + 2\delta\mu(\mathbf{x}_0)}{\Lambda_0(\mathbf{x}_0) + 2\nu_0(\mathbf{x}_0)} - \frac{(\lambda_0(\mathbf{x}_0) + 2\mu_0(\mathbf{x}_0))(\delta\Lambda(\mathbf{x}_0) + 2\delta\nu(\mathbf{x}_0))}{(\Lambda_0(\mathbf{x}_0) + 2\nu_0(\mathbf{x}_0))^2}, \\ \delta Q_S(\mathbf{x}_0) &= \frac{\delta\mu(\mathbf{x}_0)}{\nu_0(\mathbf{x}_0)} - \frac{\mu_0(\mathbf{x}_0)\delta\nu(\mathbf{x}_0)}{\nu_0^2(\mathbf{x}_0)}, \end{aligned} \tag{55}$$

recovered in each point $\mathbf{x}_0$ of the diffracted domain $\mathcal{M}$. It is an equivalent way to work with different parametrisations because it is possible to change the parameters by simple linear transformation

$$\mathbf{f} = \mathbf{M}\mathbf{f}_2, \tag{56}$$

where the matrix $\mathbf{M}$ is independent on the diffraction angle θ (see the Appendix). This matrix $\mathbf{M}$ for the transformation (55) turns out to be well conditioned for usual values of the model parameters.

Conclusion

We have developed a fast inversion technique based on both the Born approximation and the asymptotic Green functions for recovering both elastic and attenuation parameters of a medium where waves are recorded along the free surface. We solve the linearized inverse scattering problem for perturbations in different parameters treating separately the propagation modes. We haved a closed form of diffraction kernels under the assumptions of asymptotic solutions. We have derived an anelastic numerical algorithm in the frequency domain in contrast to other approaches which have developed elastic algorithms in the time domain. The separation of the propagation and attenuation parameters is possible from wave fitting. Extension to real data is the purpose of future work.

Acknowledgements

This work has been partly founded by European Commission and Norvegian Research Council in the framework of the JOULE II program (Project "Reservoir-oriented Delineation technology"). We are grateful to G. Lambaré for providing the algorithm for ray tracing and S. Operto for valuable comments. We thank U. Bruzzo and G. Caviglia of the Departement of Mathematics of the University of Genova (Italy) for interesting discussions. Publication number 103 of Géosciences-Azur.

References

Aki K., Richards P. (1980): *Quantitative seismology: Theory and methods.* (W. H. Freeman and Co, San Francisco, CA).

Beydoun W., and Mendes M. (1989): *Elastic ray-born l2 migration/inversion.* Geophys. J. Int., **97**, 151–160.

Beylkin G., and Burridge R. (1990): *Linearized inverse scattering problems in acoustics and elasticit.* Wave Motion, **12**, 15–52.

Bleistein N. (1987): *On the imaging of reflectors in the earth.* Geophysics, **52**, 931–942.

Caviglia G., Morro A., and Pagani E. (1990): *Inhomogeneous waves in viscoelastic media.* Wave Motion, **12**, 143–159.

Červený V., and Hron F. (1980): *The ray series method and dynamic ray tracing system for three dimensional inhomogeneous media.* Bull., Seis. Soc. Am., **70**, 47.

Crase E., Pica A., Noble M., McDonald J., and Tarantola A. (1990): *Robust elastic nonlinear waveform inversion: Application to real data.* Geophysics, **55**, 527–538.

Eringen A. (1980): *Mechanics of continua* (Robert E. Krieger Publishing Company).

Ferry J. (1961): *Viscoelastic properties of polymers* (John Wiley and Sons, Inc., New York).

Forgues E. (1996): *Inversion linéarisée multiparamètres via la théorie des rais* (Thèse de Doctorat de l'Universié de Paris VII., Paris).

Hatzidimitriou P. (1995): *S-wave attenuation in the crust in northern greece.* Bull., Seis. Soc. Am., **85**, 1381–1387.

Hearn D., and Krebes E. (1990a): *Complex rays applied to wave propagation in a viscoelastic medium.* Pageoph, **132**, 401–415.

—— (1990b): *On computing ray-synthetic seismograms for anelastic media using complex rays.* Geophysics, **55**, 422–432.

Jin S., Madariaga R., Virieux J., and Lambaré G. (1991): *Two-dimensional asymptotic iterative elastic inversion.* Geophys. J. Int., **108**, 1–14.

Kjartansson E. (1979): *Constant q-wave propagation and attenuation.* J. Geophys. Res., **84**, 4737–4748.

Lambaré G., Virieux J., Jin S., and Madariaga R. (1992): *Iterative asymptotic inversion in the acoustic approximation.* Geophysics, **57**, 1138–1154.

Lambaré G., Lucio, P. and Hanyga A. (1996): *Two-dimensional multivalued traveltime and amplitude maps by uniform sampling of ray fiel.* Geophys. J. Int., **125**, 584–598.

Mavko G., and Nur M. (1978): *Wave attenuation in partially satured rocks.* Geophysics, **44**, 161–178.

O'Connel R., and Budiansky B. (1978): *Measures of dissipation in viscoelastic media.* Geophys. Res. Lett., **5**, 5–8.

Ribodetti A., Virieux J., and Durand S. (1995): *Asymptotic theory for viscoacoustic seismic imaging.* 65th Ann. Internat. Mtg., , Soc. Expl. Geophys., 631–634.

Tarantola A. (1986): *A strategy for nonlinear inversion of seismic reflection data.* Geophysics, **51**, 1893–1903.

Tarantola A. (1988): *Theoretical background for the inversion of seismic waveforms including elasticity and attenuation.* Pageoph, **128**, 365–399.

Virieux J., Flores-Luna C., and Gibert D. (1994): *Asymptotic theory for diffusive electromagnetic imaging.* Geophys. J. Int., **119**, 857–868.

White J. (1965): *Seismic waves: Radiation, transmission and attenuation* (McGraw-Hill Book Co., Inc., New York).

Wu R. (1989): *The perturbation method in elastic wave scattering.* Pageoph, **131**, 605–637.

Appendix : The Ray-Born scattering matrix

We define θ_1 to be the angle at diffacting point at $\mathbf{x}$ that the slowness vector $\mathbf{p_s}(\mathbf{s}, \mathbf{x})$ makes with the vertical (parallel to z direction); θ_2 the angle at diffacting point at $\mathbf{x}$ that the slowness vector $\mathbf{p_r}(\mathbf{x}, \mathbf{r})$ makes with the vertical; γ_1 the angle that the attenuation vector $\alpha_\mathbf{s}(\mathbf{s}, \mathbf{x})$ makes with the vertical and

γ_2 the angle that the attenuation vector $\alpha_{\mathbf{r}}(\mathbf{x},\mathbf{r})$ makes with the vertical. Then we define the total angles

$$\theta = \theta_1 + \theta_2, \tag{57}$$

and

$$\gamma = \gamma_1 + \gamma_2. \tag{58}$$

Then we can write the complex components of the Ray-Born scattering matrix in term of these angles

$$W_\lambda^{PP} = \frac{1}{c_{0_P}^2}\left(1 - \frac{\cos\gamma^{PP}}{4Q_{0_P}^2} + i\left(\frac{\cos(\theta_1^{PP}-\gamma_2^{PP})}{2Q_{0_P}} + \frac{\cos(\theta_2^{PP}-\gamma_1^{PP})}{2Q_{0_P}}\right)\right),$$

$$\begin{aligned}
W_\mu^{PP} &= \frac{2}{c_{0_P}^2}\cos\theta^{PP}\left(\cos\theta^{PP} - \frac{\cos\gamma^{PP}}{4Q_{0_P}^2} + \right.\\
&\left. +i\left(\frac{\cos(\theta_1^{PP}-\gamma_2^{PP})}{2Q_{0_P}} + \frac{\cos(\theta_2^{PP}-\gamma_1^{PP})}{2Q_{0_P}}\right)\right),\\
W_\mu^{PS} &= -\frac{2}{c_{0_P}c_{0_S}}\sin\theta^{PS}\left(\cos\theta^{PS} - \frac{\cos\gamma^{PS}}{4Q_{0_P}Q_{0_S}} + \right.\\
&\left. +i\left(\frac{\cos(\theta_1^{PS}-\gamma_2^{PS})}{2Q_{0_S}} + \frac{\cos(\theta_2^{PS}-\gamma_1^{PS})}{2Q_{0_P}}\right)\right),\\
W_\mu^{SP} &= \frac{2}{c_{0_P}c_{0_S}}\sin\theta^{SP}\left(\cos\theta^{SP} - \frac{\cos\gamma^{SP}}{4Q_{0_P}Q_{0_S}} + \right.\\
&\left. +i\left(\frac{\cos(\theta_1^{SP}-\gamma_2^{SP})}{2Q_{0_P}} + \frac{\cos(\theta_2^{SP}-\gamma_1^{SP})}{2Q_{0_S}}\right)\right),\\
W_\mu^{SVSV} &= \frac{1}{c_{0_S}^2}\cos\theta^{SS}\left(\cos\theta^{SS} - \frac{\cos\gamma^{SS}}{4Q_{0_S}^2} + \right.\\
&\left. +i\left(\frac{\cos(\theta_1^{SS}-\gamma_2^{SS})}{2Q_{0_S}} + \frac{\cos(\theta_2^{SS}-\gamma_1^{SS})}{2Q_{0_S}}\right)\right) + \\
&-\left(\frac{1}{c_{0_S}^2}\sin\theta^{SS}\left(\sin\theta^{SS} - \frac{\sin\gamma^{SS}}{4Q_{0_S}^2}\right.\right.\\
&\left.\left. +i\left(\frac{\sin(\theta_1^{SS}-\gamma_2^{SS})}{2Q_{0_S}} + \frac{\sin(\theta_2^{SS}-\gamma_1^{SS})}{2Q_{0_S}}\right)\right)\right),\\
W_\mu^{SHSH} &= \frac{1}{c_{0_S}^2}\left(\cos\theta^{SS} - \frac{\cos\gamma^{SS}}{4Q_{0_S}^2} + \right.\\
&\left. +i\left(\frac{\cos(\theta_1^{SS}-\gamma_2^{SS})}{2Q_{0_S}} + \frac{\cos(\theta_2^{SS}-\gamma_1^{SS})}{2Q_{0_S}}\right)\right),
\end{aligned}$$

$W_\rho^{PP} = \cos\theta^{PP}$, $W_\rho^{PS} = -\sin\theta^{PS}$, $W_\rho^{SP} = \sin\theta^{SP}$, $W_\rho^{SVSV} = \cos\theta^{SS}$, $W_\rho^{SVSV} = \cos\theta^{SS}$, $W_\rho^{SHSH} = 1$;$W_\Lambda^{PP} = W_\lambda^{PP}$, $W_\nu^{PP} = W_\mu^{PP}$, $W_\nu^{PS} = W_\mu^{PS}$, $W_\nu^{SP} = W_\mu^{SP}$, $W_\nu^{SVSV} = W_\mu^{SVSV}$, $W_\nu^{SHSH} = W_\mu^{SHSH}$, where c_{0_P} and c_{0_S} are the bulk and shear velocities in the propagation medium and Q_{0_P} and Q_{0_S} the quality factors. We remark that when $Q_P \to \infty$ and $Q_S \to \infty$, i.e. in the elastic case, we find the classical elastic Ray-Born scattering matrix (Forgues (1996)).

An Inverse Time Domain Problem for a Stratified, Biperiodic and 2D Medium Using an Optimization Method

Stephane Alestra[1], and Eric Duceau[1]

Département Bases Physiques et Mathématiques, AEROSPATIALE CCR, F-92152 Suresnes Cedex, France

Abstract. The aim of an inverse scattering electromagnetism problem is to determine physical properties of an object or configuration, from the known scattered near-fields or far fields.
In this paper, the involved parameters (permittivity and conductivity profile, impedance operator) should be reconstructed from the knowledge of time domain datas (grating mode, reflection coefficient, far-field ...).
The problem is treated as an optimal control problem where the norm of the difference between measured and computed data is minimized, constrained to the state equation governing the system.
The original constrained optimisation problem is reduced to the stationnary point evaluation of an augmented functional, which is obtained by the method of "Lagrangian multipliers".
Profile reconstruction is carried out by a descent method (Quasi-Newton method). At each iteration, the state and adjoint state are solved by a Finite Difference Time Domain (FDTD) method. New estimates for the permittivity are obtained by a one dimensional search in a suitable descent direction.

1 Introduction

The problem of a lossless one-dimensional slab in time domain, was first theoretically treated by Kay, Sabatier, Gelfand and Levitan (1955). They reduced it to an equivalent, uniquely solvable, quantum-mechanical scattering problem with an integro-differential approach.
Concerning inverse problems for the wave equations, we refer to the geophysical litterature where different approaches and models have been studied extensively. The earth is modelled as a stratified elastic medium whose density and shear modulus vary as a function of depth. Under certain assumptions, available theoretical results about uniqueness, global and local stability estimates are obtained by Symes.
For electromagnetic medium, one has to mention the integral relation methods developped by Tabbara (1979), Lesselier (1982) and Tijhuis (1981) using Born approximation. Some other authors (Kristensson and al.,1986) have examined the numerical practicability of an integro-differential approach, and,

a more accessible procedure for arriving at the scattering kernels by using invariant embedding equations.
In higher dimensions, some interesting papers (see Colton, Kress (1993), and Kirsch (1995)) proved uniqueness and stability results in frequential domain, the time domain approach being not so investigated.
The approach used in this work is analogous to the gradient method developped by Bamberger, Chavent and Lailly (1977) in identification problems for geophysical explorations.

1.1 Mathematical introduction to the inverse problem

The admissible set of parameters will be denoted C_{adm}. Most of the electromagnetic characterization problems are usually considered to be "open problems" where the domain of the computed field is ideally unbounded. Clearly, no computer can store an unlimited amount of datas, and therefore the field computation zone must be limited in size.
The computation zone must be large enough to enclose the structure of interest (the slab), and a suitable boundary condition (Absorbing or Impedance Boundary Condition) on the computation zone board must be used to simulate the extension to infinity.
For theoretical results, we refer to Collino (1992),and Joly (1987) papers. In the following part of the paper, all the variables will be splitted into two components : the one pertaining to the volume object, the other relative to the boundary object. We define the parameter p or control variable

$$p = \begin{pmatrix} p_i \\ p_b \end{pmatrix}$$

and

$$U = \begin{pmatrix} U_i \\ U_b \end{pmatrix}$$

which is here a component of the electromagnetic field.
We note $y = U_{obs}$ the measurement data, M the Maxwell differential operator (in the volume) , B the Boundary differential operator, S_{inc} the incident excitation (plane wave).
Denote by $A: \; p \longmapsto y = A(p)$ the application (linear or non linear forward map) from the space of solutions X into the space of datas Y. One also refers to the data y as the image and to the solution p as the object or profile function.
The inverse problem consists in solving the operator equation $y = A(p)$, from the knowledge of y . According to the usual definition, inverse problem is well-posed if the three following requirements are fullfilled :

- for every $y \in Y$, a solution p exists.
- the application A is one to one (uniqueness when existence is assumed)
- When p exists then $p = A^{-1}y$ and A^{-1} is a bounded operator (stability of the solutions when observations datas vary)

Those properties are often obtained locally (by linearization) : Therefore, Newton methods can be employed with guarantees about convergence.

1.2 Least square solutions

A well-kown way to find p uses a minimization procedure (non linear least square inversion, (see Tarantola, 1984).

Introduction of variables Denote by $Res(U)$ the residual (difference between measurement datas U_{obs} and the computed value U) :

$$Res(U) = (\Sigma_{k=1}^{k=N} F(U - U_{obs})(\delta_{capt(k)})$$

where N is the number of measurement points and $\delta_{capt(k)}$ is the abscissa of the point number k. $F = Id$, for near field measurement and F is the near field-far field mapping if far field measurements are employed. Introducing a space-time scalar product on an adapted Hilbert Space, the cost function is written

$$j(p) = J(U(p)) = \frac{1}{2} < Res(U), Res(U) >$$

By the optimal control theory, finding p_{opt} as

$$j(p_{opt}) = \min j(p) \qquad \forall p \in C_{adm}$$

under the state equation constraint

$$\begin{cases} M(p_i)U_i = S_{inc}(\delta_{sinc}, t) & \text{in the volume} \\ B(p_b)U_b = 0 & \text{on the boundary} \end{cases}$$

is equivalent to search the saddle point of an associated Lagrangian defined below (see Lions, 1968)).

Introduction of the associated Lagrangian Let Q be a costate variable (dual variablo). Then the Lagrangian, function of three variables (p, U, Q) is

$$\begin{aligned} L(p, U, Q) &= J(U) + < Q_i, M(p_i)U_i - S_{inc} > \\ &\quad + < Q_b, B(p_b)U_b > \\ &= \frac{1}{2} < Res(U), Res(U) > \\ &\quad + < Q_i, M(p_i)U_i - S_{inc} > \\ &\quad + < Q_b, B(p_b)U_b > \end{aligned}$$

Remark that p and U are completely independant here, the constraint being added via the second term of Lagrangian .
Searching the saddle point of the Lagrangian is finding $(p_{opt}, U_{opt}, U^*_{opt})$ which verifies

$$L(p_{opt}, U_{opt}, Q) \leq L(p_{opt}, U_{opt}, U^*_{opt}) \leq L(p, U, U^*_{opt})$$
$$\forall p, U, Q \in C_{adm}$$

If $(p_{opt}, U_{opt}, U^*_{opt})$ is saddle point of the Lagrangian, then p_{opt} is the optimal control searched.

Definition of the costate equation Let $U^* = \begin{pmatrix} U_i^* \\ U_b^* \end{pmatrix}$ the corresponding adjoint state to U. U^* verifies the first Euler equation

$$\frac{\partial L(p, U, U^*)}{\partial U} = 0$$

then the adjoint (or dual) equations are performed by

$$\begin{cases} M^*(p_i)U_i^* = Res(U) & \text{in the volume} \\ B^*(p_b)U_b^* = 0 & \text{on the boundary} \end{cases}$$

Computation of the gradients If U and p are in relation with the state equation, denote by $U = U(p)$ the solution to

$$\begin{cases} M(p_i)U_i = S_{inc}(\delta_{sinc}, t) & \text{in the volume} \\ B(p_b)U_b = 0 & \text{on the boundary} \end{cases}$$

The Lagrangian is then simplified in

$$L(p, U, U^*) = j(p)$$

Computing the gradients of j, for the parameter p, we have

$$\nabla j(p)\delta p = \left(\frac{\partial L}{\partial p}\delta p + \frac{\partial L}{\partial U}U'\,\delta p\right)(p, U, U^*)$$

Now, with the relation

$$\frac{\partial L(p, U, U^*)}{\partial U} = 0$$

gradient formulas are obtained

$$\frac{\nabla j}{\nabla p} = \frac{\partial}{\partial p}L(p, U, U^*)$$
$$= < U^*, \frac{\partial M}{\partial p}U >$$

Therefore

$$\frac{\nabla j}{\nabla p_i} = < U_i^*, \frac{\partial M}{\partial p_i} U_i >$$

$$\frac{\nabla j}{\nabla p_b} = < U_b^*, \frac{\partial B}{\partial p_b} U_b >$$

The Optimisation processus With the initial guess p_0, a sequence of parameters, converging towards the optimal p_{opt} is built, with the properties

$$J(p_{n+1}) < J(p_n) \qquad \forall n$$

At each iteration, the algorithm needs :

- A descent direction d_n verifying

$$< \nabla J(p_n), d_n >< 0$$

- A positive λ_n solution to

$$\lambda_n = \min_{\lambda > 0} J(p_n + \lambda d_n)$$

- the following parameter is computed by

$$p_{n+1} = p_n + \lambda_n d_n$$

For a Quasi-Newton method, a choice for the descent direction d_n is

$$d_n = -H_n \nabla J(p_n)$$

where H_n is a definite positive matrix, approximation of the hessian of cost function. The optimizor used for our numerical computations is M2QN1, developped at INRIA Rocquencourt by Lemarechal (1976).

1.3 Parameters sensibility study

This part is devoted to the stability analysis for the non linear least square solution. The rate of convergence for Newton method is linked with the linearized operator boundedness. Thus, supposing that $y_0 = Ap_0$, and linearizing around p_0, one gets

$$\begin{aligned} p &= p_0 + \delta p \\ Y &= y_0 + \delta y \\ A(p) &\approx A(p_0) + A_0'(p_0)\delta p \end{aligned}$$

We obtain the associated linearized minimization problem

$$j_l(\delta p) = \min_{q \in K} j(q) = \|A_0'(q) - \delta y\|^2 \tag{1}$$

where K is a subset of X

Let $A_0'^*$ the adjoint operator of A_0', then

$$H(p_0) = A_0'^* A_0'$$

is a linear, compact, self-adjoint operator.
Denote by $\delta p = (A^+)\delta y$ the solution to (1). A^+ is called the pseudo-generalized inverse operator. The operator A^+ is not in general bounded and by hessian singular values analysis, the problem can be ill-conditionned if the condition number $\chi(A_0') = \frac{\lambda_{max}}{\lambda_{min}}$ is much more greater than one.
Therefore, regularization method have to be applied for stability problems, and a priori information is incorporated in the cost function. Let R_α be the Tichonov bounded regularization operator of A^+

$$R_\alpha = (A_0'^* A_0' + \alpha Id)^{-1} A_0'^*$$

The best choice of the α parameter must realize a compromise between stability and precision and regularization techniques have to be understood in connection with stochastic links and a priori information.

2 Application to a monodimensional electromagnetic problem

2.1 Theoretical problem

Non dispersive medium The governing time domain Maxwell equations for electromagnetic fields are

$$\begin{cases} \varepsilon(z)\frac{\partial E}{\partial t} + \frac{\partial H}{\partial z} = 0 \\ \mu\frac{\partial H}{\partial t} + \frac{\partial E}{\partial z} = 0 \end{cases}$$

with initial conditions for E and H, and with plane wave excitation.
In the way to have a well-posed inverse electromagnetic problem similar geophysics, it is better to change the depth z variable into the traveltime variable $x(z)$, such $x(z) = \int_0^z \frac{1}{c(z)}dz = \int_0^z (\varepsilon\mu(z))^{\frac{1}{2}}dz$.
The impedance $\eta(x)$ is defined by $\eta(x) = \left(\frac{\mu}{\varepsilon}(z)\right)^{\frac{1}{2}}$ and the reflectivity by $r(x) = \frac{1}{2}\frac{\partial \ln \varepsilon\mu(z)}{\partial x}$.
We refer to Symes (1983,1986) for the homeomorphism property of the forward map concerning $\eta(x)$ parameter, and also diffeomorphism property of the forward map concerning $r(x)$, for a Dirac excitation $f(t) = \delta(t)$, and for

the consequences about the inverse problem.
Concerning reflectivity equation, some analytical methods (Invariant Embedding equations) are used by Kristensson (1986) to find $r(x)$ and for lossy electromagnetic medium, we refer to Chaderjian and Bube (1993), and Marechal papers (1986), where one of the parameter is fixed (for exemple the lossy term $\sigma(x)$), and the other (impedance) is the reconstructed parameter.

Linear dispersive medium For a linear polarizable medium, we have the splitting $D(z,\omega) = \varepsilon(z)E(z,\omega)+P(z,\omega)$ with P being the polarization vector

$$P = P(z,\omega) = \frac{Q(z,\omega)}{R(z,\omega)}E$$

and Q et R are two polynomials z dependant coefficients in $i\omega$. With respect to time domain formulation, the general formulas that link Partial Differential Equations (PDE) with Ordinary Differential Equation (ODE) can be written

$$\begin{cases} \varepsilon(z)\frac{\partial E}{\partial t} + \frac{\partial H}{\partial z} + \frac{\partial P}{\partial t} = & 0 \\ \mu\frac{\partial H}{\partial t} + \frac{\partial E}{\partial z} \quad = & 0 \\ R(z,\partial t) \quad = & Q(z,\partial t)E \end{cases}$$

In the litterature, inverse problems for dispersive medium are treated by Weston (1972) with the Invariant Embedding equations model, but stability results have to be proved. Particular linear dispersive medium are Debye law for atomic dipolair relaxation, Lorentz law and Drude law.

2.2 Optimal control algorithm for a non dispersive medium

Using the previous notations, and assuming that $\mu = 1$, we define:
The control parameter $p = (\varepsilon,\sigma)$, the state variable (direct variable $U = (E,H)$), the adjoint variable (dual variable $U_a = (E_a,H_a)$, the incident excitation source at point δ_{sinc}, at time t $S_{inc}(\delta_{sinc},t) = (-j_s,j_m)$ and the Maxwell differential operator

$$M = \begin{pmatrix} \varepsilon\frac{\partial}{\partial t}+\sigma & \frac{\partial}{\partial z} \\ \frac{\partial}{\partial z} & \frac{\partial}{\partial t} \end{pmatrix}$$

We also use this formalism for dispersive medium, and gradient formulas will be given.

2.3 Numerical validation

To illustrate an exemple of reconstruction with real datas, a numerical application with a lossless three-layers dielectric (Kevlar ($\varepsilon = 3.5$), Polystyren ($\varepsilon = 1.01$) and Lecoflex ($\varepsilon = 2.9$)) is presented. The lenght of each layer is fixed.

Measurements are done in the anechoic chamber with a Radar covered range frequency band from 2 Ghz to 18 Ghz. A frequential reflection coefficient versus frequency is then constructed in phase and a discrete inverse Fourier transform is carried out to produce real impulsionnal datas (the frequential window is enlarged to respect Shannon's sampling theorem for the time domain datas, then a conjugation and a smoothing procedure are employed) . At the end of minimization procedure (15 iterations), synthetic impulsionnal datas have a good agreement with experimental datas and the three parameters $\varepsilon_1, \varepsilon_2, \varepsilon_3$ are well reconstructed.
Sensibility results for parameters will be shown and the linear locally comportment of $U(p)$ (see Alestra, 1994) will be presented. Furthermore, the optimal control approach is used to determine the material parameters of the dispersive slab's layers that minimize the reflection coefficient (optimization) over a specified range of frequencies and given the total thickness of the layers.

3 Application to a bidimensional electromagnetic problem

3.1 Theoretical problem

Maxwell's equation General Maxwell equations are :

$$\begin{cases} \partial_t H + rotE &= \quad -J_m \\ -\varepsilon\partial_t E + rotH &= J_e = J_s + J_c \end{cases}$$

Initial and boundary conditions are added to close the system. Furthermore, J_m is the fictive magnetic current, J_e J_s the fictive eletric current, J_s the source current and J_c the conduction current.

Near field-Far field mapping Of particular interest in electromagnetism are far fields measurements. Far fields are given from the knowledge of near fields by an integral representation: Stratton-Chu formulas. Electric currents $J_e = n \wedge H$ and magnetic ones are computed by the FDTD code on a equivalent Huyghens surface S. Then we denote by $M_0(x_0, y_0, z_0)$ the source point on S, and M the observation point. Then the 3D far field representation is

$$E(M,t) = \int_s \mathbf{f}(M_0, t^*) dM_0$$

with the 3D retarded potential function f given by

$$4\pi R\mathbf{f}(M_0,t) = -\frac{1}{\varepsilon_0 c} divJ_e(M_0,t) + \frac{1}{c}\mathbf{u} \wedge \frac{d}{dt} J_m(M_0,t)$$

where $\mathbf{u}$ is the unitary vector in the direction $\mathbf{MM_0}$.
The frequential relation

$$E^r_{2D} = \left(\frac{2\pi c}{j\omega}\right) E^r_{3D}$$

based on Green's kernel and available for $R >> 1$ where $R = d(M_0, M)$ will be used to compute quickly the 2D far field.

3.2 2D optimal control algorithm

State variable and adjoint one are the same as 1D case, while Maxwell Differential operator is slightly modified

$$M = \begin{pmatrix} \varepsilon\frac{\partial}{\partial t} + \sigma & -rot \\ rot & \frac{\partial}{\partial t} \end{pmatrix}$$

. With the notations of part 1.1, F is the near field-far field integro-differential mapping . Its corresponding adjoint operator F^* (far field-near field integro-differential mapping) will be introduced.

3.3 Numerical validation

An explicit and order two leapfrog scheme is employed in the FDTD code (see Yee (1966) and Taflove (1992)). For numerical dispersion and stability and moreover for Absorbing Boundary conditions, we refer to Joly (1987) and Collino (1992). Those conditions are in fact expressed by a "Dirichlet-to Neumann" T operator, described by the relation $\partial_n u + Tu = 0$ on the boundary.
A numerical illustration of the method is the case of an absorbing medium reconstruction fixed on a perfectly metallic 2D profile. The code (see Duceau, 1993 and 1996) was implemented on a parallel computer.

4 Application to an electromagnetic doubly-periodic problem

4.1 Theoretical problem

By assuming periodicity, a group of well-defined difficulties must be solved . In fact, the diffraction due to the periodic system must be taken into account (see Petit, 1980). This effect causes the appearance of waves propagating in preferential directions and apparition of evanescent modes determined by the elementary cell size and by the incident wave frequency. The problem is studied on a elementary cell and periodicity conditions are employed to

simulate the whole structure periodicity (see Delort et Duceau, 1993).
For a pseudo-periodic incident field, such

$$E_{inc}(x, y + d_1, z + d_2) = e^{ik_y d_1} e^{ik_z d_2} E_{inc}(x, y, z)$$

with $d = (d_1, d_2)$ y et z being the grating steps and $k = (k_x, k_y, k_z)$ the wave number, the scattered field is pseudo-periodic (with the same pseudoperiodicity) and Rayleigh expansion can be obtained on a tranverse plane (parallel to the grating plane) $x = x_s$

$$E(x, t + \frac{x}{c}) = \sum_{p,q} E_{pq}(x, t) e^{2i\pi(p\frac{y}{d_1} + q\frac{z}{d_2})}$$

and Fourier coefficients are given by

$$E_{pq}(x, t) = \frac{1}{d_1 d_2} \int_{d_1} \int_{d_2} \left(e^{-2i\pi(p\frac{y}{d_1} + q\frac{z}{d_2})} \right) d_y d_z$$

. Normal incidence will be used in the FDTD code.

4.2 Numerical validation

Reconstruction results of dielectric inclusions in substrate for doubly periodic grating will be given, for Mode 00 observation datas. Then, informations about other modes will be progressively incorporated.

5 Application to an equivalent impedance operator reconstruction

5.1 Theoretical problem

We consider two electromagnetic mediums. The aim of the problem is to modelize the 2nd medium by an equivalent surface condition (impedance condition). The 2nd medium is not described in the volume. Moreover, for complex medium, those conditions are used as equivalent classes (in an electromagnetic equivalent reflection sense) for complex materials. On the interface between the two medium, we have the exact frequential condition, linking the traces E and H.

$$[A] \, \mathbf{E_{tan}} + [B] \, \mathbf{n} \wedge \mathbf{H_{tan}} = 0$$

A and B are two frequential operators depending on $\hat{\varepsilon}, \omega, k$.
In time domain, this condition is described by a Dirichlet to Neumann map linking the tangential traces of E and H, with the non local space time Z operator.

In a practical standpoint, as for the CLA, the exact Z operator is approximated for high frequencies, by Z_{app} in the way to have a maniable numerical condition. Impedance operator issued from Padé developpment are obtained, and Kreiss techniques are used to prove stability.

5.2 Numerical validation

An equivalent impedance condition is searched for a grating periodic medium with dielectric and conductive inclusions. The first mode 00 (specular mode) is observed. The reflection coefficient depends on the frequency range of eigenmodes. Therefore, for each frequency range, an infinite equivalent medium is searched and results of reconstruction will be detailed.

6 Conclusion

In this paper, we showed a generic optimization technique applied to different electromagnetic problems. FIrst, in 1D, we have systematically studied the models and analyzed the measurements, as well as the parameters sensibilitty and regularization techniques.
Therefore, "real reconstructions" are available from the knowledge of impulsive datas, after filtering and treating the frequential experimental measurements.
Then the 2D problem was overviewed. The classical measurement datas were specified (Far field) and therefore the inverse problem goes on. Preliminar results on parallel computers showed good efficiency for the algorithm.
Next section, periodic grating problem was studied, and almost, classical observation datas (propagative eigenmodes). Reconstruction for dielectric and conductive inclusions for periodic gratings are simulated.
The last section concerned the equivalent impedance reconstruction for a complex medium (an approximation of the exact impedance operator). After the mathematical model choice, we established that a reflection coefficient piecewise evaluation was possible, for a periodic grating.

7 Acknowledgement

This work was carried out at AEROSPATIALE Research Center and in collaboration with INRIA ROCQUENCOURT.
The authors wish to thank especially Pr GUILLOT from Université PARIS XIII for his lecture and critical overview of this paper, Patrick Joly from INRIA ROCQUENCOURT for his many ideas, and Francois Hamel from ENS for many helpful discussions (for the dispersive part).

Thanks also to the measurement staff at AEROSPATIALE Research Center : Catherine Druez and Gérard Piau.

References

S.Alestra. (1994) : *Méthodes inverses en électromagnétisme: application aux milieux stratifiés et bipériodiques.* Rapport interne AEROSPATIALE

A.Bamberger G.Chavent and P.Lailly. (1977) : *Une Application de la théorie du contrôle optimal à un problème inverse de sismique.* Annales de Géophysique.

B. Chaderjian and K. Bube. (1993) : *Recovery of Perturbations in an acoustic medium with attenuation from several plane wave responses.* SIAM J. Appl. Math.

F.Clement and G.Chavent. (1993) : *Separating propagation and reflection parameters in the acoustic wave equation.* Rapport INRIA 1839.

F.Collino. (1992) : *Absorbing Boundary Conditions.* (Rapport INRIA).

D.Colton and R.Kress. (1993) : *Inverse Acoustic and Electromagnetic Scaterring theory.* Applied Math. Sciences Springer Verlag.

T.Delort and E.Duceau. (1993) : *Méthodes d'éléments finis et de Différences finies pour l'étude de structures 3D bipériodiques.* Communication JEE Toulouse.

D.C. Dobson. (1993) : *Exploiting ill-posedness in the design of diffractive optical structure.* SPIE vol. 1919 Mathematics in Smart Structure.

E.Duceau. (1993) : *Simulation Numérique Industrielle des équations de Maxwell par le code AS-TEMMIS.* 2ème prix Concours SEYMOUR-CRAY.

E.Duceau. (1996) : ASERIS *Documentation du logiciel développé et commercialisé par AEROSPATIALE* . AEROSPATIALE.

I.M. Gelfand and B.Levitan. (1955) : *On the determination of a differential equation equation from its spectral function.* Am. Math. Soc. trans. pp 253-304.

P.Joly. (1987) : *Analyse Numérique et mathématique de problèmes liés à la propagation d'ondes acoustiques, élastiques et électromagnétiques.* Thèse de docteur-es-sciences Univ. PARIS IX Dauphine.

A.Kirsch. (1995) *The Inverse Scattering Problem.* Proceedings of the Ecole des Ondes INRIA Problème Inverse.

G.Kristensson and R-J Krueger. (1994 1986) : *Direct and Inverse Scattering in the time domain for a dissipative wave equation Part II Simultaneous reconstruction of dissipation and phase velocity profiles.* J. Math. Phys. 27 pp 1683-1693.

C.Lemarechal. (1976) : *Nondifferentiable optimization subgradient methods.* Lect. Notes in Econ. and Math. Systems 117 Optimization and operational Research Springer Berlin.

D.Lesselier. (1982) : *Optimization techniques and Inverse problems : Reconstruction of conductivity profiles in the time domain.* IEEE Trans. Antennas Propagat. 30 pp 59-65.

J.L Lions. (1968) : *Contrôle Optimal des Systèmes gouvernés par des Équations aux Dérivées Partielles.* DUNOD PARIS.

N.J Marechal. (1986) : *Inverse problems for Lossy Hyperbolic Equations.* Ph.D. Thesis Univ. of California (UCLA).

R.Petit. (1980) : *Electromagnetic theory of grating.* Springer-Verlag, Berlin.

W.Symes. (1983) : *Impedance profile inversion via the first transport equation.* J. Math. Anal. Appl. 94.

W.Symes. (1986) : *On the relation between coefficient and boundary values for solutions of Webster's horn equaton.* SIAM J. Math. Anal. 17 pp 1400-1420.

W.Tabbara. (1979) : *Reconstruction of permittivity profiles from a spectral analaysis of the reflection coefficient.* IEEE Trans. Antennas Propagat. 27 pp 241-244.

A.Taflove. (1992) : *Basis and Application of FDTD techniques for Modeling Electromagnetic Wave Interaction.* IEEE Course.

A. Tarantola. (1984) : *The Seismic reflection inverse problem.* Inverse Problems of Elastic and Acoustic waves SIAM 104-181.

A.Tijhuis. (1981) : *Iterative determination of permittivity and conductivity profiles of a dielectric slab in the time domain.* IEEE Trans. Antennas Propagat. AP 29 pp 239-244.

V.H. Weston. (1972) : *On the Inverse problem for a hyperbolic dispersive partial differential operator equation.* J. Math Phys. 13 pp 1952-1956.

K.S. Yee. (1966) : *Numerical solution of Initial Boundary Value Problems in Isotropic Media.* IEEE Trans. Antennas Propagat. AP14 pp 302-307.

Shape Reconstruction of a Penetrable Homogeneous 3D Scattering Body via the ICBA

T. Scotti[1], A. Wirgin[1]

Laboratoire de Mécanique et d'Acoustique, 31 chemin Joseph Aiguier, 13402 Marseille cedex 09, France

Abstract. This work deals with the inverse problem of the determination of the shape of a generally non-spherical penetrable 3D body from the way it scatters incident sonic plane waves. The measurements of the diffracted field are matched to a partial wave representation involving unknown coefficients. Rather than solve for these coefficients (i.e., forward problem) by invoking the transmission conditions, it is supposed that they are locally those of the penetrable sphere of the same composition (as that of the given body) which intersects the given body at its boundary (this is the so-called ICBA, i.e., Intersecting Canonical Body Approximation). These coefficients are known explicitly to within a single parameter which is none other than the length of the position vector joining the origin of the laboratory system to the given point on the boundary of the body. By varying the locations of the measurement point and corresponding boundary point, one generates a discrete form of the parametric equation of the boundary.

1 Problem ingredients

The measured (not necessarily far) field is known in both phase and amplitude, on a part or on the totality of a sphere completely enclosing the body. The scattering surface is acoustically penetrable. The space Ω_0 (Ω_1) surrounding (within) the body is filled with a linear, homogeneous, isotropic, non-absorbing (non-absorbing) material. The incident monochromatic field is that of a plane longitudinal wave. The unknown body is bounded by the surface Γ. The $(Oxyz)$ cartesian coordinate system and the (r, ϑ, ϕ) spherical coordinates (where O is assumed, for convenience, to be located within Γ) will be used. The $e^{-i\omega t}$ time dependence is omitted. Ψ^i will represent the incident plane wave field, $\Psi^0 = \Psi^i + \Psi^d$ (Ψ^d the diffracted field) and Ψ^1 the total fields respectively in Ω_0 and Ω_1. Ψ^0 and Ψ^1 :1) are locally square integrable in Ω_0 and Ω_1, 2) are governed by the Helmholtz equations :

$$(\Delta + k_j^2)\Psi^j = 0 \quad \text{in } \Omega_j; \;\; \mathrm{j} = 0{,}1 \quad , \tag{1}$$

with k_j the (known) wave number in Ω_j, 3) satisfy the outgoing wave condition at infinity (as concerns Ψ^d) and 4) obey the transmission boundary conditions on Γ:

$$\begin{cases} \alpha_0 \Psi^0/_\Gamma = \alpha_1 \Psi^1/_\Gamma \\ \beta_0 \partial_n \Psi^0/_\Gamma = \beta_1 \partial_n \Psi^1/_\Gamma \quad , \end{cases} \tag{2}$$

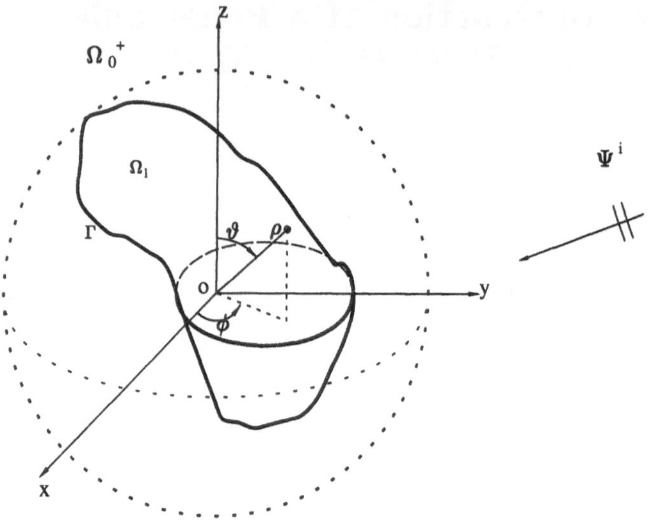

Fig. 1. scattering configuration

with α_0 , β_0 , α_1 , β_1 known constants. For the inverse problem, Ψ^0 is supposed to be known (i.e., measured or simulated) at given points on a spherical surface Γ_b (whose radius is $r_b \geq \bar{r} = Max\,[\,\rho(\vartheta,\phi)\ ;\ 0 \leq \vartheta < \pi\ ;\ 0 \leq \phi < 2\pi]$, with $\rho(\vartheta,\phi)$ the parametric equation description of Γ) enclosing the body. Thus, given Ψ^0 on Γ_b, ϑ^i, ϕ^i, (incident angles), k_j, α_j, β_j; $j = 0, 1$, the objective is to fully or partially, determine the bounding curve Γ of the body.

2 Field representation

It can be shown (Jones and Mao [1]), regardless of the shape of the scattering body, that the total velocity potential is expressible in $\Omega_0^+ = \{r \geq \bar{r}\ ;\ 0 \leq \vartheta < \pi\ ;\ 0 \leq \vartheta < 2\pi\}$ by :

$$\Psi^0(r,\vartheta,\phi) = \Psi^i(r,\vartheta,\phi) + ...$$

$$... \qquad + \sum_{n=0}^{\infty}\sum_{m=0}^{n}\left[A_{nm}\cos(m\phi) + B_{nm}\sin(m\phi)\right] P_n^m(\cos\vartheta) h_n(k_0 r) \quad (3)$$

with

$$\Psi^i(r,\vartheta,\phi) = \sum_{n=0}^{\infty}(2n+1)i^n ...$$

$$... \qquad \sum_{m=0}^{n}\epsilon_m \frac{(n-m)!}{(n+m)!} cos(m(\phi-\phi^i)) P_n^m(\cos\vartheta) P_n^m(\cos\vartheta^i) j_n(k_0 r) \quad ,$$

where j_n and h_n are the n-th order spherical Bessel and Hankel functions of the first kind respectively, P_n^m the Legendre polynomial, $\epsilon_m = 1$ for $m = 0$; $= 2$ for other m. Similarly in
$\Omega_0^- = \{r \leq \underline{r} = Min\,[\rho(\vartheta,\phi); 0 \leq \vartheta < \pi; 0 \leq \phi < 2\pi]\,; 0 \leq \vartheta < \pi; 0 \leq \phi < 2\pi\}$

$$\Psi^1(r,\vartheta,\phi) = \sum_{n=0}^{\infty}\sum_{m=0}^{n} [C_{nm}\cos(m\phi) + D_{nm}\sin(m\phi)]\, P_n^m(\cos\vartheta) j_n(k_1 r) \quad (4)$$

3 Preliminary direct problem

3.1 Particular case of a sphere with center at the origin

Let the body be a sphere of radius a. Introducing Eq.(3) and Eq.(4) into Eqs.(2), projecting successively on $\{\cos(\mu\phi)P_\nu^\mu(\cos\vartheta)\sin\vartheta\}$ and on $\{\sin(\mu\phi)P_\nu^\mu(\cos\vartheta)\sin\vartheta\}$, with $\nu = 0 \ldots N$, $\mu = 0 \ldots \nu$, gives :

$$\Psi^0(r,\vartheta,\phi) = \Psi^i(r,\vartheta,\phi) + \sum_{n=0}^{\infty}\sum_{m=0}^{n} E_{nm}(a)\cos(m(\phi-\phi^i))P_n^m(\cos\vartheta)h_n(k_0 r) \;, \quad (5)$$

$$E_{nm}(a) = \alpha_{nm}^i \frac{\beta_0\alpha_1 k_0 j_n(k_1 a) j_n'(k_0 a) - \beta_1\alpha_0 k_1 j_n'(k_1 a) j_n(k_0 a)}{\alpha_0\beta_1 k_1 j_n'(k_1 a) h_n(k_0 a) - \alpha_1\beta_0 k_0 h_n'(k_0 a) j_n(k_1 a)} \;, \quad (6)$$

$$\alpha_{nm}^i = (2n+1) i^n \epsilon_m \frac{(n-m)!}{(n+m)!} P_n^m(\cos\vartheta^i) \;, \quad (7)$$

where $j_n'(z) = dj_n/dz$, $h_n' = dh_n/dz$.

3.2 General case of a 3D body of arbitrary shape

The process is much the same as above, except that now the solution of the forward problem (i.e. A_{nm} and B_{nm}) is not known. The following approximation is made (Scotti and Wirgin [2]) : if the body is not much different from a sphere, it is assumed that, in any scattering direction ϑ^q, ϕ^p, the field can be approximated by Eq.(5) wherein E_{nm} is that of an "equivalent" sphere with center at the origin and radius η equal to the local radius of the body : $\rho(\vartheta^q, \phi^p)$. The exact expression Eq.(3) is now replaced by the approximation :

$$\Psi^0\,(r,\vartheta^q,\phi^p) \simeq \Psi^i(r,\vartheta^q,\phi^p) + \ldots$$
$$\ldots + \sum_{n=0}^{\infty}\sum_{m=0}^{n} E_{nm}(\rho(\vartheta^q,\phi^p))\cos m(\phi^i - \phi^p) P_n^m(cos\vartheta^q) h_n(kr) \;, \quad (8)$$

where the $E_{nm}(\rho(\vartheta^q,\phi^p))$ are given by Eq.(6) with a replaced by $\rho(\vartheta^q,\phi^p)$. This procedure is called the "Intersecting Canonical Body Approximation" (ICBA, Scotti and Wirgin [2]).

4 Inversion scheme

For any particular scattering direction, the expression of Ψ^0 given by Eq.(3) (in which the infinite series has been reduced to a finite series for computational purposes) is matched to the "given" data Ψ^0 :

$$\Psi^0 (r_b, \vartheta^q, \phi^p) - [\Psi^i(r_b, \vartheta^q, \phi^p) + ...$$

$$... + \sum_{n=0}^{N} \sum_{m=0}^{n} E_{nm}(\eta^{p,q}) \cos m(\phi^i - \phi^p) P_n^m(cos\vartheta^q) h_n(k_0 r_b)] \simeq 0. \quad (9)$$

E_{nm} is known analytically from Eq.(6) to within one parameter, $\eta^{p,q} = a$, so that Eq.(9) (non linear in term of $\eta^{p,q}$) enables one to determine $\eta^{p,q}$. Therefore the inverse problem reduces to : 1) determining, for each scattered direction (ϑ^q, ϕ^p), the radius η of a sphere which gives the same diffracted field as the measured field and 2) identifying η with the local radius $\eta^{p,q} = \rho(\vartheta^q, \phi^p)$ of the body. If this is done for a set of measurements in an angular sector (or all around the body) then the discretised form of the shape function $\rho(\vartheta^q, \phi^p)$ is thereby partially (or totally) obtained.

Remarks
1) If L measured samples of the diffracted field are taken at angles (ϑ^q, ϕ^p), a system of L uncoupled non-linear equations in L unknowns (one equation and one "radius" $\eta^{p,q}$ for each scattered direction) must be solved; 2) $\eta^{p,q}$ should be real, but, because of errors in using the local canonical body approximation and limiting the series to a finite number of terms, the solution $\eta^{p,q}$ of Eq.(9) is, in fact, complex; 3) only the real part of η is kept to test the results; 4) for each equation, the solution is not unique so one profile among many has to be chosen.

5 Post processing

We first reconstructed the most regular profiles, i.e., the ones for which the ϑ and ϕ derivatives are small, then eliminated profiles for which the real part of $\eta^{p,q}$ is negative or larger than r_b, and finally chose the one corresponding to the smallest imaginary part of $\int\int \rho(\vartheta, \phi) d\vartheta d\phi / \int\int d\vartheta d\phi$.

6 Numerical computation, results and conclusion

6.1 Subroutines

The spherical Bessel and Hankel functions were computed by means of the IMSL (IMSL [3]) subroutines DBSJS and DBSYS. The non linear equations (Eq.(9)) were solved by means of the IMSL subroutine DZANLY. The latter computes the complex zeros of a complex function by the Müller method.

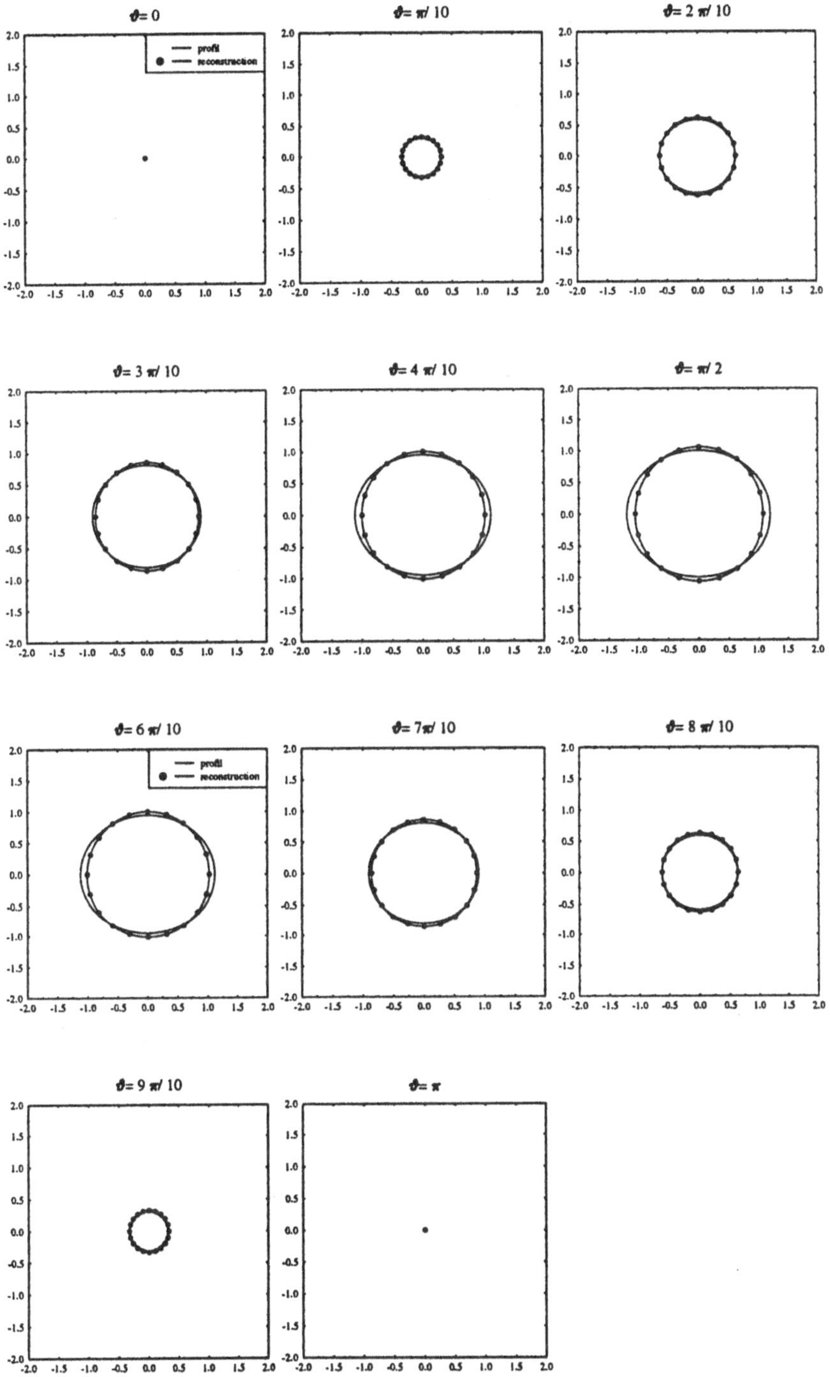

Fig. 2. Cross sections for various ϑ of an ellipsoid of semi axis $(a_x = 1.2, a_y = 1, a_z = 1)$.

6.2 Parameter choices and results

We chose : $k_0 = 1$, $k_1 = 0.5$, $\alpha_0 = \alpha_1 = 1$, $\beta_0 = \beta_1 = 1.5$, measurements in the near field ($r_b = 2, 200$ uniformly distributed scattered directions around the body), $N = 7$, and $\vartheta^i = \phi^i = \pi/3$. The body is an ellipsoide of semi axis $a_x = 1.2$, $a_y = 1$, $a_z = 1$. An example of the results of the computations are given in Figs.(2).
The computations took : for the direct problem, 8h 33mn; for the inverse problem, 16s. The direct problem was solved (to compute Ψ^0 on Γ_b) by means of the Rayleigh-Fourier method (Bolomey and Wirgin [4]).

6.3 Conclusion

The local canonical body approximation for the forward problem easily enables the location and reconstruction of the shape of 3D bodies at low cost with reasonable accuracy, even for bodies whose shape is quite different from that of a sphere.

REFERENCES

[1] Jones D.S. and Mao X.Q., 1989, The inverse problem in hard acoustic scattering. *Inverse probs.*, **5** 731-748.

[2] Scotti T. and Wirgin A., 1995, Shape reconstruction using diffracted waves and canonical solutions, *Inverse Probs.*, **11** 1097-1111.

[3] IMSL, 1991 User's manual Fortran subroutines for mathematical applications: MATH/ LIBRARY, special functions, Version 2.0 *Ref.SFLB and MALB-USM-PERFECT-EN9104-2.0, IMSL, Houston.*

[4] Bolomey J. C. and Wirgin A., 1974, Numerical comparaison of the Green's function and the Waterman and Rayleigh theories of scattering from a cylinder with arbitrary cross section, *Proc.IEE*, **121** 794-804

Program of the Conference of Aix-les-Bains (23–27 sept.1996)

Sponsors: We acknowledge *financial support of:*

DGA/DRET	Direction Générale de l'Armement Direction de la Recherche et de la Technologie
THOMSON-CSF	
USAF/EOARD	United States Air Force European Office of Aerospace Research and Development.

Conference Chairs

G.	Chavent	Université Paris-Dauphine/INRIA Rocquencourt, France
P.C.	Sabatier	Université de Montpellier II, France

Scientific Committee

M.	Bertero	Università di Genova, Italy
G.	Chavent	Université Paris-Dauphine/INRIA Rocquencourt, France
M.	Cheney	Rensselaer Polytechnic Institute, Troy, USA
D.	Colton	University of Delaware, Newart, USA
H.W.	Engl	Johannes-Kepler-Universität, Linz, Austria
R.	Ewing	Texas A & M University, College Station, USA
A.	Friedman	University of Minnesota, Minneapolis, USA
R.	Kress	Universität Göttingen, Germany
K.	Kunisch	Technische Universität Graz, Austria
A.K.	Louis	Universität Saarbrücken, Germany
W.	Rundell	Texas A & M University, College Station, USA
P.C.	Sabatier	Université de Montpellier II, France
W.	Symes	Rice University, Houston, USA

Program Committee

M.	Bertero	Università di Genova, Italy
G.	Beyklin	University of Colorado, Boulder, USA
K.	Chadan	Université d'Orsay, France
G.	Chavent	Université Paris-Dauphine/INRIA Rocquencourt, France
M.	Cheney	Rensselaer Polytechnic Institute, Troy, USA
D.	Colton	University of Delaware, Newart, USA
H.W.	Engl	Johannes-Kepler-Universität, Linz, Austria
R.	Ewing	Texas A & M University, College Station, USA
A.	Friedman	University of Minnesota, Minneapolis, USA
G.	Grunbaum	University of California, Berkeley, USA

P.	Joly	INRIA Rocquencourt, France
R.	Kleinman	University of Delaware, Newark, USA
R.	Kress	Universität Göottingen, Germany
K.	Kunisch	Technische Universität Berlin, Germany
A.K.	Louis	Universität Saarbrücken, Germany
A.	Masmoudi	Université Paul Sabatier/CERFACS, Toulouse, France
J.C.	Nedelec	Ecole Polytechnique, Paris, France
R.G.	Newton	Indiana University, Bloomington, USA
E.R.	Pike	King's College, London, UK
W.	Rundell	Texas A & M University, College Station, USA
P.C.	Sabatier	Université de Montpellier II, France
P.	Sacks	Iowa State University, Ames, USA
S.	Ström	The Royal Institute of Technology, Stockholm, Sweden
W.	Symes	Rice University, Houston, USA
W.	Tabbara	Ecole supérieure d'Electricité , Jouy-en-Josas, France

Technical Organization

M.-C. Sance INRIA Rocquencourt

Lectures Presented at the Aix Meeting

Chairmen: G. Chavent, M. Bertero

An Algorithm for 3D Ultrasound Tomography. *F Natterer* (Wilhelms-Universität, Münster, Germany)

An Inverse Time Domain Problem for a Stratified, Biperiodic and 2D Medium Using an Optimization Method. S. Alestra, *E. Duceau* (Aérospatiale CCR, Suresnes, France)

Retrieval of Object Information from Electron Diffraction as III-Posed Inverse Problem. *K. Scheerschmidt* (Max Planck Institute of Microstructure Physics, Weinberg, Germany)

Recent Developments in Numerical Methods for Time Dependent Scattering. *P. Joly* (INRIA Rocquencourt, Le Chesnay, France)

Inverse Electromagnetic Scattering from an Orthotropic Medium. R. Potthast (University of Delaware, USA)

Inverse Problems in Transport Theory. M. Mokhtar-Kharroubi (Université de Franche Comté, Besançon, France)

Chair: M. Bertero, C. de Mol
Application of the Approximate Inverse Scattering. *A.K. Louis* (Universität des Saarlandes, Saarbrücken, Germany)

The Inverse Spectral Problem for a Non-Selfadjoint Operator and Its Application to a Random Body. M. Lassas (Rolf Nevanlinna-Institute, Helsinki, Finland)

Invited Lecture:
On Inverse Scattering at Midfrequencis. V. Isakov (Wichita State University, Kansas, USA)

Study of a New Tensorial Tomographic Problem. M. Hochhold, H. Leeb (Technische Universität, Wien, Austria)

Scattering Inverse Problems and Transport Theory. S. Patch (Stanford University, USA)

Finding Sound Speed and Electric Permittivity from Interior Transmission Eigenvalues. J.R. Mc Laughlin (Rensselaer Polytechnic Institute, Troy, USA)

Ion Cyclotron Emission in Tokamak Plasmas. D. Fraboulet, (CEA, Saint Paul lez Durance, France)

Chairman: D. Colton
Invited Lecture:
Location and Reconstruction of Objects Using a Modified Gradient Approach. *R.E. Kleinman* (University of Delaware, USA)
P.M. van den Berg, (Delft University of Technology) The Netherlands
D. Lesselier, (C.N.R.S-SUPELEC, Gif-sur-Yvette France)

Inverse Obstacle Scattering Based on Resonant Frequencies. *C. Labreuche* (University of Delaware, USA and Thomson CSF, Orsay, France)

Identification of Electromagnetic Parameters for Media with Microstructure. J. Gottlieb, S. Schlaeger (Universität Karlsruhe, Germany)
S.I. Kabanikhin, V.G. Romanov (Siberian Branch of Russian Academy of Sciences, Russia)

Invited Lecture:
Recovery of Strongly Scattering Permittivity Distributions from Limited Backscattered Data Using a Nonlinear Filtering Technique. *M.A. Fiddy* (University of Massachusetts, Lowell, USA)

Inverse 3D Acoustic and Electromagnetic Obstacle Scattering by Adaptive Iteration. *M. Haas*, W. Rieger, G. Lehner (Universität Stuttgart, Germany)

Singular Behaviour of the Potential on the Boundary in Electrical Impedance Tomography. S. Ispas, S. Ciulli (CNRS/Université de Montpellier II, France) M. Pidcock (Oxford Brookes University, UK)

Chairmen: R. Kleinnan, D. Isakov

New Developments in the Applications of Inverse Scattering to Target Recognition and Remote Sensing. A. Gerard, A. Guran, G. Maze, J. Ripoche, *H. Überall* (Catholic University of America, Washington, USA)

Shape Reconstruction of an Impenetrable 2D Scattering Body Via the Rayleigh Hypothesis. A. Wirgin, T. Scotti (CNRS, Marseille, France)

Shape Reconstruction of a Penetrable Homogeneous 3D Scattering Body Via the ICBA. *T. Scotti*, A. Wirgin (CNRS, Marseille, France)

Reconstruction of an Impenetrable Obstacle Immersed in a Shallow Water Acoustic Waveguide. *C. Rozier*, D. Lesselier (CNRS-Supelec, Gif-sur-Yvette, France)
T. Angell, R. Kleinman (University of Delaware, Newark, USA)

A Unified Viewpoint of Diffraction Tomography Methods Within a Bayesian Estimation Framework. *H. Carfantan*, A. Mohammad-Djafari (CNRS, Gif-sur-Yvette, France)

Born Inversion in Realistic Backgrounds by Means of Recursive Green's Functions. *T.J. Moser*, M. Biryulina, G. Ryzhikov (University of Bergen, Norway)

Stability Estimates in an Inverse Problem for the Transfer Equation. V.G. Romanov (Institute of Mathematics Novosibirsk, Russia)

Chairman: A. Louis

Invited lecture:
A Linear Method for Solving Inverse Scattering Problems in the Resonance Rogion. *D.L. Colton* (University of Delaware, Newark, USA)

Invited lecture:
Inverse Scattering for N-Body Systems with Time-Dependent Potentials. *R. Weder* (Univ. Nacional Autonoma de Mexico, Mexico)

Invited lecture:
An Inverse Problem for a Wave Equation in a Stratified Medium. L. Fatone, P. Maponi, C. Pignotti (Università di Camerino, Italy)
F. Zirilli (Università degli Studi di Roma, Italy)

On the Approximate Solution of the Inverse Obstacle Problem in Acoustics. G.F. Crosta (Università degli Studi di Milano, Italy)

A Generalisation of Karp's Theorem in Acoustic Scattering Theory. *T. Ha-Duong* (Université de Technologie de Compiègne, France)

Chairmen: R. Kress, R. Weder
Invited Lecture:
Mathematical Programming for Positive Solutions of Ill-Conditioned Inverse Problems. *R. Pike* (King's College, London, UK)

An Application of Fourier Series in Inverse Scattering. P. Hahner (Institut für Numerische und Angewandte Mathematik, Göttingen, Germany)

Identification of Seismic Diffracted Fields Using 2D and 3D Discrete Wavenumber–Boundary Integral Equation Simulations: Methodology and Results.
S. Durand, S. Gaffet, F. Tressols, J. Virieux (UMR Géosciences Azur, Valbonne, France)

Invited lecture:
Resolution and Super-Resolution in Inverse Diffraction. *M. Bertero* (Università degli Studi di Genova, Italy)

Inverse Scattering Approach for Stratified Chiral Media. A. Boutet de Monvel (Université Paris 7 et CNRS, France),
D. Shepelsky (Institute for Temperature Physics, Kharkiv, Ukraina)

Inverse Scattering Problem, at fixed Energy, for the Class of Yukawian Potentials. G. Viano (Università di Genova, Italy)

Chairmen: G. Viano, S. Patch
Iterative Methods for the Inverse Potential Problem. F. Hettlich (Universität Erlangen-Nürnberg, Germany,
W. Rundell (Texas A&M University, College Station, USA)

New Situation in Quantum Mechanics. N. Zakhariev, V.M. Chabanov (Joint Institute for Nuclear Research, Dubna, Russia)

Two-Dimensional Exactly Solvable Models. A. Suzko (Joint Institute for Nuclear Research, Moscow, Russia)

Direct and Inverse Scattering of Guided Waves in Laterally Heterogeneous Media. *F.E. Ernst*, G.C. Herman (Delft University of Technology, The Netherlands)

Scattered Seismic Wave Fields Identification Using a Very Small Dense Array. F. Tressols, S. Gaffet, A. Deschamps (UMR Géosciences Azur, Valbonne, France)

A New Version of Wexler Algorithm in Electrical Impedance Tomography. Th. Condamines, P.-M. Marsili (Université Paul Sabatier, Toulouse, France)

Chairman: W. Rundell

Identifiability of the Diffusion Coefficient in a 1D Elliptic Equation. X. Goudou (Université de Montpellier II, France)

On The Inverse Seismic Problem for Horizontally Layered Elastic Media. Subsidiary Studies. *M. Cuer* (Université de Montpellier II, France)
J.-L. Petit(Institut Français du Pétrole, Rueil-Malmaison, France)

VSP Inversion with 3D Propagation Modelling. J.-L. Petit(Institut Français du Pétrole, Rueil-Malmaison, France)

Invited Lecture:

Numerical Methods in Inverse Obstacle Scattering. *R. Kress* (Georg-August-Universität, Göttingen, Germany)

A Quasi-tomography of a Comet Nucleus: the CONSERT Experiment. J.P. Barriot (GRGS-GTP, Toulouse, France), W. Kofman (Centre d'Etudes des Phénomènes Alétoires et Géophysiques, St. Martin d'Hères, France)
T. Hagfors (Max Plank Institut für Aeronomie, Katlenburg-Lindau, Germany,) G. Picardi (INFOCOM, Roma, Italy), J. Van Zyl (Jet Propulsion Laboratory, Pasadena, USA)

A Level-Set Approach for Eddy Current Imaging of Defects in a Conductive Half-Space. *A. Litman*, D. Lesselier (CNRS-Supelec, Gif-sur-Uvette, France), F. Santosa, (University of Minnesota, Minneapolis, USA)

Chairman: E.R. Pike

Spectral Analysis of Surface Waves: an Automated Inversion Technique Based on a Gauss-Newton Inversion Algorithm. W. Dewulf, G. Degrande, G. De

Roeck (Katholieke Universiteit te Leuven, Heverlee, Belgium)

Non-linear Inversion of Synthetic Seismic Reflection Data by Simulated Annealing. *P. Amand*, J. Virieux (IGSA, Valbonne, France)

Asymptotic Theory for Imaging the Attenuation Factor Q. *A. Ribodetti*, (ICSA, Valbonne et GEMCO, Villefranche sur Mer, France)
J. Virieux(IGSA, Valbonne, France)

Determination of Dopant Concentrations using Grazing-Emission X-Ray Fluorescence Spectrometry. H.P. Urbach (Philips Research Laboratories, Eindhoven, The Netherlands)

Springer and the environment

At Springer we firmly believe that an international science publisher has a special obligation to the environment, and our corporate policies consistently reflect this conviction.
We also expect our business partners – paper mills, printers, packaging manufacturers, etc. – to commit themselves to using materials and production processes that do not harm the environment. The paper in this book is made from low- or no-chlorine pulp and is acid free, in conformance with international standards for paper permanency.

Springer

Lecture Notes in Physics

For information about Vols. 1–455
please contact your bookseller or Springer-Verlag

Vol. 456: H. B. Geyer (Ed.), Field Theory, Topology and Condensed Matter Physics. Proceedings, 1994. XII, 206 pages. 1995.

Vol. 457: P. Garbaczewski, M. Wolf, A. Weron (Eds.), Chaos - The Interplay Between Stochastic and Deterministic Behaviour. Proceedings, 1995. XII, 573 pages. 1995.

Vol. 458: I. W. Roxburgh, J.-L. Masnou (Eds.), Physical Processes in Astrophysics. Proceedings, 1993. XII, 249 pages. 1995.

Vol. 459: G. Winnewisser, G. C. Pelz (Eds.), The Physics and Chemistry of Interstellar Molecular Clouds. Proceedings, 1993. XV, 393 pages. 1995.

Vol. 460: S. Cotsakis, G. W. Gibbons (Eds.), Global Structure and Evolution in General Relativity. Proceedings, 1994. IX, 173 pages. 1996.

Vol. 461: R. López-Pen˜a, R. Capovilla, R. Garcı´a-Pelayo, H. Waelbroeck, F. Zertuche (Eds.), Complex Systems and Binary Networks. Lectures, México 1995. X, 223 pages. 1995.

Vol. 462: M. Meneguzzi, A. Pouquet, P.-L. Sulem (Eds.), Small-Scale Structures in Three-Dimensional Hydrodynamic and Magnetohydrodynamic Turbulence. Proceedings, 1995. IX, 421 pages. 1995.

Vol. 463: H. Hippelein, K. Meisenheimer, H.-J. Röser (Eds.), Galaxies in the Young Universe. Proceedings, 1994. XV, 314 pages. 1995.

Vol. 464: L. Ratke, H. U. Walter, B. Feuerbach (Eds.), Materials and Fluids Under Low Gravity. Proceedings, 1994. XVIII, 424 pages, 1996.

Vol. 465: S. Beckwith, J. Staude, A. Quetz, A. Natta (Eds.), Disks and Outflows Around Young Stars. Proceedings, 1994. XII, 361 pages, 1996.

Vol. 466: H. Ebert, G. Schütz (Eds.), Spin - Orbit-Influenced Spectroscopies of Magnetic Solids. Proceedings, 1995. VII, 287 pages, 1996.

Vol. 467: A. Steinchen (Ed.), Dynamics of Multiphase Flows Across Interfaces. 1994/1995. XII, 267 pages. 1996.

Vol. 468: C. Chiuderi, G. Einaudi (Eds.), Plasma Astrophysics. 1994. VII, 326 pages. 1996.

Vol. 469: H. Grosse, L. Pittner (Eds.), Low-Dimensional Models in Statistical Physics and Quantum Field Theory. Proceedings, 1995. XVII, 339 pages. 1996.

Vol. 470: E. Martı´nez-González, J. L. Sanz (Eds.), The Universe at High-z, Large-Scale Structure and the Cosmic Microwave Background. Proceedings, 1995. VIII, 254 pages. 1996.

Vol. 471: W. Kundt (Ed.), Jets from Stars and Galactic Nuclei. Proceedings, 1995. X, 290 pages. 1996.

Vol. 472: J. Greiner (Ed.), Supersoft X-Ray Sources. Proceedings, 1996. XIII, 350 pages. 1996.

Vol. 473: P. Weingartner, G. Schurz (Eds.), Law and Prediction in the Light of Chaos Research. X, 291 pages. 1996.

Vol. 474: Aa. Sandqvist, P. O. Lindblad (Eds.), Barred Galaxies and Circumnuclear Activity. Proceedings of the Nobel Symposium 98, 1995. XI, 306 pages. 1996.

Vol. 475: J. Klamut, B. W. Veal, B. M. Dabrowski, P. W. Klamut, M. Kazimierski (Eds.), Recent Developments in High Temperature Superconductivity. Proceedings, 1995. XIII, 362 pages. 1996.

Vol. 476: J. Parisi, S. C. Müller, W. Zimmermann (Eds.), Nonlinear Physics of Complex Systems. Current Status and Future Trends. XIII, 388 pages. 1996.

Vol. 477: Z. Petru, J. Przystawa, K. Rapcewicz (Eds.), From Quantum Mechanics to Technology. Proceedings, 1996. IX, 379 pages. 1996.

Vol. 479: H. Latal, W. Schweiger (Eds.), Perturbative and Nonperturbative Aspects of Quantum Field Theory. Proceedings, 1996. X, 430 pages. 1997.

Vol. 480: H. Flyvbjerg, J. Hertz, M. H. Jensen, O. G. Mouritsen, K. Sneppen (Eds.), Physics of Biological Systems. From Molecules to Species. X, 364 pages. 1997.

Vol. 481: F. Lenz, H. Grießhammer, D. Stoll (Eds.), Lectures on QCD. VII, 276 pages. 1997.

Vol. 482: X.-W. Pan, D. H. Feng, M. Vallières (Eds.), Contemporary Nuclear Shell Models. Proceedings, 1996. XII, 309 pages. 1997.

Vol. 483: G. Trottet (Ed.), Coronal Physics from Radio and Space Observations. XVII, 226 pages. 1997.

Vol. 484: L. Schimansky-Geier, T. Pöschel (Eds.), Stochastic Dynamics. XVIII, 386 pages. 1997.

Vol. 486: G. Chavent, P. C. Sabatier (Eds.), Inverse Problems of Wave Propagation and Diffraction. XV, 379 pages. 1997.

Vol. 488: B. Apagyi, G. Endrédi, P. Lévay (Eds.), Inverse and Algebraic Quantum Scattering Theory. XV, 385 pages. 1997.

Vol. 489: G. M. Simnett, C. E. Alissandrakis, L. Vlahos (Eds.), Solar and Heliospheric Plasma Physics. VIII, 278 pages, 1997.

Vol. 490: O. Boratav, A. Eden, A. Erzan (Eds.), Turbulence Modeling and Vortex Dynamics. XIV, 652 pages, 1997.

New Series m: Monographs

Vol. m 1: H. Hora, Plasmas at High Temperature and Density. VIII, 442 pages. 1991.

Vol. m 2: P. Busch, P. J. Lahti, P. Mittelstaedt, The Quantum Theory of Measurement. XIII, 165 pages. 1991. Second Revised Edition: XIII, 181 pages. 1996.

Vol. m 3: A. Heck, J. M. Perdang (Eds.), Applying Fractals in Astronomy. IX, 210 pages. 1991.

Vol. m 4: R. K. Zeytounian, Mécanique des fluides fondamentale. XV, 615 pages, 1991.

Vol. m 5: R. K. Zeytounian, Meteorological Fluid Dynamics. XI, 346 pages. 1991.

Vol. m 6: N. M. J. Woodhouse, Special Relativity. VIII, 86 pages. 1992.

Vol. m 7: G. Morandi, The Role of Topology in Classical and Quantum Physics. XIII, 239 pages. 1992.

Vol. m 8: D. Funaro, Polynomial Approximation of Differential Equations. X, 305 pages. 1992.

Vol. m 9: M. Namiki, Stochastic Quantization. X, 217 pages. 1992.

Vol. m 10: J. Hoppe, Lectures on Integrable Systems. VII, 111 pages. 1992.

Vol. m 11: A. D. Yaghjian, Relativistic Dynamics of a Charged Sphere. XII, 115 pages. 1992.

Vol. m 12: G. Esposito, Quantum Gravity, Quantum Cosmology and Lorentzian Geometries. Second Corrected and Enlarged Edition. XVIII, 349 pages. 1994.

Vol. m 13: M. Klein, A. Knauf, Classical Planar Scattering by Coulombic Potentials. V, 142 pages. 1992.

Vol. m 14: A. Lerda, Anyons. XI, 138 pages. 1992.

Vol. m 15: N. Peters, B. Rogg (Eds.), Reduced Kinetic Mechanisms for Applications in Combustion Systems. X, 360 pages. 1993.

Vol. m 16: P. Christe, M. Henkel, Introduction to Conformal Invariance and Its Applications to Critical Phenomena. XV, 260 pages. 1993.

Vol. m 17: M. Schoen, Computer Simulation of Condensed Phases in Complex Geometries. X, 136 pages. 1993.

Vol. m 18: H. Carmichael, An Open Systems Approach to Quantum Optics. X, 179 pages. 1993.

Vol. m 19: S. D. Bogan, M. K. Hinders, Interface Effects in Elastic Wave Scattering. XII, 182 pages. 1994.

Vol. m 20: E. Abdalla, M. C. B. Abdalla, D. Dalmazi, A. Zadra, 2D-Gravity in Non-Critical Strings. IX, 319 pages. 1994.

Vol. m 21: G. P. Berman, E. N. Bulgakov, D. D. Holm, Crossover-Time in Quantum Boson and Spin Systems. XI, 268 pages. 1994.

Vol. m 22: M.-O. Hongler, Chaotic and Stochastic Behaviour in Automatic Production Lines. V, 85 pages. 1994.

Vol. m 23: V. S. Viswanath, G. Müller, The Recursion Method. X, 259 pages. 1994.

Vol. m 24: A. Ern, V. Giovangigli, Multicomponent Transport Algorithms. XIV, 427 pages. 1994.

Vol. m 25: A. V. Bogdanov, G. V. Dubrovskiy, M. P. Krutikov, D. V. Kulginov, V. M. Strelchenya, Interaction of Gases with Surfaces. XIV, 132 pages. 1995.

Vol. m 26: M. Dineykhan, G. V. Efimov, G. Ganbold, S. N. Nedelko, Oscillator Representation in Quantum Physics. IX, 279 pages. 1995.

Vol. m 27: J. T. Ottesen, Infinite Dimensional Groups and Algebras in Quantum Physics. IX, 218 pages. 1995.

Vol. m 28: O. Piguet, S. P. Sorella, Algebraic Renormalization. IX, 134 pages. 1995.

Vol. m 29: C. Bendjaballah, Introduction to Photon Communication. VII, 193 pages. 1995.

Vol. m 30: A. J. Greer, W. J. Kossler, Low Magnetic Fields in Anisotropic Superconductors. VII, 161 pages. 1995.

Vol. m 31: P. Busch, M. Grabowski, P. J. Lahti, Operational Quantum Physics. XI, 230 pages. 1995.

Vol. m 32: L. de Broglie, Diverses questions de mécanique et de thermodynamique classiques et relativistes. XII, 198 pages. 1995.

Vol. m 33: R. Alkofer, H. Reinhardt, Chiral Quark Dynamics. VIII, 115 pages. 1995.

Vol. m 34: R. Jost, Das Märchen vom Elfenbeinernen Turm. VIII, 286 pages. 1995.

Vol. m 35: E. Elizalde, Ten Physical Applications of Spectral Zeta Functions. XIV, 228 pages. 1995.

Vol. m 36: G. Dunne, Self-Dual Chern-Simons Theories. X, 217 pages. 1995.

Vol. m 37: S. Childress, A.D. Gilbert, Stretch, Twist, Fold: The Fast Dynamo. XI, 410 pages. 1995.

Vol. m 38: J. González, M. A. Martín-Delgado, G. Sierra, A. H. Vozmediano, Quantum Electron Liquids and High-T_c Superconductivity. X, 299 pages. 1995.

Vol. m 39: L. Pittner, Algebraic Foundations of Non-Commutative Differential Geometry and Quantum Groups. XII, 469 pages. 1996.

Vol. m 40: H.-J. Borchers, Translation Group and Particle Representations in Quantum Field Theory. VII, 131 pages. 1996.

Vol. m 41: B. K. Chakrabarti, A. Dutta, P. Sen, Quantum Ising Phases and Transitions in Transverse Ising Models. X, 204 pages. 1996.

Vol. m 42: P. Bouwknegt, J. McCarthy, K. Pilch, The W_3 Algebra. Modules, Semi-infinite Cohomology and BV Algebras. XI, 204 pages. 1996.

Vol. m 43: M. Schottenloher, A Mathematical Introduction to Conformal Field Theory. VIII, 142 pages. 1997.

Vol. m 44: A. Bach, Indistinguishable Classical Particles. VIII, 157 pages. 1997.

Vol. m 45: M. Ferrari, V. T. Granik, A. Imam, J. C. Nadeau (Eds.), Advances in Doublet Mechanics. XVI, 214 pages. 1997.

Vol. m 46: M. Camenzind, Les noyaux actifs de galaxies. XVIII, 218 pages. 1997.

Vol. m 48: P. Kopietz, Bosonization of Interacting Fermions in Arbitrary Dimensions. XII, 259 pages. 1997.